Lecture Notes in Computer Science 10490

Commenced Publication in 1973
Founding and Former Series Editors:
Gerhard Goos, Juris Hartmanis, and Jan van Leeuwen

More information about this series at http://www.springer.com/series/7407

Vladimir P. Gerdt · Wolfram Koepf
Werner M. Seiler · Evgenii V. Vorozhtsov (Eds.)

Computer Algebra in Scientific Computing

19th International Workshop, CASC 2017
Beijing, China, September 18–22, 2017
Proceedings

Editors
Vladimir P. Gerdt
Joint Institute of Nuclear Research
Dubna
Russia

Wolfram Koepf
Universität Kassel
Kassel
Germany

Werner M. Seiler
Universität Kassel
Kassel
Germany

Evgenii V. Vorozhtsov
Russian Academy of Sciences
Novosibirsk
Russia

ISSN 0302-9743 ISSN 1611-3349 (electronic)
Lecture Notes in Computer Science
ISBN 978-3-319-66319-7 ISBN 978-3-319-66320-3 (eBook)
DOI 10.1007/978-3-319-66320-3

Library of Congress Control Number: 2017950084

LNCS Sublibrary: SL1 – Theoretical Computer Science and General Issues

Printed on acid-free paper

This Springer imprint is published by Springer Nature
The registered company is Springer International Publishing AG
The registered company address is: Gewerbestrasse 11, 6330 Cham, Switzerland

Preface

The International Workshop on Computer Algebra in Scientific Computing (CASC) is a leading conference which provides the opportunity for all researchers from home and abroad to present their research results annually. CASC is the forum of excellence for the exploration of the frontiers in the field of computer algebra and its applications in scientific computing. It brings together scholars, engineers, and scientists from various disciplines including computer algebra. This workshop provides a platform for the delegates to exchange new ideas and application experiences, share research results, discuss existing issues and challenges, and explore international cooperation in cutting-edge technology face to face.

This year, the 19th CASC conference was held in Beijing (China). Study on computer algebra in China started with the work of Prof. Wen-Tsun Wu on automated geometry theorem proving and characteristic set methods for polynomial equation solving in the late 1970s. In 1990, the Research Center of Mathematics Mechanization (MMRC) was established in the Chinese Academy of Sciences, and the center runs a series of academic programs on computer algebra and related areas.

In particular, jointly with the Japanese Society for Symbolic and Algebraic Computation, the Asian Symposium on Computer Mathematics was started in 1995 and held every two years. MMRC also organized The International Symposium on Symbolic and Algebraic Computation (ISSAC) in 2005. Other major research groups include the Laboratory of Automated Reasoning at the Chongqing Branch of the Chinese Academy of Sciences led by Prof. Jingzhong Zhang, and the research group at Beihang University led by Prof. Dongming Wang. In 2007, the Computer Mathematics Society of China was established, and Prof. Xiao-Shan Gao was the founding president of the society. The society runs an annual conference with approximately 100–150 participants.

Prof. Xiao-Shan Gao has kindly agreed to be one of the General Chairs of the CASC 2017 workshop. This has affected the choice of Beijing as a venue for CASC 2017.

This volume contains 26 full papers submitted to the workshop by the participants and accepted by the Program Committee after a thorough reviewing process with usually three independent referee reports. Additionally, the volume includes two contributions corresponding to the invited talks.

Polynomial algebra, which is at the core of computer algebra, is represented by contributions devoted to the convergence conditions of interval Newton's method applied to the solution of a nonlinear system; certifying the simple real zeros of overdetermined polynomial systems with interval methods; decomposition of polynomial sets into lexicographic Gröbner bases and into normal triangular sets; computation of all the isolated solutions to a special class of polynomial systems with the aid of a special homotopy continuation method; computing real witness points of general polynomial systems with the aid of the penalty function based critical point approach; algorithms for zero-dimensional ideals using linear recurrent sequences;

finding quasihomogeneous isolated hypersurface singularities with the aid of an interface of the computer algebra system (CAS) POLYMAKE in the CAS SINGULAR; and full rank representation of real algebraic sets with applications in visualizing plane and space curves with singularities. Two papers deal with the problems arising in polynomial interpolation: one focussing on the optimal knots selection for spline interpolation in the case of sparse reduced data, and one focussing on sparse interpolation algorithms for black box univariate or multivariate polynomials whose coefficients are from a finite set.

The invited talk of Lihong Zhi is devoted to computing multiple zeros of polynomial systems. It shows how to compute the multiplicity structure of each multiple zero and the lower bound on the minimal distance between the multiple zero and other zeros of the system. The developed algorithms were implemented in the CAS Maple.

Several papers deal with using computer algebra for the investigation of various mathematical and applied topics related to ordinary differential equations (ODEs), focussing on, for example, the introduction of the concept of a Laurent Gröbner basis for the investigation of the Laurent (differential) polynomial systems, and the study of local integrability of an autonomous system of ODEs with the aid of an approach based on power geometry.

The invited talk by S. Abramov handles the problem of the solvability of linear systems of ordinary differential equations whose coefficients have the form of infinite formal power series. The problem is to decide whether the system has non-zero Laurent series, regular, or formal exponential-logarithmic solutions, and to find all such solutions if they exist. Maple-based procedures are presented for constructing local solutions.

Four papers deal with applications of symbolic and symbolic-numeric computations for investigating and solving partial differential equations (PDEs) and ODEs in mathematical physics and fluid mechanics, focussing on, for example, the symbolic-numeric integration of the dynamical Cosserat partial differential equations describing the mechanical behavior of elastic rods; the symbolic-numeric solution with Maple of the parametric self-adjoint 2D elliptic boundary-value problem with the aid of a high-accuracy finite element method; and a new symbolic-numeric preconditioned solver for incompressible Navier–Stokes equations using the integral form of collocation equations.

Applications of CASs in mechanics, physics, and biology are represented by the following themes: investigation of the asymptotic stability of a satellite with a gravitational stabilizer; satellite dynamics subject to damping torques; *Mathematica*-based analysis of the relative equilibria stability in a problem of celestial mechanics; stationary motions of the generalized Kowalewski gyrostat and their stability; and symbolic versus numerical computation and visualization of parameter regions for multistationarity of biological networks.

The remaining topics include the computation of some integer sequences in Maple; algorithms for computing the integer points of a polyhedron; a divide and conquer algorithm for sparse nonlinear interpolation; and normalization of indexed differentials based on function distance invariants.

The CASC 2017 workshop was supported financially by the National Center for Mathematics and Interdisciplinary Sciences of the Chinese Academy of Sciences, the

Academy of Mathematics and Systems Science of the Chinese Academy of Sciences, and the Key Laboratory of Mathematics Mechanization of the Chinese Academy of Sciences. We appreciate that they provided free accommodation for a number of participants.

Our particular thanks are due to the members of the CASC 2017 local organizing committee at the Key Laboratory of Mathematics Mechanization, Chinese Academy of Sciences, Jin-San Cheng, Changbo Chen, Ruyong Feng, and Zhikun She, who ably handled all the local arrangements in Beijing. In addition, Profs. Xiao-Shan Gao and Jin-San Cheng provided us with the history of computer algebra activities in China.

Furthermore, we want to thank all the members of the Program Committee for their thorough work. We are grateful to Matthias Orth (Universität Kassel) for his technical help in the preparation of the camera-ready manuscript for this volume. Finally, we are grateful to the CASC publicity chair, Andreas Weber (Rheinische Friedrich-Wilhelms-Universität Bonn), and his assistant, Hassan Errami, for the design of the conference poster and the management of the conference web page http://www.casc.cs.uni-bonn.de.

July 2017

Vladimir P. Gerdt
Wolfram Koepf
Werner M. Seiler
Evgenii V. Vorozhtsov

Organization

CASC 2017 was organized jointly by the Institute of Mathematics at Kassel University and the Key Laboratory of Mathematics Mechanization, Academy of Mathematics and System Sciences, Chinese Academy of Science, Beijing, China.

Workshop General Chairs

Xiao-Shan Gao, Beijing
Vladimir P. Gerdt, Dubna
Werner M. Seiler, Kassel

Program Committee Chairs

Wolfram Koepf, Kassel
Evgenii V. Vorozhtsov, Novosibirsk

Program Committee

Moulay Barkatou, Limoges
François Boulier, Lille
Jin-San Cheng, Beijing
Victor F. Edneral, Moscow
Matthew England, Coventry
Jaime Gutierrez, Santander
Sergey A. Gutnik, Moscow
Thomas Hahn, Munich
Jeremy Johnson, Philadelphia
Victor Levandovskyy, Aachen
Marc Moreno Maza, London, Canada
Veronika Pillwein, Linz
Alexander Prokopenya, Warsaw
Georg Regensburger, Linz
Eugenio Roanes-Lozano, Madrid
Valery Romanovski, Maribor
Doru Stefanescu, Bucharest
Thomas Sturm, Saarbrücken
Akira Terui, Tsukuba
Elias Tsigaridas, Paris
Jan Verschelde, Chicago
Stephen M. Watt, W. Ontario, Canada

Additional Reviewers

Alexander Bathkin
Russell Bradford
Martin Bromberger
Alexander Bruno
Xiaojie Dou
Bruno Grenet
Qiaolong Huang
Manuel Kauers
Denis Khmelnov
Thomas Peter
Hamid Rahkooy
Daniel Robertz
Olivier Ruatta
Vasily Shapeev
Yilei Tang
Francis Valiquette
Nathalie Verdière
Andreas Weber
Zafeirakis Zafeirakopoulos

Local Organization

Jin-San Cheng, Beijing (Chair)
Changbo Chen, Chongqing
Ruyong Feng, Beijing
Zhikun She, Beijing

Publicity Chair

Andreas Weber, Bonn

Website

http://www.casc.cs.uni-bonn.de/2017
(Webmaster: Hassan Errami)

Contents

Linear Differential Systems with Infinite Power Series Coefficients (*Invited Talk*)

S.A. Abramov(✉)

Dorodnitsyn Computing Centre, Federal Research Center Computer Science and Control of Russian Academy of Sciences, Vavilova, 40, Moscow 119333, Russia
sergeyabramov@mail.ru

Abstract. Infinite power series may appear as inputs for certain mathematical problems. This paper examines two possible solutions to the problem of representation of infinite power series: the algorithmic representation (for each series, an algorithm is specified that, given an integer i, finds the coefficient of x^i, — any such algorithm defines a so called computable, or constructive, series) and a representation in an approximate form, namely, in a truncated form.

1 Introduction

Infinite power series play an important role in mathematical studies. Those series may appear as inputs for certain mathematical problems. In order to be able to discuss the corresponding algorithms, we must agree on representation of the infinite series (algorithm inputs are always objects represented by specific finite words in some alphabet). This paper examines two possible solutions to the problem of representation of power series.

In Sect. 2, we consider the algorithmic representation. For each series in x, an algorithm is specified that, given an integer i, finds the coefficient of x^i. Any deterministic algorithms are allowed (any such algorithm defines a so called computable, or constructive series). Here there is a dissimilarity with the publications [14], [15, Chap. 10], where some specific case of input (mainly the hypergeometric type) is considered, and the coefficients of the power series which are returned by the corresponding algorithms can be given "in closed form".

For example, suppose that a linear ordinary differential system S of arbitrary order with infinite formal power series coefficients is given, decide whether the system has non-zero Laurent series, regular, or formal exponential-logarithmic solutions, and find all such solutions if they exist. If the coefficients of the original systems are arbitrary formal power series represented algorithmically (thus, we are not able, in general, to recognize whether a given series is equal to zero or not) then these three problems are algorithmically undecidable, and this can be deduced from the classical results of Turing [21]. But, it turns out that the

S.A. Abramov—Supported in part by the Russian Foundation for Basic Research, project No. 16-01-00174.

V.P. Gerdt et al. (Eds.): CASC 2017, LNCS 10490, pp. 1–15, 2017.
DOI: 10.1007/978-3-319-66320-3_1

first two problems are decidable in the case when we know in advance that a given system S is of full rank [5]. However, the third problem (finding formal exponential-logarithmic solutions) is not decidable even in this case [3]. It is shown that, despite the fact that such a system has a basis of formal exponential-logarithmic solutions involving only computable (i.e., algorithmically represented) series, there is no algorithm to construct such a basis. But, it is possible to specify a limited version of the third problem, for which there is an algorithm of the desired type: namely, if S and a positive integer d are such that for the system S the existence of at least d linearly independent solutions is guaranteed, then we can construct such d solutions [20].

It is shown also that the algorithmic problems connected with the ramification indices of irregular formal solutions of a given system are mostly undecidable even if we fix a conjectural value ρ of the ramification index [2]. However, there is nearby an algorithmically decidable problem: if a system S of full rank and positive integers ρ, d are such that for S the existence at least of d linearly independent formal solutions of ramification index ρ is guaranteed then one can compute such d solutions of S.

Thus, when we use the algorithmic way of power series representation, a neighborhood of algorithmically solvable and unsolvable problems is observed.

For the solvable problems mentioned above, a Maple implementation is proposed [9]. In Sect. 2.2, we report some experiments.

Note that the ring of computable formal power series is smaller than the ring of all formal power series because not every sequence of coefficients can be represented algorithmically. Indeed, the set of elements of the constructive formal power series is countable (each of the algorithms is a finite word in some fixed alphabet) while the set of all power series is uncountable.

In Sect. 3, we consider an "approximate" representation. A well-known example is the results [16] related to the number of terms of entries in A that can influence some components of formal exponential-logarithmic solutions of a differential system $x^s y' = Ay$, where s is a given non-negative integer, A is a matrix whose entries are power series. As a further example we consider matrices with infinite power series entries and suppose that those series are represented in an approximate form, namely, in a truncated form. Thus, it is assumed that a polynomial matrix P which is the l-truncation (l is a non-negative integer, $\deg P = l$) of a power series matrix A is given, and P is non-singular, i.e., $\det P \neq 0$. In [4], it is proven that the question of strong non-singularity, i.e., the question whether P is not the l-truncation of a singular matrix having power series entries, is algorithmically decidable. Assuming that a non-singular power series matrix A (which is not known to us) is represented by a strongly non-singular polynomial matrix P, we give a tight lower bound for the number of initial terms of A^{-1} which can be determined from P^{-1}.

We discuss the possibility of applying the proposed approach to "approximate" linear higher-order differential systems: if a system is given in the approximate truncated form and the leading matrix is strongly non-singular then the results [16,18] and their generalization can be used, and the number of reliable terms of Laurent series solution can be estimated by the algorithm proposed in [6].

Theorems are known that if a system has a solution in the form of a series, then this system also has a solution in the form of a series with some specific properties such that the initial terms of these series coincide (and estimates of the number of coinciding terms are given), see, e.g., [13]. To avoid misunderstandings, note that this is a different type of task. We are considering a situation where a truncated system is initially given, and we do not know the original system. We are trying to establish, whether it is possible to get from the solutions of this system an information on solutions of any system obtained from this system by a prolongation of the polynomial coefficients to series.

The information that can be extracted from truncated series, matrices, systems, etc. may be sufficient to obtain certain characteristics of the original (untruncated) objects. Naturally, these characteristics are incomplete, but may suffice for some purposes.

In Sect. 4, we discuss the fact that the width of a given full-rank system S with computable formal power series coefficients can be found, where the width of S is the smallest non-negative integer w such that any l-truncation of S with $l \geqslant w$ is a full-rank system. It is shown also that the above-mentioned value w exists for any full-rank system [5]. We introduce also the notion of the s-width. This is done on the base of the notion of the strong non-singularity.

2 Algorithmic Representation

Definition 1. We suppose that for each series $a(x) = \sum_{i=0}^{\infty} a_i x^i$ under consideration, an algorithm Ξ_a (a procedure, terminating in finitely many steps) such that $a(x) = \sum_{i=0}^{\infty} \Xi_a(i) x^i$, i.e., such that $a_i = \Xi_a(i)$ $\forall i$ is given. We will call such series *computable* (or *constructive*).

2.1 Computable Infinite Power Series in the Role of Coefficients of Linear Differential Systems

Let K be a field of characteristic 0. We will use the standard notation $K[x]$ for the ring of *polynomials* in x and $K(x)$ for the field of *rational functions* of x with coefficients in K. Similarly, we denote by $K[[x]]$ the ring of *formal power series* and $K((x)) = K[[x]][x^{-1}]$ its quotient field (the field of *formal Laurent series*) with coefficients in K. The ring of $n \times n$-matrices with entries belonging to a ring (a field) R is denoted by $\mathrm{Mat}_n(R)$.

Definition 2. A ring (field) is said to be *constructive* if there exist algorithms for performing the ring (field) operations and an algorithm for zero testing in the ring (field).

We suppose that the ground field K is a constructive field of characteristic 0. We write θ for $x\frac{d}{dx}$ and consider differential systems of the form

$$A_r(x)\theta^r y + A_{r-1}(x)\theta^{r-1} y + \cdots + A_0(x) y = 0 \tag{1}$$

where $y = (y_1, \ldots, y_m)^T$ is a column vector of unknown functions, and $y_1, \ldots, y_m$ are the *components* of y.

For the matrices

$$A_0(x), A_1(x), \ldots, A_r(x) \qquad (2)$$

we have $A_i(x) \in \mathrm{Mat}_m(K[[x]])$, $i = 0, 1, \ldots, r$, and $A_r(x)$ (the *leading* matrix of the system) is non-zero.

We call elements of the matrices $A_i(x)$ *system coefficients.* As the system coefficients will appear computable series.

It can be deduced from the classical results of Turing [21] that

We are not able, in general, to test whether a given computable series is equal to zero or not; for a square matrix whose entries are computable series - to test, whether this matrix is non-singular or not.

However, it turns out that the problems of finding solutions of some types are decidable in the case when we know in advance that a given differential system S is of full rank, i.e., that the equations of the system are linearly independent over $K[\theta]$. Algorithms for constructing local solutions of certain types can be proposed (the components of local solutions either are series in x, or contain such series as constituents). All the involved series are supposed to be formal.

Definition 3. The solutions whose components are formal Laurent series are *Laurent* solutions. The components of a *regular* solution are of the form

$$y_i(x) = \sum_{i=1}^{u} x^{\lambda_i} \sum_{s=0}^{k_i} g_{i,s}(x) \frac{\ln^s x}{s!}, \qquad (3)$$

where $u, k_i \in \mathbb{N}$, $\lambda_i \in \bar{K}$ and $g_{i,s}(x) \in \bar{K}((x))^m$ ($\bar{K}$ denotes the *algebraic closure* for K.)

Definition 4. A *proper formal (exponential-logarithmic) solution* of a system is a solution of the form

$$e^{Q(\frac{1}{t})} t^{\lambda} \Phi(t), \;\; x = t^{\rho}, \qquad (4)$$

where

$\lambda \in \bar{K}$;

$Q(\frac{1}{t})$ is a polynomial in $\frac{1}{t}$ over $\bar{K}$ and the constant term of this polynomial is equal to zero;

ρ is a positive integer;

$\Phi(t)$ is a column vector with components in the form $\sum_{i=0}^{k} g_i(t) \log^i(t)$, and all $g_i(t)$ are power series over $\bar{K}$.

If ρ has the minimal possible value in representation (4) of a proper formal solution then ρ is the *ramification index* of that solution.

A *formal (exponential-logarithmic) solution* is a finite linear combination with coefficients from $\bar{K}$ of proper formal solutions.

Formal exponential-logarithmicis solutions are of a special interest since, e.g., any system of the form $y' = Ay$, where A is an $m \times m$-matrix whose entries are formal Laurent series, has m linearly independent (over $\bar{K}$) formal solutions [19].

The main problems which are considered in this section are the following. Suppose that a linear ordinary differential system S of arbitrary order having the form (1) with computable formal power series coefficients (entries of the matrices $A_i(x)$) is given, test whether the system has

(1) non-zero Laurent series,
(2) regular, or
(3) formal exponential-logarithmic solutions,

and find all such solutions if they exist.

Theorem 1. (i) [5,8] *The first two problems are decidable in the case when we know in advance that a given system S is of full rank, i.e., in the case where the equations of the given system are linearly independent over the ring $K[\theta]$.*

(ii) [3] *Despite the fact that such a system has a basis of formal exponential-logarithmic solutions involving only computable series, there is no algorithm to construct such a basis.*

However, it is possible to specify a limited version of the third problem, for which there is an algorithm of the desired type:

Theorem 2 [20]. *If S and a positive integer d are such that for the system S the existence of at least d linearly independent solutions is guaranteed, we can construct such d solutions.*

It is shown also that the algorithmic problems connected with the ramification indices of irregular formal solutions of a given system are mostly undecidable even if we fix a conjectural value of the ramification index:

Theorem 3 [2]. *There exists no algorithm which, given a system S with computable power series coefficients and a positive integer ρ, tests the existence of a proper formal solution of ramification index ρ for the system S.*

Thus,

When we use the algorithmic way of power series representation, a neighborhood of algorithmically solvable and unsolvable problems is observed.

2.2 Procedures for Constructing Local Solutions

For the solvable problems mentioned above, a Maple [17] implementation as procedures of the package `EG` was proposed [9]. The package is available from http://www.ccas.ru/ca/eg.

We report some experiments (Figs. 1, 2 and 3). The degree of the truncation of the series involved in the solutions returned by our procedures is not less

```
> sys := Matrix([[Sum(f(k)*x^k,k=0..infinity),-x^2,-1-x],
                 [-x^2,-Sum(x^k,k=1..infinity),-1-x^3],
                 [-x^3,x,-Sum(x^k,k=0..infinity)]]).y(x)+
        Matrix([[x^2,0,0],[0,1,0],[0,0,1]]).theta(y(x),x,1)+
        Matrix([[x+Sum(x^k,k=3..infinity),0,0],
                 [0,Sum(x^k,k=1..infinity),0],
                 [0,0,Sum(x^k,k=1..infinity)]]).theta(y(x),x,2);
```

$$sys := \begin{bmatrix} \sum_{k=0}^{\infty} f(k)\,x^k & -x^2 & -1-x \\ -x^2 & -\left(\sum_{k=1}^{\infty} x^k\right) & -x^3-1 \\ -x^3 & x & -\left(\sum_{k=0}^{\infty} x^k\right) \end{bmatrix} . \, y(x) + \begin{bmatrix} x^2 & 0 & 0 \\ 0 & 1 & 0 \\ 0 & 0 & 1 \end{bmatrix} . \, \theta(y(x), x, 1) + \begin{bmatrix} x+\sum_{k=3}^{\infty} x^k & 0 & 0 \\ 0 & \sum_{k=1}^{\infty} x^k & 0 \\ 0 & 0 & \sum_{k=1}^{\infty} x^k \end{bmatrix} . \, \theta(y(x), x, 2)$$

Fig. 1. An example of a system ($f(k)$ is not yet defined).

than it is required by the user. That degree can be even bigger: in any case, it is big enough to represent the dimension of the space of the solutions under consideration.

Let $m = 3$ and the system be of the form presented in Fig. 1.

Suppose that we define the procedure for computing coefficients of the series $\sum_{k=0}^{\infty} f(k)x^k$ as presented in Fig. 2.

```
> f := proc(k)
        if k::integer then
            piecewise(k<0, 0, k <= 1, -1, k = 2, 1, -k^2+k)
        else
            'procname(k)'
        end if
     end proc:
>
```

Fig. 2. $f(k)$ is defined.

(Thus, $\sum_{k=0}^{\infty} f(k)x^k = -1 - x + x^2 + \sum_{k=3}^{\infty}(-k^2 + k)x^k$.) The results of the search for Laurent, regular and formal solutions are presented on Fig. 3.

The procedure of the construction of all formal solutions constructs also all regular, and in particular, all Laurent solutions. Actually, one procedure `EG[FormalSolution]` is sufficient in order to obtain solutions of all three types. However, if it is required to construct, say, only Laurent solutions, then it is advantageous to use procedure `EG[LaurentSolution]`, because it will construct them considerably faster, even if the original system has no formal solutions but the Laurent ones. For this reason, we propose three procedures for searching solutions of various types.

In conclusion of this section note that the ring of computable formal power series is smaller than the ring of all formal power series because not every sequence of coefficients can be represented algorithmically. Indeed, the set of

```
> EG:-LaurentSolution(sys, theta, y(x), 0);
        [x_c1 + O(x^2), -x_c1 + O(x^2), -x_c1 + O(x^2)]
>
> EG:-RegularSolution(sys, theta, y(x), 0);
[ln(x) (x_c1 + O(x^2)) + x_c2 + O(x^2), ln(x) (-x_c1 + O(x^2)) + _c1 + x (-_c2 + 2 _c1) + O(x^2), ln(x) (-x_c1 + O(x^2))
   - x_c2 + O(x^2)]
>
> Res := EG:-FormalSolution(sys, theta, y(x), t, 'solution_dimension' = 6):
  Res[1]; Res[2]; Res[3];
[x = t, [ln(t) (t_c1 + O(t^2)) + t_c2 + O(t^2), ln(t) (-t_c1 + O(t^2)) + _c1 + t (-_c2 + 2 _c1) + O(t^2), ln(t) (-t_c1 + O(t^2))
   - t_c2 + O(t^2)]]
[x = t, e^(1/t) [ln(t) O(t^3) - t^2 _c3 + O(t^3), ln(t) (t^2 _c3 + O(t^3)) + t^2 _c4 - t_c3 + O(t^3), ln(t) O(t^3) - t^2 _c3 - t_c3 + O(t^3)]]
        [x = t^2, e^(-2/t) [sqrt(t) (_c5 + O(t)), sqrt(t) O(t), sqrt(t) O(t)]]
> 
```

Fig. 3. Laurent, regular and formal solutions of the system.

elements of the computable formal power series is countable (each of the algorithms is a finite word in some fixed alphabet) while the set of all power series is uncountable.

3 Approximate (Truncated) Representation

Now, we consider an "approximate" representation of series.

A well-known example [16] is the result by Lutz and Schäfke. It is related to the number of terms of entries of a power series matrix A that can influence initial terms of some constituents of formal exponential-logarithmic solutions of a differential system $x^s y' = Ay$, where s is a non-negative integer.

As a further example [4], we consider matrices with infinite power series entries and suppose that those series are represented in an approximate form, namely, in a truncated form.

We start with introducing some notions.

If $l \in \mathbb{Z}, a \in K((x))$ then we define the *l-truncation* $a^{\langle l \rangle}$ which is obtained by omitting all the terms of degree larger than l in a. For a non-zero element $a = \sum a_i x^i$ of $K((x))$, we denote by $\operatorname{val} a$ the *valuation* of a defined by $\operatorname{val} a = \min\{i \text{ such that } a_i \neq 0\}$; by convention, $\operatorname{val} 0 = \infty$.

For $A \in \operatorname{Mat}_n(K((x)))$, we define $\operatorname{val} A$ as the minimum of the valuations of the entries of A. We define the *leading coefficient* of a non-zero matrix $A \in \operatorname{Mat}_n(K((x)))$ as $\operatorname{lc} A = (x^{-\operatorname{val} A} A)|_{x=0}$. For $A \in \operatorname{Mat}_n(K[x])$, we define $\deg A$ as the maximum of the degrees of the entries of A.

The notation A^T is used for the transpose of a matrix (vector) A. I_n is the *identity* $n \times n$-matrix.

Given $A \in \operatorname{Mat}_n(K((x)))$, we define the matrix $A^{\langle l \rangle} \in \operatorname{Mat}_n(K[x, x^{-1}])$ obtained by replacing the entries of A by their l-truncations (if $A \in \operatorname{Mat}_n(K[[x]])$ then $A^{\langle l \rangle} \in \operatorname{Mat}_n(K[x])$).

If $P \in \mathrm{Mat}_n(K[x])$ then any $\hat{P} \in \mathrm{Mat}_n(K[[x]])$ such that $(\hat{P})^{\langle \deg P \rangle} = P$ is a *prolongation* of P.

3.1 Strongly Non-singular Matrices

Definition 5. A polynomial matrix P which is non-singular, i.e., $\det P \neq 0$, is *strongly non-singular* if P is not the l-truncation ($l = \deg P$) of a singular matrix having power series entries; in other words, P is strongly non-singular if $\det \hat{P} \neq 0$ for any prolongation $\hat{P}$ of P.

It is proven that the question of strong non-singularity is algorithmically decidable. For the answer to this question, the number

$$h = \deg P + \mathrm{val}\, P^{-1} \tag{5}$$

plays the key role.

Theorem 4 [4]. *P is strongly non-singular if and only if*

$$\deg P + \mathrm{val}\, P^{-1} \geqslant 0, \tag{6}$$

i.e., $h \geqslant 0$.

Example 1. If P is a non-singular constant matrix then P is a strongly non-singular due to the latter proposition. However, the matrix

$$\begin{pmatrix} x & 0 \\ 1 & x \end{pmatrix}, \tag{7}$$

is not strongly non-singular:

$$\det \begin{pmatrix} x & x^2 \\ 1 & x \end{pmatrix} = 0. \tag{8}$$

This could be recognized in advance: for (7) we have $\deg P = 1, \mathrm{val}\, P^{-1} = -2$ (since $\det P = x^2$), and the inequality $h \geqslant 0$ does not hold: $1 - 2 = -1$. □

Assuming that a non-singular power series matrix A (which is not known to us) is represented by a strongly non-singular polynomial matrix P, we give a tight lower bound for the number of initial terms of entries of A^{-1} which can be determined from P^{-1}.

Theorem 5 [4]. *Let P be a polynomial matrix. If the inequality $h \geqslant 0$ holds then first, for any prolongation $\hat{P}$, the valuations of the determinant and the inverse matrix of the approximate matrix and, resp., of the determinant and the inverse of the prolonged matrix coincide. Second, in the determinants of the approximate and prolonged matrices, the coefficients coincide for $x^{\mathrm{val}\, \det P}$, as well as h subsequent coefficients (for larger degrees of x). A similar statement holds for the inverse matrix. The bound h is tight.*

Example 2. Let

$$P = \begin{pmatrix} 1+x & 0 \\ 1 & 1-x \end{pmatrix}.$$

Here $h = 1$. The matrix P is strongly non-singular.

Let

$$\hat{P} = \begin{pmatrix} 1+x+x^2+\dots & 0 \\ 1 & 1-x \end{pmatrix}.$$

We have

$$\det P = 1 - x^2 = \underline{1 + 0\cdot x} - 1\cdot x^2, \quad \det \hat{P} = \underline{1 + 0\cdot x} + 0\cdot x^2 + \dots$$

We have also:

$$P^{-1} = \begin{pmatrix} 1/(1+x) & 0 \\ -1/(1-x^2) & 1/(1-x) \end{pmatrix} = \begin{pmatrix} \underline{1-x}+x^2+\dots & \underline{0+0\cdot x} \\ \underline{-1+0\cdot x}-x^2-\dots & \underline{1+x}+x^2+\dots \end{pmatrix},$$

$$\hat{P}^{-1} = \begin{pmatrix} 1-x & 0 \\ -1 & 1/1-x \end{pmatrix} = \begin{pmatrix} \underline{1-x} & \underline{0+0\cdot x} \\ \underline{-1+0\cdot x} & \underline{1+x}+x^2+\dots \end{pmatrix}.$$

□

As a consequence of Theorem 5, if val det $P = e$ then val det $\hat{P} = e$ and

$$\det P - \det \hat{P} = O(x^{e+h+1}).$$

Similarly, if val $P^{-1} = e$ then val $(\hat{P})^{-1} = e$ and

$$P^{-1} - \hat{P}^{-1} = O(x^{e+h+1}).$$

3.2 When only a Truncated System Is Known

In this section, we are interested in the following question. Suppose that for a system S of the form (1) only a finite number of terms of the entries of $A_0(x), A_1(x), \dots, A_r(x)$ is known, i.e., we know not the system S itself but the system $S^{\langle l\rangle}$ for some non-negative integer l. Suppose that we also know that

- $\mathrm{ord} S^{\langle l\rangle} = \mathrm{ord} S$,
- $A_r(x)$ is invertible.

How many terms of Laurent series solutions of S can be determined from the given "approximate" system $S^{\langle l\rangle}$?

We first recall the following result:

Proposition 1 [6, Proposition 6]. *Let S be a system of the form (1) and*

$$\gamma = \min_i \text{val}\ \left(A_r^{-1}(x)A_i(x)\right), \quad q = \max\{-\gamma, 0\}.$$

There exists an algorithm that uses only the terms of degree less than

$$rmq + \gamma + \text{val}\ \det A_r(x) + 1 \tag{9}$$

of the entries of the matrices $A_0(x), A_1(x), \ldots, A_r(x)$, and computes a non-zero polynomial (the so called indicial *polynomial* [12, Chap. 4, Sect. 8], [10, Definition 2.1], [6, Sect. 3.2]*) $I(\lambda)$ such that:*

- *if $I(\lambda)$ has no integer root then (1) has no solution in $K((x))^m \setminus \{0\}$,*
- *otherwise, let e_*, e^* be the minimal and maximal integer roots of $I(\lambda)$; then the sequence*

$$a_k = rmq + \gamma + \text{val}\ \det A_r(x) + \max\{e^* - e_* + 1,\ k + (rm-1)q\}, \tag{10}$$

$k = 1, 2, \ldots,$ is such that for any $e \in \mathbb{Z}$, $k \in \mathbb{Z}^+$ and column vectors

$$c_e, c_{e+1}, \ldots, c_{e+k-1} \in K^m,$$

the system S possesses a solution $y(x) \in K((x))^m$ of the form

$$y(x) = c_e x^e + c_{e+1}x^{e+1} + \cdots + c_{e+k-1}x^{e+k-1} + O(x^{e+k}),$$

if and only if, the system $S^{\langle a_l \rangle}$ possesses a solution $\tilde{y}(x) \in K((x))^m$ such that $\tilde{y}(x) - y(x) = O(x^{e+k})$.

Using the latter proposition we prove

Theorem 6 [4]. *Let Σ be a system of the form*

$$P_r(x)\theta^r y + P_{r-1}(x)\theta^{r-1}y + \cdots + P_0(x)y = 0$$

with polynomial matrices $P_0(x), P_1(x), \ldots, P_r(x)$. Let its leading matrix $P_r(x)$ be strongly non-singular. Let

$$d = \deg P_r,\ \ p = -\text{val}\, P_r^{-1},\ \ h = d - p,\ \ \gamma = \min_{0 \leqslant i \leqslant r-1} (\text{val}\,(P_r^{-1}P_i))$$

be such that the inequality

$$h - p - \gamma \geqslant 0$$

holds. Let $I(\lambda)$ be the indicial polynomial of Σ. Let the set of integer roots of $I(\lambda)$ be non-empty, and e_*, e^* be the minimal and maximal integer roots of $I(\lambda)$. Let a non-negative integer k satisfy the equality

$$\max\{e^* - e_* + 1,\ k + (rm-1)q\} = l - rmq - \gamma - \text{val}\ \det P_r(x). \tag{11}$$

Let $\hat{\Sigma}$ be an arbitrary system of the form (1) such that $\hat{\Sigma}^{\langle l \rangle} = \Sigma$ for $l = \deg \Sigma$ (i.e., $\hat{\Sigma}$ is an arbitrary prolongation of Σ). Then for any $e \in \mathbb{Z}$, the system $\hat{\Sigma}$ possesses a solution

$$\hat{y}(x) \in K((x))^m, \quad \operatorname{val} \hat{y}(x) = e,$$

if and only if, the system Σ possesses a solution $y(x) \in K((x))^m$ such that

$$y(x) - \hat{y}(x) = O(x^{e+k+1}) \tag{12}$$

(evidently, the equalities $\operatorname{val} \hat{y}(x) = e$ and (12) imply that $\operatorname{val} y(x) = e$).

Example 3. Let

$$P_1 = \begin{pmatrix} 1 & 0 \\ 0 & 1-x \end{pmatrix}, \quad P_0 = \begin{pmatrix} 0 & -1 \\ -x + 2x^2 + 2x^3 + 2x^4 & -2 + 4x \end{pmatrix}.$$

For the first-order differential system Σ

$$P_1(x)\theta y + P_0(x)y = 0$$

we have

$$d = 1, \;\; p = 0, \;\; h = 1, \;\; \gamma = 0, \;\; I(\lambda) = \lambda(\lambda - 2), \;\; e^* - e_* + 1 = 3.$$

The conditions of Theorem 6 are satisfied.

The general solution of Σ is

$$y_1 = C_1 - C_1 x + C_2 x^2 - C_2 x^3 + 0x^4 + \frac{2C_1}{15}x^5 + \frac{C_1}{30}x^6 + \left(\frac{C_1}{210} + \frac{2C_2}{35}\right)x^7 + \dots,$$

$$y_2 = \quad - C_1 x + 2C_2 x^2 - 3C_2 x^3 + 0x^4 + \frac{2C_1}{3}x^5 + \frac{C_1}{5}x^6 + \left(\frac{C_1}{30} + \frac{2C_2}{5}\right)x^7 + \dots,$$

where C_1 and C_2 are arbitrary constants.

Equation (11) has the form $\max\{3, k\} = 4$, thus

$$k = 4.$$

This means that all Laurent series solutions of any system $\hat{\Sigma}$ of the form

$$A_1(x)\theta y + A_0(x)y = 0 \tag{13}$$

with non-singular matrix A_1 and such that $\hat{\Sigma}^{\langle 4 \rangle} = \Sigma$ (we have $\deg \Sigma = 4$) are power series solutions having the form

$$\hat{y}_1 = C_1 - C_1 x + C_2 x^2 - C_2 x^3 + O(x^5),$$

$$\hat{y}_2 = \quad - C_1 x + 2C_2 x^2 - 3C_2 x^3 + O(x^5),$$

where C_1, C_2 are arbitrary constants. Consider, e.g., the first-order differential system $\hat{\Sigma}$ of the form (13) with

$$A_1 = \begin{pmatrix} 1 & 0 \\ 0 & 1-x \end{pmatrix},$$

$$A_0 = \begin{pmatrix} 0 & -1 \\ -x + 2x^2 + 2x^3 + 2x^4 + 2x^5 + 2x^6 + x^7 + x^8 + \dots & -2 + 4x \end{pmatrix}.$$

Its general solution is

$$\hat{y}_1 = \underline{C_1 - C_1 x + C_2 x^2 - C_2 x^3} + 0x^4 + 0x^5 + 0x^6 + \frac{C_1}{35}x^7 + \dots,$$

$$\hat{y}_2 = \quad \underline{-C_1 x + 2C_2 x^2 - 3C_2 x^3} + 0x^4 + 0x^5 + 0x^6 + \frac{C_1}{5}x^7 + \dots,$$

what corresponds to the forecast and expectations. □

Remark 1. The latter example shows that Theorem 6 gives a tight bound for possible value of k: in that example that we cannot take $k + 1$ instead of k. Indeed, y_1 contains the term $\frac{2C_1}{15}x^5$, while $\hat{y}_1$ has factually no term of degree 5.

We see that the information that can be extracted from truncated series, matrices, systems, etc. may be sufficient to obtain certain characteristics of the original (untruncated) objects. Naturally, these characteristics are incomplete, but may suffice for some purposes.

In the context of truncated systems we considered only the problem of testing the existence and constructing Laurent series solutions, but we did not discuss similar problems related to regular and formal exponential-logarithmic solutions. We will continue to investigate this line of enquiry.

4 The Width

In conclusion, we discuss a plot which connects both thematic lines of the paper.

Definition 6 [4,5]**.** Let S be a system of full rank over $K[[x]][\theta]$. The minimal integer w such that $S^{\langle l \rangle}$ is of full rank for all $l \geqslant w$ is called the *width* of S The minimal integer w_s such that any system S_1 having power series coefficients and satisfying the condition $S_1^{\langle w_s \rangle} = S^{\langle w_s \rangle}$, is of full rank, is called the *s-width* (the *strong width*) of S.

We will use the notations $w(S), w_s(S)$ when it is convenient.

Any linear algebraic system can be considered as a linear differential system of zero order. This lets us state using the following example that for an arbitrary differential system S we have $w_s(M) \neq w(M)$ in general, however, the inequality

$$w_s(S) \geqslant w(S)$$

holds.

Example 4. Let A be

$$\begin{pmatrix} x & x^3 \\ 1 & x \end{pmatrix}, \tag{14}$$

then

$$w(A) = 1,$$

since $\det A^{\langle 0 \rangle} = 0$ and

$$A^{\langle 1 \rangle} = A^{\langle 2 \rangle} = \begin{pmatrix} x & 0 \\ 1 & x \end{pmatrix}, \ \det \begin{pmatrix} x & 0 \\ 1 & x \end{pmatrix} \neq 0,$$

and $A^{\langle l \rangle} = A$ when $l \geqslant 3$, $\det A \neq 0$. However, $w_s(A) > 1$, due to $\det \begin{pmatrix} x & x^2 \\ 1 & x \end{pmatrix} = 0$. It is easy to check that $w_s(A) = 2$. □

It was proven in [5, Theorem 2] that if a system S of the form (1) is of full rank then there exists the width w of S. The value w may be computed if the coefficients of S are represented algorithmically.

As for the idea of the proof from [5], it is shown that the rank-preserving EG-eliminations [1,7] give a confirmation for the fact that S is of full rank. That confirmation uses only a finite number of the terms of power series which are coefficients of S. For this, the induced recurrent system R is considered (such R is a specific recurrent system for the coefficients of Laurent series solutions of S). This system has polynomial coefficients of degree less than or equal to $r = \operatorname{ord} S$. The system S is of full rank if and only if R is of full rank as a recurrent system. A recurrent system of this kind can be transformed by a special version of the EG-eliminations [5, Sect. 3] into a recurrent system $\tilde{R}$ whose leading matrix is non-singular. This gives the confirmation mentioned above. It is important that only a finite number of the coefficients of R are involved in the obtained leading matrix of $\tilde{R}$ (due to some characteristic properties of the used version of the EG-eliminations). Each of polynomial coefficients of R is determined from a finite number (bounded by a non-negative integer N) of the coefficients of the power series involved in S. This proves the existence of the width and of the s-width as well. The mentioned number N can be computed algorithmically when all power series are represented algorithmically; thus, in this case we can compute the width of S since we can test [1,7,11] whether a finite order differential system with polynomial coefficients is of full rank or not. From this point we can consider step-by-step $S^{\langle N-1 \rangle}, S^{\langle N-2 \rangle}, \dots, S^{\langle 1 \rangle}, S^{\langle 0 \rangle}$ until there appears the first which is not of full rank. If all the truncated systems are of full rank then $w = 0$.

Concerning the s-width, we get the following theorem

Theorem 7 [4]. *Let S be a full rank system of the form (1). Then the s-width $w_s(S)$ is defined. If the power series coefficients of S are represented algorithmically then we can compute algorithmically a non-negative integer N such that $w_s(S) \leqslant N$.*

However, it is not exactly clear how to find the minimal value N, i.e., $w_s(S)$. Is this problem algorithmically solvable? The question is still open.

Acknowledgments. The author is thankful to M. Barkatou, D. Khmelnov, M. Petkovšek, E. Pflügel, A. Ryabenko and M. Singer for valuable discussions.

References

1. Abramov, S.: EG-eliminations. J. Differ. Eqn. Appl. **5**, 393–433 (1999)
2. Abramov, S.: On ramification indices of formal solutions of constructive linear ordinary differential systems. J. Symbolic Comput. **79**, 475–481 (2017)
3. Abramov, S.A., Barkatou, M.A.: Computable infinite power series in the role of coefficients of linear differential systems. In: Gerdt, V.P., Koepf, W., Seiler, W.M., Vorozhtsov, E.V. (eds.) CASC 2014. LNCS, vol. 8660, pp. 1–12. Springer, Cham (2014). doi:10.1007/978-3-319-10515-4_1
4. Abramov, S., Barkatou, M.: On strongly non-singular polynomial matrices. In: Schneider, C., Zima, E. (eds.) Advances in Computer Algebra: Proceedings of the Waterloo Workshop in Computer Algebra 2016. Springer, Heidelberg (2017, accepted)
5. Abramov, S., Barkatou, M., Khmelnov, D.: On full rank differential systems with power series coefficients. J. Symbolic Comput. **68**, 120–137 (2015)
6. Abramov, S.A., Barkatou, M.A., Pflügel, E.: Higher-order linear differential systems with truncated coefficients. In: Gerdt, V.P., Koepf, W., Mayr, E.W., Vorozhtsov, E.V. (eds.) CASC 2011. LNCS, vol. 6885, pp. 10–24. Springer, Heidelberg (2011). doi:10.1007/978-3-642-23568-9_2
7. Abramov, S., Bronstein, M.: Linear algebra for skew-polynomial matrices. Rapport de Recherche INRIA RR-4420 (2002). http://www.inria.fr/RRRT/RR-4420.html
8. Abramov, S.A., Khmelnov, D.E.: Regular solutions of linear differential systems with power series coefficients. Program. Comput. Softw. **40**(2), 98–106 (2014)
9. Abramov, S.A., Ryabenko, A.A., Khmelnov, D.E.: Procedures for searching local solutions of linear differential systems with infinite power series in the role of coefficients. Program. Comput. Softw. **42**(2), 55–64 (2016)
10. Barkatou, M., Pflügel, E.: An algorithm computing the regular formal solutions of a system of linear differential equations. J. Symbolic Comput. **28**, 569–588 (1999)
11. Beckermann, B., Cheng, H., Labahn, G.: Fraction-free row reduction of matrices of skew polynomials. In: Proceedings of the ISSAC 2002, pp. 8–15. ACM, New York (2002)
12. Coddington, E., Levinson, N.: Theory of Ordinary Differential Equations. McGraw-Hill, New York (1955)
13. Denef, J., Lipshitz, L.: Power series solutions of algebraic differential equations. Math. Ann. **267**, 213–238 (1984)
14. Koepf, W.: Power series in computer algebra. J. Symbolic Comput. **13**, 581–603 (1992)
15. Koepf, W.: Computeralgebra. Eine algorithmisch orientierte Einführung. Springer, Heidelberg (2006)
16. Lutz, D.A., Schäfke, R.: On the identification and stability of formal invariants for singular differential equations. Linear Algebra Appl. **72**, 1–46 (1985)
17. Maple online help. http://www.maplesoft.com/support/help/

18. Pflügel, E.: Effective formal reduction of linear differential systems. Appl. Algebra Eng. Commun. Comput. **10**(2), 153–187 (2000)
19. van der Put, M., Singer, M.F.: Galois Theory of Differential Equations. Grundlehren der mathematischen Wissenschaften, vol. 328. Springer, Heidelberg (2003)
20. Ryabenko, A.A.: On exponential-logarithmic solutions of linear differential systems with power series coefficients. Program. Comput. Softw. **41**(2), 112–118 (2015)
21. Turing, A.: On computable numbers, with an application to the Entscheidungsproblem. Proc. Lond. Math. Soc. Ser. 2 **42**, 230–265 (1936)

On the Asymptotic Stability of a Satellite with a Gravitational Stabilizer

Andrei V. Banshchikov(✉)

Matrosov Institute for System Dynamics and Control Theory
of Siberian Branch of Russian Academy of Sciences,
PO Box 292, 134, Lermontov Str., Irkutsk 664033, Russia
bav@icc.ru

Abstract. The problem of the influence of the structure of forces on the stability of the relative equilibrium of a controlled satellite with a gravitational stabilizer on the circular orbit is studied. In the space of entered parameters, the regions with different degrees of instability by Poincaré are found. Assuming an instability of a potential system, the problem of the possibility of its stabilization up to asymptotic stability is considered. A parametric analysis of the obtained inequalities with the help of "Mathematica" built-in tools for symbolic-numerical modelling is carried out.

1 Introduction

Investigation of stability and stabilization of nonlinear or linearized models of mechanical systems often leads to the problem of "parametric analysis" of the conditions (inequalities) obtained. In the case of parametric analysis, it is important to have a possibility to estimate the domain of values of the parameters under which a desired system's state is provided. Naturally, it is hard to hope for obtaining any readable analytical results for the models which have high dimensions and contain many parameters. At this stage, one can efficiently use software packages of computer algebra (SPCA) as well as the corresponding software elaborated on the basis of these software packages.

The paper considers a problem of stability of the position of relative equilibrium in the orbital coordinate system of a controlled satellite with a gravitational stabilizer. The mechanical system in question is a well-studied model (see, for example, the review [1]). To obtain sufficient stability conditions, the second Lyapunov method and the Barbashin–Krasovskii theorem were applied. As noted in [1], obtaining the necessary stability conditions (by linear equations of perturbed motion) leads to presenting very bulky calculations. In contrast to the passive stabilization and orientation systems, the possibilities of active control of a gravitational stabilizer are investigated in [2], in particular, the optimization of the system by degrees of stability and accuracy.

The application of computer algebra methods and SPCA capabilities to the problems of celestial mechanics has rich history and till today attracts academic attention (see, for example, [3,4]).

V.P. Gerdt et al. (Eds.): CASC 2017, LNCS 10490, pp. 16–26, 2017.
DOI: 10.1007/978-3-319-66320-3_2

2 Description and Construction of a Symbolical Model

The system's mass center moves along the Kepler circular orbit with constant angular velocity $\boldsymbol{\omega}$. For the description of a motion of the system, two right-handed rectangular Cartesian coordinate systems are introduced (the orbital coordinate system (OCS) and the coordinate system rigidly connected to a satellite). To define relative positioning of the axes of these coordinate systems, the directional cosines defined by the angles ψ, θ, φ of Euler's type, are used (see, for example, [2]). The stabilizer is a rigid rod with point mass at its free end. The rod is connected to the satellite with a 2-degree-of-freedom suspension. The rotation axes of the rod coincide with the direction of the axes of pitch and roll. The system is influenced by a gravitation moment. When moving undisturbed, the system's principal central axes of inertia coincide with the axes of orbital coordinate system, and the rod is oriented along the radius of the orbit. This is the equilibrium position of a satellite with the stabilizer in regard to OCS.

With the help of the developed software [5,6], the following results are obtained in a symbolic form on PC for the system of bodies in question:

- kinetic energy and force function of the approximate Newtonian field of gravitation;
- nonlinear equations of motion in Lagrange form of the 2nd kind;
- matrices of equations of perturbed motion in the first approximation in the vicinity of equilibrium position;
- coefficients of the system's characteristic equation.

Linearized in the vicinity of the equilibrium position, equations of motion for a satellite with a stabilizer are decomposed into two subsystems. Respectively, a "pitch" subsystem (θ) and a "yaw-and-roll" subsystem (ψ, φ) are:

$$\begin{cases} M_1\,\ddot{q}_1 + K_1\,q_1 = Q_1 \\ M_2\,\ddot{q}_2 + G\,\dot{q}_2 + K_2\,q_2 = Q_2, \end{cases} \tag{1}$$

where all derivatives are calculated on dimensionless time $\tau = \omega t$ ($\omega = |\boldsymbol{\omega}|$ is the module of orbital angular velocity); $q_1 = \begin{pmatrix} \theta \\ \delta \end{pmatrix}$, $q_2 = \begin{pmatrix} \psi \\ \varphi \\ \sigma \end{pmatrix}$; δ, σ are rotation angles of the rod with regard to the satellite's body; $Q_1 = \begin{pmatrix} 0 \\ Q_\delta \end{pmatrix}$, $Q_2 = \begin{pmatrix} 0 \\ 0 \\ Q_\sigma \end{pmatrix}$ are control forces;

$$M_1 = \begin{pmatrix} c & f \\ f & d \end{pmatrix}; \quad K_1 = 3\begin{pmatrix} b-a & f \\ f & f \end{pmatrix}; \quad M_2 = \begin{pmatrix} a & 0 & 0 \\ 0 & b & f \\ 0 & f & d \end{pmatrix};$$

$$K_2 = \begin{pmatrix} c-b & 0 & 0 \\ 0 & 4(c-a) & 4f \\ 0 & 4f & 3f+d \end{pmatrix}; \quad G = \begin{pmatrix} 0 & c-b-a & 0 \\ a+b-c & 0 & 0 \\ 0 & 0 & 0 \end{pmatrix}.$$

Here, we introduce the following notations:

$$a = J_y; \quad b = J_x + m\,r\,(l+r) + \frac{1}{3}m\,l^2 + m_0(l+r)^2; \quad c = b + J_z - J_x;$$
$$d = \Big(\frac{m}{3} + m_0\Big)l^2; \quad f = \Big(\frac{m}{2} + m_0\Big)rl + d; \quad c - b - a = J_z - J_x - J_y,$$

where m and m_0 are masses of the rod and the point load at the end, respectively; $l>0$ is the rod length; $r \geq 0$ is the distance from the system's mass center to the point of attachment of the rod; J_x, J_y, J_z; a, b, c are principal inertia moments of the satellite and whole system, respectively.

Taking into account the mass distribution in the system and in the ellipsoid of inertia of rigid body, the following inequalities are valid

$$\begin{gathered} b > a > 0, \quad c > a, \quad f > d > 0, \quad c > f, \quad b > f, \\ c + a - b \equiv J_z + J_y - J_x > 0, \quad b + a - c \equiv J_x + J_y - J_z > 0, \end{gathered} \tag{2}$$

Equation (1) may be interpreted as equations of oscillations of a mechanical system influenced by potential (with the matrices K_1, K_2) and gyroscopic (with the matrix G) forces. These forces are determined by gravitation forces as well as by orbital motion. The matrices M_1 and M_2 play the role of diagonal blocks of a positive definite matrix of kinetic energy.

3 Formulation of the Problem

According to Kelvin–Chetaev's theorems [7], examination of stability of trivial solution begins with the analysis of the matrix of potential forces. Let us write out the conditions of positive definiteness of matrices K_1, K_2:

$$b > a + f, \quad c > b, \quad (c-a)(3f+d) - 4f^2 > 0. \tag{3}$$

Let us assume that

(1) for the "pitch" subsystem, the values of the parameters satisfy the condition $a < b < a + f$ (i.e., the first inequality in (3) is violated);
(2) for the "yaw-and-roll" subsystem, the last inequality in (3) or simultaneously the second and third inequalities are changed to the opposite.

Taking into account the assumptions presented, the system is unstable when initial potential forces are in action. The simultaneous stabilization of the two subsystems by additional forces of different nature is required. For this purpose, control forces with the suspension of the rod are added into the right-hand sides of the motion Eq. (1) as it is shown below

$$Q_\delta = \tilde{k}^*_\theta\,\dot\theta - \tilde{k}^*_\delta\,\dot\delta + \tilde{k}_\theta\theta - \tilde{k}_\delta\delta; \qquad Q_\sigma = \tilde{k}^*_\varphi\,\dot\varphi - \tilde{k}^*_\sigma\,\dot\sigma + \tilde{k}_\varphi\varphi - \tilde{k}_\sigma\sigma, \tag{4}$$

where $\tilde{k}^*_\theta = \dfrac{k^*_\theta}{\omega}$; $\tilde{k}^*_\delta = \dfrac{k^*_\delta}{\omega}$; $\tilde{k}_\theta = \dfrac{k_\theta}{\omega^2}$; $\tilde{k}_\delta = \dfrac{k_\delta}{\omega^2}$; $\tilde{k}^*_\varphi = \dfrac{k^*_\varphi}{\omega}$; $\tilde{k}^*_\sigma = \dfrac{k^*_\sigma}{\omega}$; $\tilde{k}_\varphi = \dfrac{k_\varphi}{\omega^2}$; $\tilde{k}_\sigma = \dfrac{k_\sigma}{\omega^2}$ are constant coefficients.

The objective of the paper is to investigate the effect of the structure of forces on the stability of the equilibrium position of system (1). In addition, the problem of the possibility of ensuring the asymptotic stability of the two subsystems by a "reduced" set of forces represented in (4) is formulated.

By splitting the matrices in terms of velocities and coordinates in Eq. (1) into the symmetric and skew-symmetric parts, it is not difficult to write out the structure of the forces affecting the system. For example, concerning the "yaw-and-roll" subsystem, potential (with a matrix P_2), non-conservative (N_2), dissipative (D_2) and gyroscopic (G_2) forces are added to the initial potential (with a matrix K_2) and gyroscopic (with a matrix G) forces, where

$$P_2 = \begin{pmatrix} 0 & 0 & 0 \\ 0 & 0 & -\frac{\tilde{k}_\varphi}{2} \\ 0 & -\frac{\tilde{k}_\varphi}{2} & \tilde{k}_\sigma \end{pmatrix}; \quad N_2 = \begin{pmatrix} 0 & 0 & 0 \\ 0 & 0 & \frac{\tilde{k}_\varphi}{2} \\ 0 & -\frac{\tilde{k}_\varphi}{2} & 0 \end{pmatrix};$$

$$D_2 = \begin{pmatrix} 0 & 0 & 0 \\ 0 & 0 & -\frac{\tilde{k}^*_\varphi}{2} \\ 0 & -\frac{\tilde{k}^*_\varphi}{2} & \tilde{k}^*_\sigma \end{pmatrix}; \quad G_2 = \begin{pmatrix} 0 & 0 & 0 \\ 0 & 0 & \frac{\tilde{k}^*_\varphi}{2} \\ 0 & -\frac{\tilde{k}^*_\varphi}{2} & 0 \end{pmatrix}.$$

4 Regions of System's Instability

For the convenience of graphical representation of the regions with different degrees of instability and subsequent parametric analysis, we introduce four dimensionless parameters:

$$\alpha = \frac{c-b}{a} = \frac{J_z - J_x}{J_y}; \quad \gamma = \frac{b-a}{c}; \quad p_1 = \frac{d}{f}; \quad p_2 = \frac{f}{c}. \tag{5}$$

The physically obtainable values of the parameters, taking into account (2), lie within the intervals: $-1 < \alpha < 1, \;\; 0 < \gamma < 1, \;\; 0 < p_1 \leq 1, \;\; 0 < p_2 < 1$. It is not difficult to show that conditions (2) imply $\gamma + \alpha > 0$.

The diagonal blocks of the initial matrix of potential forces (when $Q_\delta = 0$, $Q_\sigma = 0$) in notation (5) have the form:

$$K_1 = 3\begin{pmatrix} \gamma & p_2 \\ 1 & 1 \end{pmatrix}; \quad K_2 = \begin{pmatrix} \alpha & 0 & 0 \\ 0 & 4(\gamma+\alpha) & 4p_2(\alpha+1) \\ 0 & 4 & 3+p_1 \end{pmatrix}.$$

In the space of the outlined parameters, the relations $\gamma = p_2$, $\alpha = 0$, $S \equiv (\gamma+\alpha)(3+p_1) - 4p_2(\alpha+1) = 0$ define the surfaces which separate the regions having different degrees of instability. For example, Fig. 1 shows these regions for the values of the parameters $p_1 = 4/5$, $p_2 = 5/7$.

It is known that if the equilibrium position is unstable at potential forces, Kelvin–Chetaev's theorem [7] of influence of gyroscopic forces tells us that gyroscopic stabilization is possible only for systems with an even degree of instability.

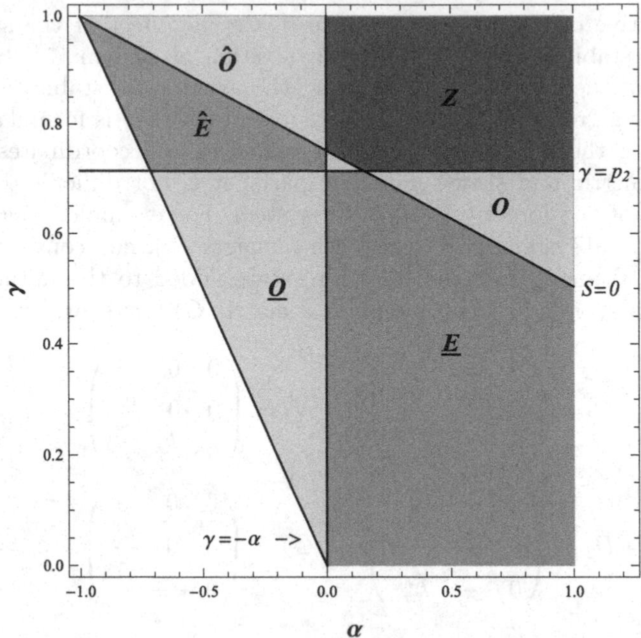

Fig. 1. Regions with different degrees of instability.

Here, respectively, instability regions for the entire system have: $\boldsymbol{Z}$ – zero degree; $\widehat{\boldsymbol{E}}$ – an even degree (when $\gamma > p_2$) and $\underline{\boldsymbol{E}}$ (when $\gamma < p_2$); $\boldsymbol{O}$, $\underline{\boldsymbol{O}}$, $\widehat{\boldsymbol{O}}$ – odd degree.

The evenness (or oddness) of the degree of instability according to Poincaré is determined by positivity (or negativity) of the determinant of the matrix of potential forces. It is necessary to emphasize that for the values of the parameters from the regions $\widehat{\boldsymbol{E}}$ and $\underline{\boldsymbol{E}}$, the unstable equilibrium position has an even degree of instability (i.e., $\det K = \det K_1 * \det K_2 > 0$). Thus, under certain conditions, equilibrium can be stabilized due to the influence of gyroscopic forces. Earlier in [8], the author has proved the stabilization of the equilibrium in the needle-shaped part (subregion) of the region $\widehat{\boldsymbol{E}}$ for an uncontrolled satellite. The matrices K_1 and K_2 are positive definitive in the region $\boldsymbol{Z}$. On the basis of another Kelvin–Chetaev's theorem, the addition to the potential forces of gyroscopic forces preserves the nature of stability of the investigated motion.

The mass distribution in the system in which the initial matrix of potential forces of the system will be positive definitive is usually given for the applied problems of spacecraft dynamics. Further, due to the addition of primarily dissipative forces, the asymptotic stability of motion is ensured by Lyapunov's theorem. However, unstable systems may also be of interest and, besides, "nonstandard" situations on the orbit are possible.

Thus, taking into account the assumptions made in the formulation of the problem in Sect. 3, we shall consider the possibility of stabilizing an unstable system in the region $\underline{\boldsymbol{O}}$ (when $\gamma < p_2,\ \alpha < 0,\ (\gamma+\alpha)(3+p_1) - 4p_2(\alpha+1) < 0$) or in the region $\underline{\boldsymbol{E}}$ (when $\gamma < p_2,\ \alpha > 0,\ (\gamma+\alpha)(3+p_1) - 4p_2(\alpha+1) < 0$) to asymptotic stability by additional forces (4).

5 Parametric Analysis of Asymptotic Stability Conditions

It is obvious that the characteristic equation of system (1) is factorized: $\Lambda(\lambda) \equiv \Lambda^{(1)} * \Lambda^{(2)} = 0$. After performing elementary transformations with the characteristic matrices (multiplying their rows by positive factors), we obtain the characteristic determinants in notation (5), respectively, in the "pitch" subsystem and in the "yaw-and-roll" subsystem:

$$\Lambda^{(1)} = \begin{vmatrix} \lambda^2 + 3\gamma & p_2(\lambda^2+3) \\ \lambda^2 - \lambda\tilde{k}^*_\theta + (3-\tilde{k}_\theta) & \lambda^2 p_1 + \lambda\tilde{k}^*_\delta + (3+\tilde{k}_\delta) \end{vmatrix} = \sum_{i=0}^{4} w_i\,\lambda^i, \qquad \text{where}$$

$$w_4 \equiv \det M_1 = p_1 - p_2,\ w_3 = \tilde{k}^*_\delta + p_2\tilde{k}^*_\theta,\ w_2 = 3\,(p_1\gamma - 2p_2 + 1) + \tilde{k}_\delta + p_2\tilde{k}_\theta,$$

$$w_1 = 3\Big(\gamma\,\tilde{k}^*_\delta + p_2\tilde{k}^*_\theta\Big),\ w_0 = 3\Big(3\,(\gamma - p_2) + \gamma\,\tilde{k}_\delta + p_2\tilde{k}_\theta\Big);$$

$$\Lambda^{(2)} = \begin{vmatrix} \lambda^2+\alpha & \lambda(\alpha-1) & 0 \\ \lambda(\alpha-1)(\gamma-1) & \lambda^2(1+\gamma\alpha)+4(\alpha+\gamma) & (\lambda^2+4)p_2(\alpha+1) \\ 0 & \lambda^2 - \lambda\tilde{k}^*_\varphi + (4-\tilde{k}_\varphi) & \lambda^2 p_1 + \lambda\tilde{k}^*_\sigma + (3+p_1+\tilde{k}_\sigma) \end{vmatrix} =$$

$$= \sum_{i=0}^{6} v_i\,\lambda^i, \qquad \text{where} \quad v_6 \equiv \det M_2 = (1+\gamma\alpha)\,p_1 - (\alpha+1)p_2,$$

$$v_5 = (1+\gamma\alpha)\,\tilde{k}^*_\sigma + (\alpha+1)\,p_2\,\tilde{k}^*_\varphi, \quad v_1 = 4\,\alpha\Big((\alpha+\gamma)\,\tilde{k}^*_\sigma + (\alpha+1)\,p_2\,\tilde{k}^*_\varphi\Big),$$

$$v_3 = (1+3\gamma+\alpha(\alpha+2\gamma+3))\,\tilde{k}^*_\sigma + (\alpha+1)(4+\alpha)\,p_2\,\tilde{k}^*_\varphi,$$

$$v_4 = (1+\gamma\alpha)(3+p_1+\tilde{k}_\sigma) + (1+3\gamma+\alpha(3+\alpha+2\gamma))p_1 - (\alpha+1)p_2(8+\alpha-\tilde{k}_\varphi),$$

$$v_2 = (1+3\gamma+\alpha\,(\alpha+2\gamma+3))\Big(3+p_1+\tilde{k}_\sigma\Big) + 4\,\alpha\,p_1\,(\gamma+\alpha) + \\ + (\alpha+1)\,p_2\Big((4+\alpha)\,\tilde{k}_\varphi - 8(\alpha+2)\Big),$$

$$v_0 = 4\,\alpha\Big((\alpha+\gamma)\Big(3+p_1+\tilde{k}_\sigma\Big) + (\alpha+1)\,p_2\Big(\tilde{k}_\varphi - 4\Big)\Big).$$

The principal diagonal minors of the Hurwitz matrix, respectively, for two subsystems

$$\Delta_3^{(1)} = w_1 w_2 w_3 - w_4 w_1^2 - w_0 w_3^2; \quad \Delta_3^{(2)} = \begin{vmatrix} v_5 & v_3 & v_1 \\ v_6 & v_4 & v_2 \\ 0 & v_5 & v_3 \end{vmatrix}; \quad \Delta_5^{(2)} = \begin{vmatrix} v_5 & v_3 & v_1 & 0 & 0 \\ v_6 & v_4 & v_2 & v_0 & 0 \\ 0 & v_5 & v_3 & v_1 & 0 \\ 0 & v_6 & v_4 & v_2 & v_0 \\ 0 & 0 & v_5 & v_3 & v_1 \end{vmatrix}$$

are analytically obtained with SPCA "Mathematica" and were used in further calculations, but due to bulkiness, their explicit form is not given here.

The fulfillment of the conditions on the existence of roots with negative real parts for the polynomial $\Lambda(\lambda)$

$$w_i > 0,\ (i = \overline{0,4});\quad \Delta_3^{(1)} > 0, \tag{6}$$

$$v_i > 0,\ (i = \overline{0,6});\quad \Delta_3^{(2)} > 0;\quad \Delta_5^{(2)} > 0 \tag{7}$$

ensures the asymptotic stability of the system's equilibrium position on the basis of Lyapunov's theorem on the first approximation.

It is worth noting that the conditions $w_4 > 0$ and $v_6 > 0$ are satisfied by virtue of the positive definiteness of the kinetic energy matrix.

5.1 Stabilization in the "Pitch" Subsystem

With the help of "Mathematica" function *Reduce* designed to find the symbolic (analytical) solution of the inequalities systems, the conditions for the control parameters $\tilde{k}_\theta^*$, $\tilde{k}_\delta^*$, $\tilde{k}_\theta$, $\tilde{k}_\delta$ (when $p_1 > p_2, \gamma < p_2$) ensuring the fulfillment of the system of inequalities (6) are obtained. Due to the solution's bulkiness, its presentation is omitted here. It is worth noting that "extra" forces entail "costs" of their technical implementation.

An analysis of the solution obtained allows us to conclude that it is possible to achieve stabilization of the subsystem to asymptotic stability by a "reduced" set of control forces in Case 1 $Q_\delta = -\tilde{k}_\delta^*\,\dot{\delta} - \tilde{k}_\delta\,\delta$ or Case 2 $Q_\delta = \tilde{k}_\theta^*\,\dot{\theta} + \tilde{k}_\theta\,\theta$. In Case 1, additional dissipative and potential forces make an impact on the subsystem, and in Case 2, all forces (potential, non-conservative, dissipative and gyroscopic) are present. As a result, the following proposition is formulated and proved.

Proposition 1. *When choosing control parameters that satisfy the conditions*

$$\tilde{k}_\delta^* > 0,\ \ \tilde{k}_\delta > 3\left(\frac{p_2}{\gamma} - 1\right) \text{ in Case 1 or } \tilde{k}_\theta^* > 0,\ \ \tilde{k}_\theta > 3\left(1 - \frac{\gamma}{p_2}\right) \text{ in}$$

Case 2, all the roots of the polynomial $\Lambda^{(1)}(\lambda)$ have negative real parts.

5.2 Stabilization in the "Yaw-and-Roll" Subsystem

We note that the control parameters $\tilde{k}_\varphi^*$ and $\tilde{k}_\sigma^*$ enter only the odd coefficients v_1, v_3, v_5 of the characteristic equation. With the above mentioned *Reduce* function, their positivity is analyzed separately for the regions $\underline{\boldsymbol{O}}$ and $\underline{\boldsymbol{E}}$. For example, for the region $\underline{\boldsymbol{O}}$, the function call and the solution have the following form:

$$\textit{Reduce}[\,\{0 < p_2 < p_1 \le 1,\ 0 < \gamma < p_2,\ -\gamma < \alpha < 0,\ S < 0,$$
$$v_1 > 0,\ v_3 > 0,\ v_5 > 0\},\{\,\tilde{k}_\sigma^*,\ \tilde{k}_\varphi^*\,\},\ \text{Reals}]$$

$$p_2 < p_1 \leq 1 \wedge \gamma < p_2 \wedge -\gamma < \alpha < 0 \wedge$$
$$\wedge\, \tilde{k}^*_\sigma > 0 \wedge -\frac{(1+3\gamma+\alpha(3+\alpha+2\gamma))\,\tilde{k}^*_\sigma}{(\alpha+1)(\alpha+4)p_2} < \tilde{k}^*_\varphi < -\frac{(\alpha+\gamma)\,\tilde{k}^*_\sigma}{(\alpha+1)p_2}.$$

Looking at the analytical solution of this system of inequalities, we note the positivity of $\tilde{k}^*_\sigma$ and the negativity of $\tilde{k}^*_\varphi$. Therefore, forces (in the matrix D_2) can only be dissipative but not accelerating. As a result, the following proposition is formulated and proved.

Proposition 2. *It is impossible to ensure the coefficients* v_1, v_3, v_5 *are simultaneously positive for the values of the parameters from the region* $\underline{\boldsymbol{O}}$ *when* $\tilde{k}^*_\sigma = 0$ *or* $\tilde{k}^*_\varphi = 0$, *but in the region* $\underline{\boldsymbol{E}}$, *this can be done.*

Thus, in order to stabilize the system in the region $\underline{\boldsymbol{O}}$, a complete set of control forces with respect to velocities is required (in contrast to the region $\underline{\boldsymbol{E}}$, where a "reduced" set of forces is sufficient).

It is not possible to obtain an analytical solution for the entire system of inequalities (7) because of the large number of parameters and the complexity of the expressions being analyzed. Therefore, to simplify the analysis, let us move on to symbolic-numerical analysis for fixed values of some parameters.

To start with, we consider the question of the possibility of asymptotic stability for the region $\underline{\boldsymbol{O}}$. Since in this region $v_0\big|_{\tilde{k}_\sigma=0,\,\tilde{k}_\varphi=0} \equiv \det K_2 > 0$, it is possible not to take into account the positional forces in Q_σ from (4) (i.e., let us add $\tilde{k}_\sigma = 0$ and $\tilde{k}_\varphi = 0$). When solving the system of inequalities (7) using *Reduce* function for the specific numerical values $\tilde{k}^*_\varphi < 0$, $\tilde{k}^*_\sigma > 0$ (for example, $\tilde{k}^*_\sigma = 1$, $\tilde{k}^*_\varphi = -\gamma/p_2$, $p_1 = 4/5$) we get the answer FALSE (i.e. the system is incompatible). The same answer was received in the case $\tilde{k}_\sigma \neq 0$, $\tilde{k}_\varphi \neq 0$ (i.e. under the action of the whole set of forces Q_σ). As a result of the analysis, the following proposition can be formulated.

Proposition 3. *For the values of the parameters in the region* $\underline{\boldsymbol{O}}$ *system (1) cannot be stabilized up to the asymptotic stability due to the control forces' effect (4).*

Now, let us consider the question of the possibility of asymptotic stability for the region $\underline{\boldsymbol{E}}$. Taking into account the second part of Proposition 2, we assume that $Q_\sigma = -\tilde{k}^*_\sigma \dot{\sigma} - \tilde{k}_\sigma \sigma$ (that is, additionally only dissipative and potential forces act). In this case, the principal diagonal minors of the third and fifth order Hurwitz matrix do not depend on the second control parameter $\tilde{k}_\sigma$ and have the form:

$$\Delta_3^{(2)} = -p_2(\alpha-1)^2(\alpha+1)(\gamma-1)(\,9\,(1-\gamma)+\alpha\,(6\,(1-\gamma)+\alpha+1)\,)(\tilde{k}^*_\sigma)^2,$$
$$\Delta_5^{(2)} = -144\,p_2^2\,\alpha\,(\alpha-1)^4(\alpha+1)^2(\gamma-1)^3(\tilde{k}^*_\sigma)^3.$$

When solving the system of inequalities (7) (where, as in Fig. 1, $p_1 = 4/5$, $p_2 = 5/7$) in relation to $\tilde{k}^*_\sigma$, $\tilde{k}_\sigma$ using function

$$\textit{Reduce}[\,\{0<\gamma<5/7,\ 0<\alpha<1,\ S<0,\ v_6>0,\ v_0>0,\ v_2>0,\ v_4>0,$$
$$v_1>0,\ v_3>0,\ v_5>0,\ \Delta_3^{(2)}>0,\ \Delta_5^{(2)}>0\,\},\{\,\tilde{k}^*_\sigma,\ \tilde{k}_\sigma\,\},\textit{Reals}],$$

we get the answer:

$$
\begin{aligned}
&\tilde{k}_\sigma^* > 0 \wedge \tilde{k}_\sigma > \frac{100 - 33\,\alpha - 133\,\gamma}{35\,(\alpha+\gamma)} \wedge \\
&\wedge \left(\left(0 < \alpha \le \frac{3}{25} \wedge 0 < \gamma < \frac{5}{7}\right) \vee \left(\frac{3}{25} < \alpha \le \frac{5}{33} \wedge \frac{25\alpha - 3}{28\alpha} < \gamma < \frac{5}{7}\right) \vee \right. \\
&\left. \vee \left(\frac{5}{33} < \alpha < \frac{19}{44} \wedge \frac{25\,\alpha - 3}{28\,\alpha} < \gamma < \frac{100 - 33\,\alpha}{133}\right)\right). \qquad (8)
\end{aligned}
$$

It is not difficult to show that in the region $\underline{\boldsymbol{E}}$, the value $\frac{100-33\,\alpha-133\,\gamma}{35\,(\alpha+\gamma)} > 0$, and, therefore, the parameter $\tilde{k}_\sigma$ in (8) is positive. We note that any positive value of the other parameter $\tilde{k}_\sigma^*$ satisfies solution (8). Thus, in the present case, Q_σ are the forces of friction and elasticity.

Let us construct the region of asymptotic stability (8) in the parameter plane α, γ using "Mathematica" function *RegionPlot*, designed for a graphical representation of the solution of the system of inequalities, with the next value of the parameter $\tilde{k}_\sigma = 10$. The result obtained is shown with a shaded region in Fig. 2. It has been found that with an increasing (decreasing) value $\tilde{k}_\sigma$, this area expands (narrows) within the limits of the borders found $v_6 = 0$, $S = 0$, $\alpha = 0$, $\gamma = 0$, $\gamma = 5/7$ (see Fig. 2) and disappears at a value $\tilde{k}_\sigma = 0$.

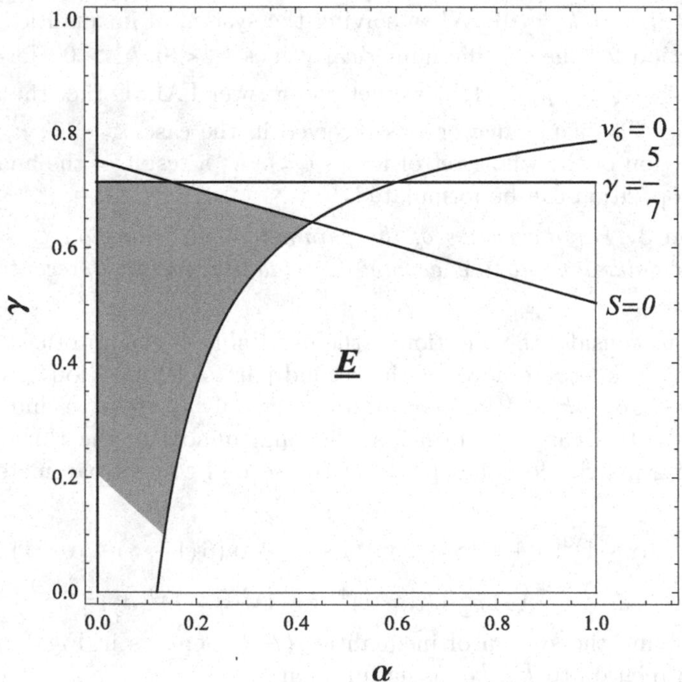

Fig. 2. Region of asymptotic stability.

A similar symbolic-numerical analysis has also been carried out for the control forces $Q_\sigma = \tilde{k}^*_\varphi \dot{\varphi} + \tilde{k}_\varphi \varphi$. As a result of the analysis, the following proposition can be formulated.

Proposition 4. *For the values of the parameters from the region* $\underline{\boldsymbol{E}}$, *system (1) can be stabilized up to an asymptotic stability thanks to the effect of control forces* $Q_\sigma = -\tilde{k}^*_\sigma \dot{\sigma} - \tilde{k}_\sigma \sigma$ *or* $Q_\sigma = \tilde{k}^*_\varphi \dot{\varphi} + \tilde{k}_\varphi \varphi$.

6 Conclusion

Based on the analogy with the parametric analysis presented above, the possibility of asymptotic stability was also investigated for other regions in Fig. 1. The study has shown that replacing the initial parameters a, b, c, f, d with the parameters α, γ, p_1, p_2 only slightly simplified the symbolic-numerical analysis. But due to the limited values of α, γ, p_1, p_2, this replacement allowed us to see a qualitative picture of the research. For a future research, the problem of the influence of the structure of forces on system's stability and its stabilization requires a more detailed study.

It is necessary to emphasize the problems of reliability and precision of computations, as well as the problems of explicitness and speeding-up of the process of investigations can be partially solved when SPCA is chosen as a software tool. Along with the application of the SPCA (as "a calculator") for solving a definite problem, the approach, which implies the elaboration of some software for solving a definite class of problems on the basis of the internal programming language of the SPCA (in our case – "Mathematica"), is quite important. Practically, the whole above analysis has been conducted using this software.

The work has been partially supported by the Russian Foundation for Basic Research (grant No. 16-07-00201). The research is partially supported by the Council for Grants of the President of Russian Federation, state support of the leading scientific schools, project No. NSh-8081.2016.9.

References

1. Sarychev, V.A.: Problems of orientation of satellites. Itogi Nauki i Tekhniki. Ser. "Space Res." **11**, 5–224 (1978). VINITI Publication, Moscow (in Russian)
2. Potapenko, E.M.: Dynamics of a spacecraft with direct active control of the gravity gradient stabilizer. Kosmicheskie Issledovaniya **26**(5), 699–708 (1988). (in Russian)
3. Gutnik, S.A., Guerman, A., Sarychev, V.A.: Application of computer algebra methods to investigation of influence of constant torque on stationary motions of satellite. In: Gerdt, V.P., Koepf, W., Seiler, W.M., Vorozhtsov, E.V. (eds.) CASC 2015. LNCS, vol. 9301, pp. 198–209. Springer, Cham (2015). doi:10.1007/978-3-319-24021-3_15
4. Prokopenya, A.N., Minglibayev, M.Z., Mayemerova, G.M.: Symbolic calculations in studying the problem of three bodies with variable masses. Program. Comput. Softw. **40**(2), 79–85 (2014)

5. Banshchikov, A.V., Burlakova, L.A., Irtegov, V.D., Titorenko, T.N.: Symbolic computation in modelling and qualitative analysis of dynamic systems. Comput. Technol. **19**(6), 3–18 (2014). (in Russian)
6. Banshchikov, A.V., Irtegov, V.D., Titorenko, T.N.: Software package for modeling in symbolic form of mechanical systems and electrical circuits. Certificate of State Registration of Computer Software No. 2016618253. Federal service for intellectual property. Issued 25 July 2016 (in Russian)
7. Chetaev, N.G.: Stability of Motion. Works on Analytical Mechanics. AS USSR, Moscow (1962). (in Russian)
8. Banshchikov, A.V.: Parametric analysis of stability conditions for a satellite with a gravitation stabilizer. In: Ganzha, V.G., et al. (ed.) CASC 2002, pp. 1–6. Technische Universität München, Munich (2002)

Sparse Interpolation, the FFT Algorithm and FIR Filters

Matteo Briani[✉], Annie Cuyt, and Wen-shin Lee

Department of Mathematics and Computer Science (Wis-Inf),
Universiteit Antwerpen, Middelheimlaan 1, B-2020 Antwerpen, Belgium
{Matteo.Briani,annie.cuyt,wen-shin.lee}@uantwerpen.be

Abstract. In signal processing, the Fourier transform is a popular method to analyze the frequency content of a signal, as it decomposes the signal into a linear combination of complex exponentials with integer frequencies. A fast algorithm to compute the Fourier transform is based on a binary divide and conquer strategy.

In computer algebra, sparse interpolation is well-known and closely related to Prony's method of exponential fitting, which dates back to 1795. In this paper we develop a divide and conquer algorithm for sparse interpolation and show how it is a generalization of the FFT algorithm.

In addition, when considering an analog as opposed to a discrete version of our divide and conquer algorithm, we can establish a connection with digital filter theory.

1 Sparse Interpolation

Let the function $\phi(t)$ be given by

$$\phi(t) = \sum_{i=1}^{n} \alpha_i \exp(2\pi \mathrm{i} \mu_i t)$$

and let us consider the general nonlinear interpolation problem of the samples $\phi(t_j)$, given by

$$\phi(t_j) = \sum_{i=1}^{n} \alpha_i \exp(2\pi \mathrm{i} \mu_i j/M), \qquad j = 0, \ldots, 2n-1, \ldots \tag{1}$$

with

$$\sqrt{-1} = \mathrm{i}, \quad \text{distinct } \mu_i \in \mathbb{C}, \quad \alpha_i \in \mathbb{C} \setminus \{0\}, \quad |\mathrm{Re}(\mu_i)| < M/2, \quad t_j = j/M,$$

where, without loss of generality, $M \in \mathbb{N}$. A solution of this interpolation problem was already presented in 1795 in [1] and can also be found in [2, pp. 378–382]. Let us denote $\Omega_i = \exp(2\pi \mathrm{i}\mu_i/M)$, with $\Omega_i \neq \Omega_k$ when $i \neq k$ because $|\mathrm{Re}(\mu_i)| < M/2$. It is apparent that the data $\phi(t_j)$ are structured, namely

M. Briani—This research is supported by the Instituut voor Wetenschap en Technology - IWT.

V.P. Gerdt et al. (Eds.): CASC 2017, LNCS 10490, pp. 27–39, 2017.
DOI: 10.1007/978-3-319-66320-3_3

$$\phi(t_j) = \sum_{i=1}^{n} \alpha_i \Omega_i^j, \qquad j = 0, \ldots, 2n-1, \ldots \tag{2}$$

We now want to obtain the values $\Omega_i, i = 1, \ldots, n$ and $\alpha_i, i = 1, \ldots, n$ from the $2n$ samples $\phi(t_j)$. From Ω_i the value μ_i can easily be deduced because $2\pi|\mathrm{Re}(\mu_i)|/M < \pi$ and hence no periodicity problem arises. Temporarily we assume that n is known. How n can be extracted from the samples is explained in Sect. 2.

Consider the polynomial

$$\prod_{i=1}^{n} (z - \Omega_i) = z^n + b_{n-1} z^{n-1} + \cdots + b_1 z + b_0 \tag{3}$$

with so far unknown coefficients $b_i, i = 1, \ldots, n$. Since the Ω_i are its zeroes, we find for $k \geq 0$,

$$\begin{aligned} 0 &= \sum_{i=1}^{n} \alpha_i \Omega_i^k (\Omega_i^n + b_{n-1} \Omega_i^{n-1} + \cdots + b_0) \\ &= \sum_{i=1}^{n} \alpha_i \Omega_i^{n+k} + \sum_{j=0}^{n-1} b_j \left(\sum_{i=1}^{n} \alpha_i \Omega_i^{j+k} \right) \\ &= \phi(t_{k+n}) + \sum_{j=0}^{n-1} b_j \phi(t_{k+j}). \end{aligned}$$

In other words, we can conclude that the structured data $\phi(t_j)$ are linearly generated,

$$\begin{pmatrix} \phi(t_0) & \ldots & \phi(t_{n-1}) \\ \vdots & \ddots & \vdots \\ \phi(t_{n-1}) & \ldots & \phi(t_{2n-2}) \end{pmatrix} \begin{pmatrix} b_0 \\ \vdots \\ b_{n-1} \end{pmatrix} = - \begin{pmatrix} \phi(t_n) \\ \vdots \\ \phi(t_{2n-1}) \end{pmatrix}. \tag{4}$$

This linear system allows us to compute the coefficients $b_i, i = 0, \ldots, n-1$ and actually compose the polynomial (3) having $\Omega_i, i = 1, \ldots, n$ as its zeroes. Let us now denote by $H_n^{(r)}$ the Hankel matrix

$$H_n^{(r)} = \begin{pmatrix} \phi(t_r) & \ldots & \phi(t_{r+n-1}) \\ \vdots & \ddots & \vdots \\ \phi(t_{r+n-1}) & \ldots & \phi(t_{r+2n-2}) \end{pmatrix}$$

and by $H_n^{(0)}(z)$ the Hankel polynomial [3, p. 625]

$$H_n^{(0)}(z) = \begin{vmatrix} \phi(t_0) & \ldots & \phi(t_{n-1}) & \phi(t_n) \\ \vdots & \ddots & \vdots & \vdots \\ \phi(t_{n-1}) & \ldots & \phi(t_{2n-2}) & \phi(t_{2n-1}) \\ 1 & \ldots & z^{n-1} & z^n \end{vmatrix}.$$

Then

$$\prod_{i=1}^{n}(z-\Omega_i) = \frac{H_n^{(0)}(z)}{|H_n^{(0)}|},$$

where $|H_n^{(0)}|$ denotes the determinant of $H_n^{(0)}$. From the matrix factorisations

$$H_n^{(0)} = V_n D_\alpha V_n^T,$$

$$H_n^{(1)} = V_n D_\alpha \begin{pmatrix} \Omega_1 & & \\ & \ddots & \\ & & \Omega_n \end{pmatrix} V_n^T,$$

where V_n and D_α respectively denote the Vandermonde matrix

$$V_n = \begin{pmatrix} 1 & 1 & \dots & 1 \\ \Omega_1 & \Omega_2 & \dots & \Omega_n \\ \vdots & \vdots & & \vdots \\ \Omega_1^{n-1} & \Omega_2^{n-1} & \dots & \Omega_n^{n-1} \end{pmatrix}$$

and the diagonal matrix

$$D_\alpha = \begin{pmatrix} \alpha_1 & & \\ & \ddots & \\ & & \alpha_n \end{pmatrix},$$

it is easy to see that the polynomial zeroes Ω_i can also be obtained as generalized eigenvalues [4, 5]. So the Ω_i also satisfy

$$\det\left(H_n^{(1)} - \Omega_i H_n^{(0)}\right) = 0, \qquad i = 1, \dots, n. \tag{5}$$

The coefficients α_i in the model (1) can be obtained from any set of n interpolation conditions taken from (2),

$$\begin{pmatrix} \Omega_1^j & \dots & \Omega_n^j \\ \vdots & & \vdots \\ \Omega_1^{j+n-1} & \dots & \Omega_n^{j+n-1} \end{pmatrix} \begin{pmatrix} \alpha_1 \\ \vdots \\ \alpha_n \end{pmatrix} = \begin{pmatrix} \phi(t_j) \\ \vdots \\ \phi(t_{j+n-1}) \end{pmatrix}, \qquad 0 \le j \le n. \tag{6}$$

With Ω_i computed as above, the remaining equations are linearly dependent.

Whether solving (4) or (5), the Hankel matrices involved tend to become quite ill-conditioned when n increases [6, 7]. So in practice, one may be interested in a divide and conquer approach where the full system is divided into several smaller systems, thus keeping the condition number under control. In Sect. 2 we present such an algorithm, which we connect to the traditional FFT in Sect. 3. Our goal is not to incorporate sparsity considerations into the FFT algorithm as in [8], but rather to add the divide and conquer approach of the FFT to sparse interpolation. Related work can be found in [9] where digital filters are used as a splitting technique and Prony's method is used to solve for the non-filtered μ_i.

So here the classical FFT algorithm will appear as a special case, when restricting the μ_i to integer values. In its most general form, with μ_i complex, our formula is related to a comb filter. The former is the subject of the Sects. 2 and 3, while the latter is discussed in the Sects. 4 and 5.

2 Divide and Conquer Approach

In this section we assume for simplicity that $\mathrm{Re}(\mu_i) \in \mathbb{Z}$ and we introduce $\omega = \exp(2\pi \mathrm{i}/N)$ with the integer $N > 0$. In addition we require that N divides M, thus guaranteeing that $M/N \in \mathbb{N}$. From our samples $\phi(t_j)$ we now deduce N linear combinations $\phi_k(t_j)$ by the construction [10, pp. 15–17]

$$\phi_k(t_j) := \frac{1}{N} \sum_{\ell=0}^{N-1} \omega^{k\ell} \phi(t_j + \ell/N), \qquad k = 0, \ldots, N-1. \tag{7}$$

These $\phi_k(t_j)$ are linear combinations of already collected samples $\phi(t_{j+M\ell/N})$ since $t_j + \ell/N$ can be expressed as $(j + M\ell/N)/M$. Figure 1 graphically illustrates formula (7). Each derived sample contains only some of the original components of (1), as can be seen from the rearrangement

$$\begin{aligned}
\phi_k(t_j) &= \frac{1}{N} \sum_{\ell=0}^{N-1} \omega^{k\ell} \phi(t_{j+M\ell/N}) \\
&= \frac{1}{N} \sum_{\ell=0}^{N-1} \omega^{k\ell} \sum_{i=1}^{n} \alpha_i \exp\left(2\pi \mathrm{i} \mu_i (j/M + \ell/N)\right) \\
&= \frac{1}{N} \sum_{\ell=0}^{N-1} \omega^{k\ell} \sum_{i=1}^{n} \alpha_i \exp(2\pi \mathrm{i} \mu_i t_j) \omega^{\ell \mu_i} \\
&= \frac{1}{N} \sum_{i=1}^{n} \alpha_i \exp(2\pi \mathrm{i} \mu_i t_j) \left(\sum_{\ell=0}^{N-1} \omega^{\ell(k+\mu_i)} \right).
\end{aligned} \tag{8}$$

We remark that

$$\begin{aligned}
&\sum_{\ell=0}^{N-1} \omega^{\ell(k+\mu_i)} = N \text{ if } \operatorname{mod}(k + \mu_i, N) = 0, \\
&\sum_{\ell=0}^{N-1} \omega^{\ell(k+\mu_i)} = 0 \text{ otherwise.}
\end{aligned} \tag{9}$$

So actually, every component of the original exponential sum (1) is present in one and only one linear combination ϕ_k. When $\mathrm{Re}(\mu_i) \in \mathbb{Z}$ formula (7) allows a perfect split of (1) over N smaller sized problems. Since each ϕ_k has the same

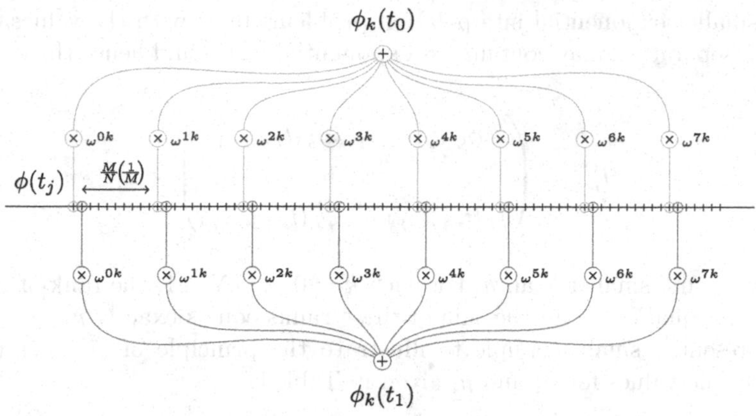

Fig. 1. *Formula (7) with $M = 80$ and $N = 8$.*

exponential structure as (1), we can apply (4) or (5) to it and identify the parameters α_i and μ_i present in ϕ_k from the values $\phi_k(t_j)$. And this for each smaller exponential sum $\phi_k, k = 0, \ldots, N-1$.

But (7) also remains valid for general $\mu_i \in \mathbb{C}$ as it is merely a linear combination of the samples taken at equidistant points. In Sect. 4 we see that, what changes when going from $\mathrm{Re}(\mu_i) \in \mathbb{Z}$ to $\mu_i \in \mathbb{C}$, is that the factor

$$\sum_{\ell=0}^{N-1} \omega^{\ell(k+\mu_i)}$$

that accompanies each term in a particular $\phi_k(t_j)$ is replaced by expression (13) of which the behaviour is illustrated in Fig. 2.

Let us now discuss the number of terms in each of the ϕ_k and for this we first consider the detection of n in (1) which we didn't touch in Sect. 1. In an exact (noisefree) context, the value of n can simply be detected from the theorems given in [3, p. 603] and [11, pp. 20–31]:

$$\begin{aligned} \det H_n^{(r)} &\neq 0, \\ \det H_\nu^{(r)} &= 0, \qquad \nu > n, \end{aligned}$$

It is analyzed in [12] that when $\nu < n$, the value $\det H_\nu^{(r)}$ is not guaranteed zero as for $\nu > n$, or guaranteed nonzero as for $\nu = n$, but can vanish accidentally when by the choice of M and r one hits a zero of this expression. From these statements the number of components n can be obtained as the rank of $H_\nu^{(r)}$ for $\nu > n$. In order to inspect $|H_\nu^{(r)}|$ for $\nu > n$, additional samples up to $t_{r+2\nu-2}$ need to be provided, in other words at least the additional sample $\phi(t_{2n})$ in case $r = 0$ and $\nu = n + 1$.

The smaller exponential interpolation problems built with the values $\phi_k(t_j)$ for each k separately, may contain less exponential terms and hence their Hankel matrices

$$H_{n,k}^{(r)} = \begin{pmatrix} \phi_k(t_r) & \dots & \phi_k(t_{r+n-1}) \\ \vdots & \ddots & \vdots \\ \phi_k(t_{r+n-1}) & \dots & \phi_k(t_{r+2n-2}) \end{pmatrix}$$

may have a rank smaller than n. For each $k = 0, \dots, N-1$, the rank of $H_{\nu,k}^{(r)}$ is less than or equal to n and the sum of these ranks equals exactly n.

We present a small example to illustrate the principle of (7). Let (1) be defined by the values for α_i and μ_i given in Table 1.

Table 1. *Ill-conditioned example of* (1).

$\mathrm{Re}(\mu_i)$	5	6	7	8	9	45	−10	−33
$\mathrm{Im}(\mu_i)$	0	0	0	0	0	0	0	0
$\mid \alpha_i \mid$	1	1	1	1	1	1	1	1
$\arg(\alpha_i)$	0	$\pi/4$	$\pi/2$	$3\pi/4$	π	$5\pi/4$	$3\pi/2$	$7\pi/4$

With $M = 100$ and $n = 8$ the Hankel matrix $H_n^{(0)}$ has a condition number of the magnitude 7.7×10^9! In [13] oversampling is used as a means to reduce the condition number. Here we use (8) to split the exponential analysis problem and bring the condition number down. We take $N = 5$. Each of the samples $\phi_k(t_j)$ for $k = 0, \dots, 4$ involves only a subset of the original components $\exp(2\pi i \mu_i t_j)$, as detailed in Table 2.

Table 2. *Example from Table 1 split into* $N = 5$ *subsets.*

k	$\mathrm{Re}(\mu_i)$			Condition nr
0	5	45	−10	2.2×10^0
1	9			1.0×10^0
2	8			1.0×10^0
3	7	−33		1.4×10^0
4	6			1.0×10^0

The major improvement in the conditioning is not only due to the reduction in size of the Hankel matrices involved, but also to a much better disposition in the complex plane of the frequencies μ_i per subsum.

3 The FFT Algorithm

An algorithm related to formula (7) is the FFT algorithm which retrieves the coefficients α_i from a set of samples $\phi(t_j), j = 0, \ldots, M-1$ given by

$$\phi(t_j) = \sum_{i=1}^{M} \alpha_i \exp(2\pi \mathrm{i} ij/M). \tag{10}$$

The difference between (10) and (1) is that now all integer frequencies appear, so $\mu_i = i$, and that therefore the number of terms in the sum equals M, which is also the number of samples. The coefficients α_i in (10) are called Fourier coefficients. In a way, (7) is a generalization of the FFT to sparse interpolation or Prony's algorithm as we now explain in some more detail.

Let $M = N_1 \times \cdots \times N_m$ with all $N_k \in \mathbb{N}$. Then the FFT algorithm breaks down the set of samples (10) into new different sets as follows. We detail the first divide of $\phi(t_j)$ into N_1 smaller exponential sums, starting from (8). For $\mu_i = i$ and $n = M$, we find from (8):

$$\phi_k(t_j) = \frac{1}{N_1} \sum_{i=1}^{M} \alpha_i \exp(2\pi \mathrm{i} ij/M) \left(\sum_{\ell=0}^{N_1-1} \omega^{\ell(k+i)} \right), \qquad k = 0, \ldots, N_1 - 1$$

where

$$\frac{1}{N_1} \sum_{\ell=0}^{N_1-1} \omega^{\ell(k+i)}$$

evaluates to either 0 or 1. Preserving only the terms that are not multiplied by zero leads to

$$\begin{aligned} \phi_k(t_j) &= \sum_{i=1}^{M/N_1} \alpha_{1+(i-1)N_1+k} \exp(2\pi \mathrm{i} j(1 + (i-1)N_1 + k)/M) \\ &= \sum_{i=1}^{M/N_1} \alpha_{1+(i-1)N_1+k} \exp(2\pi \mathrm{i} jiN_1/M) \exp(2\pi \mathrm{i} j(1 - N_1 + k)/M) \\ &= \sum_{i=1}^{M/N_1} \alpha_{1+(i-1)N_1+k} \exp(2\pi \mathrm{i} ij/(M/N_1)) \exp(2\pi \mathrm{i} j(1 - N_1 + k)/M) \\ & \qquad\qquad k = 0, \ldots, N_1 - 1 \end{aligned} \tag{11}$$

The subsequent step in which each smaller sum is divided into N_2 new smaller sums is obvious for $k = N_1 - 1$, but the other ϕ_k first need to be multiplied by the so-called twiddle factor $\exp(-2\pi \mathrm{i} j(1 - N_1 + k)/M)$ in order to bring them in the correct form (1). For the subdivision of each of the N_1 sums into N_2 yet smaller sums, one substitutes in (11) and the expression for the twiddle factors, M by M/N_1 and N_1 by N_2. In this way one continues until the algorithm has created

M sums each containing only one component of the form $\alpha_i \exp(2\pi \mathrm{i} i j/M)$. Thus at the final stage each single component immediately reveals the coefficient α_i.

The case where $M = 2^m$ is of particular interest because then (8) and (11) simplify even further ($\omega = \exp(\pi \mathrm{i}) = -1$) into

$$\phi_k(t_j) = \frac{1}{2}\sum_{\ell=0}^{1}(-1)^{\ell k}\phi(t_j + \ell/2), \qquad k = 0, 1.$$

4 An Analog Version of the Splitting Technique

We now consider a generalization of (7) when it does not make sense to require that the $\mathrm{Re}(\mu_i)$ be integer, as we did in the discrete case. To this end we introduce, in addition to $\omega = \exp(2\pi \mathrm{i}/N)$,

$$\Omega = \omega\kappa, \qquad ||\kappa|| = 1.$$

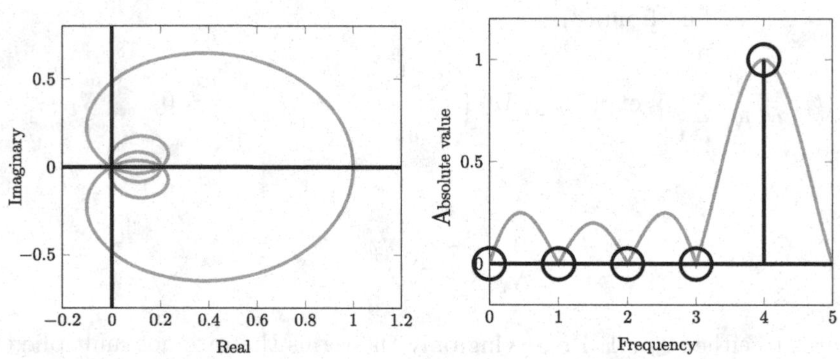

Fig. 2. *The functions $\mathcal{M}_4(1,\mu)/N$ (left) and* $|\mathcal{M}_4(1,\mu)/N|$ (right) for $\mu \in [0, 5]$.

The samples $\phi_k(t_j)$ derived from the samples $\phi(t_j)$ are then defined by the following continuous analogon of (7):

$$\begin{aligned}
\phi_k(t_j) &= \frac{1}{N}\sum_{\ell=0}^{N-1}\Omega^{k\ell}\phi(t_j + \ell/N) \\
&= \frac{1}{N}\sum_{\ell=0}^{N-1}\Omega^{k\ell}\sum_{i=1}^{n}\alpha_i \exp\left(2\pi \mathrm{i}\mu_i j/M + 2\pi \mathrm{i}\mu_i \ell/N\right) \\
&= \frac{1}{N}\sum_{\ell=0}^{N-1}\Omega^{k\ell}\sum_{i=1}^{n}\alpha_i \exp(2\pi \mathrm{i}\mu_i j/M)\omega^{\ell\mu_i} \\
&= \frac{1}{N}\sum_{i=1}^{n}\alpha_i \exp(2\pi \mathrm{i}\mu_i j/M)\sum_{\ell=0}^{N-1}\omega^{\ell(k+\mu_i)}\kappa^{\ell k} \\
&= \frac{1}{N}\sum_{i=1}^{n}\alpha_i \exp(2\pi \mathrm{i}\mu_i j/M)\mathcal{M}_k(\kappa, \mu_i), \qquad k = 0, \ldots, N-1, \qquad (12)
\end{aligned}$$

where $\mathcal{M}_k(\kappa, \mu)$, for fixed N, is defined by

$$\mathcal{M}_k(\kappa, \mu) := \frac{1 - \left(\omega^{k+\mu}\kappa^k\right)^N}{1 - \omega^{k+\mu}\kappa^k}. \tag{13}$$

In case $\kappa = 1$ formula (12) coincides with (8). However, the value of (13) does not reduce to 0 or N as in (9). By (12) all integer frequencies $\mathrm{Re}(\mu_i)$ are either zeroed or copied to ϕ_k, as in (9), while the non-integer frequencies inbetween are amplified as in Fig. 2, where we illustrate (12) for $\kappa = 1, N = 5$ and $\mathrm{Re}(\mu_i) \in [0, 5]$. The function $\mathcal{M}_k(\kappa, \mu)$ is periodic, and in Fig. 2 the period equals 5. The effect on the integer frequencies $\mu = i, i = 0, \ldots, 5$ is accentuated in the graph at the bottom in Fig. 2.

The complex number $\kappa = \exp(2\pi i\theta)$ on the unit circle acts as a continuous shifter of $\mathrm{Re}(\mu_i)$, as shown in Fig. 3. Increasing k to $k + 1$ in (7) can also be achieved by choosing $\kappa = \exp(2\pi i/N)$ in (12).

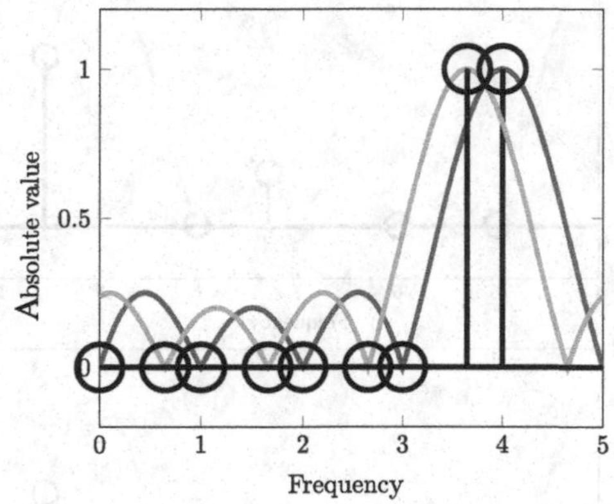

Fig. 3. *Influence of the parameter κ while N and ω are kept equal in both $\mathcal{M}_k$ graphs.*

Table 3. *Analog divide and conquer illustration.*

$\mathrm{Re}(\mu_i)$	5	6	7.3	8	9.5	45	−10	−33
$\mathrm{Im}(\mu_i)$	0	0	−0.1	0	−0.001	0	0	0
$\mid \alpha_i \mid$	1	1	1	1	1	1	1	1
$\arg(\alpha_i)$	0	$\pi/4$	$\pi/2$	$3\pi/4$	π	$5\pi/4$	$3\pi/2$	$7\pi/4$

We repeat the example of Sect. 2 where the data have now been altered so that $\mathrm{Re}(\mu_i) \notin \mathbb{Z}$ and $\mathrm{Im}(\mu_i) \neq 0$. The new data can be found in Table 3. We take a look at $\phi_4(t_j)$ given by (7) and (12) but with the μ_i from Table 3 and with $\kappa = 1$. The components in $\phi_4(t_j)$ are now multiplied by $\mathcal{M}_4(1, \mu_i)/N$. So none of the non-integer frequencies is annihilated. The μ_i with non-integer

Table 4. *Analysis of $\phi_4(t_j)$ for μ_i from Table 3.*

$\mathrm{Re}(\mu_i)$	7.3	9.5
$\mathrm{Im}(\mu_i)$	−0.1	−0.001
$\mid \alpha_i \mathcal{M}_4(1, \mu_i)/N \mid$	0.1361	0.6456

real parts are weakened in modulus as indicated in Table 4. By repeating the multiplication with $\mathcal{M}_4(1, \mu_i)/N$ this effect is strengthened. In order to retrieve the correct α_i, the coefficient of $\exp(2\pi \mathrm{i} \mu_i j/M)$ in $\phi_4(t_j)$ which can be obtained using a standard exponential analysis needs to be multiplied by $N/\mathcal{M}_4(\kappa, \mu_i)$. The effect of $\mathcal{M}_4(1, \mu)$ is graphically illustrated in Fig. 4.

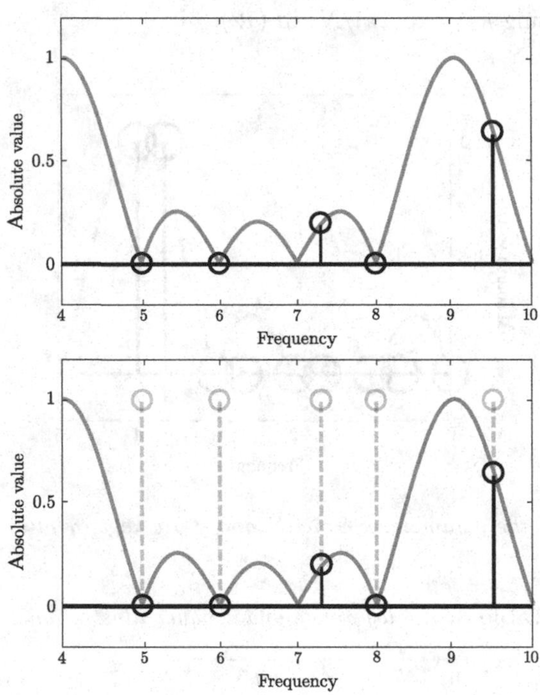

Fig. 4. *Effect of the function $\mathcal{M}_4(1, \mu)/N$ on the frequencies in Table 3.*

5 Connection to FIR Filters

We want to illustrate how formula (12) can be interpreted as the result of a digital filter. In general, a digital filter takes a set of samples as input, applies a transform and delivers another set of samples as output. In a finite impulse response or FIR filter the output samples are a linear combination of the present and previous input samples. If we denote the filter coefficients by β_ℓ and the sampling distance is $1/M$, then the filtered signal $\psi(t_j)$ equals

$$\psi(t_j) = \sum_{\ell=0}^{L-1} \beta_\ell \phi(t_j - \ell/M).$$

When the input signal is the unit impulse $\delta(\cdot)$ where δ is the Kronecker delta function, then the output signal is called the impulse response $h(t_j)$ given by

$$h(t_j) = \sum_{\ell=0}^{L-1} \beta_\ell \delta(t_j - \ell/M) = \beta_j, \qquad t_j = j/M.$$

The transfer function associated with the FIR filter ψ equals

$$H(z) = \sum_{\ell=0}^{L-1} \beta_\ell z^{-\ell}.$$

In order to establish a link with formula (12), we define for k fixed and $\ell = 0, \ldots, L-1 = M-1$, (remember that N divides M),

$$\beta_{\ell k} := \begin{cases} \dfrac{1}{N} \Omega^{k(N-(\ell+1)/(M/N))}, & (\ell+1)/(M/N) \in \mathbb{N} \\ 0, & \text{otherwise.} \end{cases}$$

When putting the $\beta_{\ell k}$ for fixed k in a vector, they are structured in N blocks of size M/N, each block containing $M/N - 1$ zeroes and one power of Ω^k:

$$\frac{1}{N}\left(0, \ldots, 0, \Omega^{(N-1)k}, 0, \ldots, 0, \Omega^k, 0, \ldots, 0, \Omega^0\right)$$

Since formula (12) is based on the current and future samples, we also need to shift the signal in order to fit the filter description:

$$\overline{\phi}(t_j) := \phi(t_j + (1 - 1/M)).$$

Then

$$\psi(t_j) = \phi_k(t_j) = \sum_{\ell=0}^{M-1} \beta_{\ell k} \overline{\phi}(t_j - \ell/M). \tag{14}$$

The impulse response of the filter (12), rewritten as (14), is given by

$$h_k(t_j) = \beta_{jk}.$$

The filter (12) gets a crisper look, meaning that it is flatter in the neighborhood of the zeroes and exhibits a sharper peak where it attains one, when applied iteratively. In Fig. 5 we show the result of (12) applied once (as in Fig. 2), twice and five times, reminding us more and more of a comb filter [14, p. 474].

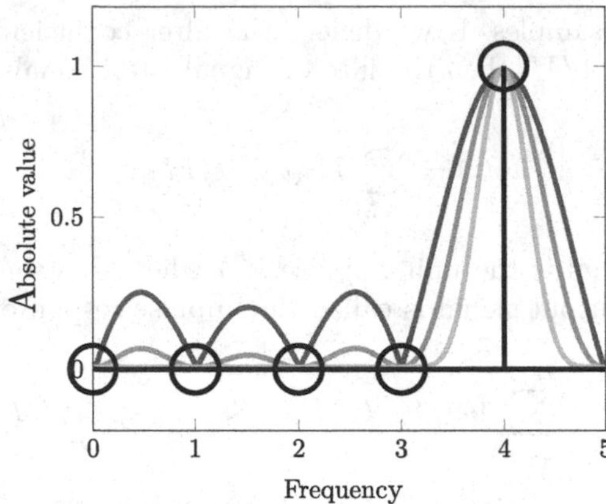

Fig. 5. *FIR filter* (12) *applied once, twice and five times.*

6 Conclusion

Sparse interpolation, which is a special case of multi-exponential analysis, can be combined with a divide and conquer technique which is a direct generalization of the fast Fourier transform algorithm in case the frequencies belong to a discrete set. This connection opens up new computational possibilities in the fitting of sparse models to data.

An analog version of our general divide and conquer method is related to digital filter theory, more precisely FIR filter theory.

References

1. de Prony, R.: Essai expérimental et analytique sur les lois de la dilatabilité des fluides élastiques et sur celles de la force expansive de la vapeur de l'eau et de la vapeur de l'alkool, à différentes températures. J. Ec. Poly. **1**, 24–76 (1795)
2. Hildebrand, F.: Introduction to Numerical Analysis. McGraw Hill, New York (1956)
3. Henrici, P.: Applied and computational complex analysis I. Wiley, New York (1974)
4. Hua, Y., Sarkar, T.K.: Matrix pencil method for estimating parameters of exponentially damped/undamped sinusoids in noise. IEEE Trans. Acoust. Speech Signal Process. **38**(5), 814–824 (1990)
5. Golub, G., Milanfar, P., Varah, J.: A stable numerical method for inverting shape from moments. SIAM J. Sci. Comput. **21**, 1222–1243 (1999)
6. Beckermann, B.: The condition number of real Vandermonde, Krylov and positive definite Hankel matrices. Numer. Math. **85**, 553–577 (2000)
7. Beckermann, B., Golub, G., Labahn, G.: On the numerical condition of a generalized Hankel eigenvalue problem. Numer. Math. **106**(1), 41–68 (2007)
8. Potts, D., Tasche, M., Volkmer, T.: Efficient spectral estimation by MUSIC and ESPRIT with application to sparse FFT. Front. Appl. Math. Stat. **2**, 1–16 (2016). Article 1

9. Heider, S., Kunis, S., Potts, D., Veit, M.: A sparse prony FFT. In: Proceedings of the 10th International Conference on Sampling Theory and Applications (SAMPTA), pp. 572–575 (2013)
10. Cuyt, A., Lee, W.-s.: Smart sampling and sparse reconstruction. GB Priority 1114255.1, (filed on 18.08.2011, published on 21.02.2013). WIPO Patentscope. https://patentscope.wipo.int/search/docservicepdf_pct/id00000020191665/PDOC/WO2013024177.pdf
11. Baker Jr., G., Graves-Morris, P.: Padé Approximants. Encyclopedia of Mathematics and Its Applications, vol. 59, 2nd edn. Cambridge University Press, Cambridge (1996)
12. Kaltofen, E., Lee, W.-s., Lobo, A.A.: Early termination in Ben-Or/Tiwari sparse interpolation and a hybrid of Zippel's algorithm. In: Proceedings of the 2000 International Symposium on Symbolic and Algebraic Computation, pp. 192–201. ACM, New York (2000)
13. Potts, D., Tasche, M.: Parameter estimation for nonincreasing exponential sums by Prony-like methods. Linear Algebra Appl. **439**(4), 1024–1039 (2013). 17th Conference of the International Linear Algebra Society, Braunschweig, Germany, August 2011
14. Schlichthärle, D.: Digital Filters. Springer, Heidelberg (2011). doi:10.1007/978-3-642-14325-0

On New Integrals of the Algaba-Gamero-Garcia System

Alexander D. Bruno[1], Victor F. Edneral[2,3(✉)], and Valery G. Romanovski[4,5,6]

[1] Keldysh Institute of Applied Mathematics of RAS, Miusskaya Sq. 4, Moscow 125047, Russia
abruno@keldysh.ru

[2] Skobeltsyn Institute of Nuclear Physics, Lomonosov Moscow State University, Leninskie Gory 1(2), Moscow 119991, Russia
edneral@theory.sinp.msu.ru

[3] Peoples' Friendship University of Russia (RUDN University), 6 Miklukho-Maklaya Street, Moscow 117198, Russian Federation
edneral_vf@rudn.university

[4] Faculty of Electrical Engineering and Computer Science, University of Maribor, Smetanova 17, SI-2000 Maribor, Slovenia

[5] CAMTP – Center for Applied Mathematics and Theoretical Physics, University of Maribor, Krekova 2, SI-2000 Maribor, Slovenia

[6] Faculty of Natural Science and Mathematics, University of Maribor, Koroška cesta 160, SI-2000 Maribor, Slovenia
valery.romanovsky@uni-mb.si

Abstract. We study local integrability of a plane autonomous polynomial system of ODEs depending on five parameters with a degenerate singular point at the origin. The approach is based on making use of the Power Geometry Method and the computation of normal forms. We look for the complete set of necessary conditions on parameters of the system under which the system is locally integrable near the degenerate stationary point. We found earlier that the sets of parameters satisfying these conditions consist of four two-parameter subsets in the full five-parameter co-space. Now we consider the special subcase of the case $b^2 = 2/3$ and separate subsubcases when additional first integrals can exist. Here we have found two such integrals.

Keywords: Ordinary differential equations · Integrability · Resonant normal form · Power Geometry · Computer algebra

1 Introduction

We consider an autonomous system of ordinary differential equations of the form

$$dx_i/dt \stackrel{\text{def}}{=} \dot{x}_i = \varphi_i(X), \quad i = 1, 2, \tag{1}$$

where $X = (x_1, x_2) \in \mathbb{C}^2$ and $\varphi_i(X)$ are polynomials.

V.P. Gerdt et al. (Eds.): CASC 2017, LNCS 10490, pp. 40–50, 2017.
DOI: 10.1007/978-3-319-66320-3_4

A method of the analysis of integrability of system (1) based on power transformations [5] and computation of normal forms near stationary solutions of transformed systems (see [3] and Chap. 2 in [4]) was proposed in [6–8].

In a neighborhood of the stationary point $X = 0$, system (1) can be written in the form

$$\dot{X} = A\,X + \tilde{\Phi}(X), \tag{2}$$

where $\tilde{\Phi}(X)$ has no terms linear in X.

Let λ_1 and λ_2 be the eigenvalues of the matrix A. If at least one of them is different from zero, then the stationary point $X = 0$ is called an *elementary* stationary point. In this case, system (2) has a normal form (see, e.g. Chap. 2 in [4]). If both eigenvalues vanish, then the stationary point $X = 0$ is called a *nonelementary* stationary point. In this case, there is no normal form for system (2). But using power transformations, we can split the nonelementary stationary point $X = 0$ to a set of elementary stationary points [5]. For each of these elementary stationary points, we can compute the normal form and write the conditions of local integrability.

In the present paper, we demonstrate how this approach can be applied to study the local and global integrability in the case of planar system near a stationary point $X^0 = 0$ of high degeneracy in the case of the system

$$\begin{aligned} \dot{x} &= \alpha\,y^3 + \beta\,x^3\,y + (a_0\,x^5 + a_1\,x^2 y^2), \\ \dot{y} &= \gamma\,x^2\,y^2 + \delta\,x^5 + (b_0\,x^4 y + b_1\,x\,y^3). \end{aligned} \tag{3}$$

This system is a subfamily of the system introduced in [1].

Systems with a nilpotent matrix of the linear part were thoroughly studied by Lyapunov and others. In system (3), there is no linear part, and the first approximation is not homogeneous. This is the simplest case of a planar system without linear part and with Newton's open polygon (see Chap. 2 in [4]) consisting of a single edge. In generic case, such problems have not been studied. However, the system with such support was considered in [1], where the authors set $-\alpha = \delta = 1$ and $3\,\beta + 2\,\gamma = 0$. Further, the authors of [1] studied the Hamiltonian subcase of this system under the additional assumption that the Hamiltonian function is expandable into the product of only square-free factors.

We study the following problem: what are the conditions on parameters under which system (3) is locally or globally integrable when the system is non-Hamiltonian.

We discuss the first quasi-homogeneous approximation of system (3) and the necessary conditions of local integrability. First, we calculate the sets of these conditions. Then, we prove that these conditions are sufficient for the global integrability by computing the corresponding first integrals of motion. We do it first for the case when $b^2 \neq 2/3$ in system (6), and after that for a subcase of the case $b^2 = 2/3$. The search for the necessary conditions of local integrability by this approach is impossible without computer algebra methods.

2 Problem Statement

We start from the study of the case when the first quasi-homogeneous approximation of (3) considered in [7–9] has the form

$$\dot{\tilde{x}} = \alpha\,\tilde{y}^3 + \beta\,\tilde{x}^3\,\tilde{y}, \quad \dot{\tilde{y}} = \gamma\,\tilde{x}^2\,\tilde{y}^2 + \delta\,\tilde{x}^5, \tag{4}$$

where $\alpha \neq 0$ and $\delta \neq 0$. Using the linear transformation $x = \sigma\tilde{x}$, $y = \tau\tilde{y}$ we can fix two nonzero parameters in (4) and obtain the system

$$\dot{x} = -y^3 - b\,x^3\,y, \quad \dot{y} = c\,x^2\,y^2 + x^5. \tag{5}$$

Each autonomous planar quasi-homogeneous system (4) has an integral, but it is not necessarily analytic. We are interested to have the analytic integrability of (4), so we look for conditions on parameters under which system (3) is locally or globally analytically integrable.

The following result was proven in [7–9]:

Theorem 1. *In the case* $D \stackrel{\mathrm{def}}{=} (3\,b + 2\,c)^2 - 24 \neq 0$, *system (5) is locally integrable if and only if the number* $N = (3\,b - 2\,c)/\sqrt{D}$ *is rational.*

In this paper, we will study a simple particular case $c = 1/b$, then $N = 1$ and $D = (3b - 2/b)^2$. In view of Theorem 1, the first quasi-homogeneous approximation has an analytic integral if $b^2 \neq 2/3$ but it is not a Hamiltonian system.

We will study the integrability problem for the entire system (3) with the first quasi-homogeneous approximation (5) writing the system in the form

$$\begin{aligned} dx/dt &= -y^3 - b\,x^3 y + a_0\,x^5 + a_1\,x^2 y^2, \\ dy/dt &= (1/b)\,x^2 y^2 + x^5 + b_0\,x^4 y + b_1\,x\,y^3. \end{aligned} \tag{6}$$

Thus, we consider the system with five arbitrary parameters a_i, b_i, $(i = 0, 1)$ and $b \neq 0$.

3 Necessary Conditions of Local Integrability

The rationality of the ratio λ_1/λ_2 and the condition A (see [3,4,7,8]) are the necessary and sufficient conditions for local analytical integrability of a planar system near an elementary stationary point. The condition A is a strong algebraic condition on coefficients of the normal form. For local integrability of the original system (1) near a degenerate (nonelementary) stationary point, it is necessary to have local integrability near each of the elementary stationary points, which are produced by the blowing up process described below.

The algorithm for calculation of the normal form and the normalizing transformation together with the corresponding computer program are briefly described in [11].

At the first step, we should rewrite (6) in a non-degenerate form. It can be done using the power transformation (see Chap. 1, \$1.8 in [4])

$$x = u\,v^2, \quad y = u\,v^3 \tag{7}$$

and the time rescaling $u^2 v^7 dt = d\tau$. As a result, we obtain system (6) in the form

$$\begin{aligned} du/d\tau &= -3\,u - [3\,b + (2/b)]u^2 - 2\,u^3 + (3\,a_1 - 2\,b_1)u^2 v + \\ &\quad (3\,a_0 - 2\,b_0)u^3 v, \\ dv/d\tau &= v + [b + (1/b)]u\,v + u^2 v + (b_1 - a_1)u\,v^2 + (b_0 - a_0)u^2 v^2. \end{aligned} \tag{8}$$

Under the power transformation (7), the point $x = y = 0$ blows up into two straight invariant lines $u = 0$ and $v = 0$. Along the line $u = 0$, system (8) has a single stationary point $u = v = 0$. Along the second line $v = 0$, this system has four elementary stationary points

$$u = 0, \quad u = -\frac{1}{b}, \quad u = -\frac{3b}{2}, \quad u = \infty. \tag{9}$$

The necessary condition of local integrability of system (6) near the point $x = y = 0$ is local integrability near all stationary points of system (8).

Lemma 1. *Near the points $u = v = 0$ and $u = \infty, v = 0$, system (8) is locally integrable.*

This lemma was proven in [7–9].

Thus, we must find conditions of local integrability at two other stationary points (9). Then we will have the conditions of local integrability of system (6) near the original point.

Let us consider the stationary point $u = -1/b, v = 0$. First we restrict ourselves to the case $b^2 \neq 2/3$ when the linear part of system (8), after the shift $u = w - 1/b$, has non-vanishing eigenvalues. At $b^2 = 2/3$, the matrix of the linear part of the shifted system in new variables w and v has the Jordan cell with both zero eigenvalues (17). This case will be studied by means of one more power transformation below. In papers [3,4], the condition A was formulated and applied to the considered problem in [6]. For two-dimensional systems, it is a sufficient condition of their integrability. It is an algebraic condition on the coefficients of the normal form. In our case, it can be written as a system of algebraic equations. We have computed the condition A with the program described in [11]. There are two solutions of the corresponding subset of equations from the condition A [9] at $b \neq 0$:

$$a_0 = 0, \quad a_1 = -b_0\,b, \quad b_1 = 0, \quad b^2 \neq 2/3 \tag{10}$$

and

$$a_0 = a_1\,b, \quad b_0 = b_1\,b, \quad b^2 \neq 2/3. \tag{11}$$

The consideration of the stationary point $u = -3\,b/2, v = 0$ under condition (11) gives three more two-parameter (depending on a_1 and b) solutions

$$\begin{array}{llll}
(1)\ \ b_1 = -2\,a_1, & a_0 = a_1 b, & b_0 = b_1 b, & b^2 \neq 2/3, \\
(2)\ \ b_1 = (3/2)\,a_1, & a_0 = a_1 b, & b_0 = b_1 b, & b^2 \neq 2/3, \\
(3)\ \ b_1 = (8/3)\,a_1, & a_0 = a_1 b, & b_0 = b_1 b, & b^2 \neq 2/3.
\end{array} \tag{12}$$

Thus, we have proved.

Theorem 2. *Conditions (10) and (12) form the set of necessary conditions of local integrability of system (8) near all its stationary points and the local integrability of system (6) at the stationary point $x = y = 0$.*

This theorem was proven in [7–9].

4 Sufficient Conditions of Integrability

The conditions presented in Theorem 2 as the necessary and sufficient conditions for the local integrability of system (6) at the stationary point at the origin can be considered as good candidates for sufficient conditions of global integrability. However, it is necessary to prove the sufficiency of these conditions by independent methods. It is necessary to do it for each of four conditions (10) and (12) at each of the stationary points $u = -3b/2, v = 0$ and $u = -1/b, v = 0$, for $b^2 \neq 2/3$.

In [12], we found first integrals of system (8) for all cases (10), (12) (mainly by the Darboux method, see, e.g., [13]).

We found four families of solutions which exhausted all cases mentioned above:

1. At $a_0 = 0, \quad a_1 = -b_0\, b, \quad b_1 = 0$:

$$\begin{aligned} I_{1uv} &= u^2(3\,b + 2\,u)v^6, \\ I_{1xy} &= 2\,x^3 + 3\,b\,y^2. \end{aligned} \tag{13}$$

2. At $b_1 = -2a_1$, $a_0 = a_1 b$, $b_0 = b_1 b$:

$$\begin{aligned} I_{2uv} &= u^2\, v^6\,(3\,b + u\,(2 - 6\,a_1\,b\,v)), \\ I_{2xy} &= 2\,x^3 - 6\,a_1\,b\,x^2\,y + 3\,b\,y^2. \end{aligned} \tag{14}$$

3. At $b_1 = 3a_1/2$, $a_0 = a_1 b$, $b_0 = b_1 b$:

$$\begin{aligned} &I_{3uv} = [4 - 4a_1\,u\,v + 3^{5/6} a_1 \times {}_2F_1\,(2/3, 1/6; 5/3; -2u/(3b)) \times \\ &u\,v\,(3 + 2u/b)^{1/6}]/[u^{1/3} v\,(3b + 2u)^{1/6}], \\ &I_{3xy} = [a_1 x^2(-4 + 3^{5/6}\,{}_2F_1\,(2/3, 1/6; 5/3; -2\,x^3/(3\,b\,y^2)) \times \\ &(3 + 2x^3/(b\,y^2))^{1/6}) + 4y]/[y^{4/3}(3\,b + 2\,x^3/y^2)^{1/6}], \end{aligned} \tag{15}$$

4. At $b_1 = 8a_1/3$, $a_0 = a_1 b$, $b_0 = b_1 b$:

$$\begin{aligned} I_{4u,v} &= [u\,(3 + 2\,a_1^2 b u) + 6\,a_1\,b\,v]/ \\ &[3\,u\,[u^3(6 + a_1^2 b\,u) + 6\,a_1^2 b\,u^2 v + 9\,b\,v^2]^{1/6}] - \\ &8\,a_1\sqrt{-b}/3^{5/3} B_{6+a_1\sqrt{-6\,b\,u}+3\,v\sqrt{-6\,b/u^3}}(5/6, 5/6), \end{aligned} \tag{16}$$

where $B_t(a, b)$ is the incomplete beta function and ${}_2F_1(a, b; c; z)$ is the hypergeometric function [2].

The first integrals and solutions do not have any singularities for the values $b^2 = 2/3$, but the approach with the aid of which these solutions were found has the limitation $b^2 \neq 2/3$, so there are possible additional integrals at these values. Thus, we need to study the case $b^2 = 2/3$ separately.

5 Case $b^2 = 2/3$, Subcase $3a_0 - 2b_0 = b(3a_1 - 2b_1)$

Let us consider the case $b = \sqrt{2/3}$. At these values b, both stationary points $u = -3b/2, v = 0$ and $u = -1/b, v = 0$ are collapsing, and after the shift $u \to w - 1/b$, we have instead of (8) the degenerate system

$$\begin{aligned} &\tfrac{dw}{d\tau} = v(-\tfrac{9}{2}\sqrt{\tfrac{3}{2}}\,a_0 + \tfrac{9}{2}\,a_1 + 3\sqrt{\tfrac{3}{2}}\,b_0 - 3\,b_1) + \\ &wv(\tfrac{27}{2}\,a_0 - 3\sqrt{6}\,a_1 - 9\,b_0 + 2\sqrt{6}\,b_1) + \\ &\sqrt{6}\,w^2 + w^2 v(-9\sqrt{\tfrac{3}{2}}\,a_0 + 3\,a_1 + 3\sqrt{6}\,b_0 - 2\,b_1) - \\ &2\,w^3 + w^3 v(3\,a_0 - 2\,b_0), \\ &\tfrac{dv}{d\tau} = -\tfrac{\sqrt{6}}{6} wv + v^2(-\tfrac{3}{2}\,a_0 + \sqrt{\tfrac{3}{2}}\,a_1 + \tfrac{3}{2}\,b_0 - \sqrt{\tfrac{3}{2}}\,b_1) + \\ &w^2 v + wv^2((\sqrt{6}\,a_0 - a_1 - \sqrt{6}\,b_0 + b_1) + \\ &+w^2 v^2(-a_0 + b_0). \end{aligned} \tag{17}$$

This system has zero eigenvalues at the stationary point $w = v = 0$, so we should apply a power transformation once again. In [9], we used the transformation

$$v \to r^2 w, \quad \dot{v} \to 2\dot{r}rw + r^2\dot{w}, \tag{18}$$

and obtained the systems with resonances of 19th and 27th orders. We calculated the corresponding normal form with 4 free parameters till 19th order, but for finding new solutions we should have more equations, and we need to compute the 27th order resonance. This resonance exists only if $3a_0 - 2b_0 \neq b(3a_1 - 2b_1)$, $b^2 = 2/3$, and its calculation is very hard. We postpone the investigation of this subcase.

But if $b^2 = 2/3$, equation (17) can be rewritten as

$$\begin{aligned}
&\tfrac{dw}{d\tau} = -3v/(2b)[(3a_0 - 2b_0) - b(3a_1 - 2b_1)]+\\
&wv(\tfrac{27}{2}\,a_0 - 3\sqrt{6}\,a_1 - 9\,b_0 + 2\sqrt{6}\,b_1)+\\
&\sqrt{6}\,w^2 + w^2v(-9\sqrt{\tfrac{3}{2}}\,a_0 + 3\,a_1 + 3\sqrt{6}\,b_0 - 2\,b_1)-\\
&2\,w^3 + w^3v(3\,a_0 - 2\,b_0),\\
&\tfrac{dv}{d\tau} = -\tfrac{\sqrt{6}}{6}wv + v^2(-\tfrac{3}{2}\,a_0 + \sqrt{\tfrac{3}{2}}\,a_1 + \tfrac{3}{2}\,b_0 - \sqrt{\tfrac{3}{2}}\,b_1)+\\
&w^2v + wv^2((\sqrt{6}\,a_0 - a_1 - \sqrt{6}\,b_0 + b_1)+\\
&+w^2v^2(-a_0 + b_0).
\end{aligned} \tag{19}$$

We see that in systems (17) and (19), the coefficient of v in the linear part of the first equation is zero if $3a_0 - 2b_0 = b(3a_1 - 2b_1)$. So we have the special subcase

$$3a_0 - 2b_0 = b(3a_1 - 2b_1), \quad b^2 = 2/3.$$

For this subcase, we use the transformation

$$u \to w - 1/b, \quad v \to rw, \quad \dot{v} \to \dot{r}w + r\dot{w}, \tag{20}$$

with the time scaling by division of the equations by $w/\sqrt{6}$, so $\tilde{\tau} = w\tau/\sqrt{6}$. Then, from (8) we have

$$\begin{aligned}
\tfrac{dw}{d\tilde{\tau}} &= 6w + 3(3a_1 - 2b_1)rw - 2\sqrt{6}w^2 - 2\sqrt{6}(3a_1 - 2b_1)rw^2+\\
&\quad 2(3a_1 - 2b_1)rw^3,\\
\tfrac{dr}{d\tilde{\tau}} &= -7r - (9a_1 - \sqrt{\tfrac{3}{2}}b_0 - 5b_1)r^2 + 3\sqrt{6}rw+\\
&\quad (7\sqrt{6}a_1 - 2b_0 - 13\sqrt{\tfrac{2}{3}}b_1)r^2w - (8a_1 - \sqrt{\tfrac{3}{2}}b_0 - \tfrac{16}{3}b_1)r^2w^2.
\end{aligned} \tag{21}$$

This is a three-parameter system with the resonance of the 13th order at the stationary point $w = 0, r = 0$ on the invariant line $w = 0$. Along this line, there is also another stationary point. It is possible to prove the integrability of the system there, and this point does not supply any additional restriction on the parameters.

We have calculated the normal form for (21) till the 26th order and obtained two equations for the condition A. They are $a13 = 0$ and $a26 = 0$, where $a13$ and $a26$ are given in [14]. Each of these equations is homogeneous in parameters a_1, b_0, b_1 of system (6) of sixth and twelfth orders, for example, $a13$ is

```
a13 =
77591416320*a1^6*s6+65110407552*a1^5*b0-343384549344*a1^5*b1*s6-
214574033664*a1^4*b0^2*s6-1084658542848*a1^4*b0*b1+
495240044652*a1^4*b1^2*s6-618953467392*a1^3*b0^3+
59995851552*a1^3*b0^2*b1*s6+1782026653968*a1^3*b0*b1^2-
325584668628*a1^3*b1^3*s6-8037029376*a1^2*b0^4*s6+
642627782784*a1^2*b0^3*b1+230489977896*a1^2*b0^2*b1^2*s6-
1080958485096*a1^2*b0*b1^3+105084809187*a1^2*b1^4*s6-
```

```
29504936448*a1*b0^5+95627128896*a1*b0^4*b1*s6-
130189857408*a1*b0^3*b1^2-155744503512*a1*b0^2*b1^3*s6+
270984738720*a1*b0*b1^4-15802409798*a1*b1^5*s6+
19669957632*b0^5*b1-20179406208*b0^4*b1^2*s6-
15425489664*b0^3*b1^3+25998124528*b0^2*b1^4*s6-
22559067296*b0*b1^5+882415736*b1^6*s6,
```

where $s6 = \sqrt{6}$. Both $a13$ and $a26$ are equal to zero at solutions (10) and (12).

Homogeneous algebraic equations in three variables can be rewritten as inhomogeneous equations in two variables. If we suppose that $a_1 = 0$, we get only one and zero dimensional solutions in the parametric cospace. Let us postpone the consideration of these cases and suppose that $a_1 \neq 0$. In this case, we substitute $b_0 = c_0\, a_1, b_1 = c_1\, a_1$ and obtain the system of two equations in two variables $a13(c_0, c_1) = 0, \quad a26(c_0, c_1) = 0$. The resultant of two corresponding polynomials in each of two variables is identically equal to zero. It is interesting that the condition A of the 19th order from [9,15] is *identically equal* to $a13$ up to multiplication by a constant. So it is enough to solve only equation $a13(c_0, c_1) = 0$. This equation can be factorized as the product of four factors including a_1^6:

$$
\begin{aligned}
a13 = {} & 48(c_1 - 3/2)\times \\
& (c_0 - 1/12\sqrt{6}c_1 + 1/2\sqrt{6})^2\times \\
& [409790784c_0^3 - 104\sqrt{6}c_0^2(-9152256 + 3385633c_1)- \\
& 208c_0(-10917702 + c_1(-360720 + 3319927c_1))+ \\
& \sqrt{6}(-718439040 + c_1(2461047528+ \\
& c_1(-1944898681 + 441207868c_1)))]\times \\
& a_1^6.
\end{aligned} \tag{22}
$$

From the first two factors, we get two two-parametric solutions $c_1 = 3/2$ and $c_1 = 6 + 2\sqrt{6}c_0$ or

$$
\begin{aligned}
b_1 &= 3a_1/2, & a_0 &= (2b_0 + b(3a_1 - 2b_1))/3, & b &= \sqrt{2/3}, \\
b_1 &= 6a_1 + 2\sqrt{6}b_0, & a_0 &= (2b_0 + b(3a_1 - 2b_1))/3, & b &= \sqrt{2/3}.
\end{aligned} \tag{23}
$$

For solutions (23), we calculate the normal form of (21) till the 36th order and obtain that for each solution it is a diagonal linear system.

The use of general roots of the polynomial, which is a cubic factor in (22), is too complicated. There are some partial one-parametric solutions which yield vanishing $a13$ at (10), (12), for instance:

$$
\begin{aligned}
b_1 &= 5a_1/3, & b_0 &= -5a_1/(12\sqrt{6}), & a_0 &= (2b_0 + b(3a_1 - 2b_1))/3, & b &= \sqrt{2/3}, \\
b_1 &= 8a_1/3, & b_0 &= 8\sqrt{6}a_1/9, & a_0 &= (2b_0 + b(3a_1 - 2b_1))/3, & b &= \sqrt{2/3}.
\end{aligned}
$$

We note that the first solution from given above is a new solution which does not intersect with (10) and (12), but we do not see here one-parameter solutions.

For each set of parameters (23), one can find the Darboux integrating factor $\mu = f_1^a \cdot f_2^d \cdot f_3^c$ (see e.g. [12,13]). In both cases, system (21) has invariant lines $f_1 = r, f_2 = w, f_3 = 1 - \sqrt{2/3}w$.

In the first case (when $b_1 = 3/2a_1$)

$$\mu_1 = r^a w^d f_3^c,$$

where

$$a = -2, \quad d = -\frac{13}{6}, \quad c = -\frac{4}{3}.$$

In the second case (when $b_1 = 6a_1 + 2\sqrt{6}b_0$)

$$\mu_2 = r^a w^d f_3^c,$$

where

$$a = \frac{3a_1 + 2\sqrt{6}b_0}{3a_1 + \sqrt{6}b_0}, \quad d = \frac{8a_1 + 5\sqrt{6}b_0}{6a_1 + 2\sqrt{6}b_0}, \quad c = \frac{-a_1}{3a_1 + \sqrt{6}b_0}.$$

The corresponding first integrals of (21) are

$$\begin{aligned} I_{1rw} = {} & w^{-7/6}(1 - \sqrt{\tfrac{2}{3}}w)^{-1/3}[-9a_1 + 3\sqrt{6}b_0 - \tfrac{42}{r} - 6(\sqrt{6}a_1 + 5b_0)w + \\ & 2(9a_1 + 4\sqrt{6}b_0)w^2 - 2^{1/6}(9\sqrt{2}a_1 + 8\sqrt{3}b_0)w^{5/3}(-\sqrt{6} + 2w)^{1/3} \times \\ & {}_2F_1(-1/2, 1/3; 1/2; \sqrt{2/3}/w)], \\ I_{2rw} = {} & r^{3\frac{3a_1}{3a_1+\sqrt{6}b_0}} \cdot w^{7/3+\frac{7b_0}{3\sqrt{6}a_1+6b_0}} \cdot (1 - \sqrt{2/3}w)^{\frac{-a_1}{3a_1+\sqrt{6}b_0}} \times \\ & \{\tfrac{-6+2\sqrt{6}w}{6a_1+3\sqrt{6}b_0} + r[3 + 2w(-\sqrt{6} + w)]\}. \end{aligned} \tag{24}$$

In the original variables x, y of Eq. (6), these integrals up to multiplication by a number have the form:

$$\begin{aligned} I_{1xy} = {} & (y/x^2)(\sqrt{6} + 2x^3/y^2)^{-7/6}(x^3/y^2)^{2/3} \cdot \{42\sqrt{6} + \\ & 1/(xy^3)[-36a_1x^6 - 16\sqrt{6}b_0x^6 + 84x^4y24\sqrt{6}a_1x^3y^2 - \\ & 36b_0x^3y^2 + 2^{1/3}(x^3/y^2)^{1/3}y^2(\sqrt{6} + (x^3/y^2)^{2/3} \times \\ & (2(9a_1 + 4\sqrt{6}b_0)x^3 + 3(3\sqrt{6}a_1 + 8b_0)y^2) \times \\ & {}_2F_1(-1/2, 1/3; 1/2; \tfrac{3y^2}{3y^2+\sqrt{6}x^3})]\}, \\ I_{2xy} = {} & y(\sqrt{2/3} + x^3/y^2)^{-1/2+\frac{a_1}{-6a_1-2\sqrt{6}b_0}}(x^2/y)^{-\frac{a_1}{3a_1+\sqrt{6}b_0}} \times \\ & \{3 + (x^2/y^2)[\sqrt{6}x + 3(2a_1 + \sqrt{6}b_0)y]\}. \end{aligned} \tag{25}$$

In the case $b = -\sqrt{2/3}$, we obtain a similar formula.

6 Analytical Properties of the Integrals

We should check analyticity the obtained first integrals of (25) near the origin $x = y = 0$.

We note that by Theorem 4.13 of [10], if a system has a Darboux integrating factor of the form

$$\mu = f_1^{\beta_1} f_2^{\beta_2} (1 + \text{h.o.t})^{\beta}$$

then it has an analytic first integral except of the case when both β_1 and β_2 are integer numbers greater than 1. In both cases, the above orders a and b of the integrating factor $\mu_{1,2}$ are not integer simultaneously in the generic case.

7 Conclusions

For a five-parameter non-Hamiltonian planar system (6), we have found for the case $b^2 \neq 2/3$ four sets of two-parametric necessary conditions on parameters under which the system is locally integrable near the degenerate point $x = y = 0$. These sets of conditions are also sufficient for local and global integrability of system (6). For the subcase $b^2 = 2/3$ and $3a_0 - 2b_0 = b(3a_1 - 2b_1)$, we have found two more first integrals. For our search of additional first integrals, we need to calculate the condition A at the point with the resonance of the 27th order for the subcase $b^2 = 2/3$, $3a_0 - 2b_0 \neq b(3a_1 - 2b_1)$ [9].

We have used Standard Lisp for the normal forms calculations. The integrating factors and integrals were calculated using the computer algebra system Mathematica.

Acknowledgements. Victor F. Edneral was supported by the grant NSh-7989.2016.2 of the President of Russian Federation and by the Ministry of Education and Science of the Russian Federation (Agreement number 02 A03.21.0008), Valery G. Romanovski was supported by the Slovenian Research Agency (research core funding No. P1-0306).

References

1. Algaba, A., Gamero, E., Garcia, C.: The integrability problem for a class of planar systems. Nonlinearity **22**, 395–420 (2009)
2. Bateman, H., Erdêlyi, A.: Higher Transcendental Functions, vol. 1. McGraw-Hill Book Company, New York (1953)
3. Bruno, A.D.: Analytical form of differential equations (I, II). Trudy Moskov. Mat. Obsc. **25**, 119–262 (1971), **26**, 199–239 (1972) (Russian). Trans. Moscow Math. Soc. **25**, 131–288 (1971), **26**, 199–239 (1972) (English)
4. Bruno, A.D.: Local Methods in Nonlinear Differential Equations. Nauka, Moscow (1979) (Russian). Springer, Berlin (1989) (English)
5. Bruno, A.D.: Power Geometry in Algebraic and Differential Equations. Fizmatlit, Moscow (1998) (Russian). Elsevier Science, Amsterdam (2000) (English)
6. Bruno, A.D., Edneral, V.F.: Algorithmic analysis of local integrability. Dokl. Akad Nauk **424**(3), 299–303 (2009) (Russian). Doklady Mathem. **79**(1), 48–52 (2009) (English)
7. Bruno, A.D., Edneral, V.F.: On integrability of a planar ODE system near a degenerate stationary point. In: Gerdt, V.P., Mayr, E.W., Vorozhtsov, E.V. (eds.) CASC 2009. LNCS, vol. 5743, pp. 45–53. Springer, Heidelberg (2009). doi:10.1007/978-3-642-04103-7_4
8. Bruno, A.D., Edneral, V.F.: On integrability of a planar system of ODEs near a degenerate stationary point. J. Math. Sci. **166**(3), 326–333 (2010)
9. Bruno, A.D., Edneral, V.F.: On possibility of additional solutions of the degenerate system near double degeneration at the special value of the parameter. In: Gerdt, V.P., Koepf, W., Mayr, E.W., Vorozhtsov, E.V. (eds.) CASC 2013. LNCS, vol. 8136, pp. 75–87. Springer, Cham (2013). doi:10.1007/978-3-319-02297-0_6
10. Christopher, C., Mardešić, P., Rousseau, C.: Normalizable, integrable, and linearizable saddle points for complex quadratic systems in $\mathbf{C}^2$. J. Dyn. Control Syst. **9**, 311–363 (2003)

11. Edneral, V.F.: An algorithm for construction of normal forms. In: Ganzha, V.G., Mayr, E.W., Vorozhtsov, E.V. (eds.) CASC 2007. LNCS, vol. 4770, pp. 134–142. Springer, Heidelberg (2007). doi:10.1007/978-3-540-75187-8_10
12. Edneral, V., Romanovski, V.G.: On sufficient conditions for integrability of a planar system of ODEs near a degenerate stationary point. In: Gerdt, V.P., Koepf, W., Mayr, E.W., Vorozhtsov, E.V. (eds.) CASC 2010. LNCS, vol. 6244, pp. 97–105. Springer, Heidelberg (2010). doi:10.1007/978-3-642-15274-0_9
13. Romanovski, V.G., Shafer, D.S.: The Center and Cyclicity Problems: A Computational Algebra Approach. Birkhäuser, Boston (2009)
14. http://theory.sinp.msu.ru/~edneral/CASC2017/a13-26.txt
15. http://theory.sinp.msu.ru/~edneral/CASC2017/a19.txt

Full Rank Representation of Real Algebraic Sets and Applications

Changbo Chen[1,2], Wenyuan Wu[1,2](✉), and Yong Feng[1,2]

[1] Chongqing Key Laboratory of Automated Reasoning and Cognition, Chongqing Institute of Green and Intelligent Technology, Chinese Academy of Sciences, Chongqing, China
changbo.chen@hotmail.com, {wuwenyuan,yongfeng}@cigit.ac.cn
[2] University of Chinese Academy of Sciences, Beijing, China

Abstract. We introduce the notion of the full rank representation of a real algebraic set, which represents it as the projection of a union of real algebraic manifolds $V_{\mathbb{R}}(F_i)$ of $\mathbb{R}^m$, $m \geq n$, such that the rank of the Jacobian matrix of each F_i at any point of $V_{\mathbb{R}}(F_i)$ is the same as the number of polynomials in F_i.

By introducing an auxiliary variable, we show that a squarefree regular chain T can be transformed to a new regular chain C having various nice properties, such as the Jacobian matrix of C attains full rank at any point of $V_{\mathbb{R}}(C)$. Based on a symbolic triangular decomposition approach and a numerical critical point technique, we present a hybrid algorithm to compute a full rank representation.

As an application, we show that such a representation allows to better visualize plane and space curves with singularities. Effectiveness of this approach is also demonstrated by computing witness points of polynomial systems having rank-deficient Jacobian matrices.

1 Introduction

Numerical algebraic geometry [12,33], which solves problems arising from algebraic geometry by numerical computation techniques, has made great progress in the past decades with the advent of homotopy continuation methods [24,34]. For well-posed problems, such algorithms can often provide good approximated answers in a shorter time than methods based on symbolic computation. On the other hand, different from symbolic approaches, the accuracy of computation with a geometric object may depend greatly on the algebraic representations.

Rank deficiency of the Jacobian matrix is a typical factor affecting the performance of numerical root finding procedures like Newton iteration. The iteration process may not converge or converge only linearly to points not meeting the full rank condition. To handle this problem, approaches such as deflation [20,22,23,26,28] have been proposed. In the particular case of real varieties, algebraic representations such as being implicitly sum of squares of polynomials bring another level of difficulty for numerical computation since a slight

V.P. Gerdt et al. (Eds.): CASC 2017, LNCS 10490, pp. 51–65, 2017.
DOI: 10.1007/978-3-319-66320-3_5

deformation may change the dimension. Such ill-posed problems might be handled to some extent by computing polynomial SOS [29] and real radicals [3,19,25] based on semidefinite programming.

Different from the above approaches, to handle the singularities of a real algebraic set V caused by self-intersections, non-radicalness, sum of squares and others, we lift V to a higher dimensional space such that the variety V becomes the union of projections of real algebraic manifolds $M_1, \ldots, M_s$. Each M_i is represented by a set of polynomials attaining full rank at any point of M_i. See Sect. 2 for illustrative examples. Such a representation allows some operations like existential quantifier elimination [13], witness points computing [11,30,35], curve tracing and border curve generating [7] to be accomplished by firstly applying them to a smooth geometric object with full rank algebraic representation, which allows efficient algorithms with good complexity [32], and next making use of the projection operator, which costs nothing if the object is represented by points.

In this paper, we provide a proof-of-concept algorithm, see Sect. 3, for computing such a full rank representation based on a symbolic triangular decomposition method [6] and a numeric critical point technique [35]. The full rank condition is achieved by introducing an auxiliary variable to a squarefree regular chain. The resulting new object is still a squarefree regular chain but has extra nice properties, such as the ideal generated by it being the same as its saturated ideal (importance of such regular chains are discussed in [1]). Such a representation is partly motivated by the work [5] of decomposing a semi-algebraic system into regular semi-algebraic systems and the work [31] for computing witness points via triangular decompositions. Related theoretical work on lifting plane curves as projections of non singular space curves can be found in [4].

We provide two applications of our methods. The first is to plot singular plane and space curves [8,14–18], detailed in Sect. 4. The second is to compute witness points of systems for which classical Lagrange multiplier method fails due to rank-deficiency of Jacobian matrices, presented in Sect. 5. Effectiveness of this method is justified by examples in the two applications. In Sect. 6, we summarize the main results of the paper and discuss some possible future research directions.

2 Full Rank Representation of Real Algebraic Sets

In this section, we introduce the notion of full rank representation of a real algebraic set and provide several examples to illustrate the notion.

Definition 1. *Let $F \subseteq \mathbb{R}[x_1, \ldots, x_n]$ and $V_{\mathbb{R}}(F)$ be the zero set of F in $\mathbb{R}^n$. Let $m \geq n$ and $\pi : \mathbb{R}^m \to \mathbb{R}^n$ be the projection function which sends a point $(x_1, \ldots, x_n, x_{n+1}, \ldots, x_m)$ in $\mathbb{R}^m$ to the point $(x_1, \ldots, x_n)$ in $\mathbb{R}^n$. A* full rank representation *of $V_{\mathbb{R}}(F)$ is a sequence of polynomial sets $F_i \subseteq \mathbb{R}[x_1, \ldots, x_m]$, $i = 1, \ldots, s$, such that*

(i) We have $V_{\mathbb{R}}(F) = \pi(\cup_{i=1}^s V_{\mathbb{R}}(F_i))$, where $V_{\mathbb{R}}(F_i)$ is the zero set of F_i in $\mathbb{R}^m$.
(ii) For any i, we have $V_{\mathbb{R}}(F_i) \neq \emptyset$.

(iii) For any i, the Jacobian matrix $\mathcal{J}_{F_i}$ with respect to the variables of F_i attains full rank $|F_i|$, where $|F_i|$ is the number of elements of F_i, at any point of $V_{\mathbb{R}}(F_i)$.

Geometrically, a full rank representation represents a real algebraic set $V_{\mathbb{R}}(F)$ as the projection of a union of possibly several smooth manifolds. To make it precise, we recall the regular level set theorem, see pages 113–114 of [21].

Definition 2. *Let M and N be smooth manifolds. If $F : M \to N$ is a smooth map, a point $p \in M$ is said to be a regular point of F if the induced map between tangent spaces $F_* : T_pM \to T_{F(p)}N$ is surjective; it is called a critical point otherwise. A point $c \in N$ is said to be a regular value of F if every point of the level set $F^{-1}(c)$ is a regular point, and a critical value otherwise. A level set $F^{-1}(c)$ is called a regular level set if c is a regular value.*

Theorem 1. *Every regular level set of a smooth map is a closed embedded submanifold whose codimension is equal to the dimension of the range.*

Proposition 1. *Let $G \subseteq \mathbb{R}[x_1, \ldots, x_m]$ and k be the number of elements in G. Assume that $V_{\mathbb{R}}(G) \neq \emptyset$ and at any point of $V_{\mathbb{R}}(G)$, the Jacobian matrix $\mathcal{J}_G$ has full rank k. Then $V_{\mathbb{R}}(G)$ is a smooth submanifold of $\mathbb{R}^m$ with codimension k.*

Proof. Consider the map $G : \mathbb{R}^m \to \mathbb{R}^k$. Since at any point of $V_{\mathbb{R}}(G)$, the Jacobian matrix $\mathcal{J}_G$ has full rank k, we know that any point of $V_{\mathbb{R}}(G)$ is a regular point of the map G by Definition 2. Thus $V_{\mathbb{R}}(G) = G^{-1}(0)$ is a smooth submanifold of $\mathbb{R}^m$ with codimension k by Theorem 1.

Example 1. *Consider the real algebraic set defined by the polynomial*

$$f := z^4 + 2\,z^2y^2 + 2\,z^2x^2 + 2\,y^4 + 4\,y^2x^2 + 2\,x^4 - 4\,z^2 - 6\,y^2 - 6\,x^2 + 5.$$

Then $G := \{g_1, g_2\}$, where $g_1 := x^2 + y^2 - 1, g_2 := x^2 + y^2 + z^2 - 2$, is a full rank representation of $V_{\mathbb{R}}(f)$. Indeed, we have $f = g_1^2 + g_2^2$. The Jacobian matrix $\mathcal{J}_f$ has rank 0 at any point of $V_{\mathbb{R}}(f)$ whereas $\mathcal{J}_G$ has rank 2 at any point of $V_{\mathbb{R}}(G)$. An alternative full rank representation of f is as below:

$$\left\{\begin{matrix} y^2 + x^2 - 1, \\ z + 1, \\ wy - 1 \end{matrix}\right\}, \left\{\begin{matrix} y^2 + x^2 - 1, \\ z - 1, \\ wy - 1 \end{matrix}\right\}, \left\{\begin{matrix} x - 1, \\ y, \\ z + 1 \end{matrix}\right\}, \left\{\begin{matrix} x + 1, \\ y, \\ z + 1 \end{matrix}\right\}, \left\{\begin{matrix} x - 1, \\ y, \\ z - 1 \end{matrix}\right\}, \left\{\begin{matrix} x + 1, \\ y, \\ z - 1 \end{matrix}\right\}.$$

Example 2. *Consider the Motzkin polynomial $M := x^4y^2 + x^2y^4 - 3\,y^2x^2 + 1$. It is known that M can not be written as sum of squares of polynomials in $\mathbb{R}[x, y]$. A full rank representation of it is: $\{x-1, y-1\}, \{x-1, y+1\}, \{x+1, y+1\}, \{x+1, y-1\}$.*

Example 3. *Consider the lemniscate of Gerono defined by the polynomial $y^2 + x^4 - x^2$. It has a singular point $(0, 0)$. A full rank representation of it is:*

$$\left\{\begin{matrix} x^4 - x^2 + y^2, \\ zy - 1 \end{matrix}\right\}, \left\{\begin{matrix} x - 1, \\ y, \\ z - 1 \end{matrix}\right\}, \left\{\begin{matrix} x, \\ y, \\ z - 1 \end{matrix}\right\}, \left\{\begin{matrix} x + 1, \\ y, \\ z - 1 \end{matrix}\right\}.$$

The zero sets of the four polynomial systems are the union of four smooth curves and three points in $\mathbb{R}^3$*, as illustrated in Fig. 2 (one red point is hidden), whose projection in* (x, y) *space is exactly the lemniscate in Fig. 1.*

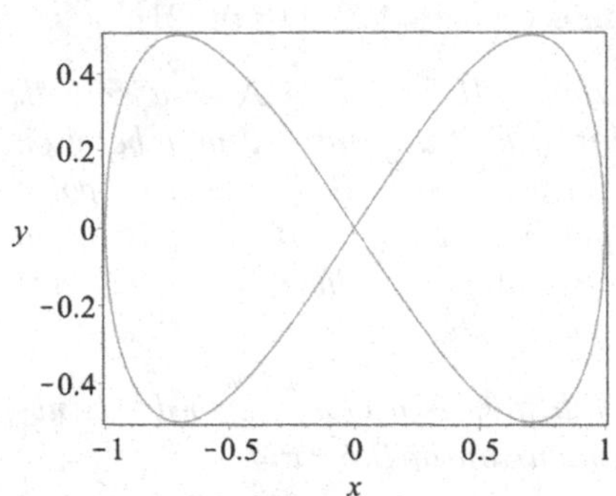

Fig. 1. Lemniscate of Gerono.

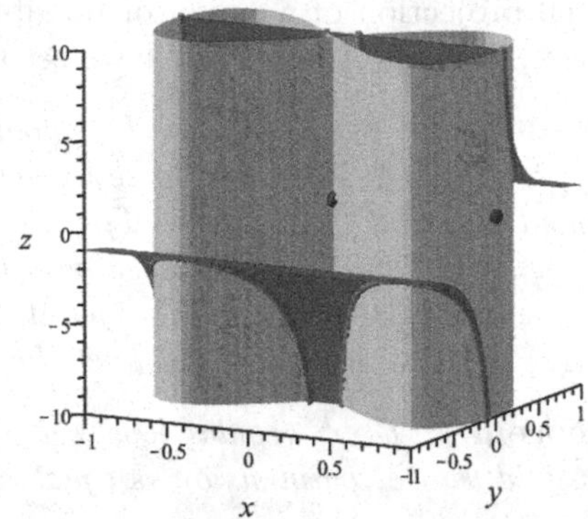

Fig. 2. A lifting of the lemniscate. (Color figure online)

3 Compute Full Rank Representation

In this section, we show that after introducing an auxiliary variable, a squarefree regular chain can be transformed to a new one having various nice properties, based on which we present a hybrid algorithm to compute a full rank representation of a given real algebraic set.

Throughout this section, let $\mathbf{k}$ be a field of characteristic 0 and $\mathbf{K}$ be the algebraic closure of $\mathbf{k}$. We refer to [2,6] for basic notions on regular chains.

Lemma 1 [2]. *Let* T *be a regular chain in* $\mathbf{k}[x_1, \ldots, x_n]$. *Let* $\mathrm{init}(T)$ *be the squarefree part of the product of the initials of polynomials in* T. *Let* $\mathrm{sat}(T) := \langle T \rangle : \mathrm{init}(T)^\infty$. *Let* $W(T)$ *be the quasi-component of* T, *that is* $W(T) = V(T) \setminus V(\mathrm{init}(T))$. *Then* $V(\mathrm{sat}(T)) = \overline{W(T)}$ *holds. Moreover,* $\mathrm{sat}(T)$ *is an unmixed ideal with dimension* $n - |T|$.

Definition 3 ([10], **page** 31). *Let* $V \subseteq \mathbf{K}^n$ *be an equidimensional variety of dimension* d. *Let* $F \subseteq \mathbf{k}[x_1, \ldots, x_n]$ *be a set of generators of its defining ideal* $\mathcal{I}(\mathcal{V})$. V *is nonsingular at a point* $p \in V$ *if the rank of the Jacobian matrix of* F *at* p *is* $n - d$. V *is* nonsingular *if it is nonsingular at every point.*

Lemma 2. *Let* $F = \{f_1, \ldots, f_k\} \subseteq \mathbb{C}[x_1, \ldots, x_n]$. *Assume that at any point* p *of* $V_{\mathbb{C}}(F)$, *the Jacobian matrix* $\mathcal{J}_F$ *has rank* k. *Then* $V_{\mathbb{C}}(F)$ *is a complex submanifold of* $\mathbb{C}^n$ *with codimension* k.

Proof. It follows directly from the definition of complex submanifold (see Definition 2.8 in [27]).

Theorem 2 (Jacobian Criterion [9], **page 402).** *Let $R := \mathbf{k}[x_1, \ldots, x_n]$ be a polynomial ring. Let $\mathcal{I} := \langle F \rangle \subseteq \mathbf{k}[x_1, \ldots, x_n]$ be an ideal. Let $\mathfrak{p}$ be a prime ideal of R containing $\mathcal{I}$. Let c be the codimension of $\mathcal{I}_\mathfrak{p}$ in $R_\mathfrak{p}$. Then:*

- *The Jacobian matrix $\mathcal{J}_F$ modulo $\mathfrak{p}$ has rank $\leq c$.*
- *$(R/\mathcal{I})_\mathfrak{p}$ is a regular local ring iff the matrix $\mathcal{J}$ modulo $\mathfrak{p}$ has rank $= c$.*

Lemma 3. *Let $F = \{f_1, \ldots, f_k\} \subseteq \mathbb{C}[x_1, \ldots, x_n]$. Assume that at any point p of $V_\mathbb{C}(F)$, the Jacobian matrix $\mathcal{J}_F$ has rank k. Then the ideal generated by F is an equidimensional radical ideal of codimension k.*

Proof. By Lemma 2, the ideal $\langle F \rangle$ is an equidimensional ideal of codimension k. Next we show it is radical (the argument below is similar to those used in the proof of Lemma 7 in [13]). Let $R := \mathbb{C}[x_1, \ldots, x_n]$ and $\mathcal{I} := \langle F \rangle$. Let $\mathfrak{p}$ be an associated prime ideal of $\mathcal{I}$. Since $\langle F \rangle$ is an equidimensional ideal of codimension k, the codimension of $\mathcal{I}_\mathfrak{p}$ in $R_\mathfrak{p}$ is also k no matter whether $\mathfrak{p}$ is an isolated or embedded prime. By Theorem 2, $(R/\mathcal{I})_\mathfrak{p}$ is a regular local ring, which implies that $\mathcal{I}_\mathfrak{p}$ is prime. On the other hand, $\mathcal{I}_\mathfrak{p} = (\cap Q_i)_\mathfrak{p}$, where Q_i are components of a minimal primary decomposition of $\mathcal{I}$ such that $\sqrt{Q_i} = \mathfrak{p}$. Thus $\cap Q_i$ is prime, which implies that $\mathcal{I}$ is radical.

Theorem 3. *Let $T = \{t_1, \ldots, t_k\}$ be a squarefree regular chain of $\mathbb{Q}[x_1 < \cdots < x_n]$. Let $d = n - k$. Let $\mathrm{init}(T)$ be the squarefree part of the product of the initials of polynomials in T and $\mathrm{sep}(T)$ be the squarefree part of the product of the separants of polynomials in T. Set h to be the least common multiple of $\mathrm{init}(T)$ and $\mathrm{sep}(T)$. We introduce a new variable w and let $C := \{T, hw - 1\}$. Then we have*

(i) C is a squarefree regular chain of $\mathbb{Q}[x_1 < \cdots < x_n < w]$.
(ii) At any point p of $V_\mathbb{C}(C)$, the rank of the Jacobian matrix of C is $k + 1$.
(iii) The ideal $\langle C \rangle$ is radical and $C = \mathrm{sat}(C)$ holds.
(iv) $V_\mathbb{C}(C)$ is a nonsingular variety and a complex submanifold of $\mathbb{C}^{n+1}$ with dimension d.
(v) Assume $V_\mathbb{R}(C) \neq \emptyset$ holds. Then $V_\mathbb{R}(C)$ is a smooth manifold of dimension d.

Proof. Note that $\mathrm{sep}(hw - 1) = h$. Since T is a squarefree regular chain and h is regular modulo $\mathrm{sat}(T)$, we deduce (i).

To prove (ii), let $f = hw - 1$, note that we have

$$\mathcal{J}_C := \begin{pmatrix} \frac{\partial t_1}{\partial x_1} \cdots \frac{\partial t_1}{\partial x_d} & \frac{\partial t_1}{\partial x_{d+1}} & & & & \\ \frac{\partial t_2}{\partial x_1} \cdots \frac{\partial t_2}{\partial x_d} & \frac{\partial t_2}{\partial x_{d+1}} & \frac{\partial t_2}{\partial x_{d+2}} & & & \\ \cdots & \cdots & & & & \\ \frac{\partial t_k}{\partial x_1} \cdots \frac{\partial t_k}{\partial x_d} & \frac{\partial t_k}{\partial x_{d+1}} & \frac{\partial t_k}{\partial x_{d+2}} & \cdots & \frac{\partial t_k}{\partial x_n} & \\ \frac{\partial f}{\partial x_1} \cdots \frac{\partial f}{\partial x_d} & \frac{\partial f}{\partial x_{d+1}} & \frac{\partial f}{\partial x_{d+2}} & \cdots & \frac{\partial f}{\partial x_n} & \frac{\partial f}{\partial w} \end{pmatrix}.$$

Consider the minor being the product of $\frac{\partial t_1}{\partial x_{d+1}} \cdots \frac{\partial t_k}{\partial x_n} \frac{\partial f}{\partial w}$, which is the polynomial $\mathrm{sep}(t_1) \cdots \mathrm{sep}(t_k) h$. Thus the minor is nonzero at any point of $V_\mathbb{C}(C)$. Therefore the rank of the Jacobian matrix of C is $k + 1$ at any point of $V_\mathbb{C}(C)$.

Next we prove (*iii*). By Lemma 3, the ideal $\langle C\rangle$ is radical. Since $W(C) \subseteq V(C) \subseteq \overline{W(C)}$ holds, we have $V(C) = \overline{W(C)} = V(\mathrm{sat}(C))$ by Lemma 1. Since both $\langle C\rangle$ and $\mathrm{sat}(C)$ are radical, we have $\langle C\rangle = \sqrt{\langle C\rangle} = \sqrt{\mathrm{sat}(C)} = \mathrm{sat}(C)$. Thus (*iii*) is proved.

Property (*iv*) follows directly from (*ii*), Definition 3 and Lemma 2 while Property (*v*) follows directly from (*ii*) and Proposition 1.

Remark 1. *In Theorem 3, if we replace h by the border polynomial of T, then the conclusion still holds.*

Next we present a hybrid algorithm to compute the full rank representation of a real variety. The algorithm relies on the following subroutines:

- $\mathsf{Triangularize}(F)$ [6]: given a set of polynomials F, it returns a set of squarefree regular chains $T_1, \ldots, T_s$ such that $V(F) = \cup_{i=1}^{s} W(T_i)$.
- $\mathsf{Intersect}(p, T)$ [6]: given a polynomial p and a squarefree regular chain T, it returns a set of squarefee regular chains $T_1, \ldots, T_s$ such that $V(p) \cap W(T) \subseteq \cup_{i=1}^{s} W(T_i) \subseteq V(p) \cap \overline{W(T)}$.
- $\mathsf{WitnessPoints}(F)$ [35]: given a set of polynomials F whose Jacobian matrix has full rank at any regular point of $V_{\mathbb{R}}(F)$, it numerically computes a finite set $W \subseteq V_{\mathbb{R}}(F)$ such that W meets every connected component of $V_{\mathbb{R}}(F)$.

Proposition 2. *Algorithm 1 terminates and correctly computes a full rank representation of $V_{\mathbb{R}}(F)$.*

Proof. In the first for loop of Algorithm 1, each polynomial $h \in H$ is regular modulo $\mathrm{sat}(T)$, which implies that for any $C \in \mathsf{Intersect}(h, T)$, we have $\dim(C) < \dim(T)$. So the algorithm terminates. Its correctness follows directly from Theorem 3.

Remark 2. *In Algorithm 1, if we replace H by the border polynomial set [5] of T, then the witness points of G (same as that of $[G, H]$) can be computed symbolically by combining open cylindrical algebraic decomposition and real root isolation of zero-dimensional regular chains (see* [5]*). However, this approach tends to produce polynomials of larger sizes due to computations of iterated resultants, and thus makes the full rank representation less suitable as the input of numerical algorithms.*

Next we illustrate the main steps of Algorithms 1 by an example.

Example 4. *Consider the following polynomial system*

$$F := \left\{ \begin{array}{l} {x_2}^2 x_1 - {x_3}^2 x_1 + {x_1}^2 + 2\,x_2\,x_1 + 2\,{x_2}^2 + 2\,x_3\,x_4 + {x_4}^2 - x_1 - 1, \\ -{x_3}^2 x_1 + {x_2}^2 x_1 + {x_4}^2 + 2\,x_3\,x_4 + 2\,{x_3}^2 + 2\,x_2\,x_1 + {x_1}^2 - x_1 + 1 \end{array} \right\}.$$

1. *The command* $\mathsf{Triangularize}$ *returns a single squarefree regular chain:*

$$T := \left\{ \begin{array}{l} {x_4}^2 + 2\,x_3\,x_4 + 2\,{x_2}^2 + 2\,x_2\,x_1 + {x_1}^2 - 1 \\ {x_3}^2 - {x_2}^2 + 1 \end{array} \right.$$

Algorithm 1. FullRankDecompose(F)

Input: A polynomial system $F \subseteq \mathbb{Q}[x_1 < \cdots < x_n]$.
Output: A full rank representation of $V_{\mathbb{R}}(F)$ (with witness points information).
begin
  introduce a new variable w; set $S_1 :=$ Triangularize(F), $S_2 := \emptyset$ and $S_3 := \emptyset$;
  while $S_1 \neq \emptyset$ **do**
    choose and remove a regular chain T from S_1;
    let H_1 (resp. H_2) be the set of nonconstant irreducible factors of the initials (resp. separants) of polynomials in T;
    $H := H_1 \cup H_2$;
    $S_2 := S_2 \cup \{[T, H]\}$;
    for $h \in H_2 \setminus H_1$ **do**
      $S_1 := S_1 \cup$ Intersect(h, T);
  for $[T, H] \in S_2$ **do**
    let h be the product of polynomial in H; set $g := hw - 1$; $G := T \cup \{g\}$;
    let $S :=$ WitnessPoints(G);
    if $S \neq \emptyset$ **then**
      $S_3 := S_3 \cup \{[G, S]\}$
  return the sequence of elements in S_3

2. *By computing the initials and separants of polynomials in T, we have $H_1 = \emptyset$ and $H = H_2 = \{x_4 + x_3, x_3\}$.*
3. *Calling* Intersect($x_4 + x_3, T$) *returns $T_1 := \{x_2 + x_1, x_3^2 - x_1^2 + 1, x_4 + x_3\}$ while calling* Intersect(x_3, T) *returns $T_2 := \{x_2 - 1, x_3, x_4^2 + x_1^2 + 2x_1 + 1\}$ and $T_3 := \{x_2 + 1, x_3, x_4^2 + x_1^2 - 2x_1 + 1\}$.*
4. *In the next iterations of the while loop, we call successively* Intersect(x_3, T_1), Intersect(x_4, T_2) *and* Intersect(x_4, T_3) *and obtain the following two regular chains (by removing duplicated ones): $T_4 := \{x_1 + 1, x_2 - 1, x_3, x_4\}$ and $T_5 := \{x_1 - 1, x_2 + 1, x_3, x_4\}$.*
5. *We have $S_2 := \{[T, H], [T_1, \{x_3\}], [T_2, \{x_4\}], [T_3, \{x_4\}], [T_4, \emptyset], [T_5, \emptyset]\}$ after the while loop terminates.*
6. *Consider the first iteration of the second for loop. We have $h = (x_3 + x_4)(x_3)$, $g = hw - 1$ and $G = T \cup \{g\}$. Calling* WitnessPoints(G) *returns $\emptyset$.*
7. *When the for loop terminates, we have the final full rank representation, built on top of the regular chains T_1, T_4 and T_5:*

```
[[[x_2+x_1, x_3^2-x_1^2+1, x_4+x_3, _w*x_3-1],
   [[x_1 = -1.356357868, x_2 = 1.356357868, x_3 = .9163550989,
     x_4 = -.9163550989, _w = 1.091280008],
    [x_1 = 1.356357868, x_2 = -1.356357868, x_3 = -.9163550989,
     x_4 = .9163550989, _w = -1.091280008]]],
  [[x_1-1, x_2+1, x_3, x_4, _w-1],
    [[x_1 = 1.000000000, x_2 = -1.000000000, x_3 = 0., x_4 = 0., _w = 1.000000000]]],
  [[x_1+1, x_2-1, x_3, x_4, _w-1],
    [[x_1 = -1.000000000, x_2 = 1.000000000, x_3 = 0., x_4 = 0., _w = 1.000000000]]]]
```

4 Applications on Plotting Singular Plane and Space Curves

In this section, we present applications of the full rank representation on visualizing singular plane and space curves.

Algorithm PlotSingularCurve
Input: a plane or space curve defined by $F(x_1, \ldots, x_r) = 0$ and a box B of $\mathbb{R}^r$, where $r = 2, 3$.
Output: the plotting of the curve inside the box B.
Steps:

1. Let $S :=$ FullRankDecompose(F).
2. For each (G, W) of S, let W' be the union of points of W inside B and the intersection points of $V_{\mathbb{R}}(G)$ with the fibers of the boundaries of B.
3. For each (G, W'), trace the smooth curve $V_{\mathbb{R}}(G)$ by a prediction- projection method using points in W'.
4. Plot the projection of the traced points in B.

Paper [18] presents a list of challenges for real algebraic plane curve visualization software. Among them, we choose two irreducible algebraic curves which softwares like Maple have difficulties to correctly visualize near singular points.

The first example is defined in Challenge 15 of [18].

$$\tilde{Y}_r^{--} := -(x^2 - y^2)^2 + ax^r.$$

Here we choose $a = 7/6, r = 8$. A full rank representation of it computed by Algorithm 1 is as below, where some of the witness points are omitted for brevity. To have better numerical stability when tracing the curve, we have rescaled the coefficients of the polynomials, which is a typical technique for dealing with machine epsilon, and increased the degree of the auxiliary variable $\_w$, which makes $\_w$ converge faster to infinity when its coefficient approaches to zero.

```
[[[5.000000*10^6*y^4-1.0000000*10^7*x^2*y^2-5.833333333*10^6*x^8+5.000000*10^6*x^4,
   (1000.*x^2*y-1000.*y^3)*_w^3-1],
  [[x = -.8361036479, y = -.4137816431, _w = -.1660497798],
   [x = .8361036479, y = .4137816431, _w = .1660497798],
   [x = .9575968845, y = 0.9351001503e-1, _w = .2274988437],
   [x = .4174607241, y = .3761233446, _w = .4327575059],
   ...  ...  ... ]],
  [[10000.*x^4-8571.428571, 10.*y, 1.*_w-1],
   [[x = -.9621954582, y = 0., _w = 1.000000000],
   [x = .9621954582, y = 0., _w = 1.000000000]]],
  [[10.*x, 10.*y, 1.*_w-1], [[x = 0., y = 0., _w = 1.000000000]]]]
```

Figure 3 shows the plotting of the curve by the command plots:- implicitplot in Maple, where the curve is not completely plotted near the singular point $(0, 0)$. Figure 4 illustrates a correct visualization by PlotSingularCurve.

A second example is defined in Challenge 17. The polynomial is:

$$SA_{k,\ell} := (y - 1 - x^k)^\ell (y - x^k)^\ell + (y - 1)^{k\ell+1} y^{k\ell}.$$

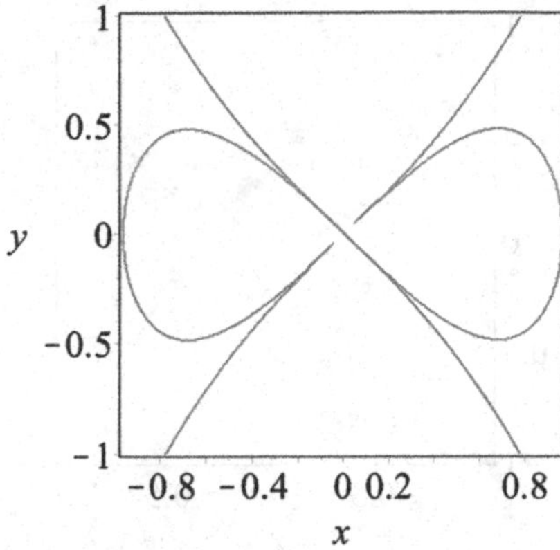

Fig. 3. By plots:- implicitplot in Maple.

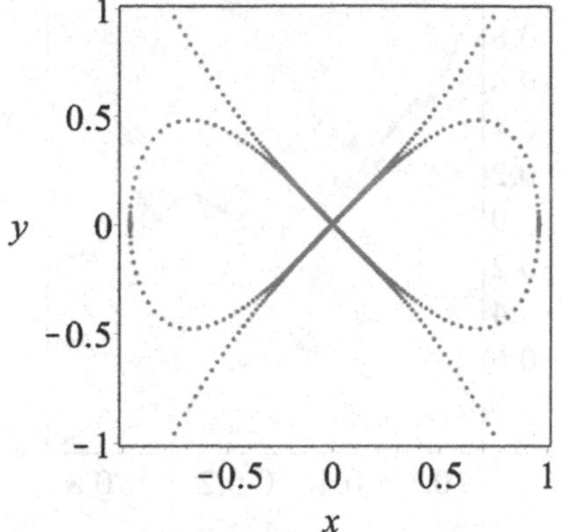

Fig. 4. By PlotSingularCurve.

Here we choose $k = \ell = 2$. A full rank representation of it computed by Algorithm 1 is as below, where some of the witness points are omitted for brevity. Again, we have rescaled the coefficients of the polynomials and increased the degree of the auxiliary variable $\_w$ in the representation.

```
[[[1.000000*10^6*y^9+1.000000*10^6*x^8-5.000000*10^6*y^8-4.000000*10^6*y*x^6
   +1.0000000*10^7*y^7+2.000000*10^6*x^6+6.000000*10^6*y^2*x^4-1.0000000*10^7*y^6
   -6.000000*10^6*y*x^4-4.000000*10^6*y^3*x^2+5.000000*10^6*y^5+1.000000*10^6*x^4
   +6.000000*10^6*y^2*x^2-2.000000*10^6*y*x^2-2.000000*10^6*y^3+1.000000*10^6*y^2,
   (-1.285714286*10^6*y^8+5.714285714*10^6*y^7-1.0000000*10^7*y^6
   +5.714285714*10^5*x^6+8.571428571*10^6*y^5-1.714285714*10^6*y*x^4
   -3.571428571*10^6*y^4+1.714285714*10^6*y^2*x^2+8.571428571*10^5*x^4
   -1.714285714*10^6*y*x^2+8.571428571*10^5*y^2+2.857142857*10^5*x^2
   -2.857142857*10^5*y)*_w^3-1],
  [[x = -.3683735588, y = .1510941675, _w = -0.6572659595e-1],
   [x = .7826945944, y = -.8765869978, _w = -0.4645572677e-2],
   ... ... ... ]],
 [[10.*x-10., 10.*y-10., 1.*_w-1],
  [[x = 1.000000000, y = 1.000000000, _w = 1.000000000]]],
 [[10.*x, 10.*y-10., 1.*_w-1], [[x = 0., y = 1.000000000, _w = 1.000000000]]],
  [[10.*x, 10.*y, 1.*_w-1], [[x = 0., y = 0., _w = 1.000000000]]],
 [[10.*x+10., 10.*y-10., 1.*_w-1],
  [[x = -1.000000000, y = 1.000000000, _w = 1.000000000]]]]
```

Figure 5 shows the plotting of the curve by the command plots:-implicitplot in Maple, where the curve is not completely plotted near the singular point $(0, 0)$. The solitary point $(0, 1)$ is also missing. Figure 6 illustrates a correct visualization by PlotSingularCurve.

Next we present an application on plotting a space curve. The example comes from [8, 16]. The curve is the zero set of the following polynomial system F.

$$F := \{(x - y + z)^2 + y^2 - 2(x - y + z), ((x - y + z)^2 + y^2 + z^2)^2 - 4((x - y + z)^2 + y^2)\}.$$

A visualization of $V_{\mathbb{R}}(F)$ by PlotSingularCurve is shown in Fig. 7.

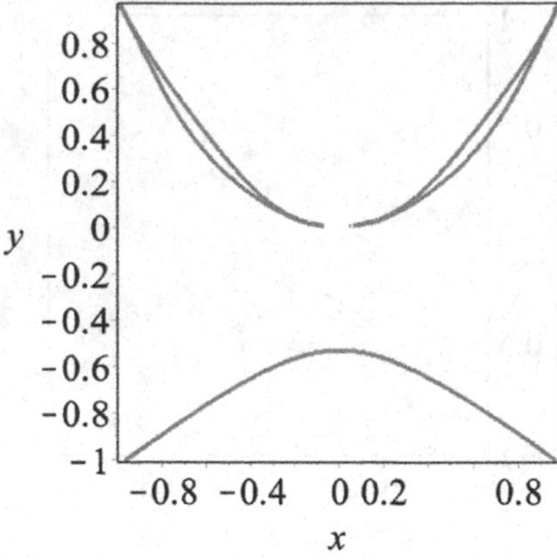

Fig. 5. By plots:- implicitplot in Maple.

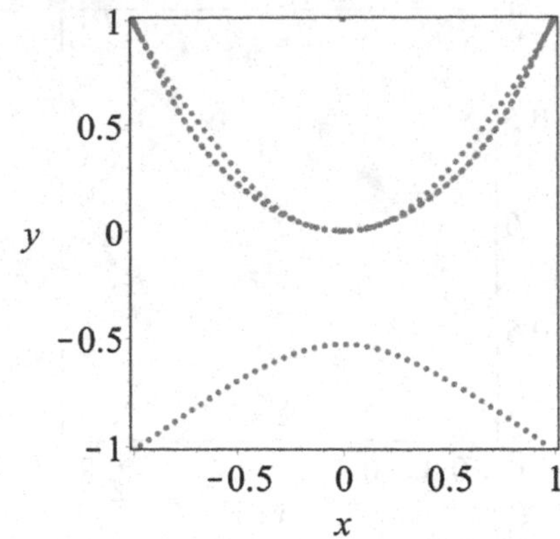

Fig. 6. By PlotSingularCurve.

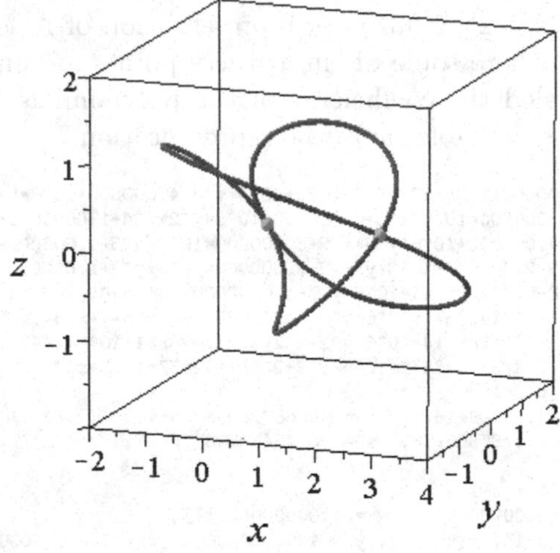

Fig. 7. A visualization of $V_{\mathbb{R}}(F)$ by PlotSingularCurve with green points being self intersection points. (Color figure online)

5 Experimentation

In this section, we report on the experimental results of a preliminary implementation of our algorithm in Maple. The test examples are created from random polynomials with some transformation such that the Jacobian matrix is rank deficient and polynomial SOS methods do not directly apply. We compare with the RealTriangularize command [5] in the REGULARCHAINS library of MAPLE, which can also compute real witness points and real dimensions of polynomial systems with rank-deficient Jacobian matrices.

Example 5. *Consider a set of polynomials* $G := \{g_1, g_2, g_3, g_4, g_5\}$, *where* $g_1 := -9\,x_4 - 4\,x_3 + 10\,x_1 + 1, g_2 := -8 + 4\,x_3 + 10\,x_5, g_3 := -7\,x_5\,x_2 - 5\,x_2 - 8\,x_5 + 1, g_4 := 7\,x_4\,x_3 - 10\,{x_2}^2 + 1, g_5 := -6\,x_5\,x_2 - 3\,{x_4}^2 + 1$. *Let* M *be a* 2×2 *invertible matrix*

$$M := \begin{pmatrix} 1 & x_1 \\ x_2 & x_1 x_2 - 1 \end{pmatrix}.$$

Set $(f_1, f_2)^T := M(g_1^2 + g_2^2, g_3)^T$ *and* $F := \{f_1, f_2, g_4, g_5\}$. *That is*

$$F = \left\{ \begin{array}{l} -7\,x_5\,x_2\,x_1 + 100\,{x_5}^2 + 80\,x_5\,x_3 - 8\,x_5\,x_1 + 81\,{x_4}^2 + 72\,x_4\,x_3 \\ \qquad - 180\,x_4\,x_1 + 32\,{x_3}^2 - 80\,x_3\,x_1 - 5\,x_2\,x_1 + 100\,{x_1}^2 \\ \qquad - 160\,x_5 - 18\,x_4 - 72\,x_3 + 21\,x_1 + 65, \\ -7\,x_5\,{x_2}^2 x_1 + 100\,{x_5}^2 x_2 + 80\,x_5\,x_3\,x_2 - 8\,x_5\,x_2\,x_1 + 81\,{x_4}^2 x_2 \\ \qquad + 72\,x_4\,x_3\,x_2 - 180\,x_4\,x_2\,x_1 + 32\,{x_3}^2 x_2 - 80\,x_3\,x_2\,x_1 \\ \qquad - 5\,{x_2}^2 x_1 + 100\,x_2\,{x_1}^2 - 153\,x_5\,x_2 - 18\,x_4\,x_2 - 72\,x_3\,x_2 \\ \qquad + 21\,x_2\,x_1 + 8\,x_5 + 70\,x_2 - 1, \\ \quad 7\,x_4\,x_3 - 10\,{x_2}^2 + 1, \\ -6\,x_5\,x_2 - 3\,{x_4}^2 + 1 \end{array} \right\}.$$

A full rank representation of $V_{\mathbb{R}}(F)$ *(after normalizing the coefficients) is*

```
[[[0.4183612727e-1*x1^7-.2285796501*x1^6+.5030640417*x1^5-.3897793981*x1^4-.4559214459*x1^3
    +1.000000000*x1^2-.6441493178*x1+.1609043497,
  .1529373720*x2-0.5369788910e-1*x1^6+.2699715843*x1^5-.5451541805*x1^4+.2573550740*x1^3
    +.7684914458*x1^2-1.000000000*x1+.3914206422,
  0.6037479144e-1*x3-0.6628199824e-1*x1^6+.3082918410*x1^5-.5584479577*x1^4+.2069073607*x1^3
    +.8725734034*x1^2-1.000000000*x1+.2543012124,
  .1556812072*x4+0.7596151567e-1*x1^6-.3533133600*x1^5+.6400011227*x1^4-.2371231577*x1^3
    -1.000000000*x1^2+.9730562705*x1-.3087361007,
  .1509369786*x5+0.6628199820e-1*x1^6-.3082918408*x1^5+.5584479574*x1^4-.2069073606*x1^3
    -.8725734030*x1^2+.9999999997*x1-.3750507950, 1.0*_w-1],
 [[x1 = -1.234303648, x2 = -.5486535450, x3 = -.2498708024, x4 = -1.149283697,
  x5 = .8999483232, _w = 1.000000000],
  [x1 = 2.176655446, x2 = 1.618025934, x3 = 2.917166233, x4 = 1.233098803,
  x5 = -.3668665021, _w = 1.000000000],
  [x1 = .8850340374, x2 = -.4777062337, x3 = .1805787394, x4 = 1.014225043,
  x5 = .7277685022, _w = 1.000000000]]]]
```

Let F^* *be the polynomial system in the output. It is a square system consisting of* 6 *polynomials and defines a zero-dimensional real variety. The smallest singular values of* $\mathcal{J}_{F^*}$ *at the three zeros of* $V_{\mathbb{R}}(F^*)$ *are respectively:*

$$0.0454144731863706450, 0.0351114360453165381, 0.0176834495031580330.$$

Example 6. *Consider the polynomial system* F*:*

$$\left\{ \begin{array}{l} 64\,{x_3}^4 - 16\,x_1\,{x_3}^3 + 50\,{x_3}^2{x_1}^2 - 3\,{x_3}^2 x_5 - 3\,{x_4}^3 + 42\,{x_3}^2 x_1 - 126\,x_3\,x_2\,x_1 \\ \qquad - 2\,{x_2}^3 + 41\,{x_3}^2 - 54\,x_3\,x_2 - 18\,x_3\,x_1 + 81\,{x_2}^2 - 6\,x_3 + 18\,x_2 + 6, \\ 64\,{x_3}^4 x_1 - 16\,{x_3}^3{x_1}^2 + 50\,{x_3}^2{x_1}^3 - 3\,x_5\,{x_3}^2 x_1 \\ \qquad - 3\,{x_4}^3 x_1 + 64\,{x_3}^4 - 16\,x_1\,{x_3}^3 + 92\,{x_3}^2{x_1}^2 - 126\,x_3\,x_2\,{x_1}^2 \\ \qquad - 2\,{x_2}^3 x_1 + 3\,{x_3}^2 x_5 + 3\,{x_4}^3 + 83\,{x_3}^2 x_1 - 180\,x_3\,x_2\,x_1 \\ \qquad - 18\,x_3\,{x_1}^2 + 2\,{x_2}^3 + 81\,{x_2}^2 x_1 + 41\,{x_3}^2 - 54\,x_3\,x_2 - 24\,x_3\,x_1 \\ \qquad + 81\,{x_2}^2 + 18\,x_2\,x_1 - 6\,x_3 + 18\,x_2 + 6\,x_1 + 4 \end{array} \right\}.$$

Set the variable order to be $x_1 < x_2 < x_3 < x_4 < x_5$. *A full rank representation of* $V_{\mathbb{R}}(F)$ *computed by Algorithm 1 consists of* 3 *components, with dimensions being respectively* $2, 1, 1$, *numbers of polynomials being respectively* $4, 5, 5$, *and total degrees of polynomials in them being respectively*

$$(3, 2, 5, 5), (1, 1, 1, 3, 1), (1, 1, 1, 3, 1).$$

The number of computed witness points are respectively $9, 1, 3$. *The maximum residuals after substituting the witness points into the three components are respectively* $5.34 \times 10^{-7}, 5. \times 10^{-10}, 6. \times 10^{-10}$. *The smallest singular values of the Jacobian matrices evaluated at the witness points are respectively*

$$\begin{array}{l} 0.217390601993606e-1, 0.217565704741585e-1, 3.32650846923473, \\ 1.32160172547595, 0.184438528285523e-1, 0.184312217006799e-1, \\ .472264278357534, 1.13453515931668, 1.14395048107664, .125000000000000, \\ 0.340696090687040e-1, 0.340696090687040e-1, .125000000000000. \end{array}$$

Example 7. *Consider the following polynomial system*

$$F := \left\{ \begin{array}{r} 3\,x_4\,x_1^3 + 100\,x_2^4 - 119\,x_3^2x_2^2 - 6\,x_4^2x_3\,x_2 + 36\,x_3^4 + 9\,x_4^4 + 3\,x_4\,x_1^2 \\ - 40\,x_2^2x_1 - 16\,x_3\,x_2\,x_1 + 24\,x_3^2x_1 + 48\,x_4^2x_1 + 71\,x_1^2 \\ - 7\,x_2\,x_1 - 20\,x_2^2 + 2\,x_3\,x_2 + 12\,x_3^2 - 6\,x_4^2 - 8\,x_1 - 7\,x_2 + 3, \\ 9\,x_4^4 - 6\,x_4^2x_3\,x_2 + 3\,x_4\,x_1^3 + 36\,x_3^4 - 119\,x_3^2x_2^2 + 100\,x_2^4 + 48\,x_4^2x_1 \\ - 3\,x_4\,x_1^2 + 24\,x_3^2x_1 - 16\,x_3\,x_2\,x_1 - 40\,x_2^2x_1 - 6\,x_4^2 + 12\,x_3^2 \\ + 2\,x_3\,x_2 - 20\,x_2^2 - 7\,x_2\,x_1 + 71\,x_1^2 + 7\,x_2 - 14\,x_1 + 1 \end{array} \right\}.$$

Set the variable order as $x_1 < x_2 < x_3 < x_4$. *A full rank representation of* $V_{\mathbb{R}}(F)$ *computed by Algorithm 1 consists of* 3 *components, with dimensions being respectively* $1, 0, 0$, *numbers of polynomials being respectively* $4, 5, 5$, *and total degrees of polynomials in them being respectively* $(12, 19, 3, 35), (8, 9, 9, 7, 1), (20, 19, 19, 19, 1)$. *The number of computed witness points are respectively* $4, 4, 4$. *The maximum residuals after substituting the witness points into the three components are respectively* $0.1419818555e-4, 3.09 \times 10^{(-8)}, 0.2263585e-5$. *The smallest singular values of the Jacobian matrices evaluated at the witness points are respectively*

$$\begin{array}{l} 1.01136778729643, 1.54387370543690, 1.51495924713963, 1.52421428042282, \\ 0.778955506429449e-1, .660455232353615, .317809084873550, .105611642184416, \\ 0.000221843023788232e-3, 0.615594132791045e-3, 0.165300475257133e-3, \\ 0.596754059706055e-3. \end{array}$$

Example 8. *Consider the following system* F:

$$\left\{ \begin{array}{l} x_1^3 - 16\,x_3\,x_1^2 + 100\,x_2^2x_1 + 160\,x_4\,x_2\,x_1 + 64\,x_3^2x_1 + 64\,x_4^2x_1 - 14\,x_1^2 \\ \quad + 20\,x_2\,x_1 + 112\,x_3\,x_1 + 16\,x_4\,x_1 - 4\,x_3\,x_5 - 3\,x_4\,x_5 + 50\,x_1 - 8\,x_3 + 1, \\ x_2\,x_1^3 - 16\,x_3\,x_2\,x_1^2 + 100\,x_2^3x_1 + 160\,x_4\,x_2^2x_1 + 64\,x_3^2x_2\,x_1 \\ \quad + 64\,x_4^2x_2\,x_1 - 14\,x_2\,x_1^2 + 20\,x_2^2x_1 + 112\,x_3\,x_2\,x_1 + 16\,x_4\,x_2\,x_1 \\ \quad - 4\,x_2\,x_3\,x_5 - 3\,x_2\,x_4\,x_5 - x_1^2 + 50\,x_2\,x_1 + 16\,x_3\,x_1 - 100\,x_2^2 - 8\,x_3\,x_2 \\ \quad - 160\,x_4\,x_2 - 64\,x_3^2 - 64\,x_4^2 + 14\,x_1 - 19\,x_2 - 112\,x_3 - 16\,x_4 - 50, \\ -6\,x_4\,x_1 + 2\,x_2 - 7\,x_5 + 1, \\ -2\,x_6\,x_1 - 4\,x_4\,x_3 + 3\,x_5 + 1 \end{array} \right\}.$$

Set the variable order to be $x_1 < \cdots < x_6$. *A full rank representation of* $V_{\mathbb{R}}(F)$ *computed by Algorithm 1 consists of* 6 *components, with dimensions being respectively* 1, 1, 0, 0, 0, 0, *numbers of polynomials being respectively* 6, 6, 7, 7, 7, 7, *and total degrees of polynomials in them being respectively*

$$(4,2,3,2,2,9),(3,13,3,2,2,19),(5,4,1,4,4,4,1),$$
$$(1,1,1,1,1,1,1),(5,4,1,4,4,4,1),(4,3,1,3,2,3,1).$$

The number of computed witness points are respectively 2, 3, 3, 1, 1, 2. *The maximum residuals after substituting the witness points into the six components are respectively* $0.67701e-4, 3.8\times10^{-8}, 1.17\times10^{-8}, 0, 2.8\times10^{-9}, 4.\times10^{-10}$. *The smallest singular values of the Jacobian matrices of polynomial systems in the output evaluated at the witness points are respectively*

$0.243258982897929e-5, 0.243028372806473e-5, .129125424091993,$
$.678574657147939, 0.295982421600441e-1, 0.735206574833336e-3,$
$0.712398160744306e-3, 0.467778814287120e-2, 0.571345331300000e-2,$
$0.457875824937666e-4, 0.301415284267564e-2, 0.306478932392468e-2.$

In all four examples, the smallest singular values are far from machine epsilon, which indicates that the Jacobian matrices have full (numerical) rank. Table 1 summarizes an experimental comparison with the RealTriangularize command. The column *in-deg* and *out-deg* denote respectively the maximum total degrees of polynomials appearing in the input and output of FullRankDecompose. The experimentation was conducted on an Ubuntu Laptop (Intel i7-4700MQ CPU @ 2.40GHz, 8.0GB memory). The memory usage was restricted to 60% of the total memory. It is interesting to notice that FullRankDecompose outperforms RealTriangularize for systems having larger degrees.

Table 1. Summary of experimentation results on four examples.

Sys	in-deg	out-deg	Complex-dim	Real-dim	RealTriangularize	FullRankDecompose
Ex 5	4	7	1	0	$> 1800(s)$	$275(s)$
Ex 6	3	5	3	2	$1(s)$	$2(s)$
Ex 7	4	35	2	1	$> 1800(s)$	$92(s)$
Ex 8	4	19	2	1	$> 1800(s)$	$320(s)$

6 Conclusion and Future Work

In this paper, we introduced the notion of the full rank representation of a real algebraic set to remove various singularities, which often lead to ill-posedness in numerical computation. A proof-of-concept hybrid algorithm was proposed to compute it. Such a representation was successfully applied to visualize some plane and space curves with singularities. Effectiveness of the hybrid algorithm was also illustrated by computing full rank representations of some nontrivial examples with rank-deficient Jacobian matrices.

Nevertheless, we notice that the current algorithm computes more than necessary (like in a triangular shape) for a full rank representation. How to reduce the size (including degrees and coefficients) of polynomials in the representation remains a challenge, which is crucial for the stability of numerical procedures taking such representation as input. In a future work, we will also investigate the possibility of combining with other numerical methods, such as sum of squares based on semidefinite programming.

Acknowledgements. The authors would like to thank Hoon Hong and anonymous reviewers for their helpful comments. This work is partially supported by the projects NSFC (11471307, 11671377, 61572024), cstc2015jcyjys40001, and the Key Research Program of Frontier Sciences of CAS (QYZDB-SSW-SYS026).

References

1. Alvandi, P., Chen, C., Hashemi, A., Maza, M.M.: Regular chains under linear changes of coordinates and applications. In: Gerdt, V.P., Koepf, W., Seiler, W.M., Vorozhtsov, E.V. (eds.) CASC 2015. LNCS, vol. 9301, pp. 30–44. Springer, Cham (2015). doi:10.1007/978-3-319-24021-3_3
2. Aubry, P., Lazard, D., Moreno Maza, M.: On the theories of triangular sets. J. Symb. Comput. **28**(1–2), 105–124 (1999)
3. Brake, D., Hauenstein, J., Liddell, A.: Validating the completeness of the real solution set of a system of polynomial equations. ISSAC **2016**, 143–150 (2016)
4. Caire, L.: Plane curves as projections of non singular space curves. Manuscripta Math. **67**(1), 433–450 (1990)
5. Chen, C., Davenport, J., May, J., Moreno Maza, M., Xia, B., Xiao, R.: Triangular decomposition of semi-algebraic systems. J. Symb. Comput. **49**, 3–26 (2013)
6. Chen, C., Moreno Maza, M.: Algorithms for computing triangular decomposition of polynomial systems. J. Symb. Comput. **47**(6), 610–642 (2012)
7. Chen, C., Wu, W.: A numerical method for computing border curves of bi-parametric real polynomial systems and applications. In: Gerdt, V.P., Koepf, W., Seiler, W.M., Vorozhtsov, E.V. (eds.) CASC 2016. LNCS, vol. 9890, pp. 156–171. Springer, Cham (2016). doi:10.1007/978-3-319-45641-6_11
8. Daouda, D., Mourrain, B., Ruatta, O.: On the computation of the topology of a non-reduced implicit space curve. ISSAC **2008**, 47–54 (2008)
9. Eisenbud, D.: Commutative Algebra: With a View Toward Algebraic Geometry. Graduate Texts in Mathematics, vol. 150. Springer, Heidelberg (2013). doi:10.1007/978-1-4612-5350-1
10. Hartshorne, R.: Algebraic Geometry. Graduate Texts in Mathematics, vol. 52. Springer, Heidelberg (1997). doi:10.1007/978-1-4757-3849-0
11. Hauenstein, J.D.: Numerically computing real points on algebraic sets. Acta Appl. Math. **125**(1), 105–119 (2012)
12. Hauenstein, J., Sommese, A.: What is numerical algebraic geometry. J. Symb. Comp. **79**, 499–507 (2017). Part 3
13. Hong, H., El Din, M.S.: Variant quantifier elimination. J. Symb. Comp. **47**(7), 883–901 (2012)
14. Hong, H.: An efficient method for analyzing the topology of plane real algebraic curves. Math. Comput. Simul. **42**(4), 571–582 (1996)

15. Imbach, R., Moroz, G., Pouget, M.: Numeric and certified isolation of the singularities of the projection of a smooth space curve. MACIS **2015**, 78–92 (2016)
16. Jin, K., Cheng, J.: Isotopic epsilon-meshing of real algebraic space curves. SNC **2014**, 118–127 (2014)
17. Jin, K., Cheng, J.-S., Gao, X.-S.: On the topology and visualization of plane algebraic curves. In: Gerdt, V.P., Koepf, W., Seiler, W.M., Vorozhtsov, E.V. (eds.) CASC 2015. LNCS, vol. 9301, pp. 245–259. Springer, Cham (2015). doi:10.1007/978-3-319-24021-3_19
18. Labs, O.: A list of challenges for real algebraic plane curve visualization software. In: Emiris, I.Z., Sottile, F., Theobald, T. (eds.) Nonlinear Computational Geometry, pp. 137–164. Springer, New York (2010)
19. Lasserre, J., Laurent, M., Rostalski, P.: Semidefinite characterization and computation of zero-dimensional real radical ideals. Found. Comput. Math. **8**(5), 607–647 (2008)
20. Lecerf, G.: Quadratic newton iteration for systems with multiplicity. Found. Comput. Math. **2**(3), 247–293 (2002)
21. Lee, J.M.: Introduction to Smooth Manifolds. Graduate Texts in Mathematics, vol. 218. Springer, Heidelberg (2003). doi:10.1007/978-1-4419-9982-5
22. Leykin, A.: Numerical primary decomposition. ISSAC **2008**, 165–172 (2008)
23. Leykin, A., Verschelde, J., Zhao, A.: Newton's method with deflation for isolated singularities of polynomial systems. TCS **359**(1), 111–122 (2006)
24. Li, T.Y.: Numerical solution of multivariate polynomial systems by homotopy continuation methods. Acta Numerica **6**, 399–436 (1997)
25. Ma, Y., Wang, C., Zhi, L.: A certificate for semidefinite relaxations in computing positive-dimensional real radical ideals. J. Symb. Comput. **72**, 1–20 (2016)
26. Mantzaflaris, A., Mourrain, B.: Deflation and certified isolation of singular zeros of polynomial systems. ISSAC **2011**, 249–256 (2011)
27. Morrow, J.A., Kodaira, K.: Complex Manifolds, vol. 355. American Mathematical Society, Providence (1971)
28. Ojika, T., Watanabe, S., Mitsui, T.: Deflation algorithm for the multiple roots of a system of nonlinear equations. J. Math. Anal. Appl. **96**(2), 463–479 (1983)
29. Parrilo, P.: Semidefinite programming relaxations for semialgebraic problems. Math. Program. **96**(2), 293–320 (2003)
30. Rouillier, F., Roy, M.F., El Din, M.S.: Finding at least one point in each connected component of a real algebraic set defined by a single equation. J. Complex. **16**(4), 716–750 (2000)
31. El Din, M.S., Schost, É.: Properness defects of projections and computation of at least one point in each connected component of a real algebraic set. Discrete Comput. Geom. **32**(3), 417–430 (2004)
32. El Din, M.S., Spaenlehauer, P.: Critical point computations on smooth varieties: degree and complexity bounds. In: ISSAC 2016, pp. 183–190 (2016)
33. Sommese, A., Verschelde, J., Wampler, C.: Numerical decomposition of the solution sets of polynomial systems into irreducible components. SIAM J. Numer. Anal. **38**(6), 2022–2046 (2001)
34. Sommese, A., Wampler, C.: The Numerical Solution of Systems of Polynomials Arising in Engineering and Science. World Scientific Press, Singapore (2005)
35. Wu, W., Reid, G.: Finding points on real solution components and applications to differential polynomial systems. ISSAC **2013**, 339–346 (2013)

Certifying Simple Zeros of Over-Determined Polynomial Systems

Jin-San Cheng(✉) and Xiaojie Dou(✉)

Key Lab of Mathematics Mechanization, Institute of Systems Science, Academy of Mathematics and Systems Science, CAS, Beijing, China
{jcheng,xjdou}@amss.ac.cn

Abstract. We construct a real square system related to a given over-determined real system. We prove that the simple real zeros of the over-determined system are the simple real zeros of the related square system and the real zeros of the two systems are one-to-one correspondence with the constraint that the value of the sum of squares of the polynomials in the over-determined system at the real zeros is identically zero. After certifying the simple real zeros of the related square system with the interval methods, we assert that the certified zero is a local minimum of the sum of squares of the input polynomials. If the value of the sum of the squares of the input polynomials at the certified zero is equal to zero, then it is a zero of the input system. Notice that a complex system with complex zeros can be transformed into a real system with real zeros.

Keywords: Over-determined polynomial system · Simple real zeros · Sum of squares · Minimum point · Interval methods

1 Introduction

Finding zeros of polynomial systems is a fundamental problem in scientific computing. Newton's method is widely used to solve this problem. For a fixed approximate solution of a system, we can use the α-theory [3,10,31], the interval methods or the optimization methods [11,16,20,23,28,32] to completely determine whether it is related to a zero of the system. However, the α-theory or the interval methods focus mainly on a simple zero of a square system, that is, a system with n equations and n unknowns.

Some special certifications of a rational solution of rational polynomials with certified sum of squares decompositions are considered [2,13,15,22,26,27,30].

How about singular zeros of a well-constrained polynomial system? Usually, an over-determined system which contains the same zero as a simple one is constructed by introducing new equations. The basic idea are the deflation techniques [1,5,6,8,9,24,25,33]. In some references [4,12,17,18,21,29], new variables are also included. Moreover, some authors verify that a perturbed system possesses an isolated singular solution within a narrow and computed error bound. The multiplicity structures of singular zeros of a polynomial system are also

V.P. Gerdt et al. (Eds.): CASC 2017, LNCS 10490, pp. 66–76, 2017.
DOI: 10.1007/978-3-319-66320-3_6

studied [5,9,21]. Especially in [1,9], singular zeros of the input systems are transformed into simple zeros of the new systems when the coefficients are rational.

For the deflation methods mentioned above, on one hand, to be a zero of the perturbed systems does not mean being a zero of the input system considering the difference between the two systems; on the other hand, although the over-determined systems without introducing new variables have the same zeros as the input systems, the verification methods, such as the α-theory or the interval methods, could not be used directly on the over-determined systems in general.

In [7], the authors study computing simple zeros of over-determined polynomial systems with Newton's method in theory. They also extend the α-theory from well-constrained systems to over-determined systems. A main result about Newton's method given in their paper is Theorem 4 [7], which says that under the condition of $2\alpha_1(g,\zeta) < 1$, where g is an analytic function $g : \mathbb{E} \to \mathbb{F}$, with $\mathbb{E}$ and $\mathbb{F}$ two real or complex Hilbert spaces, ζ is an attractive fixed point for Newton's method and simultaneously, a strict local minimum for $\|g\|^2$. However, as they stated, whether the attracting fixed points for Newton's method are always local minima of $\|g\|^2$, or the zeros of the input system, is unknown.

In this paper, we consider the problem of certifying the simple real zeros of an over-determined polynomial system. After transforming the input over-determined system into a square one, we can use both the α-theory and the interval methods to certify its simple zeros. In this paper we only consider using the interval methods to certify the simple real zeros of the over-determined system. We prove that the simple real zeros of the input system are local minima of the sum of squares of the input polynomials. We also give the condition that the local minimum is a simple zero of the input system.

Let $\mathbb{R}$ be the field of real numbers. Denote $\mathbb{R}[\mathbf{x}] = \mathbb{R}[x_1,\ldots,x_n]$ as the polynomial ring. Let $\mathbf{F} = \{f_1,\ldots,f_m\} \subset \mathbb{R}[\mathbf{x}]$ be a polynomial system. Let $\mathbf{p} = (p_1,\ldots,p_n) \in \mathbb{R}^n$.

The following theorem is our main result of this paper.

Theorem 1. *Let $\Sigma = \{f_1,\ldots,f_m\} \subset \mathbb{R}[\mathbf{x}]$ $(m \geq n)$ and $f = \sum_{i=1}^{m} f_i^2$. If $\mathbf{p} \in \mathbb{R}^n$ is a simple real zero of Σ, then, we have:*

1. *$\mathbf{p}$ is a local minimum of f;*
2. *$\mathbf{p}$ is a simple real zero of Σ if and only if $(\mathbf{p}, 0)$ is a simple real zero of the square system $\Sigma_r = \{\mathbf{J}_1(f),\ldots,\mathbf{J}_n(f), f - r\}$, where $\mathbf{J}_i(f) = \frac{\partial f}{\partial x_i}$.*

In the above theorem, we get a necessary and sufficient condition to certify the simple real zeros of the input system Σ by certifying the simple real zeros of the square system Σ_r. Therefore, to certify that $\mathbf{p}$ is a simple real zero of Σ, the key point is verifying that $f(\mathbf{p}) = 0$.

However, it is difficult to decide numerically if a point is a zero of a polynomial. Thus we can not use the necessary and sufficient condition to certify the simple real zeros of Σ by certifying the simple real zeros of Σ_r.

As an alterative, we refine and certify the simple real zeros of Σ by refining and certifying a new square system $\Sigma' = \{\mathbf{J}_1(f),\ldots,\mathbf{J}_n(f)\}$ with the interval methods and get a verified inclusion $\mathbf{X}$, which contains a unique simple real zero $\hat{\mathbf{x}}$ of Σ'. In fact, $\hat{\mathbf{x}}$ is a local minimum of f. On one hand, if $f(\hat{\mathbf{x}}) = 0$, by Theorem 1, $(\hat{\mathbf{x}}, 0)$ is a simple real zero of Σ_r, and then $\hat{\mathbf{x}}$ is a simple real zero of Σ. Thus, we certified the input system Σ. On the other hand, if $f(\hat{\mathbf{x}}) \neq 0$, we will have a very small positive value $f(\hat{\mathbf{x}})$. At this point, We assert that Σ_r has a unique zero in the verified inclusion $\mathbf{X} \times [0, f(\hat{\mathbf{x}})]$, which means we certified the system Σ_r.

The paper is organized as below. We will introduce some notations and preliminaries in the next section. In Sect. 3, we will give a method to show how to transform an over-determined system into a square one. The interval verification method on the obtained square system is considered in Sect. 4.

2 Preliminaries

Let $\mathbb{C}$ be the field of complex numbers. Denote $\mathbb{C}[\mathbf{x}] = \mathbb{C}[x_1, \ldots, x_n]$ as the polynomial ring. Let $\mathbf{F} = \{f_1, \ldots, f_m\} \subset \mathbb{C}[\mathbf{x}]$ be a polynomial system. Let $\mathbf{p} = (p_1, \ldots, p_n) \in \mathbb{C}^n$. $\mathbf{F}(\mathbf{p}) = \mathbf{0}$ denotes that $\mathbf{p}$ is a zero of $\mathbf{F}(\mathbf{x}) = \mathbf{0}$.

Let A be a matrix. Denote A^T as the transpose of A and $rank(A)$ as the rank of A. Let $Mat(a_{i,j})$ denote the matrix whose i-th row j-th column element is $a_{i,j}$.

Let $\Sigma = \{f_1, \ldots, f_m\} \subset \mathbb{C}[\mathbf{x}]$ be a polynomial system. Denote $\mathbf{J}(\Sigma)$ as the Jacobian matrix of Σ. That is,

$$\mathbf{J}(\Sigma) = \begin{pmatrix} \frac{\partial f_1}{\partial x_1} & \cdots & \frac{\partial f_1}{\partial x_n} \\ \vdots & \ddots & \vdots \\ \frac{\partial f_m}{\partial x_1} & \cdots & \frac{\partial f_m}{\partial x_n} \end{pmatrix}.$$

For a polynomial $f \in \mathbb{C}[\mathbf{x}]$, let $\mathbf{J}(f)$ denote $(\frac{\partial f}{\partial x_1}, \frac{\partial f}{\partial x_2}, \ldots, \frac{\partial f}{\partial x_n})$, $\mathbf{J}_i(f) = \frac{\partial f}{\partial x_i}$ and $\mathbf{J}_{i,j}(f) = \mathbf{J}_j(\mathbf{J}_i(f)) = \frac{\partial^2 f}{\partial x_j \partial x_i}$. Denote $\Sigma_r = \{\mathbf{J}_1(f), \ldots, \mathbf{J}_n(f), f - r\}$ with $f = \sum_{j=1}^{m} f_j^2$.

We denote the value of a function matrix A at a point $\mathbf{p} \in \mathbb{C}^n$ as $A(\mathbf{p})$. Let $\mathbf{J}(\mathbf{F})(\mathbf{p})$ denote the value of a function matrix $\mathbf{J}(\mathbf{F})$ at a point $\mathbf{p}$, similarly for $\mathbf{J}(f)(\mathbf{p})$.

Definition 1. *An* **isolated solution** *of* $\mathbf{F}(\mathbf{x}) = \mathbf{0}$ *is a point* $\mathbf{p} \in \mathbb{C}^n$ *which satisfies:*

$$\exists\, \varepsilon > 0 : \{\mathbf{y} \in \mathbb{C}^n : \|\mathbf{y} - \mathbf{p}\| < \varepsilon\} \cap \mathbf{F}^{-1}(\mathbf{0}) = \{\mathbf{p}\}.$$

Definition 2. *We call an isolated solution* $\mathbf{p} \in \mathbb{C}^n$ *of* $\mathbf{F}(\mathbf{x}) = \mathbf{0}$ *a* **singular solution** *if and only if*

$$rank(\mathbf{J}(\mathbf{F})(\mathbf{p})) < n.$$

Else, we call $\mathbf{p}$ *a* **simple solution**.

Definition 3. *A* **stationary point** *of a polynomial function* $f(\mathbf{x}) \in \mathbb{C}[\mathbf{x}]$ *is a point* $\mathbf{p} \in \mathbb{C}^n$*, which satisfies:*

$$\frac{\partial f}{\partial x_i}(\mathbf{p}) = 0, \ \forall\, i = 1, \ldots, n.$$

We can find the following lemma in many undergraduate text books about linear algebra (see Example 7 on page 224 in [19]).

Lemma 1. *Let* $A \in \mathbb{R}^{m\times n}$ *be a real matrix with* $m \geq n$ *and* $B = A^T A$*. Then the ranks of* A *and* B *are the same, especially for the case that* A *is of full rank.*

In the following, we will consider the real zeros of the systems with real coefficients. It is reasonable since for a system (m equations and n unknowns) with complex coefficients, we can rewrite the system into a new one with $2\,m$ equations and $2\,n$ unknowns by splitting the unknowns $x_i = x_{i,1} + \mathbf{i}\, x_{i,2}$ and equations $f_j(x_1, \ldots, x_n) = g_{j,1}(x_{1,1}, x_{1,2}, \ldots, x_{n,1}, x_{n,2}) + \mathbf{i}\, g_{j,2}(x_{1,1}, x_{1,2}, \ldots, x_{n,1}, x_{n,2})$, where $\mathbf{i}^2 = -1$, $f_j \in \mathbb{C}[\mathbf{x}], g_{j,1}, g_{j,2} \in \mathbb{R}[\mathbf{x}]$, $j = 1, \ldots, m$, and find out the complex zeros of the original system by finding out the real zeros of the new system.

3 Transforming Over-Determined Polynomial Systems into Square Ones

In this section, we will show how to transform an over-determined polynomial system into a square one with their zeros having a one-to-one correspondence, especially for the simple zeros.

By Definition 3, we have the following lemma:

Lemma 2. *Given a polynomial system* $\Sigma = \{f_1, \ldots, f_m\} \subset \mathbb{R}[\mathbf{x}]$ $(m \geq n)$*. Let* $f = \sum_{i=1}^{m} f_i^2$ *and* $\Sigma' = \{\mathbf{J}_1(f), \mathbf{J}_2(f), \ldots, \mathbf{J}_n(f)\}$*. If* $\mathbf{p} \in \mathbb{R}^n$ *is a real zero of* Σ'*, then* $\mathbf{p}$ *is a stationary point of* f*.*

Lemma 3. *Let a polynomial system* $\Sigma = \{f_1, \ldots, f_m\} \subset \mathbb{R}[\mathbf{x}]$ $(m \geq n)$*,* $f = \sum_{i=1}^{m} f_i^2$ *and* $\Sigma' = \{\mathbf{J}_1(f), \mathbf{J}_2(f), \ldots, \mathbf{J}_n(f)\}$*. If* $\mathbf{p} \in \mathbb{R}^n$ *is a real zero of* Σ*, then we have:*

1. $\mathbf{p}$ *is a real zero of* Σ'*;*
2. $rank(\mathbf{J}(\Sigma)(\mathbf{p})) = rank(\mathbf{J}(\Sigma')(\mathbf{p}))$*.*

Proof. It is clear that $\mathbf{p}$ is a real zero of Σ' providing that $\mathbf{p}$ is a real zero of Σ, since $\mathbf{J}_i(f) = 2 \sum_{k=1}^{m} f_k\, \mathbf{J}_i(f_k)$.

To prove the second part of this lemma, we rewrite $\mathbf{J}_i(f)$ as follows.

$$\mathbf{J}_i(f) = 2\, \langle f_1, \ldots, f_m \rangle\, \langle \mathbf{J}_i(f_1), \ldots, \mathbf{J}_i(f_m) \rangle^T, \tag{1}$$

where $\langle \cdot \rangle^T$ is the transpose of a vector or a matrix $\langle \cdot \rangle$. Then

$$\mathbf{J}_{i,j}(f) = \mathbf{J}_j(\mathbf{J}_i(f)) = \mathbf{J}_j(2\sum_{k=1}^{m} f_k\,\mathbf{J}_i(f_k)) = 2\sum_{k=1}^{m}(\mathbf{J}_j(f_k)\mathbf{J}_i(f_k) + f_k\,\mathbf{J}_{i,j}(f_k))$$
$$= 2\,\langle \mathbf{J}_j(f_1),\ldots,\mathbf{J}_j(f_m)\rangle\,\langle \mathbf{J}_i(f_1),\ldots,\mathbf{J}_i(f_m)\rangle^T + 2\sum_{k=1}^{m} f_k\,\mathbf{J}_{i,j}(f_k). \tag{2}$$

Then the Jacobian matrix of Σ' is

$$\mathbf{J}(\Sigma') = \begin{pmatrix} \mathbf{J}_{1,1}(f) & \ldots & \mathbf{J}_{1,n}(f) \\ \vdots & \ddots & \vdots \\ \mathbf{J}_{n,1}(f) & \ldots & \mathbf{J}_{n,n}(f) \end{pmatrix} = \mathrm{Mat}(\mathbf{J}_{i,j}(f)).$$

We rewrite

$$\mathrm{Mat}(\mathbf{J}_{i,j}(f)) = 2\,A^T A + 2\,\mathrm{Mat}(\sum_{k=1}^{m} f_k\,\mathbf{J}_{i,j}(f_k)), \tag{3}$$

where

$$A = \begin{pmatrix} \mathbf{J}_1(f_1) & \ldots & \mathbf{J}_n(f_1) \\ \vdots & \ddots & \vdots \\ \mathbf{J}_1(f_m) & \ldots & \mathbf{J}_n(f_m) \end{pmatrix}$$

is an $m \times n$ matrix which is exactly the Jacobian matrix of Σ, that is, $\mathbf{J}(\Sigma) = A$. Then we have

$$\mathbf{J}(\Sigma')(\mathbf{p}) = A(\mathbf{p})^T A(\mathbf{p}). \tag{4}$$

By Lemma 1, the second part of the lemma is true. This ends the proof. □

Remark 1. In our construction of f and Σ', the degree of the polynomials may be doubled. However, it has no influence on our actual computation, if you notice Eq. (4) in the above proof. In fact, to get $\mathbf{J}(\Sigma')(\mathbf{p})$, we only need to compute $A(\mathbf{p})$, which does not increase our actual computing degree.

As a byproduct, thanks to the doubled degree of the polynomials, our final certifying accuracy is also improved in Lemma 4.

The following is the proof of Theorem 1

Proof. In fact, by fixing the real zero $\mathbf{p}$ as a simple zero in Lemma 3, we have $\mathbf{p}$ is a simple real zero of $\Sigma' = \{\mathbf{J}_1(f),\ldots,\mathbf{J}_n(f)\}$. Since $\mathbf{p}$ is a simple zero of Σ, $A(\mathbf{p})$ is a column full rank matrix. Therefore, it's easy to verify that $\mathbf{J}(\Sigma')(\mathbf{p}) = A(\mathbf{p})^T A(\mathbf{p})$ is a positive definite matrix. Thus, $\mathbf{p}$ is a local minimum of f and the first part of the theorem is true. Now we consider the second part.

First, it's easy to verify that $\mathbf{p}$ is the real zero of Σ if and only if $(\mathbf{p}, 0)$ is the real zero of Σ_r. With the same method as proving Lemma 3, we can get

$$rank(\mathbf{J}(\Sigma)(\mathbf{p})) = rank(\mathbf{J}(\Sigma_r)(\mathbf{p}, 0)) - 1, \tag{5}$$

which means that $\mathbf{J}(\Sigma_r)(\mathbf{p}, 0)$ is of full rank if and only if $\mathbf{J}(\Sigma)(\mathbf{p})$ is of full rank. Thus, $\mathbf{p}$ is the simple zero of Σ if and only if $(\mathbf{p}, 0)$ is the simple zero of Σ_r. The second part is true. We have finished the proof. □

From Theorem 1, we can know that the simple real zeros of Σ and Σ_r are in one to one correspondence with the constraint that the value of the sum of squares of the polynomials in Σ at the simple real zeros is identically zero. Thus we can transform an over-determined polynomial system into a square system Σ_r.

We will show a simple example to illustrate the theorem below.

Example 1. The simple zero $\mathbf{p} = (0,0)$ of the over-determined system $\Sigma = \{f_1, f_2, f_3\}$ corresponds to a simple zero of a square system $\Sigma_r = \{\mathbf{J}_1(f), \mathbf{J}_2(f), f - r\}$, where $f = f_1^2 + f_2^2 + f_3^2$ with

$$f_1 = x^2 - 2\,y, f_2 = y^2 - x, f_3 = x^2 - 2\,x + y^2 - 2\,y.$$

We can verify simply that $(\mathbf{p}, 0)$ is a simple zero of Σ_r.

Though the simple real zeros of Σ and Σ_r have a one to one correspondence, it can not be used directly to do certification of the simple zeros of Σ since we can not certify $r = 0$ numerically. But we can certify the zeros of $\Sigma' = \{\mathbf{J}_1(f), \mathbf{J}_2(f), \ldots, \mathbf{J}_n(f)\}$ as an alternative. By Theorem 1 and Lemma 2, when the value of f is zero at the certified zero, the certified zero is the very zero of the system Σ. We will discuss it in next section.

4 Certifying Simple Zeros of Over-Determined Systems

In this section, we consider to certify the over-determined system with the interval methods. We will prove the same local minimum result as [7].

The classical interval verification methods are based on the following theorem:

Theorem 2 [16,20,28,29]. *Let $f : \mathbb{R}^n \to \mathbb{R}^n$ with $f = (f_1, \ldots, f_n) \in C^1$, $\tilde{x} \in \mathbb{R}^n$, real interval vector $X \in \mathbb{IR}^n$ with $\tilde{x} \in X$ and real matrix $R \in \mathbb{R}^{n \times n}$ be given. Let an interval matrix $M \in \mathbb{IR}^{n \times n}$ be given whose i-th row M_i satisfies*

$$\{\nabla f_i(\zeta) : \zeta \in \tilde{x} + X\} \subseteq M_i.$$

Denote by I the $n \times n$ identity matrix and assume

$$-Rf(\tilde{x}) + (I - RM)X \subseteq int(X),$$

where $int(X)$ denotes the interior of X. Then, there is a unique $\hat{x} \in \tilde{x} + X$ with $f(\hat{x}) = 0$. Moreover, every matrix $\tilde{M} \in M$ is nonsingular. In particular, the Jacobian $\mathbf{J}(f)(\hat{x})$ is nonsingular.

About interval matrices, there is an important property in the following theorem.

Theorem 3 [14]. *A symmetric interval matrix A^I is positive definite if and only if it is regular and contains at least one positive definite matrix.*

Given an over-determined polynomial system $\Sigma = \{f_1, \ldots, f_m\} \subset \mathbb{R}[\mathbf{x}]$ with a simple real zero, we can compute a related square system

$$\Sigma' = \{\frac{\partial f}{\partial x_1}, \frac{\partial f}{\partial x_2}, \ldots, \frac{\partial f}{\partial x_n}\} \text{ with } f = \sum_{j=1}^{m} f_j^2.$$

Based on Lemma 3, a simple zero of Σ is a simple zero of Σ'. Thus, we can compute the approximate simple zero of Σ by computing the approximate simple zero of Σ'. Using Newton's method, we can refine these approximate simple zeros with quadratic convergence to a relative higher accuracy. Then, we can certify them with the interval method mentioned before and get a verified inclusion $\mathbf{X}$, which possesses a unique certified simple zero of the system Σ' by Theorem 2, denoted as $\hat{\mathbf{x}} \in \mathbf{X}$.

However, even though we get a certified zero $\hat{\mathbf{x}}$ of the system Σ', considering Lemma 2, we cannot say $\hat{\mathbf{x}}$ is a zero of the input system Σ, because the certified zero $\hat{\mathbf{x}}$ is just a stationary point of f. Considering Theorem 1 and the difference between Σ' and Σ_r, we have the following theorem.

Theorem 4. *Let Σ, Σ', Σ_r, f, $\hat{\mathbf{x}}$ and the interval $\mathbf{X}$ be given as above. Then, we have:*

1. *$\hat{\mathbf{x}}$ is a local minimum of f;*
2. *there exists a verified inclusion $\mathbf{X} \times [0, f(\hat{\mathbf{x}})]$, which possesses a unique simple zero of the system Σ_r. Especially, if $f(\hat{\mathbf{x}}) = 0$, the verified inclusion $\mathbf{X}$ possesses a unique simple zero of the input system Σ.*

Proof. First, it's easy to see that computing the value of the matrix $\mathbf{J}(\Sigma')$ at the interval $\mathbf{X}$ will give a symmetric interval matrix, denoted as $\mathbf{J}(\Sigma')(\mathbf{X})$. By Theorem 2, we know that for every matrix $M \in \mathbf{J}(\Sigma')(\mathbf{X})$, M is nonsingular. Therefore, the interval matrix $\mathbf{J}(\Sigma')(\mathbf{X})$ is regular. Especially, the matrix $\mathbf{J}(\Sigma')(\hat{\mathbf{x}})$, which is the Hessian matrix of f, is of full rank and therefore is positive definite. Thus, $\hat{\mathbf{x}}$ is a local minimum of f. By Theorem 3, we know that $\mathbf{J}(\Sigma')(\mathbf{X})$ is positive definite. Thus, for every point $\mathbf{q} \in \mathbf{X}$, $\mathbf{J}(\Sigma')(\mathbf{q})$ is a positive definite matrix. Considering Theorem 2, it's trivial that for the verified inclusion $\mathbf{X} \times [0, f(\hat{\mathbf{x}})]$, there exists a unique simple zero of the system Σ_r. If $f(\hat{\mathbf{x}}) = 0$, by Theorem 1, the verified inclusion $\mathbf{X}$ of the system Σ' is a verified inclusion of the original system Σ. □

Remark 2. In the above proof, we know that for every point $\mathbf{q} \in \mathbf{X}$, $\mathbf{J}(\Sigma')(\mathbf{q})$ is a positive definite matrix.

By Theorem 2, we know that there is a unique $\hat{\mathbf{x}} \in \mathbf{X}$ with $\Sigma'(\hat{\mathbf{x}}) = \mathbf{0}$. However, we could not know what the exact $\hat{\mathbf{x}}$ is. According to the usual practice, in actual computation, we will take the midpoint $\hat{\mathbf{p}}$ of the inclusion $\mathbf{X}$ as $\hat{\mathbf{x}}$ and verify whether $f(\hat{\mathbf{p}}) = 0$ or not. Considering the uniqueness of $\hat{\mathbf{x}}$ in $\mathbf{X}$, therefore, if $f(\hat{\mathbf{p}}) = 0$, we are sure that the verified inclusion $\mathbf{X}$ possesses a unique simple zero of the input system Σ. If $f(\hat{\mathbf{p}}) \neq 0$, we can only claim that there is a local minimum of f in the inclusion $\mathbf{X}$ and $\mathbf{X} \times [0, f(\hat{\mathbf{p}})]$ is a verified inclusion for the system Σ_r.

Considering the expression of Σ and f and for the midpoint $\hat{\mathbf{p}}$ of $\mathbf{X}$, we have a trivial result below.

Lemma 4. *Denote* $\epsilon = \max_{j=1}^{m} |f_j(\hat{\mathbf{p}})|$. *Under the conditions of Theorem 4, we have* $|f(\hat{\mathbf{p}})| \leq m\epsilon^2$.

Based on the above idea, we give an algorithm below. In the verification steps, we will apply the algorithm **verifynlss** in INTLAB [29], which is based on Theorem 2, to compute a verified inclusion $\mathbf{X}$ for the related square system Σ'. For simplicity, denote the interval $\mathbf{X} = [\underline{x}_1, \overline{x}_1], \cdots, [\underline{x}_m, \overline{x}_m]$ and the midpoint of $\mathbf{X}$ as $\hat{\mathbf{p}} = [(\underline{x}_1 + \overline{x}_1)/2, \ldots, (\underline{x}_m + \overline{x}_m)/2]$.

Algorithm 1. VSPS: verifying a simple zero of a polynomial system

Input: an over-determined polynomial system $\Sigma := \{f_1, \cdots, f_m\} \subset \mathbb{R}[\mathbf{x}]$ and an approximate simple zero $\tilde{\mathbf{p}} = (\tilde{p}_1, \cdots, \tilde{p}_n) \in \mathbb{R}^n$.

Output: a verified inclusion $\mathbf{X}$ and a small non-negative number.

Compute f and Σ';
Compute $\tilde{\mathbf{p}}' :=$ **Newton**$(\Sigma', \tilde{\mathbf{p}})$;
Compute $\mathbf{X} :=$ **verifynlss**$(\Sigma', \tilde{\mathbf{p}}')$ and $f(\hat{\mathbf{p}})$;
if $f(\hat{\mathbf{p}}) = 0$, **then**
    return $(\mathbf{X}, 0)$;
else
    return $(\mathbf{X}, f(\hat{\mathbf{p}}))$.
end if

The correctness and the termination of the algorithm is obvious by the above analysis.

We give two examples to illustrate our algorithm.

Example 2. Continue Example 1. Given an approximate zero $\tilde{\mathbf{p}} = (0.0003528, 0.0008131)$. Using Newton's method, we will get a high accuracy approximate zero

$$\tilde{\mathbf{p}}' = 10^{-11} \cdot (-0.104224090958505, -0.005858368844383).$$

Compute $f = f_1^2 + f_2^2 + f_3^2$ and $\Sigma' = \{\mathbf{J}_1(f), \mathbf{J}_2(f)\}$. After applying the algorithm **verifynlss** on Σ', we have a verified inclusion:

$$\mathbf{X} = \begin{pmatrix} [-0.11330049261083, \ 0.11330049261083] \\ [-0.08866995073891, \ 0.08866995073891] \end{pmatrix} \cdot 10^{-321}.$$

Based on Theorem 2, we know that there exists a unique $\hat{\mathbf{x}} \in \mathbf{X}$, s.t. $\Sigma'(\hat{\mathbf{x}}) = \mathbf{0}$.

Let $\Sigma_r = \{\mathbf{J}_1(f), \mathbf{J}_2(f), f - r\}$. By Theorem 1, we can certify the simple zero of Σ by certifying the simple zero of Σ_r. Considering the difference between Σ' and Σ_r, we check first whether the value of f at some point in the interval $\mathbf{X}$ is zero. According to the usual practice, we consider the midpoint $\hat{\mathbf{p}}$ of $\mathbf{X}$, which equals $(0, 0)$ and further, $f(\hat{\mathbf{p}})$ is zero. Therefore, we are sure that there exists a unique $\hat{\mathbf{x}} = (\hat{x}, \hat{y}) \in \mathbf{X}$, s.t. $\Sigma_r((\hat{\mathbf{x}}, 0)) = \mathbf{0}$ and then, there exists a unique simple zero $(\hat{x}, \hat{y}), |\hat{x}|, |\hat{y}| \leq 10^{-321}$, of the input system Σ in the interval $\mathbf{X}$, which means we certified the input system Σ.

Example 3. Let $\Sigma = \{f_1 = x_1^2 + 3\,x_1x_2 + 3\,x_1x_3 - 3\,x_3^2 + 2\,x_2 + 2\,x_3, f_2 = -3\,x_1x_2 + x_1x_3 - 2\,x_2^2 + x_3^2 + 3\,x_1 + x_2, f_3 = 2\,x_2x_3 + 3\,x_1 - 3\,x_3 + 2, f_4 = -6\,x_2^2x_3 + 2\,x_2x_3^2 + 6\,x_2^2 + 15\,x_2x_3 - 6\,x_3^2 - 9\,x_2 - 7\,x_3 + 6\}$ be an over-determined system. Given an approximate zero

$$\tilde{\mathbf{p}} = (-1.29655, 0.47055, -0.91761).$$

Using Newton's method, we will get a high accuracy zero

$$\tilde{\mathbf{p}}' = (-1.296687216045438, 0.470344502045004, -0.917812633399457).$$

Compute

$$f = f_1^2 + f_2^2 + f_3^2 + f_4^2 \text{ and } \Sigma' = \{\mathbf{J}_1(f), \mathbf{J}_2(f), \mathbf{J}_3(f)\}.$$

After applying the algorithm **verifynlss** on Σ', we have a verified inclusion:

$$\mathbf{X} = \begin{pmatrix} [-1.29668721603974, & -1.29668721603967] \\ [\ \ 0.47034450205107, & 0.47034450205114] \\ [-0.91781263339256, & -0.91781263339247] \end{pmatrix}.$$

Similarly, based on Theorem 2, we know that there exists a unique $\hat{\mathbf{x}} \in \mathbf{X}$, s.t. $\Sigma'(\hat{\mathbf{x}}) = \mathbf{0}$.

Proceeding as in the above example, we consider the midpoint $\hat{\mathbf{p}}$ of $\mathbf{X}$ and compute $f(\hat{\mathbf{p}}) = 3.94 \cdot 10^{-31} \neq 0$. Thus, by Theorem 4, we get a verified inclusion $\mathbf{X} \times [0, f(\hat{\mathbf{p}})]$, which contains a unique simple zero of the system Σ_r. It means that $\mathbf{X}$ may contain a zero of Σ. Even if $\mathbf{X}$ does not contain a zero of Σ, it contains a local minimum of f, which has a minimum value no larger than $f(\hat{\mathbf{p}})$.

Acknowledgement. The work is partially supported by NSFC Grants 11471327.

References

1. Akogul, T.A., Hauenstein, J.D., Szanto, A.: Certifying solutions to overdetermined and singular polynomial systems over $\mathbb{Q}$, 12 August 2014. arXiv: 1408.2721v1 [cs.SC]
2. Allamigeon, X., Gaubert, S., Magron, V., Werner, B.: Formal proofs for nonlinear optimization (2014). arXiv:1404.7282
3. Blum, L., Cucker, F., Shub, M., Smale, S.: Complexity and real computation. Springer, New York (1998)
4. Dayton, B., Li, T., Zeng, Z.: Multiple zeros of nonlinear systems. Math. Comp. **80**, 2143–2168 (2011)
5. Dayton, B., Zeng, Z.: Computing the multiplicity structure in solving polynomial systems. In: Kauers, M. (ed.) Proceedings of ISSAC 2005, pp. 116–123. ACM, New York (2005)

6. Giusti, M., Lecerf, G., Salvy, B., Yakoubsohn, J.C.: On location and approximation of clusters of zeros: case of embedding dimension one. Found. Comput. Math. **7**, 1–58 (2007)
7. Dedieu, J.P., Shub, M.: Newton's method for overdetermined systems of equations. Math. Comput. **69**(231), 1099–1115 (1999)
8. Hauenstein, J.D., Wampler, C.W.: Isosingular sets and deflation. Found. Comput. Math. **13**(3), 371–403 (2013)
9. Hauenstein, J.D., Mourrain, B., Szanto, A.: Certifying isolated singular points and their multiplicity structure. In: Proceedings of ISSAC 2015, pp. 213–220 (2015)
10. Hauenstein, J.D., Sottile, F.: Algorithm 921: alphaCertified: certifying solutions to polynomial systems. ACM Trans. Math. Softw. **38**(4), 28 (2012)
11. Kanzawa, Y., Kashiwagi, M., Oishi, S.: An algorithm for finding all solutions of parameter-dependent nonlinear equations with guaranteed accuracy. Electron. Commun. Jpn. (Part III: Fundam. Electron. Sci.) **82**(10), 33–39 (1999)
12. Kanzawa, Y., Oishi, S.: Approximate singular solutions of nonlinear equations and a numerical method of proving their existence. Sūrikaisekikenkyūsho Kōkyūroku **990**, 216–223 (1997). Theory and application of numerical calculation in science and technology, II (Japanese) Kyoto (1996)
13. Kaltofen, E., Li, B., Yang, Z., Zhi, L.: Exact certification of global optimality of approximate factorizations via rationalizing sums-of-squares with floating point scalars. In: Proceedings of ISSAC 2008, pp. 155–164, New York. ACM (2008)
14. Rohn, J.: Positive definiteness and stability of interval matrices. SIAM J. Matrix Anal. Appl. **15**, 175–184 (1994)
15. Kaltofen, E.L., Li, B., Yang, Z., Zhi, L.: Exact certification in global polynomial optimization via sums-of-squares of rational functions with rational coefficients. J. Symb. Comput. **47**(1), 1–15 (2012)
16. Krawczyk, R.: Newton-Algorithmen zur Bestimmung von Nullstellen mit Fehlerschranken. Computing **4**, 247–293 (1969)
17. Leykin, A., Verschelde, J., Zhao, A.: Newton's method with deflation for isolated singularities of polynomial systems. Theor. Comput. Sci. **359**, 111–122 (2006)
18. Li, N., Zhi, L.: Verified error bounds for isolated singular solutions of polynomial systems. SIAM J. Numer. Anal. **52**(4), 1623–1640 (2014)
19. Li, S.: Linear Algebra. Higher Education Press (2006). ISBN 978-7-04-019870-6
20. Moore, R.E.: A test for existence of solutions to nonlinear systems. SIAM J. Numer. Anal. **14**, 611–615 (1977)
21. Mantzaflaris, A., Mourrain, B.: Deflation and certified isolation of singular zeros of polynomial systems. In: Proceedings ISSAC 2011, pp. 249–256 (2011)
22. Monniaux, D., Corbineau, P.: On the generation of positivstellensatz witnesses in degenerate cases. In: Eekelen, M., Geuvers, H., Schmaltz, J., Wiedijk, F. (eds.) ITP 2011. LNCS, vol. 6898, pp. 249–264. Springer, Heidelberg (2011). doi:10.1007/978-3-642-22863-6_19
23. Nakaya, Y., Oishi, S., Kashiwagi, M., Kanzawa, Y.: Numerical verification of nonexistence of solutions for separable nonlinear equations and its application to all solutions algorithm. Electron. Commun. Jpn. (Part III: Fundam. Electron. Sci.) **86**(5), 45–53 (2003)
24. Ojika, T.: A numerical method for branch points of a system of nonlinear algebraic equations. Appl. Numer. Math. **4**, 419–430 (1988)
25. Ojika, T., Watanabe, S., Mitsui, T.: Deflation algorithm for the multiple roots of a system of nonlinear equations. J. Math. Anal. Appl. **96**, 463–479 (1983)

26. Peyrl, H., Parrilo, P.A.: A Macaulay2 package for computing sum of squares decompositions of polynomials with rational coefficients. In: Proceeding of SNC 2007, pp. 207–208 (2007)
27. Peyrl, H., Parrilo, P.A.: Computing sum of squares decompositions with rational coefficients. Theor. Comput. Sci. **409**(2), 269–281 (2008)
28. Rump, S.M.: Solving algebraic problems with high accuracy. In: Proceedings of the Symposium on A New Approach to Scientific Computation, San Diego, CA, USA, pp. 51–120. Academic Press Professional Inc. (1983)
29. Rump, S.M., Graillat, S.: Verified error bounds for multiple roots of systems of nonlinear equations. Numer. Algorithms **54**(3), 359–377 (2010)
30. El Din, M.S., Zhi, L.: Computing rational points in convex semialgebraic sets and sum of squares decompositions. SIAM J. Optim. **20**(6), 2876–2889 (2010)
31. Smale, S.: Newton's method estimates from data at one point. In: Ewing, R.E., Gross, K.I., Martin, C.F. (eds.) The Merging of Disciplines : New Directions in Pure, Applied and Computational Mathematics, pp. 185–196. Springer, Heidelberg (1986). doi:10.1007/978-1-4612-4984-9_13
32. Yamamura, K., Kawata, H., Tokue, A.: Interval solution of nonlinear equations using linear programming. BIT Numer. Math. **38**(1), 186–199 (1998)
33. Zeng, Z.: Computing multiple roots of inexact polynomials. Math. Comput. **74**, 869–903 (2005)

Decomposing Polynomial Sets Simultaneously into Gröbner Bases and Normal Triangular Sets

Rina Dong and Chenqi Mou(✉)

LMIB–SKLSDE–School of Mathematics and Systems Science, Beihang University, Beijing 100191, China
{rina.dong,chenqi.mou}@buaa.edu.cn

Abstract. In this paper we focus on the algorithms and their implementations for decomposing an arbitrary polynomial set simultaneously into pairs of lexicographic Gröbner bases and normal triangular sets with inherent connections in between and associated zero relationship with the polynomial set. In particular, a method by temporarily changing the variable orderings to handle the failure of the variable ordering assumption is proposed to ensure splitting needed for characteristic decomposition. Experimental results of our implementations for (strong) characteristic decomposition with comparisons with available implementations of triangular decomposition are also reported.

Keywords: Normal triangular set · Gröbner basis · Characteristic decomposition · Variable ordering

1 Introduction

Polynomial elimination theory, a classical branch of algebra, mainly studies the variable elimination, ordering, and decomposition of polynomial systems to reduce them into new ones with specific algebraic structures like being triangularized [28]. There are three main elimination methods based respectively on resultants, triangular sets [24,28,34], and Gröbner bases [6,9], the last two of which can be considered as generalizations of the well-known method of Gaussian elimination for linear systems.

Triangular decomposition is the process to decompose a polynomial set into finitely many triangular sets, which are ordered triangularized polynomial sets with respect to (w.r.t.) the variable ordering, such that the zero set of the polynomial set is equal to the union of those of the triangular sets. With continuous development on the theory, methods, and algorithms for triangular decomposition (see, e.g., [2–4,8,14,18,26,27,34] and references therein), triangular sets have become a computational tool for polynomial elimination and polynomial

This work was partially supported by the National Natural Science Foundation of China (NSFC 11401018).

V.P. Gerdt et al. (Eds.): CASC 2017, LNCS 10490, pp. 77–92, 2017.
DOI: 10.1007/978-3-319-66320-3_7

system solving. Currently there are effective algorithms for decomposing polynomial sets of moderate size into triangular sets of various kinds [7,8,16,21,28]. In this paper we are particularly interested in one kind of triangular sets, namely the normal sets, which are also called normalized triangular sets [18] and p-chain [14]. Normal sets, the initials of whose polynomials involve only the parameters, are convenient for dealing with parametric polynomial systems [7,14]. Algorithms have been proposed to normalize triangular sets or to decompose arbitrary polynomial sets into normal sets [28,32].

The Gröbner basis is a set of special generators of the ideal generated by a polynomial set w.r.t. a certain term ordering. Since its introduction by Buchberger in his Ph.D. thesis [6], the Gröbner basis has gained extensive study on the theory, methods, and algorithms [12,13,15,25,33] and become a powerful tool for computational commutative algebra and algebraic geometry with diverse applications. The elimination method with Gröbner bases is mainly based on the lexicographic (LEX) term ordering because of the good structures and rich properties of LEX Gröbner bases, for example their elimination property for elimination ideals. The structures of LEX Gröbner bases were studied first for bivariate ideals [17] and then extended to general zero-dimensional polynomial ideals [10,22]. Furthermore, the relationships between triangular decomposition of a polynomial set and the LEX Gröbner basis of the ideal generated by the polynomial set have also been studied. For zero-dimensional polynomial ideals, the relationship has been studied in [19] and algorithms for computing triangular decomposition from LEX Gröbner bases have been proposed in [10,19].

The relationship between Ritt characteristic sets and LEX Gröbner bases is investigated in [30] for polynomial ideals of arbitrary dimension via the so-called W-characteristic sets which are the smallest triangular sets contained in the LEX Gröbner bases. In particular, it is shown in [30] that when the W-characteristic set is abnormal, some certain polynomial in it is pseudo-divisible by another polynomial. By using such pseudo-divisibility, an algorithm is proposed in [31] to decompose an arbitrary polynomial set simultaneously into pairs of LEX Gröbner bases and normal sets (called characteristic decomposition) which have rich interconnections and provide two kinds of representations for the zeros of the polynomial set, and the structures and properties of characteristic decomposition are also studied.

As the follow-up work of [31], this paper focuses on the algorithms for (strong) characteristic decomposition and their implementations. In particular, an assumption on the variable ordering for the pseudo-divisibility to occur, the failure of which is not touched in [31] but happens indeed in our experiments (in 8 out of 35 positive-dimensional test examples, with more details in Table 1), is further handled by temporarily changing the variable orderings. We also make an enriched comparison on the performances of our implementation for characteristic decomposition with some available implementations for triangular decomposition via normal decomposition. As shown by the experimental results, our implementation performs comparably well with other similar implementations, but with richer output.

After a brief review of triangular sets, Gröbner bases, and (strong) characteristic decomposition in Sect. 2, we present the method to handle the variable ordering condition and recall the algorithms for (strong) characteristic decomposition in detail, followed by an illustrative example, in Sect. 3. Then the experimental results with our implementations for (strong) characteristic decomposition are reported in Sect. 4.

2 Preliminaries

In this section some basic notions and notations used in the sequel are recalled. The reader is referred to [2,5,9,28] and references therein for more details on the theories of Gröbner bases and triangular sets and to [30,31] for the definitions and properties of characteristic decomposition.

2.1 Triangular Set and Triangular Decomposition

Let $\mathbb{K}$ be a field and $\mathbb{K}[x_1, \ldots, x_n]$ be the ring of polynomials in n ordered variables $x_1 < \cdots < x_n$ with coefficients in $\mathbb{K}$. For the sake of simplicity, we denote $\mathbb{K}[x_1, \ldots, x_n]$ by $\mathbb{K}[\boldsymbol{x}]$.

Let F be a polynomial in $\mathbb{K}[\boldsymbol{x}] \setminus \mathbb{K}$. With respect to the variable ordering, the greatest variable effectively appearing in F is called the *leading variable* of F and denoted by $\mathrm{lv}(F)$. Let $\mathrm{lv}(F) = x_i$. Then F can be written as $F = I x_i^k + R$ with $I \in \mathbb{K}[x_1, \ldots, x_{i-1}]$, $R \in \mathbb{K}[x_1, \ldots, x_i]$, and $\deg(R, x_i) < k = \deg(F, x_i)$. The polynomial I is called the *initial* of F, denoted by $\mathrm{ini}(F)$. For any polynomial set $\mathcal{F} \subseteq \mathbb{K}[\boldsymbol{x}]$, we use $\mathrm{ini}(\mathcal{F})$ to denote the set $\{\mathrm{ini}(F) \mid F \in \mathcal{F}\}$.

Definition 1. A finite, nonempty, ordered set $\mathcal{T} = [T_1, \ldots, T_r]$ of polynomials in $\mathbb{K}[\boldsymbol{x}] \setminus \mathbb{K}$ is called a *triangular set* if $\mathrm{lv}(T_1) < \cdots < \mathrm{lv}(T_r)$.

The *saturated ideal* of a triangular set $\mathcal{T} = [T_1, \ldots, T_r]$ is defined as $\mathrm{sat}(\mathcal{T}) = \langle \mathcal{T} \rangle : J^\infty$, where $J = \mathrm{ini}(T_1) \cdots \mathrm{ini}(T_r)$. For a triangular set $\mathcal{T} \subset \mathbb{K}[\boldsymbol{x}]$, the variables in $\{x_1, \ldots, x_n\} \setminus \{\mathrm{lv}(T_1), \ldots, \mathrm{lv}(T_r)\}$ are called its *parameters*. $\mathcal{T}$ is said to be *zero-dimensional* if it has no parameter, and *positive-dimensional* otherwise.

Definition 2. A triangular set $\mathcal{T} = [T_1, \ldots, T_r] \subseteq \mathbb{K}[\boldsymbol{x}]$ is said to be a *normal set* if $\mathrm{ini}(T_1), \ldots, \mathrm{ini}(T_r)$ only involve the parameters of $\mathcal{T}$.

One of the most commonly used triangular sets are the so-called *regular sets* or *regular chains* [16,27]. Any normal set is regular by definition. For any two polynomial sets $\mathcal{F}, \mathcal{G} \subset \mathbb{K}[\boldsymbol{x}]$, we denote by $\mathsf{Z}(\mathcal{F}/\mathcal{G})$ the set

$$\mathsf{Z}(\mathcal{F}/\mathcal{G}) := \{\bar{\boldsymbol{x}} \in \bar{\mathbb{K}}^n : F(\bar{\boldsymbol{x}}) = 0, G(\bar{\boldsymbol{x}}) \neq 0, \forall F \in \mathcal{F}, G \in \mathcal{G}\},$$

where $\bar{\mathbb{K}}$ is the algebraic closure of $\mathbb{K}$. In particular, $\mathsf{Z}(\mathcal{F}) := \mathsf{Z}(\mathcal{F}/\{\})$.

Definition 3. Let $\mathcal{F} \subset \mathbb{K}[\boldsymbol{x}]$ be a polynomial set. A finite number of triangular sets $\mathcal{T}_1, \mathcal{T}_2, \ldots, \mathcal{T}_t \subset \mathbb{K}[\boldsymbol{x}]$ are called a triangular decomposition of $\mathcal{F}$ if $\mathsf{Z}(\mathcal{F}) = \bigcup_{i=1}^t \mathsf{Z}(\mathcal{T}_i / \mathrm{ini}(\mathcal{T}_i))$.

2.2 Gröbner Basis and W-Characteristic Set

A *term ordering* $<_t$ is a total and well ordering on all the terms in $\mathbb{K}[\boldsymbol{x}]$. With a fixed term ordering $<_t$, the greatest term in a polynomial $F \in \mathbb{K}[\boldsymbol{x}]$ w.r.t. $<_t$ is called the *leading term* of F and denoted by $\mathrm{lt}(F)$.

In this paper the LEX term ordering $<_{\mathrm{LEX}}$ is of our main concern. For two different terms $\boldsymbol{x}^{\boldsymbol{\alpha}}$ and $\boldsymbol{x}^{\boldsymbol{\beta}}$ in $\mathbb{K}[\boldsymbol{x}]$ with $\boldsymbol{\alpha} = (\alpha_1, \ldots, \alpha_n)$ and $\boldsymbol{\beta} = (\beta_1, \ldots, \beta_n)$, we say that $\boldsymbol{x}^{\boldsymbol{\alpha}} <_{\mathrm{LEX}} \boldsymbol{x}^{\boldsymbol{\beta}}$ if there exists an integer i $(1 \le i \le n)$ such that $\alpha_i < \beta_i$ and for $j = i+1, \ldots, n, \alpha_j = \beta_j$.

Definition 4. Let $\mathfrak{I} \subseteq \mathbb{K}[\boldsymbol{x}]$ be an ideal. A finite set $\{G_1, \ldots, G_s\}$ of polynomials in $\mathfrak{I}$ is called a *Gröbner basis* of $\mathfrak{I}$ w.r.t. the term ordering $<_t$ if $\langle \mathrm{lt}(G_1), \ldots, \mathrm{lt}(G_s)\rangle = \langle \mathrm{lt}(\mathfrak{I})\rangle$, where $\langle \mathrm{lt}(\mathfrak{I})\rangle$ denotes the ideal generated by the leading terms of all the polynomials in $\mathfrak{I}$.

A Gröbner basis $\mathcal{G}$ is said to be *reduced* if for each $G \in \mathcal{G}$, its coefficient w.r.t. $\mathrm{lt}(G)$ is 1 and any term in G is not divisible by $\mathrm{lt}(G')$ for any $G' \in \mathcal{G} \setminus \{G\}$. The reduced Gröbner basis of any ideal w.r.t. a fixed term ordering is unique. In particular, from the reduced LEX Gröbner basis of an ideal, one can extract the W-characteristic set of this ideal as defined below.

Definition 5 [30, Definition 3.1]**.** Let $\mathcal{G}$ be the reduced LEX Gröbner basis of the ideal $\langle \mathcal{P}\rangle \subseteq \mathbb{K}[\boldsymbol{x}]$. Denote $\mathcal{G}^{(i)} = \{G \in \mathcal{G} \mid \mathrm{lv}(G) = x_i\}$. Then the ordered set of all the smallest polynomials w.r.t. $<_{\mathrm{LEX}}$ in every set $\mathcal{G}^{(i)}$ for $i = 1, \ldots, n$ is called the *W-characteristic set* of $\langle \mathcal{P}\rangle$.

Obviously, a W-characteristic set is also a triangular set. Basic properties of W-characteristic sets and the pseudo-divisibility relationship between polynomials in W-characteristic sets are recalled respectively in Proposition 1 and Theorem 1 below. We would like to mention here that it is the relationship in Theorem 1 that enables us to adopt an effective splitting strategy in algorithms for characteristic decomposition of polynomial sets.

Proposition 1 [30, Proposition 3.1]**.** *Let $\mathcal{C}$ be the W-characteristic set of $\langle \mathcal{P}\rangle \subseteq \mathbb{K}[\boldsymbol{x}]$. Then (a) For any $P \in \langle \mathcal{P}\rangle$, $\mathrm{prem}(P, \mathcal{C}) = 0$; (b) $\langle \mathcal{C}\rangle \subseteq \langle \mathcal{P}\rangle \subseteq \mathrm{sat}(\mathcal{C})$; (c) $\mathsf{Z}(\mathcal{C}/\mathrm{ini}(\mathcal{C})) \subseteq \mathsf{Z}(\mathcal{P}) \subseteq \mathsf{Z}(\mathcal{C})$.*

Theorem 1 [30, Theorem 3.9]**.** *Let $\mathcal{C} = [C_1, \ldots, C_r]$ be the W-characteristic set of $\langle \mathcal{P}\rangle \subseteq \mathbb{K}[\boldsymbol{x}]$. If $\mathcal{C}$ is not normal, then there exists an integer k $(1 \le k \le r)$ such that $[C_1, \ldots, C_k]$ is normal and $[C_1, \ldots, C_{k+1}]$ is not regular.*

Assume that the variables $x_1, \ldots, x_n$ are ordered such that the parameters of $\mathcal{C}$ are all smaller than the other variables and let $I_{k+1} = \mathrm{ini}(C_{k+1})$ and l be the integer such that $\mathrm{lv}(I_{k+1}) = \mathrm{lv}(C_l)$.

(a) *If I_{k+1} is not R-reduced w.r.t. C_l, then*

$$\mathrm{prem}(I_{k+1}, [C_1, \ldots, C_l]) = 0, \qquad \mathrm{prem}(C_{k+1}, [C_1, \ldots, C_k]) = 0.$$

(b) *If* I_{k+1} *is R-reduced w.r.t.* C_l, *then* $\mathrm{prem}(C_l, [C_1, \ldots, C_{l-1}, I_{k+1}]) = 0$ *and either* $\mathrm{res}(\mathrm{ini}(I_{k+1}), [C_1, \ldots, C_{l-1}]) = 0$ *or* $\mathrm{prem}(C_{k+1}, [C_1, \ldots, C_{l-1}, I_{k+1}, C_{l+1}, \ldots, C_k]) = 0$.

For a triangular set $\mathcal{T} \subset \mathbb{K}[\boldsymbol{x}]$ and a variable ordering $<$, we say that the variable ordering condition is satisfied for $\mathcal{T}$ w.r.t. $<$ if all the parameters of $\mathcal{T}$ are ordered before the leading variables of polynomials in $\mathcal{T}$ in $<$. As a counter-example [30, Example 3.1(b)] shows, when the variable ordering condition is not satisfied for a W-characteristic set, Theorem 1 does not hold in general. We will work on the case when the variable ordering condition is not satisfied in Algorithm 1 in Sect. 3.

2.3 (Strong) Characteristic Decomposition and Characterizable Gröbner Basis

Definition 6 [31, Definition 3.1]. A pair $(\mathcal{G}, \mathcal{C})$ with $\mathcal{G}, \mathcal{C} \subseteq \mathbb{K}[\boldsymbol{x}]$ is called a *characteristic pair* if $\mathcal{G}$ is a reduced LEX Gröbner basis, $\mathcal{C}$ is the W-characteristic set of $\langle \mathcal{G} \rangle$, and $\mathcal{C}$ is normal.

For any polynomial set $\mathcal{F} \subseteq \mathbb{K}[\boldsymbol{x}]$, we want to compute finitely many characteristic pairs $(\mathcal{G}_1, \mathcal{C}_1), \ldots, (\mathcal{G}_t, \mathcal{C}_t)$ such that

$$\mathsf{Z}(\mathcal{F}) = \bigcup_{i=1}^{t} \mathsf{Z}(\mathcal{G}_i) = \bigcup_{i=1}^{t} \mathsf{Z}(\mathcal{C}_i / \mathrm{ini}(\mathcal{C}_i)) = \bigcup_{i=1}^{t} \mathsf{Z}(\mathrm{sat}(\mathcal{C}_i)). \tag{1}$$

A finite number of characteristic pairs $(\mathcal{G}_1, \mathcal{C}_1), \ldots, (\mathcal{G}_t, \mathcal{C}_t)$ are said to be a *characteristic decomposition* of $\mathcal{F}$ if the zero relationship (1) holds.

Remark 1. As can be seen from the definition of characteristic decomposition above, from a characteristic decomposition $(\mathcal{G}_1, \mathcal{C}_1), \ldots, (\mathcal{G}_t, \mathcal{C}_t)$ of a polynomial set $\mathcal{F}$ one can easily extract a normal decomposition $\mathcal{C}_1, \ldots, \mathcal{C}_t$ of $\mathcal{F}$.

Theorem 2 [31, Theorem 4.1]. *For any finite, nonempty polynomial set* $\mathcal{F} \subseteq \mathbb{K}[\boldsymbol{x}]$, *its characteristic decomposition can be computed within a finite number of operations if the variable ordering condition is satisfied.*

Definition 7 [31, Definitions 3.7 and 3.8]. A reduced LEX Gröbner basis $\mathcal{G}$ is said to be *characterizable* if $\langle \mathcal{G} \rangle = \mathrm{sat}(\mathcal{C})$, where $\mathcal{C}$ is the W-characteristic set of $\mathcal{G}$. A characteristic pair $(\mathcal{G}, \mathcal{C})$ is said to be *strong* if $\mathrm{sat}(\mathcal{C}) = \langle \mathcal{G} \rangle$.

Clearly the reduced LEX Gröbner basis in a strong characteristic pair is characterizable. Furthermore, it is proved that the W-characteristic set of a characterizable Gröbner basis is also normal [31, Proposition 3.9], and thus a characterizable Gröbner basis and its W-characteristic set form a strong characteristic pair.

A characteristic decomposition is said to be *strong* if each characteristic pair within is strong. For any characteristic decomposition $\Psi = \{(\mathcal{G}_1, \mathcal{C}_1), \ldots, (\mathcal{G}_t, \mathcal{C}_t)\}$

of $\mathcal{F} \subseteq \mathbb{K}[\boldsymbol{x}]$, by [31, Theorem 3.22] one can explicitly transform Ψ into a strong characteristic decomposition $\bar{\Psi} = \{(\bar{\mathcal{G}}_1, \bar{\mathcal{C}}_1), \ldots, (\bar{\mathcal{G}}_t, \bar{\mathcal{C}}_t)\}$ such that the following zero relationships hold.

$$\mathsf{Z}(\mathcal{F}) = \bigcup_{i=1}^{t} \mathsf{Z}(\bar{\mathcal{G}}_i) = \bigcup_{i=1}^{t} \mathsf{Z}(\bar{\mathcal{C}}_i / \operatorname{ini}(\bar{\mathcal{C}}_i)) = \bigcup_{i=1}^{t} \mathsf{Z}(\operatorname{sat}(\bar{\mathcal{C}}_i)). \tag{2}$$

3 Algorithm for (Strong) Characteristic Decomposition

In this section we first handle the variable ordering condition in Theorem 1 by temporarily changing the variable orderings, then we incorporate this process into the proposed algorithm (Algorithm 1 in [31]) for characteristic decomposition and represent it as Algorithm 2 for self-containedness. The algorithm to transform a characteristic decomposition into a strong one by using [31, Theorem 3.22] is formulated as Algorithm 3. The algorithms for (strong) characteristic decomposition are able to decompose an arbitrary polynomial set into simultaneously (characterizable) Gröbner bases and normal triangular sets.

3.1 Algorithm to Handle the Variable Ordering Condition

For a polynomial set $\mathcal{P} \subset \mathbb{K}[\boldsymbol{x}]$ and a variable ordering $<$, let $\mathcal{G}$ be the reduced LEX Gröbner basis of $\langle \mathcal{P} \rangle$ w.r.t. $<$ and $\mathcal{C}$ be the W-characteristic set extracted from $\mathcal{G}$. If $\mathcal{C}$ is abnormal but the variable ordering condition is not satisfied for $\mathcal{C}$ w.r.t. $<$, then Theorem 1 does not hold in general, which means that the psuedo-divisibility between the polynomials in $\mathcal{C}$ may not be found and thus the splitting needed in algorithms for characteristic decomposition is not guaranteed. In order to make the psuedo-divisibility happen, the following procedure is proposed.

(1) Take a new ordering $<'$ on the variables $x_1, \ldots, x_n$.
(2) Compute the reduced LEX Gröbner basis $\mathcal{G}'$ of $\langle \mathcal{P} \rangle$ w.r.t. $<'$ and extract the new W-characteristic set $\mathcal{C}'$ of $\langle \mathcal{P} \rangle$ from $\mathcal{G}'$.
(3) Check whether the variable ordering condition is satisfied for $\mathcal{C}'$ w.r.t. $<'$ and $\mathcal{C}'$ is abnormal.

If the variable ordering condition is verified to be satisfied for $\mathcal{C}'$ w.r.t. $<'$ and $\mathcal{C}'$ is abnormal, then we stop with $\mathcal{G}'$, $\mathcal{C}'$, and $<'$, otherwise the above steps (1)–(3) are repeated until the variable ordering condition is satisfied or the repetition ends after the finite choices of possible variable orderings. The procedure described above is formulated into Algorithm 1 below.

When the Gröbner basis $\mathcal{G} = \{1\}$, the W-characteristic $\mathcal{C}' = [1]$. In this case the variable ordering condition is assumed to be satisfied for $\mathcal{C}'$, and $(\{1\}, [1], <)$ is returned as in Algorithm 1.

Algorithm 1 may return a new variable ordering $<'$ which is different from the original one, but we make use of the fact that w.r.t. $<'$ the abnormal W-characteristic set $\mathcal{C}'$ of the considered ideal $\langle \mathcal{P} \rangle$ satisfies the variable ordering

Algorithm 1. $((\mathcal{G}', \mathcal{C}'), <') := \mathsf{VOC}(\mathcal{P}, <)$ (algorithm for ensuring the variable ordering condition)

Input: $\mathcal{P}$, a finite, nonempty set of nonzero polynomials in $\mathbb{K}[\boldsymbol{x}]$;
$<$, a variable ordering.
Output: either $(\mathcal{G}', \mathcal{C}')$, a pair of reduced LEX Gröbner basis of $\langle \mathcal{P} \rangle$ and its W-characteristic set, and $<'$, a variable ordering such that $\mathcal{C}'$ satisfies the variable ordering condition w.r.t. $<'$.
or "ERROR"

1 Compute the reduced LEX Gröbner basis $\mathcal{G}$ of $\langle \mathcal{P} \rangle$ w.r.t $<$;
2 Extract the W-characteristic set $\mathcal{C}$ from $\mathcal{G}$ w.r.t $<$;
3 **if** $\mathcal{C}$ *satisfies the variable ordering condition w.r.t.* $<$ **then**
4   **return** $((\mathcal{G}, \mathcal{C}), <)$
5 **else**
6     $\Psi := \{$all possible variable orderings on $x_1, \ldots, x_n\} \setminus \{<\}$;
7   **for** $<' \in \Psi$ **do**
8   Compute the reduced LEX Gröbner basis $\mathcal{G}'$ of $\langle \mathcal{P} \rangle$ w.r.t $<'$;
9   Extract the W-characteristic set $\mathcal{C}'$ from $\mathcal{G}'$ w.r.t $<'$;
10   **if** $\mathcal{C}'$ *is abnormal and satisfies variable ordering condition w.r.t.* $<'$ **then**
11   **return** $((\mathcal{G}', \mathcal{C}'), <')$
12 **return** "ERROR"

condition, and thus Theorem 1 ensures psuedo-divisibility between polynomials in $\mathcal{C}'$ and the splitting needed is also guaranteed. We would like to emphasize that whether Theorem 1 holds or not depends on the variable ordering, but the resultant polynomial sets after splitting do not. This means that after the splitting we can continue the computation for characteristic decomposition w.r.t. the original variable ordering (see the descriptions of Algorithm 2 below for more details), and thus this change of variable orderings is only temporary.

In Line 7 of Algorithm 1, we use the heuristic to choose first the new orderings w.r.t. which $\mathcal{C}$ satisfies the variable ordering condition, namely the parameters of $\mathcal{C}$ are ordered smaller than the other variables in such orderings, and then the other orderings in Ψ. This strategy works well to succeed with the first pick of such orderings in most cases.

Even with the change of variable orderings in Algorithm 1 one may fail to find a proper abnormal W-characteristic set which satisfies the variable ordering condition. Below is one simple example for such failure, which is also the only one we have found out of over one hundred examples in our experiments. The characterization of this phenomenon and the way to continue the process of characteristic decomposition when such phenomenon occurs are our future work.

Let $\mathcal{P} = \{x^2, (x+y)z + x\} \subseteq \mathbb{K}[x, y, z]$ with $x < y < z$. The reduced LEX Gröbner basis of $\langle \mathcal{P} \rangle$ is $\mathcal{G} = \{x^2,\ xz + yz + x\}$ and the W-characteristic set is $\mathcal{C} = [x^2, (x+y)z + x]$. Clearly $\mathcal{C}$ does not satisfy the variable ordering condition w.r.t. $x < y < z$. But for all the other possible variable orderings, the new W-characteristic sets are either normal or do not satisfy the variable ordering condition w.r.t. the new variable ordering.

3.2 Algorithms for Characteristic Decomposition

Following the overall splitting strategies sketched in [30], an algorithm for characteristic decomposition is proposed in [31]. For the purpose of clarity, this algorithm is recalled and represented here, but updated with the addition of Algorithm 1 for handling the variable ordering condition.

Let $\mathcal{F} \subseteq \mathbb{K}[\boldsymbol{x}]$ be the input polynomial set and $<$ be a variable ordering. We use a set Φ to store the polynomial sets which need further computation and a set Ψ to store the characteristic pairs already computed.

(1) Pick up a polynomial set $\mathcal{P} \in \Phi$ and remove it from Φ. Then we use Algorithm 1, if it succeeds, to find a proper variable ordering $<'$ w.r.t. which the W-characteristic set $\mathcal{C}$ of the ideal $\langle \mathcal{P} \rangle$ satisfies the variable ordering condition, where $\mathcal{C}$ is extracted from the reduced LEX Gröbner basis $\mathcal{G}$ of $\langle \mathcal{P} \rangle$.

(2) If $\mathcal{C}$ is a normal set, then one knows that no change of variable orderings occurs, namely in this case $<'=<$, and a characteristic pair $(\mathcal{G}, \mathcal{C})$ is found and adjoined to Ψ. Splitting for this case follows the strategy proposed in Algorithm 1 in [31].

Otherwise, the variable ordering condition is satisfied for $\mathcal{C}$ w.r.t. $<'$ (different from $<$) and $\mathcal{C}$ is abnormal, and Theorem 1 furnishes pseudo-divisibility between polynomials in $\mathcal{C}$. By using the same splitting strategies in [31], but w.r.t. the new variable ordering $<'$, we are able to split $\mathcal{P}$ into $\mathcal{G} \cup \{H_1\}, \ldots, \mathcal{G} \cup \{H_s\}$, where $H_1, \ldots, H_s$ are polynomials reduced w.r.t. $\mathcal{G}$. Then the new polynomial sets $\mathcal{G} \cup \{H_1\}, \ldots, \mathcal{G} \cup \{H_s\}$ are adjoined to Φ.

(3) After the splitting another polynomial set $\mathcal{P}' \in \Phi$ is picked up and steps (1)–(2) are repeated for $\mathcal{P}'$ w.r.t. the original variable ordering $<$.

The above steps are repeated until Φ becomes empty, when we will get finitely many characteristic pairs $(\mathcal{G}_1, \mathcal{C}_1), \ldots, (\mathcal{G}_t, \mathcal{C}_t)$ which form a characteristic decomposition of the input polynomial set $\mathcal{F}$. Following the same proving strategies as in [31], one can show that after the addition of Algorithm 1 for handling the variable ordering condition, the algorithm for characteristic decomposition remains correct and to terminate. The method of characteristic decomposition, whose main steps are outlined above, is described formally as Algorithm 2.

The complexity of Algorithm 2 for characteristic decomposition is not touched in this paper partially due to the same underlying difficulty as in the complexity analyses of algorithms for triangular decomposition: the very complicated behaviors of splittings.

3.3 Algorithm for Strong Characteristic Decomposition

In [31, Theorem 3.22] it is proved that for any characteristic pair $(\mathcal{G}, \mathcal{C})$, one can explicitly construct a strong characteristic pair $(\bar{\mathcal{G}}, \bar{\mathcal{C}})$, where $\bar{\mathcal{G}}$ is the reduced LEX Gröbner basis of $\mathrm{sat}(\mathcal{C})$ and $\bar{\mathcal{C}}$ is the W-characteristic set of $\bar{\mathcal{G}}$. Furthermore, a characteristic decomposition $(\mathcal{G}_1, \mathcal{C}_1), \ldots, (\mathcal{G}_t, \mathcal{C}_t)$ of a polynomial set $\mathcal{F}$ can be transformed into a strong one $(\bar{\mathcal{G}}_1, \bar{\mathcal{C}}_1), \ldots, (\bar{\mathcal{G}}_t, \bar{\mathcal{C}}_t)$ of $\mathcal{F}$, without further

Algorithm 2. $\Psi := \mathsf{CharPair}(\mathcal{F}, <)$ (algorithm for characterstic decomposition)

Input: a finite, nonempty set $\mathcal{F}$ of nonzero polynomials in $\mathbb{K}[\boldsymbol{x}]$ and a variable ordering $<$.

Output: either a characteristic decomposition Ψ of $\mathcal{F}$ such that $\mathsf{Z}(\mathcal{F}) = \cup_{(\mathcal{G},\mathcal{C})\in\Psi} \mathsf{Z}(\mathcal{G}) = \cup_{(\mathcal{G},\mathcal{C})\in\Psi} \mathsf{Z}(\mathcal{C}/\operatorname{ini}(\mathcal{C}))$,
or the empty set meaning that $\mathsf{Z}(\mathcal{F}) = \emptyset$,
or the message "The variable ordering condition is not satisfied."

```
Ψ := ∅, Φ := {F};
while Φ ≠ ∅ do
    Choose P from Φ and set Φ := Φ \ {P};
    if VOC(P, <) = "ERROR" then
        return "The variable ordering condition is not satisfied.";
    else
        ((G, C), <') := VOC(P, <);
        if G ≠ {1} then
            if C is normal then
                Ψ := Ψ ∪ {(G, C)};
                Φ := Φ ∪ {G ∪ {ini(C)} | ini(C) is the initial of C w.r.t <',
                    ini(C) ∉ K, C ∈ C};
            else
                C := first polynomial in C such that [T ∈ C | lv(T) ≤' lv(C)],
                    ordered as a triangular set, is abnormal;
                I := ini(C); y := lv(I);
                C̄ := the polynomial in C whose leading variable is y;
                if I is not reduced w.r.t. C̄ then
                    Φ := Φ ∪ {G ∪ {ini(T)} | lv(T) ≤' y, T ∈ C} ∪ {G ∪ {I}};
                else
                    Q := pquo(C̄, I);
                    if prem(ini(Q), [T ∈ C | lv(T) <' y]) = 0 then
                        Φ := Φ∪{G∪{ini(T)} | lv(T) <' y, T ∈ C}∪{G∪{ini(I)}};
                    else
                        Φ := Φ ∪ {G ∪ {ini(T) | lv(T) <' y, T ∈ C}}
                            ∪{G ∪ {prem(Q, [T ∈ C | lv(T) <' y])}, G ∪ {I}};
return Ψ
```

splitting to induce additional branches. An algorithm for strong characteristic decomposition based on Algorithm 2 and the above observations is formulated as Algorithm 3.

Remark 2. Algorithm 2 decomposes any polynomial set not only into normal triangular sets, but also into reduced LEX Gröbner bases at one stroke. These two different objects which have their own structures and properties (see [31, Sect. 3.2]) are interconnected. This makes our algorithm distinct from other existing ones for triangular decomposition like RegSer and Triangularize and so

Algorithm 3. $\Sigma := \mathsf{SCharPair}(\mathcal{F}, <)$ (algorithm for strong characterstic decomposition)

$\Sigma := \emptyset$;
$\Psi := \mathsf{CharPair}(\mathcal{F}, <)$;
for $(\mathcal{G}, \mathcal{C}) \in \Psi$ **do**
  Compute the reduced LEX Gröbner basis $\bar{\mathcal{G}}$ of $\mathrm{sat}(\mathcal{C})$;
  if $\langle \bar{\mathcal{G}} \rangle = \langle \mathcal{G} \rangle$ **then**
    $\Sigma := \Sigma \cup \{(\mathcal{G}, \mathcal{C})\}$;
  else
    Extract the W-characteristic set $\bar{\mathcal{C}}$ of from $\bar{\mathcal{G}}$;
    $\Sigma := \Sigma \cup \{(\bar{\mathcal{G}}, \bar{\mathcal{C}})\}$;
return Σ

on. Algorithm 3 which decomposes polynomial sets into characterizable Gröbner bases and normal triangular sets has richer properties (see [31, Sect. 3.3]). In particular, $\mathrm{sat}(\mathcal{T}) = \langle \mathcal{T} \rangle$ holds naturally for any zero-dimensional triangular set $\mathcal{T}$, and thus in this case a characteristic decomposition is also a strong one.

3.4 An Illustrative Example

Let $\mathcal{P} = \{ax^2y + 3b^2 + a, a(b-c)xy + abx + 5c\} \subseteq \mathbb{K}[c, b, a, y, x]$ with $c < b < a < y < x$. Part of computation of strong characteristic decomposition of $\mathcal{P}$ with Algorithm 3 is recorded below to illustrate how Algorithms 1 and 3 in Sect. 3 work.

In a certain step of computation of characteristic decomposition of $\mathcal{P}$, the polynomial set

$$\begin{aligned}\mathcal{P}_1 =& \{b(ax^2 - 3b^2 - a), b(y+1)(3b^2 + a), ab(y+1)(3b^2 + a - 5x),\\ & a(3b^2 + a)b(y+1), ax^2y + 3b^2 + a, c\}\end{aligned}$$

is chosen, with the reduced LEX Gröbner basis of $\langle \mathcal{P}_1 \rangle$ computed as

$$\mathcal{G}_1 = \{c, 3b^3y + aby + 3b^3 + ab, b^3xy + b^3x, abx^2 - 3b^3 - ab, ax^2y + 3b^2 + a\}$$

and the W-characteristic set as $\mathcal{C}_1 = [c, 3b^3y + aby + 3b^3 + ab, b^3xy + b^3x]$. One can check that the W-characteristic set $\mathcal{C}_1$ is abnormal, and it does not satisfy the variable ordering condition w.r.t. $c < b < a < y < x$.

Then a new variable ordering $b < a < c < y < x$ is chosen, and the reduced LEX Gröbner basis

$$\mathcal{G}_1' = \{c, 3b^3y + aby + 3b^3 + ab, b^3xy + b^3x, abx^2 - 3b^3 - ab, ax^2y + 3b^2 + a\}$$

of $\mathcal{P}_1$ is computed and its W-characteristic set $\mathcal{C}_1' = [c, 3b^3y + aby + 3b^3 + ab, b^3xy + b^3x]$ is extracted. At this point one can find that $\mathcal{C}_1'$ satisfies the variable ordering condition and is abnormal w.r.t. the new variable ordering.

With the set of initials of the polynomials $C'_{11}, C'_{12}, C'_{13}$ w.r.t. $b < a < c < y < x$ being $\{b^3y + b^3, 3b^3 + ab\}$, the polynomial sets $\mathcal{P}_2 = \mathcal{G}'_1 \cup \{3b^3 + ab\}$ and $\mathcal{P}_3 = \mathcal{G}'_1 \cup \{b^3y + b^3\}$ are adjoined to Φ, and the computation continues w.r.t. the original variable ordering $c < b < a < y < x$ until the characteristic decomposition of $\mathcal{P}$ is computed.

For one characteristic pair $(\mathcal{G}, \mathcal{C}) = (\{c, 3b^2 + a, bx\}, [c, 3b^2 + a, bx])$ in the characteristic decomposition, the inequality $\mathrm{sat}(\mathcal{C}) \neq \langle \mathcal{G} \rangle$ is confirmed. Then the reduced LEX Gröbner basis $\bar{\mathcal{G}} = \{c, 3b^2 + a, x\}$ of $\mathrm{sat}(\mathcal{C})$ is computed and its W-characteristic set $\bar{\mathcal{C}}' = [c, 3b^2 + a, x]$ is extracted, forming a strong characteristic pair $(\bar{\mathcal{G}}, \bar{\mathcal{C}})$ in the strong characteristic decomposition of $\mathcal{P}$.

4 Implementation and Experimental Results

We have implemented Algorithms 1, 2, and 3 in MAPLE 17 and made experiments with the implementation on an Intel(R) Core(TM) i5-4210U CPU at 2.39 GHz with 8.00 GB RAM under Windows 8. The implementation is based on the functions for Gröbner basis computation available in the FGb package [11] shipped with MAPLE 17 and MAPLE's built-in packages.

In our implementation polynomial factorization is used heuristically to enhance the performance of algorithms for triangular decomposition. In our algorithm, the splitting of a polynomial set $\mathcal{P}$ into $\mathcal{G} \cup \{H_1\}, \ldots, \mathcal{G} \cup \{H_s\}$ is an essential step. Instead of $H_1, \ldots, H_s$, we can use any set of polynomials $L_1, \ldots, L_t$ which are irreducible factors w.r.t. $\mathcal{G}$ such that $\mathsf{Z}(H_1 \cdots H_s) = \mathsf{Z}(L_1 \cdots L_t)$. In order to further simplify the computation, we can also use $L_1{}' = \mathrm{nform}(L_1, \mathcal{G}), \ldots, L_t{}' = \mathrm{nform}(L_t, \mathcal{G})$ which are all reduced w.r.t. $\mathcal{G}$ instead of $L_1, \ldots, L_t$. The polynomials $L_1{}', \ldots, L'_t$ are usually smaller in size compared with $H_1, \ldots, H_s$. Our experiments show that this simple strategy is rather effective.

In the zero-dimensional case where the W-characteristic sets do not involve any parameters, the variable ordering condition, which is required in Theorem 1 to ensure the pseudo-divisibility relationship, holds naturally. However, the condition does not necessarily hold in general in the positive-dimensional case. Among 35 test examples in our experiments in which the ideals are positive-dimensional, the variable ordering condition is not satisfied for 8 of them (marked with † in Table 1). We would like to remark that Algorithm 1 succeeds for all of these 8 test examples with only one time of changing the variable orderings whenever the failure of satisfaction of the variable ordering condition happens.

The experimental results with our implementation of Algorithm 2 on 50 examples are reported in Table 1. Among these 50 examples, Ex 1–13 are from the Epsilon package, Ex 14–16 from [28], Ex 17–18 from [1], Ex 19–33 from [23], Ex 34–36 from [7], Ex 37–47 from the FGb library, and Ex 48–50 can be found with this link[1].

The computational performances of our implementation are compared with existing implementations of methods for triangular decomposition by means of

[1] http://www.lifl.fr/~lemaire/BCLM09/BCLM09-systems.txt.

normal decomposition. A normal decomposition is extracted out of the computed characteristic decomposition for our implementation and for other implementations it is computed by normalization of regular decomposition: the function Triangularize (in the RegularChains package [20] shipped in MAPLE 17) and the function RegSer (in the Epsilon package [29] for MAPLE) are used for regular decomposition, and the function normat (in Epsilon) is used for normalization of the computed regular sets.

Remark 3. In spite of the comparison made by computing normal decomposition with different implementations, we would like to emphasize that the normal sets are only part of the computation output of Algorithm 2, for reduced LEX Gröbner bases and normal W-characteristic sets are computed simultaneously and they enjoy remarkable interconnecting properties.

In Table 1, "Label" indicates the label in the above-cited references and "Var," "Eqs," and "Dim" denote the number of variables, the number of equations, and the dimension of the ideal in the examples, respectively. "Total" and "GB" under CharPair record respectively the total time (followed by the number of pairs in parenthesis) for normal decomposition using Algorithm 2 and the time for computing all the reduced LEX Gröbner bases; "Total" and "Regular" under "RegSer" and "Triangularize" record the total time for normal decomposition and the time for regular decomposition (followed by the numbers of components in parenthesis) respectively. The marks "lost" and ">4000" in the columns mean that MAPLE reports "lost kernel connections" (which persists with several repeated attempts) and that the computation does not terminate within 4000 seconds respectively.

The experimental results in Table 1 also show that for most of the examples in which the ideals are zero-dimensional, the number of normal components in the characteristic decomposition computed by Algorithm 2 is smaller than that in the normal decomposition computed by RegSer or Triangularize with normalization. This happens because the initials of polynomials in the normal W-characteristic sets of zero-dimensional ideals do not involve any variables or parameters such that no initials cause any splitting and no more polynomial set is adjoined into Φ in Algorithm 2.

It is reasonable to claim that the built-in implementation in MAPLE for algorithms to change the term orderings, especially that for the Gröbner walk in the positive-dimensional case, is the bottleneck of our current implementation in terms of efficiency. With our comparisons between implementations of algorithms for the Gröbner walk in MATHEMATICA and MAPLE, we predict that the total time for normal decomposition with our implementation can be greatly reduced for the positive-dimensional case if the step of Gröbner walk is performed with the corresponding built-in command in MATHEMATICA. For this reason, we are working on an interface to call the Gröbner walk function in MATHEMATICA from MAPLE. According to our preliminary experiments, this may bring a speedup of about 10 times for conversion of Gröbner bases in the positive-dimensional case (the total time decreases from 27.2 s to 2.2 s for Ex 32, for example).

Table 1. Timings for characteristic decomposition (in second)

					Algorithm 2		RegSer		Triangularize	
Ex	Label	Var	Eqs	Dim	Total	GB	Total	Regular	Total	Regular
1	E1	10	10	1	0.844(5)	0.736	0.234(2)	0.187(2)	2.187(13)	2.109(13)
2	E5	15	17	4	2.187(7)	2.091	0.453(4)	0.328(4)	6.156(7)	6.078(7)
3	E14	4	3	1	0.656(10)	0.452	1.875(7)	0.125(1)	2.547(7)	0.156(1)
4	E20	4	4	1	0.094(2)	0.079	—	0.016(1)	—	0.031(1)
5	E22	3	3	0	0.046(1)	0.046	0.312(2)	0.250(2)	0.140(3)	0.109(3)
6	E23†	9	5	4	0.469(11)	0.249	—	0.094(8)	—	0.094(1)
7	E28	4	4	0	0.015(1)	0.015	0.109(1)	0.093(1)	0.141(3)	0.125(3)
8	E32†	8	6	2	0.234(6)	0.157	0.110(3)	0.110(3)	—	0.110(1)
9	E33†	13	11	2	20.313(8)	18.924	0.765(5)	0.656(5)	1.687(1)	1.594(1)
10	E35†	8	8	3	0.406(11)	0.234	0.626(5)	0.532(5)	0.781(7)	0.688(7)
11	E40	6	6	0	0.344(1)	0.344	>4000	>4000	>4000	>4000
12	E47	5	4	1	1.234(8)	1.076	47.766(18)	1.297(12)	1.953(8)	0.547(2)
13	E48	7	3	4	0.453(2)	0.407	—	0.031(2)	—	0.015(1)
14	N4	5	3	3	0.094(6)	0.094	—	0.016(4)	—	0.047(3)
15	N6	4	3	2	0.063(2)	0.047	0.016(3)	0.016(3)	—	0.031(3)
16	N20	4	3	2	0.078(3)	0.047	—	<0.01(3)	—	0.016(2)
17	F_1	3	2	1	0.047(3)	0.047	—	0.016(3)	—	0.062(2)
18	F_2	4	3	1	1.531(13)	1.157	—	0.015(1)	0.281(6)	0.188(5)
19	S1	4	3	2	0.047(3)	0.047	—	0.015(3)	—	0.046(3)
20	S2	4	9	0	0.032(1)	0.032	—	0.015(1)	—	0.031(1)
21	S5	8	4	4	0.187(8)	0.125	3.188(31)	0.344(19)	1.513(9)	0.124(1)
22	S6	4	3	2	0.062(4)	0.062	—	0.016(3)	—	0.046(3)
23	S7	4	3	1	0.281(8)	0.156	0.156(7)	0.047(4)	0.249(5)	0.109(1)
24	S8	4	3	2	0.031(2)	0.015	0.062(3)	0.062(3)	0.156(2)	0.141(2)
25	S9	6	4	2	0.375(12)	0.236	0.483(21)	0.14(8)	0.188(6)	0.094(1)
26	S10	7	4	3	0.594(16)	0.313	0.438(7)	0.172(5)	0.360(3)	0.235(1)
27	S12	8	4	4	0.140(1)	0.140	—	0.016(1)	—	0.047(1)
28	S13†	5	2	3	0.296(10)	0.187	0.171(8)	0.109(8)	0.125(2)	0.094(1)
29	S14	5	3	2	0.203(8)	0.156	0.14(6)	0.109(6)	0.157(8)	0.125(8)
30	S15	12	7	5	0.344(1)	0.344	—	0.031(1)	—	0.093(1)
31	S16	16	14	3	0.640(6)	0.344	0.703(7)	0.609(7)	—	4.609(8)
32	S17	8	3	6	lost	lost	lost	lost	lost	lost
33	S18	5	3	3	lost	lost	>4000	>4000	2058.235(56)	2.641(10)
34	maclane†	10	6	5	476.766(335)	433.867	>4000	>4000	10.937(21)	2.969(11)
35	nueral	4	3	1	1.844(13)	1.611	>4000	>4000	0.250(6)	0.125(5)
36	Leykin_1†	8	6	4	184.890(116)	169.983	>4000	>4000	5.625(15)	5.578(15)
37	Rose	3	3	0	0.859(1)	0.859	1.872(1)	1.607(1)	1.125(2)	0.875(2)
38	F663†	10	9	2	2.969(6)	2.328	1.935(16)	1.202(15)	1.607(6)	1.045(4)
39	Dessin2	10	10	0	26.718(1)	26.718	>4000	>4000	>4000	>4000
40	Liu	6	4	2	203.125(26)	192.531	>4000	>4000	18.907(20)	0.469(8)
41	Wang16	4	4	0	0.125(1)	0.109	14.555(1)	0.437(1)	16.047(1)	0.172(1)
42	Cyclic5	5	5	0	0.453(11)	0.281	lost	lost	1.906(15)	1.531(15)
43	lichtblau	3	2	1	lost	lost	>4000	>4000	>4000	0.109(1)
44	filter9	9	9	0	0.515(1)	0.515	>4000	>4000	lost	lost
45	fabrice24	9	9	0	436.7(1)	436.7	lost	lost	lost	lost
46	uteshev b	4	4	0	3.438(1)	3.438	lost	lost	>4000	>4000
47	cyclic6	6	6	0	1.922(25)	1.095	lost	lost	>4000	>4000
48	4-body-h	3	3	0	5.953(3)	5.531	655.406(2)	15.922(2)	439.547(5)	0.359(5)
49	5-body-h	3	3	0	12.297(3)	11.250	1990.656(2)	107.891(2)	1560.984(5)	0.437(5)
50	fabfaux	3	3	0	0.219(1)	0.219	14.016(1)	0.188(1)	54.000(1)	0.172(1)

Table 2. Timings for strong characteristic decomposition (in second)

Ex	Label	Total	Transform	Branches
4	E20	0.125	0.078	2
6	E23	0.953	0.547	10
9	E33	>4000	—	—
10	E35	0.828	0.422	10
12	E47	>4000	—	—
14	N4	0.281	0.141	6
15	N6	0.297	0.235	2
16	N20	0.171	0.062	3
17	F_1	0.219	0.125	3
18	F_2	2.985	1.469	13
19	S1	0.140	0.093	3
21	S5	0.578	0.422	8
22	S6	0.266	0.188	4
24	S8	0.078	0.047	2
25	S9	0.735	0.391	12
28	S13	24.937	24.672	10
29	S14	35.437	35.219	8
35	Neural	3.094	1.516	13
38	f663	>4000	—	—

The experimental results of our implementation of Algorithm 3 are presented in Table 2 for 19 test examples selected from those in Table 1. In Table 2, the columns "Total" and "Transform" record the total time for computing strong characteristic decomposition using Algorithm 3 and the time for transforming a characteristic decomposition into a strong one, and "Branches" denotes the number of branches in the strong characteristic pairs computed.

As one finds from Algorithm 3, there is no more splitting in the transformation from a characteristic decomposition into a strong one, but in practice the number of branches in the characteristic decomposition may be strictly greater than that in the strong one, as the example E23 in Tables 1 and 2 shows. In the computed characteristic decomposition for the example E23, there are two different characteristic pairs

$$\begin{aligned}&(\{a, c-1, bd, -b^2+r^2, -d^2+s^2, -b^2-d^2+t^2-1, bx, dx-d\},\\&\quad [a, c-1, bd, -b^2+r^2, -d^2+s^2, -b^2-d^2+t^2-1, bx])\end{aligned}$$

and

$$\begin{aligned}&(\{a, bc-b, bd, -b^2+r^2, -c^2-d^2+s^2+2c-1, -b^2-c^2-d^2+t^2, bx, -cy+dx\\&\quad -d+y\}, [a, bc-b, bd, -b^2+r^2, -c^2-d^2+s^2+2c-1, -b^2-c^2-d^2+t^2, bx])\end{aligned}$$

which lead to the same strong characteristic pair

$$(\{a, c-1, d, -b^2+r^2, s^2, -b^2+t^2-1, x\}, [a, c-1, d, -b^2+r^2, s^2, -b^2+t^2-1, x])$$

after the transformation.

Acknowledgements. The authors would like to thank the reviewers for their detailed comments which have led to effective improvements on this paper.

References

1. Alvandi, P., Chen, C., Marcus, S., Moreno Maza, M., Schost, É., Vrbik, P.: Doing algebraic geometry with the RegularChains library. In: Hong, H., Yap, C. (eds.) ICMS 2014. LNCS, vol. 8592, pp. 472–479. Springer, Heidelberg (2014). doi:10.1007/978-3-662-44199-2_71
2. Aubry, P., Lazard, D., Moreno Maza, M.: On the theories of triangular sets. J. Symb. Comput. **28**(1–2), 105–124 (1999)
3. Aubry, P., Moreno Maza, M.: Triangular sets for solving polynomial systems: a comparative implementation of four methods. J. Symb. Comput. **28**(1), 125–154 (1999)
4. Bächler, T., Gerdt, V., Lange-Hegermann, M., Robertz, D.: Algorithmic Thomas decomposition of algebraic and differential systems. J. Symb. Comput. **47**(10), 1233–1266 (2012)
5. Becker, T., Weispfenning, V., Kredel, H.: Gröbner Bases: A Computational Approach to Commutative Algebra. Graduate Texts in Mathematics. Springer, New York (1993)
6. Buchberger, B.: Ein Algorithmus zum Auffinden der Basiselemente des Restklassenrings nach einem nulldimensionalen Polynomideal. Ph.D. thesis, Universität Innsbruck, Austria (1965)
7. Chen, C., Golubitsky, O., Lemaire, F., Moreno Maza, M., Pan, W.: Comprehensive triangular decomposition. In: Ganzha, V.G., Mayr, E.W., Vorozhtsov, E.V. (eds.) CASC 2007. LNCS, vol. 4770, pp. 73–101. Springer, Heidelberg (2007). doi:10.1007/978-3-540-75187-8_7
8. Chen, C., Moreno Maza, M.: Algorithms for computing triangular decompositions of polynomial systems. J. Symb. Comput. **47**(6), 610–642 (2012)
9. Cox, D., Little, J., O'Shea, D.: Ideals, Varieties, and Algorithms: An Introduction to Computational Algebraic Geometry and Commutative Algebra. Undergraduate Texts in Mathematics. Springer, New York (1997)
10. Dahan, X.: On lexicographic Gröbner bases of radical ideals in dimension zero: interpolation and structure. Preprint at arXiv:1207.3887 (2012)
11. Faugère, J.-C.: FGb: a library for computing Gröbner bases. In: Fukuda, K., Hoeven, J., Joswig, M., Takayama, N. (eds.) ICMS 2010. LNCS, vol. 6327, pp. 84–87. Springer, Heidelberg (2010). doi:10.1007/978-3-642-15582-6_17
12. Faugère, J.-C., Gianni, P., Lazard, D., Mora, T.: Efficient computation of zero-dimensional Gröbner bases by change of ordering. J. Symb. Comput. **16**(4), 329–344 (1993)
13. Gao, S., Volny, F., Wang, M.: A new framework for computing Gröbner bases. Math. Comput. **85**(297), 449–465 (2016)

14. Gao, X.-S., Chou, S.-C.: Solving parametric algebraic systems. In: Proceedings of ISSAC 1992, pp. 335–341. ACM Press (1992)
15. Gianni, P., Trager, B., Zacharias, G.: Gröbner bases and primary decomposition of polynomial ideals. J. Symb. Comput. **6**(2), 149–167 (1988)
16. Kalkbrenner, M.: A generalized Euclidean algorithm for computing triangular representations of algebraic varieties. J. Symb. Comput. **15**(2), 143–167 (1993)
17. Lazard, D.: Ideal bases and primary decomposition: case of two variables. J. Symb. Comput. **1**(3), 261–270 (1985)
18. Lazard, D.: A new method for solving algebraic systems of positive dimension. Discrete Appl. Math. **33**(1–3), 147–160 (1991)
19. Lazard, D.: Solving zero-dimensional algebraic systems. J. Symb. Comput. **13**(2), 117–131 (1992)
20. Lemaire, F., Moreno Maza, M., Xie, Y.: The RegularChains library in Maple 10. In: Kotsireas, I. (ed.) Maple Conference 2005, pp. 355–368. Maplesoft, Waterloo (2005)
21. Li, B., Wang, D.: An algorithm for transforming regular chain into normal chain. In: Kapur, D. (ed.) ASCM 2007. LNCS, vol. 5081, pp. 236–245. Springer, Heidelberg (2008). doi:10.1007/978-3-540-87827-8_20
22. Marinari, M.G., Mora, T.: A remark on a remark by Macaulay or enhancing Lazard structural theorem. Bull. Iran. Math. Soc. **29**(1), 1–45 (2003)
23. Mou, C., Wang, D., Li, X.: Decomposing polynomial sets into simple sets over finite fields: the positive-dimensional case. Theoret. Comput. Sci. **468**, 102–113 (2013)
24. Ritt, J.F.: Differential Algebra. American Mathematical Society, New York (1950)
25. Shimoyama, T., Yokoyama, K.: Localization and primary decomposition of polynomial ideals. J. Symb. Comput. **22**(3), 247–277 (1996)
26. Wang, D.: Decomposing polynomial systems into simple systems. J. Symb. Comput. **25**(3), 295–314 (1998)
27. Wang, D.: Computing triangular systems and regular systems. J. Symb. Comput. **30**(2), 221–236 (2000)
28. Wang, D.: Elimination Methods. Springer, Wien (2001)
29. Wang, D.: Elimination Practice: Software Tools and Applications. Imperial College Press, London (2004)
30. Wang, D.: On the connection between Ritt characteristic sets and Buchberger-Gröbner bases. Math. Comput. Sci. **10**, 479–492 (2016)
31. Wang, D., Dong, R., Mou, C.: Decomposition of polynomial sets into characteristic pairs. arXiv:1702.08664 (2017)
32. Wang, D., Zhang, Y.: An algorithm for decomposing a polynomial system into normal ascending sets. Sci. China Ser. A **50**(10), 1441–1450 (2007)
33. Weispfenning, V.: Comprehensive Gröbner bases. J. Symb. Comput. **14**(1), 1–29 (1992)
34. Wu, W.-T.: Basic principles of mechanical theorem proving in elementary geometries. J. Autom. Reason. **2**(3), 221–252 (1986)

Symbolic Versus Numerical Computation and Visualization of Parameter Regions for Multistationarity of Biological Networks

Matthew England[1], Hassan Errami[2], Dima Grigoriev[3], Ovidiu Radulescu[4], Thomas Sturm[5,6](✉), and Andreas Weber[2]

[1] Fac. Engineering, Environment & Computing, Coventry University, Coventry, UK
Matthew.England@coventry.ac.uk

[2] Institut für Informatik II, Universität Bonn, Bonn, Germany
{errami,weber}@cs.uni-bonn.de

[3] CNRS, Mathématiques, Université de Lille, Villeneuve d'Ascq, France
Dmitry.Grigoryev@math.univ-lille1.fr

[4] DIMNP UMR CNRS/UM 5235, University of Montpellier, Montpellier, France
ovidiu.radulescu@umontpellier.fr

[5] University of Lorraine, CNRS, Inria, and LORIA, Nancy, France
thomas.sturm@loria.fr

[6] MPI Informatics and Saarland University, Saarbrücken, Germany
sturm@mpi-inf.mpg.de

Abstract. We investigate models of the mitogenactivated protein kinases (MAPK) network, with the aim of determining where in parameter space there exist multiple positive steady states. We build on recent progress which combines various symbolic computation methods for mixed systems of equalities and inequalities. We demonstrate that those techniques benefit tremendously from a newly implemented graph theoretical symbolic preprocessing method. We compare computation times and quality of results of numerical continuation methods with our symbolic approach before and after the application of our preprocessing.

1 Introduction

The mathematical modelling of intra-cellular biological processes has been using nonlinear ordinary differential equations since the early ages of mathematical biophysics in the 1940s and 50s [28]. A standard modelling choice for cellular circuitry is to use chemical reactions with mass action law kinetics, leading to polynomial differential equations. Rational functions kinetics (for instance the Michaelis-Menten kinetics) can generally be decomposed into several mass action steps. An important property of biological systems is their multistationarity which means having multiple stable steady states. Multistationarity is instrumental to cellular memory and cell differentiation during development or regeneration of multicellular organisms and is also used by micro-organisms in survival strategies. It is thus important to determine the parameter values for which a biochemical model is multistationary. With mass action reactions, testing for multiple steady states boils down to counting real positive solutions of algebraic systems.

V.P. Gerdt et al. (Eds.): CASC 2017, LNCS 10490, pp. 93–108, 2017.
DOI: 10.1007/978-3-319-66320-3_8

The models benchmarked in this paper concern intracellular signaling pathways. These pathways transmit information about the cell environment by inducing cascades of protein modifications (phosphorylation) all the way from the plasma membrane via the cytosol to genes in the cell nucleus. Multistationarity of signaling usually occurs as a result of activation of upstream signaling proteins by downstream components [2]. A different mechanism for producing multistationarity in signaling pathways was proposed by Kholodenko [26]. In this mechanism the cause of multistationarity are multiple phosphorylation/dephosphorylation cycles that share enzymes. A simple, two steps phosphorylation/dephosphorylation cycle is capable of ultrasensitivity, a form of all or nothing response with no multiple steady states (Goldbeter–Koshland mechanism). In multiple phosphorylation/dephosphorylation cycles, enzyme sharing provides competitive interactions and positive feedback that ultimately leads to multistationarity [23, 26].

Our study is complementary to works applying numerical methods to ordinary differential equations models used for biology applications. Gross et al. [18] used polynomial homotopy continuation methods for global parameter estimation of mass action models. Bifurcations and multistationarity of signaling cascades was studied with numerical methods based on the Jacobian matrix [30]. Other symbolic approaches to multistationarity either propose necessary conditions or work for particular networks [8,9,20,27].

Our work here follows [5], where it was demonstrated that determination of multistationarity of an 11-dimensional model of a mitogen-activated protein kinases (MAPK) cascade can be achieved by currently available symbolic methods when numeric values are known for all but potentially one parameter. We show that the symbolic methods used in [5], viz. real triangularization and cylindrical algebraic decomposition, and also polynomial homotopy continuation methods, benefit tremendously from a graph theoretical symbolic preprocessing method. This method has been sketched by Grigoriev et al. [17] and has been used for a "hand computation," but had not been implemented before. For our experiments we use the model already investigated in [5] and a higher dimensional model of the MAPK cascade.

2 The Systems for the Case Studies

For our investigations we use models of the MAPK cascade that can be found in the Biomodels database[1] as numbers 26 and 28 [24]. We refer to those models as Biomod-26 and Biomod-28, respectively.

2.1 Biomod-26

Biomod-26, which we have studied also in [5], is given by the following set of differential equations. We have renamed the species names as $x_1, \ldots, x_{11}$ and the rate constants as $k_1, \ldots, k_{16}$ to facilitate reading:

[1] http://www.ebi.ac.uk/biomodels-main/

$$
\begin{aligned}
\dot{x}_1 &= k_2x_6 + k_{15}x_{11} - k_1x_1x_4 - k_{16}x_1x_5 \\
\dot{x}_2 &= k_3x_6 + k_5x_7 + k_{10}x_9 + k_{13}x_{10} - x_2x_5(k_{11} + k_{12}) - k_4x_2x_4 \\
\dot{x}_3 &= k_6x_7 + k_8x_8 - k_7x_3x_5 \\
\dot{x}_4 &= x_6(k_2 + k_3) + x_7(k_5 + k_6) - k_1x_1x_4 - k_4x_2x_4 \\
\dot{x}_5 &= k_8x_8 + k_{10}x_9 + k_{13}x_{10} + k_{15}x_{11} - x_2x_5(k_{11} + k_{12}) - k_7x_3x_5 - k_{16}x_1x_5 \\
\dot{x}_6 &= k_1x_1x_4 - x_6(k_2 + k_3) \\
\dot{x}_7 &= k_4x_2x_4 - x_7(k_5 + k_6) \\
\dot{x}_8 &= k_7x_3x_5 - x_8(k_8 + k_9) \\
\dot{x}_9 &= k_9x_8 - k_{10}x_9 + k_{11}x_2x_5 \\
\dot{x}_{10} &= k_{12}x_2x_5 - x_{10}(k_{13} + k_{14}) \\
\dot{x}_{11} &= k_{14}x_{10} - k_{15}x_{11} + k_{16}x_1x_5 \qquad (1)
\end{aligned}
$$

The Biomodels database also gives us meaningful values for the rate constants, which we generally substitute into the corresponding systems for our purposes here:

$$
\begin{aligned}
&k_1 = 0.02, && k_2 = 1, && k_3 = 0.01, && k_4 = 0.032, \\
&k_5 = 1, && k_6 = 15, && k_7 = 0.045, && k_8 = 1, \\
&k_9 = 0.092, && k_{10} = 1, && k_{11} = 0.01, && k_{12} = 0.01, \\
&k_{13} = 1, && k_{14} = 0.5, && k_{15} = 0.086, && k_{16} = 0.0011. \qquad (2)
\end{aligned}
$$

Using the left-null space of the stoichiometric matrix under positive conditions as a conservation constraint [14] we obtain three linear conservation laws:

$$
\begin{aligned}
x_5 + x_8 + x_9 + x_{10} + x_{11} &= k_{17}, \\
x_4 + x_6 + x_7 &= k_{18}, \\
x_1 + x_2 + x_3 + x_6 + x_7 + x_8 + x_9 + x_{10} + x_{11} &= k_{19}, \qquad (3)
\end{aligned}
$$

where k_{17}, k_{18}, k_{19} are new constants computed from the initial data. Those constants are the parameters that we are interested in here.

The steady state problem for the MAPK cascade can now be formulated as a real algebraic problem as follows. We replace the left hand sides of all equations in (1) with 0 and substitute the values from (2). This together with (3) yields a system of parametric polynomial equations with polynomials in $\mathbb{Z}[k_{17}, k_{18}, k_{19}][x_1, \ldots, x_{11}]$. Since all entities in our model are strictly positive, we add to our system positivity conditions $k_{17} > 0$, $k_{18} > 0$, $k_{19} > 0$ and $x_1 > 0$, $\ldots$, $x_{11} > 0$. In terms of first-order logic the conjunction over our equations and inequalities yields a quantifier-free Tarski formula.

2.2 Biomod-28

The system with number 28 in the Biomodels database is given by the following set of differential equations. Again, we have renamed the species names into $x_1, \ldots, x_{16}$ and the rate constants into $k_1, \ldots, k_{27}$ to facilitate reading:

$$
\begin{aligned}
\dot{x}_1 &= k_2x_9 + k_8x_{10} + k_{21}x_{15} + k_{26}x_{16} - k_1x_1x_5 - k_7x_1x_5 - k_{22}x_1x_6 - k_{27}x_1x_6 \\
\dot{x}_2 &= k_3x_9 + k_5x_7 + k_{24}x_{12} - k_4x_2x_5 - k_{23}x_2x_6 \\
\dot{x}_3 &= k_9x_{10} + k_{11}x_8 + k_{16}x_{13} + k_{19}x_{14} - k_{10}x_3x_5 - k_{17}x_3x_6 - k_{18}x_3x_6 \\
\dot{x}_4 &= k_6x_7 + k_{12}x_8 + k_{14}x_{11} - k_{13}x_4x_6 \\
\dot{x}_5 &= k_2x_9 + k_3x_9 + k_5x_7 + k_6x_7 + k_8x_{10} + k_9x_{10} + k_{11}x_8 + k_{12}x_8 - \\
&\qquad k_1x_1x_5 - k_4x_2x_5 - k_7x_1x_5 - k_{10}x_3x_5 \\
\dot{x}_6 &= k_{14}x_{11} + k_{16}x_{13} + k_{19}x_{14} + k_{21}x_{15} + k_{24}x_{12} + k_{26}x_{16} - \\
&\qquad k_{13}x_4x_6 - k_{17}x_3x_6 - k_{18}x_3x_6 - k_{22}x_1x_6 - k_{23}x_2x_6 - k_{27}x_1x_6 \\
\dot{x}_7 &= k_4x_2x_5 - k_6x_7 - k_5x_7 \\
\dot{x}_8 &= k_{10}x_3x_5 - k_{12}x_8 - k_{11}x_8 \\
\dot{x}_9 &= k_1x_1x_5 - k_3x_9 - k_2x_9 \\
\dot{x}_{10} &= k_7x_1x_5 - k_9x_{10} - k_8x_{10} \\
\dot{x}_{11} &= k_{13}x_4x_6 - k_{15}x_{11} - k_{14}x_{11} \\
\dot{x}_{12} &= k_{23}x_2x_6 - k_{25}x_{12} - k_{24}x_{12} \\
\dot{x}_{13} &= k_{15}x_{11} - k_{16}x_{13} + k_{17}x_3x_6 \\
\dot{x}_{14} &= k_{18}x_3x_6 - k_{20}x_{14} - k_{19}x_{14} \\
\dot{x}_{15} &= k_{20}x_{14} - k_{21}x_{15} + k_{22}x_1x_6 \\
\dot{x}_{16} &= k_{25}x_{12} - k_{26}x_{16} + k_{27}x_1x_6
\end{aligned}
$$

The estimates of the rate constants given in the Biomodels database are:

$$
\begin{array}{llll}
k_1 = 0.005, & k_2 = 1, & k_3 = 1.08, & k_4 = 0.025, \\
k_5 = 1, & k_6 = 0.007, & k_7 = 0.05, & k_8 = 1, \\
k_9 = 0.008, & k_{10} = 0.005, & k_{11} = 1, & k_{12} = 0.45, \\
k_{13} = 0.045, & k_{14} = 1, & k_{15} = 0.092, & k_{16} = 1, \\
k_{17} = 0.01, & k_{18} = 0.01, & k_{19} = 1, & k_{20} = 0.5, \\
k_{21} = 0.086, & k_{22} = 0.0011, & k_{23} = 0.01, & k_{24} = 1, \\
k_{25} = 0.47, & k_{26} = 0.14, & k_{27} = 0.0018. &
\end{array}
$$

Again, using the left-null space of the stoichiometric matrix under positive conditions as a conservation constraint [14] we obtain the following:

$$
\begin{aligned}
x_6 + x_{11} + x_{12} + x_{13} + x_{14} + x_{15} + x_{16} &= k_{28}, \\
x_5 + x_7 + x_8 + x_9 + x_{10} &= k_{29}, \\
x_1 + x_2 + x_3 + x_4 + x_7 + x_8 + x_9 + x_{10} + x_{11} + & \\
x_{12} + x_{13} + x_{14} + x_{15} + x_{16} &= k_{30},
\end{aligned}
$$

where k_{28}, k_{29}, k_{30} are new constants computed from the initial data. We formulate the real algebraic problem as described at the end of Sect. 2.1. In particular, note that we need positivity conditions for all variables and parameters.

3 Graph-Theoretical Symbolic Preprocessing

The complexity, primarily in terms of dimension, of polynomial systems obtained with steady-state approximations of biological models plus conservation laws is comparatively high for the application of symbolic methods. It is therefore highly relevant for the success of such methods to identify and exploit particular structural properties of the input. Our models have remarkably low total degrees with many linear monomials after some substitutions for rate constants. This suggests to preprocess with essentially Gaussian elimination in the sense of solving single suitable equations with respect to some variable and substituting the corresponding solution into the system.

Generalizing this idea to situations where linear variables have parametric coefficients in the other variables requires, in general, a parametric variant of Gaussian elimination, which replaces the input system with a finite case distinction with respect to the vanishing of certain coefficients and one reduced system for each case. With Biomod-26 and Biomod-28 considered here it turns out that the positivity assumptions on the variables are strong enough to effectively guarantee the non-vanishing of all relevant coefficients so that case distinctions are never necessary. On the other hand, those positivity conditions establish an apparent obstacle, because we are formally not dealing with a parametric system of linear equations but with a parametric linear programming problem. However, here the theory of real quantifier elimination by virtual substitution tells us that it is sufficient that the inequality constraints play a passive role. Those constraints must be considered when substituting Gauss solutions from the equations, but otherwise can be ignored [22,25].

Parametric Gaussian elimination can increase the degrees of variables in the parametric coefficient, in particular destroying their linearity and suitability to be used for further reductions. As an example consider the steady-state approximation, i.e., all left hand sides replaced with 0, of the system in (1), solving the last equation for x_5, and substituting into the first equation. The natural question for an optimal strategy to Gauss-eliminate a maximal number of variables has been answered positively only recently [17]: draw a graph, where vertices are variables and edges indicate multiplication between variables within some monomial. Then one can Gauss-eliminate a *maximum independent set*, which is the complement of a *minimum vertex cover*. Figure 1 shows that graph for Biomod-26, where $\{x_4, x_5\}$ is a minimal vertex cover, and all other variables can be linearly eliminated. Similarly, for Biomod-28 we find $\{x_5, x_6\}$ as a minimum vertex cover. Recall that minimum vertex cover is one of Karp's 21 classical NP complete problems [21]. However, our instances considered here and instances to be expected from other biological models are so small that the use of existing approximation algorithms [16] appears unnecessary. We have used real quantifier elimination, which did not consume measurable CPU time; alternatively one could use integer linear programming or SAT-solving.

It is a most remarkable fact that a significant number of biological models in the databases have that property of loosely connected variables. This phenomenon resembles the well-known *community structure* of propositional

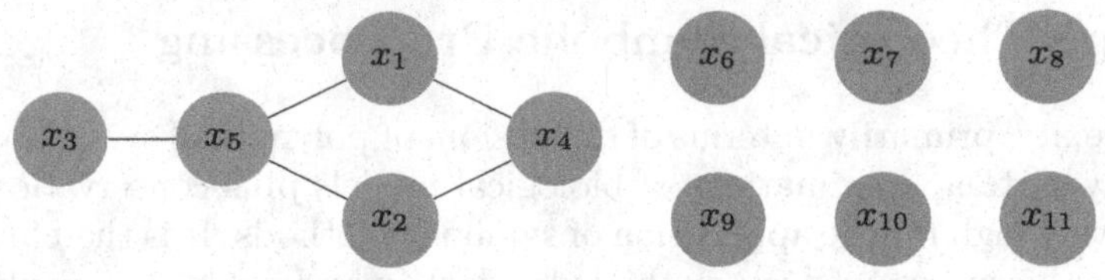

Fig. 1. The graph for Biomod-26 is loosely connected. Its minimum vertex cover $\{x_4, x_5\}$ is small. All other variables form a maximum independent set, which can be eliminated with linear methods.

satisfiability problems, which has been identified as one of the key structural reasons for the impressive success of state-of-the-art CDCL-based SAT solvers [15].

We conclude this section with the reduced systems as computed with our implementation in Redlog [11]. For Biomod-26 we obtain $x_5 > 0$, $x_4 > 0$, $k_{19} > 0$, $k_{18} > 0$, $k_{17} > 0$ and

$$\begin{aligned}
&1062444k_{18}x_4^2x_5 + 23478000k_{18}x_4^2 + 1153450k_{18}x_4x_5^2 + 2967000k_{18}x_4x_5\\
&\quad + 638825k_{18}x_5^3 + 49944500k_{18}x_5^2 - 5934k_{19}x_4^2x_5 - 989000k_{19}x_4x_5^2\\
&\quad - 1062444x_4^3x_5 - 23478000x_4^3 - 1153450x_4^2x_5^2 - 2967000x_4^2x_5\\
&\quad - 638825x_4x_5^3 - 49944500x_4x_5^2 = 0,\\
&1062444k_{17}x_4^2x_5 + 23478000k_{17}x_4^2 + 1153450k_{17}x_4x_5^2 + 2967000k_{17}x_4x_5\\
&\quad + 638825k_{17}x_5^3 + 49944500k_{17}x_5^2 - 1056510k_{19}x_4^2x_5 - 164450k_{19}x_4x_5^2\\
&\quad - 638825k_{19}x_5^3 - 1062444x_4^2x_5^2 - 23478000x_4^2x_5 - 1153450x_4x_5^3\\
&\quad - 2967000x_4x_5^2 - 638825x_5^4 - 49944500x_5^3 = 0.
\end{aligned}$$

For Biomod-28 we obtain $x_6 > 0$, $x_5 > 0$, $k_{30} > 0$, $k_{29} > 0$, $k_{28} > 0$ and

$$\begin{aligned}
&3796549898085k_{29}x_5^3x_6 + 71063292573000k_{29}x_5^3 + 106615407090630k_{29}x_5^2x_6^2\\
&\quad + 479383905861000k_{29}x_5^2x_6 + 299076127852260k_{29}x_5x_6^3\\
&\quad + 3505609439955600k_{29}x_5x_6^2 + 91244417457024k_{29}x_6^4\\
&\quad + 3557586742819200k_{29}x_6^3 - 598701732300k_{30}x_5^3x_6\\
&\quad - 83232870778950k_{30}x_5^2x_6^2 - 185019487578700k_{30}x_5x_6^3\\
&\quad - 3796549898085x_5^4x_6 - 71063292573000x_5^4 - 106615407090630x_5^3x_6^2\\
&\quad - 479383905861000x_5^3x_6 - 299076127852260x_5^2x_6^3 - 3505609439955600x_5^2x_6^2\\
&\quad - 91244417457024x_5x_6^4 - 3557586742819200x_5x_6^3 = 0,\\
&3796549898085k_{28}x_5^3x_6 + 71063292573000k_{28}x_5^3 + 106615407090630k_{28}x_5^2x_6^2\\
&\quad + 479383905861000k_{28}x_5^2x_6 + 299076127852260k_{28}x_5x_6^3\\
&\quad + 3505609439955600k_{28}x_5x_6^2 + 91244417457024k_{28}x_6^4\\
&\quad + 3557586742819200k_{28}x_6^3 - 3197848165785k_{30}x_5^3x_6\\
&\quad - 23382536311680k_{30}x_5^2x_6^2 - 114056640273560k_{30}x_5x_6^3\\
&\quad - 91244417457024k_{30}x_6^4 - 3796549898085x_5^3x_6^2 - 71063292573000x_5^3x_6\\
&\quad - 106615407090630x_5^2x_6^3 - 479383905861000x_5^2x_6^2 - 299076127852260x_5x_6^4\\
&\quad - 3505609439955600x_5x_6^3 - 91244417457024x_6^5 - 3557586742819200x_6^4 = 0.
\end{aligned}$$

Notice that no complex positivity constraints come into existence with these examples. All corresponding substitution results are entailed by the other constraints, which is implicitly discovered by using the standard simplifier from [12] during preprocessing.

4 Determination of Multiple Steady States

We aim to identify via grid sampling regions of parameter space where multistationarity occurs. Our focus is on the identification of regions with multiple positive real solutions for the parameters introduced with the conservation laws. We will encounter one or three such solutions and allow ourselves for biological reasons to assume monostability or bistability, respectively. Furthermore, a change in the number of solutions between one and three is indicative of a saddle-node bifurcation between a monostable and a bistable case. A mathematically rigorous treatment of stability would, possibly symbolically, analyze the eigenvalues of the Jacobian of the respective polynomial vector field. We consider two different approaches: first a polynomial homotopy continuation method implemented in Bertini, and second a combination of symbolic computation methods implemented in Maple. We compare the approaches with respect to performance and quality of results for both the reduced and the unreduced systems.

4.1 Numerical Approach

We use the homotopy solver Bertini [1] in its standard configuration to compute complex roots. We parse the output of Bertini using Python, and determine numerically, which of the complex roots are real and positive using a threshold of 10^{-6} for positivity. Computations are done in Python with Bertini embedded.

For System Biomod-26 we produced the two plots in Fig. 2 using the original system and the two in Fig. 3 using the reduced system. The sampling range for k_{19} was from 200 to 1000 by 50. In the left plots the sampling range for k_{17} is from 80 to 200 by 10 with k_{18} fixed at 50. In the right plots the sampling range for k_{18} is 5 to 75 by 5 with k_{17} fixed to 100. We see two regions forming according to the number of fixed points: yellow discs indicate one fixed point and blue boxes three. The diamonds indicate numerical errors where zero (red) or two (green) fixed states were identified. We analyse these further in Sect. 4.3.

For Biomod-28 we produced the two plots in Fig. 5 using the original system. The sampling range for k_{30} was from 100 to 1600 by 100. In the left plots the sampling range for k_{28} is from 40 to 160 by 10 with k_{29} fixed at 180. In the right plots the sampling range for k_{29} is from 120 to 240 by 10 with k_{28} fixed to 100. The colours and shapes indicate the number of fixed points as before. For the reduced system Bertini (wrongly) could not find any roots (not even complex ones) for any of the parameter settings. The situation did not change when going from adaptive precision to a very high fixed precision. However, we have not attempted more sophisticated techniques like providing user homotopies. We analyse these results further in Sect. 4.3.

4.2 Symbolic Approach

Our next approach will still use grid sampling, but each sample point will undergo a symbolic computation. The result will still be an approximate identification of the region (since the sampling will be finite) but the results at those sample points will be guaranteed free of numerical errors. The computations follow the strategy introduced in [5, Sect. 2.1.2]. This combined tools from the Regular Chains Library[2] available for use in Maple. Regular chains are the triangular decompositions of systems of polynomial equations (triangular in terms of the variables in each polynomial). Highly efficient methods for working in complex space have been developed based on these (see [29] for a survey).

We make use of recent work by Chen et al. [6] which adapts these tools to the real analogue: semi-algebraic systems. They describe algorithms to decompose any real polynomial system into finitely many regular semi-algebraic systems: both directly and by computation of components by dimension. The latter (the so called *lazy* variant) was key to solving the 1-parameter MAPK problem in [5]. However, for the zero dimensional computations of this paper there is only one solution component and so no savings from lazy computations.

For a given system and sample point we apply the real triangularization (RT) on the quantifier-free formula (as described at the end of Sect. 2.1: a quantifier free conjunction of equalities and inequalities) evaluated with the parameter estimates and sample point values. This produces a simplified system in several senses. First, as guaranteed by the algorithm, the output is triangular according to a variable ordering. So there is a univariate component, then a bivariate component introducing one more variable and so on. Secondly, for all the MAPK models we have studied so far, all but the final (univariate) of these equations has been linear in its main variable. This thus allows for easy back substitution. Thirdly, most of the positivity conditions are implied by the output rather than being an explicit part of it, in which case a simpler sub-system can be solved and back substitution performed instantly.

Biomod-26. For the original version of Biomod-26 the output of RT was a component consisting of 11 equations and a single inequality. The equations were in ascending main variable according to the provided ordering (same as the labelling). All but the final equation is linear in its main variable, with the final equation being univariate and degree 6 in x_1. The output of the triangularization requires that this variable be positive, $x_1 > 0$, with the positivity of the other variables implied by solutions to the system. So to proceed we must find the positive real roots of the degree 8 univariate polynomial in x_1: counting these will imply the number of real positive solutions of the parent system. We do this using the root isolation tools in the Regular Chains Library. This whole process was performed iteratively for the same sampling regime as Bertini used to produce Fig. 4.

We repeated the process on the reduced version of the system. The triangularization again reduced the problem to univariate real root isolation, this

[2] http://www.regularchains.org/

time with only one back substitution step needed. As to be expected from a fully symbolic computation, the output is identical and so again represented by Fig. 4. However, the computation was significantly quicker with this reduced system. More details are given in the comparison in Sect. 4.3.

Biomod-28. The same process was conducted on Biomod-28. As with Biomod-26 the system was triangular with all but the final equation linear in its main variable; this time the final equation is degree 8. However, unlike Biomod-26 two positivity conditions were returned in the output meaning we must solve a bivariate problem before we can back substitute to the full system. Rather than just perform univariate real root isolation we must build a Cylindrical Algebraic Decomposition (CAD) (see, e.g., [4] and the references within) sign invariant for the final two equations and interrogate its cells to find those where the equations are satisfied and variable positive. Counting these we find always 1 or 3 cells, with the latter indicating bistability. This is similar to the approach used in [5], although in that case the 2D CAD was for one variable and one parameter. We used the implementation of CAD in the Regular Chains Library [3,7] with the results producing the plots in Fig. 6.

For the reduced system we proceeded similarly. A 2D CAD still needed to be produced after triangularization and so in this case there was no reduction in the number of equations to study with CAD via back substitution. However, it was still beneficial to pre-process CAD with real triangularization: the average time per sample point with pre-processing (and including time taken to pre-process) was 0.485 s while without it was 3.577 s.

4.3 Comparison

Figures 2, 3, and 4 all refer to Biomod-26. The latter, produced using the symbolic techniques in Maple, is guaranteed free of numerical error. We see that

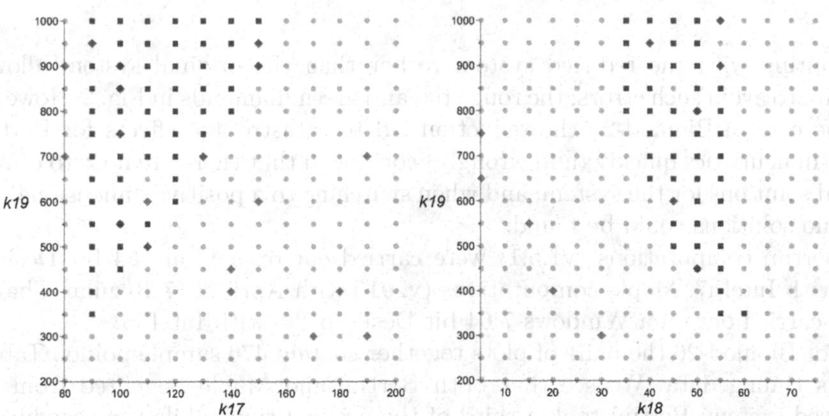

Fig. 2. Bertini grid sampling on the original version of Biomod-26 (see Sect. 4.1). The online version of this article contains colored figures

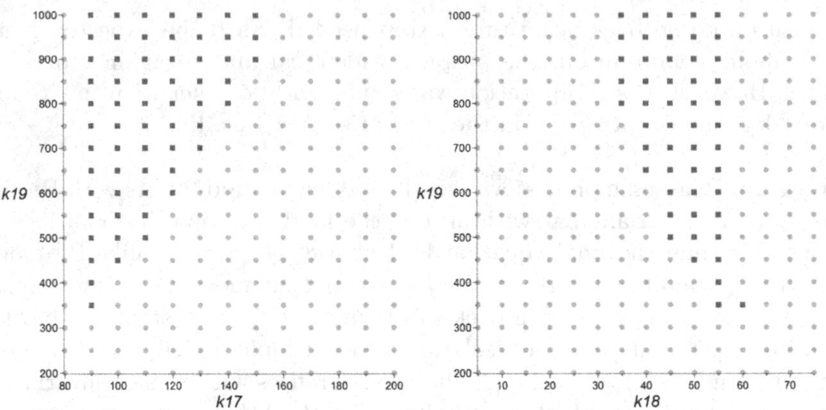

Fig. 3. Bertini grid sampling on the reduced version of Biomod-26 (see Sect. 4.1)

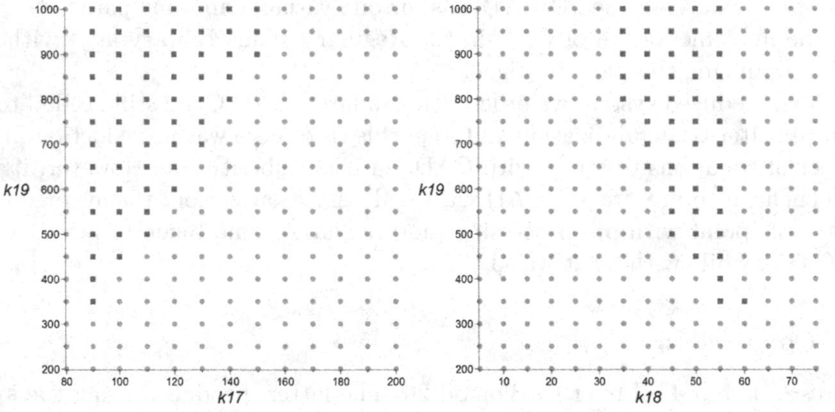

Fig. 4. Maple grid sampling on Biomod-26 (see Sect. 4.2)

computing with the reduced system rather than the original system allowed Bertini to avoid such errors: the rouge red and green diamonds in Fig. 2. However, in the case of Biomod-28 the reduction led to catastrophic effects for Bertini: built-in heuristics quickly (and wrongly) concluded that there are no zero dimensional solutions for the system, and when switching to a positive dimensional run also no solutions could be found.

Bertini computations (v1.5.1) were carried out on a Linux 64 bit Desktop PC with Intel i7. Maple computations (v2016 with April 2017 Regular Chains) were carried out on a Windows 7 64 bit Desktop PC with Intel i5.

For Biomod-26 the pairs of plots together contain 476 sample points. Table 1 shows timing data. We see that both Bertini and Maple benefited from the reduced system: Bertini took a third of the original time while the speedup for Maple was even greater: a tenth of the original. Also, perhaps surprisingly, the

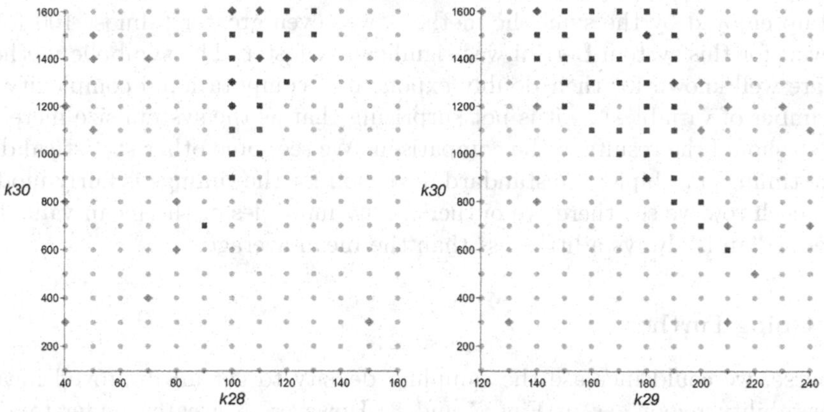

Fig. 5. Bertini grid sampling on the original version of Biomod-28 (see Sect. 4.1)

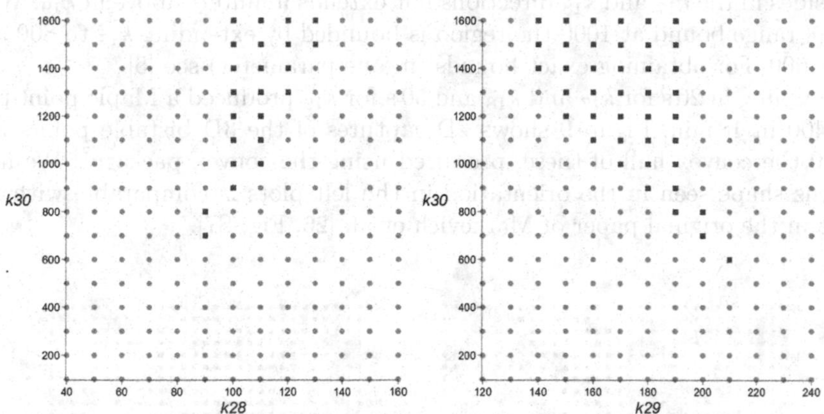

Fig. 6. Maple grid sampling on Biomod-28 (see Sect. 4.2)

Table 1. Timing data (in seconds) of the grid samplings described in Sect. 4. Numerical computation is using Bertini; Symbolic computation is using Maple Regular Chains

	Numerical	Symbolic			
	Mean	Mean	Median	StdDev	Maximum
026 – Original	2.4	0.568	0.530	0.107	0.905
026 – Reduced	0.85	0.053	0.047	0.036	0.343
028 – Original	16.57	42.430	40.529	8.632	84.116
028 – Reduced	⊥	0.485	0.468	0.119	0.796

symbolic methods were quicker than the numerical ones here. For Biomod-28 the speed-up enjoyed by the symbolic methods was even greater (almost 100 fold). However, for this system Bertini was significantly faster. The symbolic methods used are well known for their doubly exponential computational complexity (in the number of variables) so it is not surprising that as the system size increases there so should the results of the comparison. We see some other statistical data for the timings in Maple: the standard deviation for the timings is fairly modest but in each row we see there are outliers many multiples of the mean value and so the median is always a little less than the mean average.

4.4 Going Further

Of course, we could increase the sampling density to get an improved idea of the bistability region, as in Figs. 7 and 8. However, a greater understanding comes with 3D sampling. We have performed this using the symbolic approach described above, at a linear cost proportional to the increased number of sample points. This was completed for Biomod-26: the region in question is bounded to both sides in the k_{17} and k_{18} directions but extends infinitely above in k_{19}. With the k_{19} range bound at 1000 the region is bounded by extending k_{17} to 800 and k_{18} to 600. For obtaining exact bounds (in one parameter) see [5].

Sampling in 20 s for k_{17} and k_{18} and 50 s for k_{19} produced a Maple point plot of 20400 in 18 min. Figure 9 shows 2D captures of the 3D bistable points and Fig. 10 the convex hull of these, produced using the convex package[3]. We note the lens shape seen in the orientation in the left plots is comparable with the image in the original paper of Markevich et al. [26, Fig. S7].

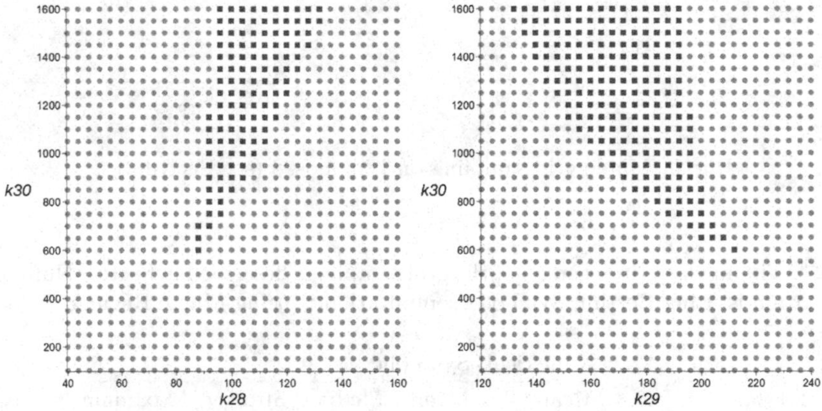

Fig. 7. As Fig. 6 but with a higher sampling rate

[3] http://www-home.math.uwo.ca/~mfranz/convex/

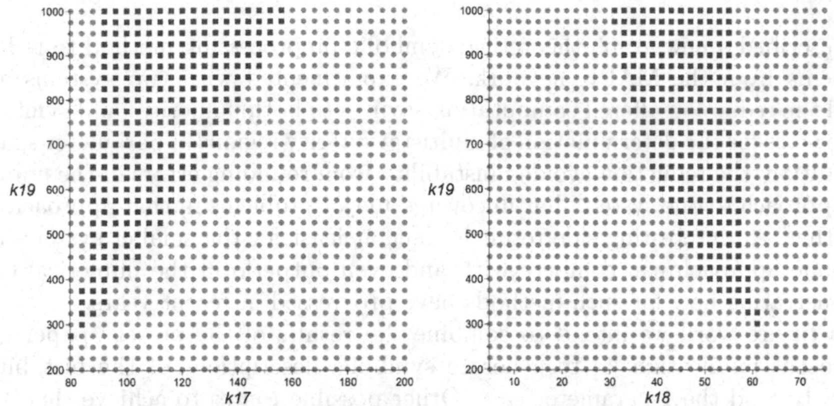

Fig. 8. As Fig. 4 but with a higher sampling rate

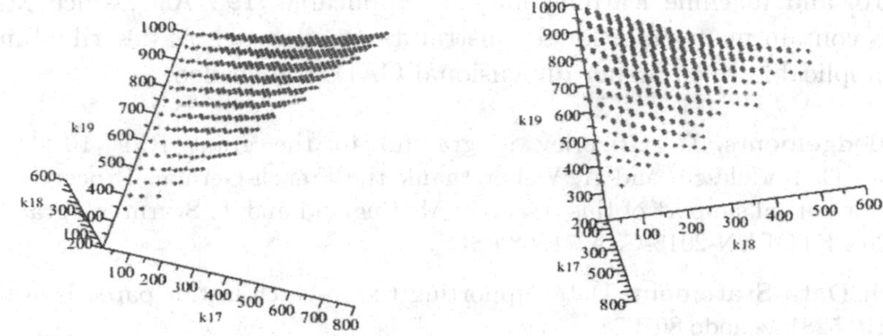

Fig. 9. 3D Maple Point Plot produced grid sampling on Biomod-26 (see Sect. 4.4)

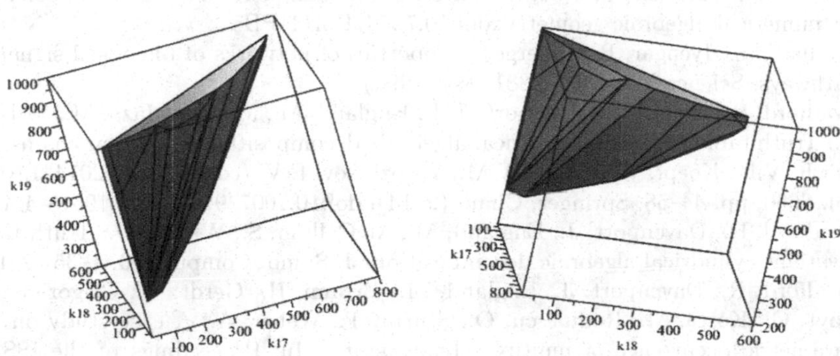

Fig. 10. Convex Hull of the bistable points in Fig. 9

5 Conclusion and Future Work

We described a new graph theoretical symbolic preprocessing method to reduce problems from the MAPK network. We experimented with two systems and found the reduction offered computation savings to both numerical and symbolic approaches for the determination of multistationarity regions of parameter space. In addition, the reduction avoided instability from rounding errors in the numerical approach to one system, but uncovered major problems in that approach for the other. An interesting side result is that, at least for the smaller system, the symbolic approach can compete with and even outperform the numerical one, demonstrating how far such methods have progressed in recent years.

In future work we intend to combine the results of the present paper and our recent publication [5] to generate symbolic descriptions of the bistability region beyond the 1-parameter case. Other possible routes to achieve this is to consider the effect of the various degrees of freedom with the algorithms used. For example, we have a free choice of variable ordering: Biomod-26 has 11 variables corresponding to 39 916 800 possible orderings while Biomod-28 has 16 variables corresponding to more than 10^{13} orderings. Heuristics exist to help with this choice [10] and machine learning may be applicable [19]. Also, since MAPK problems contain many equational constraints an approach as described in [13] may be applicable when higher dimensional CADs are needed.

Acknowledgements. D. Grigoriev is grateful to the grant RSF 16-11-10075. H. Errami, O. Radulescu, and A. Weber thank the French-German Procope-DAAD program for partial support of this research. M. England and T. Sturm are grateful to EU H2020-FETOPEN-2015-CSA 712689 SC^2.

Research Data Statement: Data supporting the research in this paper is available from doi:10.5281/zenodo.807678.

References

1. Bates, D.J., Hauenstein, J.D., Sommese, A.J., Wampler, C.W.: Bertini: software for numerical algebraic geometry. doi:10.7274/R0H41PB5
2. Bhalla, U.S., Iyengar, R.: Emergent properties of networks of biological signaling pathways. Science **283**(5400), 381–387 (1999)
3. Bradford, R., Chen, C., Davenport, J.H., England, M., Moreno Maza, M., Wilson, D.: Truth table invariant cylindrical algebraic decomposition by regular chains. In: Gerdt, V.P., Koepf, W., Seiler, W.M., Vorozhtsov, E.V. (eds.) CASC 2014. LNCS, vol. 8660, pp. 44–58. Springer, Cham (2014). doi:10.1007/978-3-319-10515-4_4
4. Bradford, R., Davenport, J., England, M., McCallum, S., Wilson, D.: Truth table invariant cylindrical algebraic decomposition. J. Symb. Comput. **76**, 1–35 (2016)
5. Bradford, R., Davenport, J., England, M., Errami, H., Gerdt, V., Grigoriev, D., Hoyt, C., Kosta, M., Radulescu, O., Sturm, T., Weber, A.: A case study on the parametric occurrence of multiple steady states. In: Proceedings of the ISSAC 2017, pp. 45–52. ACM (2017)
6. Chen, C., Davenport, J., May, J., Moreno Maza, M., Xia, B., Xiao, R.: Triangular decomposition of semi-algebraic systems. J. Symb. Comput. **49**, 3–26 (2013)

7. Chen, C., Moreno Maza, M., Xia, B., Yang, L.: Computing cylindrical algebraic decomposition via triangular decomposition. In: Proceedings of the ISSAC 2009, pp. 95–102. ACM (2009)
8. Conradi, C., Mincheva, M.: Catalytic constants enable the emergence of bistability in dual phosphorylation. J. Roy. Soc. Interface **11**(95) (2014)
9. Conradi, C., Flockerzi, D., Raisch, J.: Multistationarity in the activation of a MAPK: parametrizing the relevant region in parameter space. Math. Biosci. **211**(1), 105–31 (2008)
10. Dolzmann, A., Seidl, A., Sturm, T.: Efficient projection orders for CAD. In: Proceedings of the ISSAC 2004, pp. 111–118. ACM (2004)
11. Dolzmann, A., Sturm, T.: Redlog: computer algebra meets computer logic. ACM SIGSAM Bull. **31**(2), 2–9 (1997)
12. Dolzmann, A., Sturm, T.: Simplification of quantifier-free formulae over ordered fields. J. Symb. Comput. **24**(2), 209–231 (1997)
13. England, M., Bradford, R., Davenport, J.: Improving the use of equational constraints in cylindrical algebraic decomposition. In: Proceedings ISSAC 2015, pp. 165–172. ACM (2015)
14. Famili, I., Palsson, B.Ø.: The convex basis of the left null space of the stoichiometric matrix leads to the definition of metabolically meaningful pools. Biophys. J. **85**(1), 16–26 (2003)
15. Girvan, M., Newman, M.E.J.: Community structure in social and biological networks. Proc. Natl. Acad. Sci. USA **99**(12), 7821–7826 (2002)
16. Grandoni, F., Könemann, J., Panconesi, A.: Distributed weighted vertex cover via maximal matchings. ACM Trans. Algorithms **5**(1), 1–12 (2008)
17. Grigoriev, D., Samal, S.S., Vakulenko, S., Weber, A.: Algorithms to study large metabolic network dynamics. Math. Model. Nat. Phenom. **10**(5), 100–118 (2015)
18. Gross, E., Davis, B., Ho, K.L., Bates, D.J., Harrington, H.A.: Numerical algebraic geometry for model selection and its application to the life sciences. J. Roy. Soc. Interface **13**(123) (2016)
19. Huang, Z., England, M., Wilson, D., Davenport, J.H., Paulson, L.C., Bridge, J.: Applying machine learning to the problem of choosing a heuristic to select the variable ordering for cylindrical algebraic decomposition. In: Watt, S.M., Davenport, J.H., Sexton, A.P., Sojka, P., Urban, J. (eds.) CICM 2014. LNCS, vol. 8543, pp. 92–107. Springer, Cham (2014). doi:10.1007/978-3-319-08434-3_8
20. Joshi, B., Shiu, A.: A survey of methods for deciding whether a reaction network is multistationary. Math. Model. Nat. Phenom. **10**(5), 47–67 (2015)
21. Karp, R.M.: Reducibility among combinatorial problems. In: Complexity of Computer Computations, pp. 85–103. Plenum Press, New York (1972)
22. Košta, M.: New concepts for real quantifier elimination by virtual substitution. Doctoral dissertation, Saarland University, Germany, December 2016
23. Legewie, S., Schoeberl, B., Blüthgen, N., Herzel, H.: Competing docking interactions can bring about bistability in the MAPK cascade. Biophys. J. **93**(7), 2279–2288 (2007)
24. Li, C., Donizelli, M., Rodriguez, N., Dharuri, H., Endler, L., Chelliah, V., Li, L., He, E., Henry, A., Stefan, M.I., Snoep, J.L., Hucka, M., Le Novère, N., Laibe, C.: BioModels database: an enhanced, curated and annotated resource for published quantitative kinetic models. BMC Syst. Biol. **4**, 92 (2010)
25. Loos, R., Weispfenning, V.: Applying linear quantifier elimination. Comput. J. **36**(5), 450–462 (1993)

26. Markevich, N.I., Hoek, J.B., Kholodenko, B.N.: Signaling switches and bistability arising from multisite phosphorylation in protein kinase cascades. J. Cell Biol. **164**(3), 353–359 (2004)
27. Pérez Millán, M., Turjanski, A.G.: MAPK's networks and their capacity for multistationarity due to toric steady states. Math. Biosci. **262**, 125–37 (2015)
28. Rashevsky, N.: Mathematical Biophysics: Physico-Mathematical Foundations of Biology. Dover, New York (1960)
29. Wang, D.: Elimination Methods. Springer, Heidelberg (2000)
30. Zumsande, M., Gross, T.: Bifurcations and chaos in the MAPK signaling cascade. J. Theoret. Biol. **265**(3), 481–491 (2010)

The Polymake Interface in Singular and Its Applications

Raul Epure[1], Yue Ren[2](✉), and Hans Schönemann[1]

[1] Department of Mathematics, University of Kaiserslautern, Kaiserslautern, Germany
{epure,hannes}@mathematik.uni-kl.de

[2] Max Planck Institute for Mathematics in the Sciences, Leipzig, Germany
yueren@mis.mpg.de

Abstract. SINGULAR and POLYMAKE are computer algebra systems for research in algebraic geometry and polyhedral geometry respectively. We illustrate the implementation and the functionality of the POLYMAKE-interface in SINGULAR and exhibit its application to the arithmetic of polyhedral divisors and the reconstruction of hypersurface singularities from the Milnor algebra.

Keywords: Computer algebra · Algebraic geometry · Convex geometry

1 Introduction

SINGULAR [6] is a computer algebra system for polynomial computations with particular emphasis on applications in algebraic geometry, commutative algebra and singularity theory. POLYMAKE [9] is a software for research in polyhedral geometry. It deals with polytopes, polyhedra and fans as well as simplicial complexes, matroids, graphs and tropical hypersurfaces.

Since the initial release of SINGULAR, algebro geometric topics in which polyhedral methods play a crucial role have gained in importance (prominent examples are toric and tropical geometry). In an effort to do justice to these subjects, which lie in the intersection of algebraic and convex geometry, SINGULAR now features an interface to POLYMAKE. This gives SINGULAR users access to complex algorithms in convex geometry, some of which in turn rely on other third party software such as lattice point enumeration in NORMALIZ [3][1].

In this article, we describe the interface to polymake, and we will briefly comment on its implementation and its features. More importantly, we will show how it can be used to produce a framework for polyhedral divisors, and, on

[1] NORMALIZ is a tool for computations in affine monoids, vector configurations, lattice polytopes, and rational cones. Normaliz computes normalizations of affine monoids, integer hulls and triangulations of vector configurations, lattice points, Hilbert-/Ehrhart series and polynomials of rational polytopes, duals and Hilbert bases of rational cones.

V.P. Gerdt et al. (Eds.): CASC 2017, LNCS 10490, pp. 109–117, 2017.
DOI: 10.1007/978-3-319-66320-3_9

this example, how to create user-defined data types in SINGULAR in general. Furthermore, we will discuss the reconstruction of hypersurface singularities from their Milnor algebra, to give a glimpse on ongoing research projects which were made viable through the interface.

The interface was created in the DFG priority programme SPP 1489 in a collaboration of several big open source computer algebra systems. Its development continues in the DFG collaborative research centre SFB-TRR 195, which aims at taking a leading role in driving the development of interdisciplinary, open source infrastructure. The authors would like to thank Michael Joswig, Ewgenij Gawrilow, Benjamin Lorenz and Lars Kastner from the POLYMAKE team for their advice and continued technical support.

2 An Interface to Polymake

The POLYMAKE interface in SINGULAR is made possible through:

1. The POLYMAKE callable library functionality [10], which allows people to use POLYMAKE as a C++ callable library.
2. The SINGULAR blackbox functionality [14,15], which streamlines the integration of third party libraries into SINGULAR.

Thanks to the the aforementioned functionalities, its actual implementation is very simple, see Fig. 1. Each function merely requires a wrapper which:

Lines 3–6: checks the data type of the SINGULAR input and reads the data,
Lines 7–9: converts the SINGULAR input to a POLYMAKE object and calls the respective function in POLYMAKE,
Lines 10–12: converts the POLYMAKE output to a SINGULAR type and passes it on.

The wrapper returns `TRUE` if an error occurred and `FALSE` otherwise. This tells SINGULAR whether or not to abort the procedures in which the interface function was called.
Finally, there needs to be a line (Line 22), which tells SINGULAR:

1. the SINGULAR interpreter library containing the documentation (here: `polymake.lib`),
2. the name of the function in the SINGULAR interpreter (here: `latticePoints`),
3. whether the function is static, i.e. only to be used in SINGULAR interpreter libraries but invisible to the SINGULAR user (here: `FALSE`),
4. the name of the C++ wrapper function (here: `PMlatticePoints`).

For a more detailed exposition on the integration of third party C++ libraries into SINGULAR, please check the in-depth report on the GFANLIB[2] SINGULAR-interface [14,15].

[2] GFANLIB is a C++-library for basic manipulations of convex polyhedral cones and polyhedral fans. It also contains a fast algorithm for computing mixed volumes using homotopy methods.

```
BOOLEAN PMlatticePoints(leftv res, leftv args)
{
  leftv u = args;
  if ((u != NULL) && (u->Typ() == polytopeID) && (u->next == NULL))
  {
    gfan::ZCone* p0 = (gfan::ZCone*) u->Data();
    polymake::perl::Object* p1 = ZPolytope2PmPolytope(p0);
    polymake::Matrix<polymake::Integer> lp1 =
      p->CallPolymakeMethod("LATTICE_POINTS");
    intvec* lp0 = PmMatrixInteger2Intvec(&lp1,ok);
    res->rtyp = INTMAT_CMD;
    res->data = (char*) lp0;
    return FALSE;
  }
  WerrorS("latticePoints:␣unexpected␣parameters");
  return TRUE;
}

extern "C" int SI_MOD_INIT(polymake)(SModulFunctions* p)
{
  [...]
  p->iiAddCproc("polymake.lib","latticePoints",FALSE,PMlatticePoints);
  [...]
}
```

Fig. 1. C++ code for `blackbox` wrapper for `latticePoints`

Note that some POLYMAKE functionality depends on third party software, e.g. visualization through JREALITY[3] [16], which is why some functions in the POLYMAKE interface can only be called if the necessary third party software is installed.

The POLYMAKE interface is publicly available as part of the official SINGULAR release and a complete list of its functionality can be found in the documentation. The three main functions and their respective third party software used in the examples in this article are:

- computation of Minkowski sums,
- optimization of linear functionals via LRS[4] [1] or CDD[5] [8],
- enumeration of lattice points using LATTE[6] [4] or NORMALIZ [3].

[3] JREALITY is a Java based full-featured 3D scene graph package designed for 3D visualization and specialized in mathematical visualization. It provides several backends and can export graphics in several formats, allowing users for example to create interactive 3D elements in pdf-files.

[4] LRS is a C implementation of the reverse search algorithm for vertex enumeration and convex hull problems. All computations are done exactly in either multiple precision or fixed integer arithmetic, and the output is not stored in memory, so even problems with very large output sizes can sometimes be solved.

[5] CDDLIB is a C implementation of the Double Description Method of Motzkin et al. for generating all vertices of a general convex polyhedron in $\mathbb{R}^d$ given by a system of linear inequalities and vice versa.

[6] LATTE is a computer software dedicated to the problems of counting lattice points and integration inside convex polytopes. It contains the first ever implementation of Barnivok's algorithm.

3 User Defined Types in Singular: Polyhedral Divisors

SINGULAR offers the possibility to create user defined types from already known types and overload operators, such as +, ^ or &&, in cases in which the first operand is of a user defined type. In this section, we describe how this is done on the example of polyhedral divisors as in `divisors.lib` [2], relying on the interface to POLYMAKE for the computation of Minkowski sums and optimal values.

Polyhedral divisors were introduced by Altmann and Hausen, and they represent affine algebraic varieties with torus action in a way which encodes the torus action in a purely combinatorial fashion. To be more precise, any affine variety of dimension n with an effective action of an algebraic torus of dimension k corresponds to a polyhedral divisor living on a semiprojective variety of dimension $n-k$. They generalize the concept of affine toric varieties and are the building blocks for so-called divisorial fans, which generalize the concept of toric varieties. Of particular interest are the evaluation maps, which are used to characterize properness of polyhedral divisors.

Definition 1. *Let N be a lattice and $\sigma \subset N_{\mathbb{Q}}$ a pointed cone. Then* $\mathrm{Pol}_\sigma(N_{\mathbb{Q}})$ *is the set of all polyhedra in $N_{\mathbb{Q}}$ with tail cone σ. It has a natural semigroup structure under the Minkowski sum with neutral element σ.*

For a normal algebraic variety Y the group of rational polyhedral (Weil) divisors is *defined to be*

$$\mathrm{Div}_{\mathbb{Q}}(Y,\sigma) = \mathrm{Pol}_\sigma(N_{\mathbb{Q}}) \otimes_{\mathbb{Z}} \mathrm{Div}(Y),$$

so that its elements are formal sums $\mathfrak{D} = \sum_{i=1}^{k} \Delta_i \otimes D_i$, where $\Delta_i \subset N_{\mathbb{Q}}$ are polytopes with tail cone σ and D_i are Weil divisors on Y.

For any $u \in \sigma^\vee$ there exists a natural evaluation map

$$\begin{aligned} \mathrm{Div}_{\mathbb{Q}}(Y,\sigma) &\longrightarrow \mathrm{Div}_{\mathbb{Q}}(Y), \\ \mathfrak{D} = \sum_{i=1}^{k} \Delta_i \otimes D_i &\longmapsto \mathfrak{D}(u) = \sum_{i=1}^{k} eval_u(\Delta_i) D_i, \end{aligned}$$

where $eval_u(\Delta_i) = \max\{\langle u, v\rangle \mid v \in \sigma\}$.

New types can be created using the command `newstruct`. To call `newstruct`, one must specify a name for the new type, as well as types and names for the subobjects of which it should consist. In Fig. 2, lines 4–5 show the declaration of the type `pdivisor`, which consists of a list `summands` representing the formal sum which makes up a polyhedral divisor plus the fixed tail cone `tail`. The list `summands` will in turn consist of pairs of polytopes with tail cone `tail` and a Weil divisor.

By default, copy and print are the only two operations defined for the new type, the first copying, the second printing each subobject in order of their declaration. However, it is possible to overload more operators, using `system(''install'',...)`, with a user defined procedure with matching number of input parameters (of which the first must be of the user defined type).

Lines 9–12 are overloading the + and the * operators with the procedures defined in Lines 15–49.

The procedure `pdivmult` simply iterates through all elements of `summands`, which represent the summands in the formal sum, and scale each polytope by a given factor. The procedure `pdivplus` merges the `summands` of its two inputs, relying on POLYMAKE to compute the Minkowski sum of polytopes whose Weil divisors appear in both `summands`.

Moreover, POLYMAKE was used for implementing the evaluation of polyhedral divisors.

```
1  proc mod_init()
2  {
3    LIB "polymake.lib";
4    newstruct("pdivisor",
5      "list␣summands,␣cone␣tail");
6
7    [...]
8
9    system("install","pdivisor",
10     "+",pdivplus,2);
11   system("install","pdivisor",
12     "*",pdivmult,2);
13 }
14
15 proc pdivmult(pdivisor A, int l)
16 {
17   list LA = A.summands;
18   for (int i=1; i<=size(LA); i++)
19   {
20     LA[i,1] = scale(LA[i,1],l);
21   }
22   pDivisor Al;
23   Al.sum = LA;
24   Al.tail = A.tail;
25   return (Al);
26 }
27 proc pdivplus(pdivisor A, pdivisor B)
28 {
29   list LAB = A.summands;
30   list LB  = B.summands;
31   for (int i=1; i<=size(LB); i++)
32   {
33     p = findIndex(LAB,LB[i][2]);
34     if (p>0)
35     {
36       LAB[p][1] =
37         minkowskiSum(LAB[p][1],
38                      LB[i][1]);
39     }
40     else
41     {
42       LAB[size(LAB)+1] = LB[i];
43     }
44   }
45   pdivisor C;
46   C.summands = LAB;
47   C.tail = A.tail;
48   return(C);
49 }
```

Fig. 2. Singular code for polyhedral divisors using `newstruct`

4 Quasihomogeneous Isolated Hypersurface Singularities

In this section we present two applications regarding quasihomogeneous isolated hypersurface singularities. Before we start we need some basic definitions. In order to avoid unnecessary terminology, we give slightly restrictive definitions compared to the standard literature.

Definition 2. *Denote by* $\mathbb{C}\{x_1,\ldots,x_n\}$[7] *the ring of convergent power series over* $\mathbb{C}$ *and denote by* $\mathfrak{m}$ *its maximal ideal. Let* $f \in \mathbb{C}\{x_1,\ldots,x_n\}$*. We say* f *defines an* isolated hypersurface singularity*, if there exists some* $k \in \mathbb{N}$ *such that* $\mathfrak{m}^k \subseteq J(f) := \langle \partial_{x_1} f,\ldots,\partial_{x_n} f\rangle$*. We say* f *is a* quasihomogeneous isolated

[7] Note that while we are working mathematically over the algebraically closed field $\mathbb{C}$, our purposes allow us to restrict ourselves to rational numbers, provided our initial data is rational.

hypersurface singularity *(qhis), if there exist weights* $w_1, \dots, w_n \in \mathbb{Z}_{>0}$ *such that* $\gcd(w_1, \dots, w_n) = 1$ *and, for some fixed* $d \in \mathbb{N}$, $\sum_{i=1}^{n} w_i m_i = d$, *for all monomials* $x_1^{m_1} \cdot \ldots \cdot x_n^{m_n}$ *in the support of* f. *We refer to* d *as the* weighted degree *of* f. *We call* $M_f = \mathbb{C}\{x_1, \dots, x_n\}/J(f)$ *the* Milnor algebra *and* $\mu_f = \dim_{\mathbb{C}} M_f$ *the* Milnor number *of* f.

The following lemma gives an intrinsic characterization of quasi-homogeneity in terms of derivations, adapted to Definition 2.

Lemma 1 [17, Lemma 2.3]. *Let* $f \in \mathbb{C}\{x_1, \dots, x_n\}$ *define an isolated hypersurface singularity. Then* f *is a qhis with weights* $w_1, \dots, w_n \in \mathbb{N}$ *if and only if* $\gcd(w_1, \dots, w_n) = 1$ *and, after a suitable coordinate change,* $w_1 x_1 \partial_{x_1} f + \ldots + w_n x_n \partial_{x_n} f = df$ *for some* $d \in \mathbb{N}$.

Although qhis are defined as power series, the following lemma states that they can be considered as polynomials after a suitable coordinate change.

Theorem 1 [5, Theorem 9.1.4]. *Let* $f \in \mathbb{C}\{x_1, \dots, x_n\}$ *define a qhis and denote by* f_k *its truncation up to degree* $k \in \mathbb{N}$. *If* $k \geq \mu_f + 1$, *then there exists an automorphism* φ *of* $\mathbb{C}\{x_1, \dots, x_n\}$ *such that* $\varphi(f) = f_k$.

Theorem 1 gives us a bound for the degree of the possible monomials of f. The next theorem gives a bound on the weighted degree d of f and a formula for μ_f in terms of the weights $w_1, \dots, w_n$ of f and d.

Theorem 2 [11, Theorem 4.3]. *Let* $f \in \mathbb{C}\{x_1, \dots, x_n\}$ *define a qhis with weights* $w_1, \dots, w_n \in \mathbb{N}$ *and weighted degree* $d \in \mathbb{N}$. *Then the following hold:*

1. $d \leq C\mu_f$ *for some explicitly given* $C \in \mathbb{N}$.
2. $\mu_f = \prod_{j=1}^{n} \left(\frac{d}{w_j} - 1\right)$.

4.1 Finding Quasihomogeneous Isolated Hypersurface Singularities

A simple and nice application of the POLYMAKE interface in SINGULAR is the construction of examples of quasihomogeneous isolated hypersurface singularities. The question is, which possible weights are possible for qhis, in case a bound μ_f for the Milnor number is given. This has already been investigated by Hertling and Kurbel in [11], though they are not explicit in their constructions.

We will explain how to tackle the question in SINGULAR constructively on a specific example. We will be using their results and lattice point enumeration, which is done by NORMALIZ [3] through the POLYMAKE interface.

Example 1. Suppose $f \in \mathbb{C}\{x, y, z\}$ with $\mu_f = 14$. Theorem 2 (a) then provides a bound for the weighted degree of f and thus also a bound for all weights, as we have $w_i \leq \frac{d}{2} + 1$ using [17, Satz 1.3]. In our case, we have $d \leq 42$ so that $w_i \leq 22$ for $i = 1, 2, 3$. In Fig. 3 this is done in Lines 3 by the function `wdeg_bound`, which reads of the number of variables from the active basering and has the bound on the Milnor number as input.

In the Lines 4–15 we then construct the bounded polytope PB cut out by our inequalities. It is the intersection of the positive orthant P and the polytope B given by the weight bounds. Note that all coordinates are homogenized as in the convention of POLYMAKE.

```
1  > ring R = 0,(x,y,z),ds;
2  > int mu = 14;
3  > int d = wdeg_bound(mu); //=42
4  > intmat p[5][4] = 0,1,0,0,
5  .                  0,0,1,0,
6  .                  0,0,0,1;
7  > polytope P =
8  .   polytopeViaInequalities(p);
9  > intmat b[3][4] = 22,-1,0,0,
10 .                  22,0,-1,0,
11 .                  22,0,0,-1;
12 > polytope B =
13 .   polytopeViaInequalities(b);
14 > polytope PB =
15 .   convexIntersection(P,B);
16 > bigintmat candidates =
17 .   latticePoints(PB);
18 > nrows(candidates);
19 12167
20 > matrix checkedCandidates =
21 .   find_all_weights(candidates,mu,d);
22 > nrows(checkedCandidates);
23 156
```

Fig. 3. Computing a list of candidates

Using the POLYMAKE interface, we then see that PB contains 12167 lattice points of which only 156 satisfy Theorem 2(b), checked by **find_all_weights**. It is known that any generic linear combination of monomials of suitable weighted degree result in a qhis, allowing us to find explicit qhis with the given properties.

4.2 Reconstruction of QHIS from the Milnor Algebra

Let $f, g \in \mathbb{C}\{x_1, \ldots, x_n\}$ define qhis. The famous Mather–Yau Theorem [5, Theorem 9.1.8] states that $\mathbb{C}\{x_1, \ldots, x_n\}/\langle f\rangle \cong \mathbb{C}\{x_1, \ldots, x_n\}/\langle g\rangle$ if and only if $\mathbb{C}\{x_1, \ldots, x_n\}/J(f) \cong \mathbb{C}\{x_1, \ldots, x_n\}/J(g)$. Hence it implies that it should be possible to reconstruct f from M_f up to isomorphism, though its proof offers no way on how to do so. For homogeneous polynomials, this has been addressed in [12]. We tackle the problem in the quasihomogeneous case through methods in convex geometry. The following results are part of ongoing research.

Example 2. Consider $f = x^3 + y^3 + z^3 y \in \mathbb{C}\{x, y, z\}$, which is quasihomogeneous and defines an isolated hypersurface singularity (see Fig. 4, generated using SURFER [18]) with $J(f) = \langle x^2, 3y^2 + z^3, yz^2\rangle$. We will now attempt to reconstruct f from its Milnor algebra.

For the sake of simplicity, our example is chosen such that f and $J(f)$ are quasihomogeneous in the coordinates x, y, z. Hence, a reduced Gröbner basis computation yields that the space of weights, under whom $J(f)$ is quasihomogeneous, is spanned by the rows of

$$A = \begin{pmatrix} 1 & 0 & 0 \\ 0 & 3 & 2 \end{pmatrix}.$$

Fig. 4. The isolated hypersurface singularity of $x^3 + y^3 + z^3 y$

If this were not the case, we would have to use results from [7] regarding the structure of the derivation module of M_f and an adapted version of the coordinate changes as in the proof of [17, Lemma 2.4]. By [19, Theorem 1.2] f is quasihomogeneous if and only if $J(f)$ is weighted homogeneous with respect to a positive weight. This is clearly the case in our example, confirming that M_f does come from a qhis.

To see under which weight f is quasihomogeneous, we use the fact that $J(f)$ is weighted homogeneous with respect to all the weights under whom f is quasihomogeneous (though $J(f)$ may admit many more weights than f). A quick computation reveals the Milnor number $\mu_f = 14$, for which we have already determined a polytope PB of possible candidates in Example 1. Intersecting PB with the subspace generated by the rows of A reduces the number of lattice points from 12167 to 184.

In a separate computation we determine the socle of M_f to be the equivalence class of xz^4, pinpointing the weighted degree of f. Checking the equation in Theorem 2(b) then reduces the number of lattice points from 184 to 2 (Fig. 5).

Using syzygies we can construct examples of qhis with these weights and the corresponding degrees under certain polynomial constraints using the same idea as in Example 1. In this case our algorithm returns $g = 2x^3 + 3(y^3 + z^3 y)$ with $M_g \cong M_f$.

```
> //Computing weight cone
> cone C=positive_cone(A);
> //Bound for degree d
> int d=wdeg_bound(J);
> d;
42
> //List of weights
> bigintmat L=
. find_weights_by_bound(A,d);
```

```
> nrows(L);
184
> //Candidates after reduction
. matrix LL=
. find_weights_by_formula(L,14,s);
> print(LL);
3,3,2,9,
4,3,2,10
```

Fig. 5. Restricting the number of weight candidates using POLYMAKE

References

1. Avis, D.: Living with *lrs*. In: Akiyama, J., Kano, M., Urabe, M. (eds.) JCDCG 1998. LNCS, vol. 1763, pp. 47–56. Springer, Heidelberg (2000). doi:10.1007/978-3-540-46515-7_4
2. Böhm, J., Kastner, L., Lorenz, B., Schönemann, H., Ren, Y.: A SINGULAR 4-1-0 library for divisors and p-divisors (2016). www.singular.uni-kl.de
3. Bruns, W., Ichim, B., Römer, T., Sieg, R., Söger, C.: Normaliz 3.0. algorithms for rational cones and affine monoids (2016). https://www.normaliz.uni-osnabrueck.de
4. De Loera, J.A., Hemmecke, R., Tauzer, J., Yoshida, R.: Effective lattice point counting in rational convex polytopes. J. Symb. Comput. **38**(4), 1273–1302 (2005)
5. de Jong, T., Pfister, G.: Local Analytic Geometry. Basic Theory and Application. Advanced Lectures in Mathematics. Friedrich Vieweg & Sohn, Braunschweig (2000)
6. Decker, W., Greuel, G.-M., Pfister, G., Schönemann, H.: SINGULAR 4-1-0 – a computer algebra system for polynomial computations (2016). http://www.singular.uni-kl.de
7. Epure, R.-P.: Homogeneity and derivations on analytic algebras. Master's thesis. University of Kaiserslautern, Germany (2015)
8. Fukuda, K.: Cddlib, a C implementation of the double description method of Motzkin et al. (2016). https://www.inf.ethz.ch/personal/fukudak/cdd_home/
9. Gawrilow, E., Joswig, M.: Polymake: a framework for analyzing convex polytopes. In: Polytopes–Combinatorics and Computation (Oberwolfach, 1997), DMV Seminar, vol. 29, pp. 43–73. Birkhäuser, Basel (2000)
10. Polymake team: Polymake callable library. https://polymake.org/doku.php/reference/callable
11. Hertling, C., Kurbel, R.: On the classification of quasihomogeneous singularities. J. Singul. **4**, 131–153 (2012)
12. Isaev, A.V., Kruzhilin, N.G.: Explicit reconstruction of homogeneous isolated hypersurface singularities from their Milnor algebras. Proc. Amer. Math. Soc. **142**(2), 581–590 (2014)
13. Jensen, A.N.: Gfan 0.5, a software system for Gröbner fans and tropical varieties (2011). http://home.math.au.dk/jensen/software/gfan/gfan.html
14. Jensen, A., Ren, Y., Seelisch, F.: `gfan.lib`- A Singular 4-1-0 interface to GFANLIB (2017)
15. Jensen, A., Ren, Y., Schönemann, H.: Blackbox types in Singular and GFANLIB interface. to appear
16. Jreality: a Java library for real-time interactive 3D graphics and audio. http://www3.math.tu-berlin.de/jreality/jrealityStatic/index.php
17. Saito, K.: Quasihomogene isolierte Singularitäten von Hyperflächen. Invent. Math. **14**, 123–142 (1971)
18. Surfer: a software for visualizing real algebraic geometry in real-time. https://imaginary.org/program/surfer
19. Xu, Y.-J., Yau, S.-T.: Micro-local characterization of quasi-homogeneous singularities. Amer. J. Math. **118**(2), 389–399 (1996)

Computation of Some Integer Sequences in Maple

W.L. Fan[1(✉)], D.J. Jeffrey[1], and Erik Postma[2]

[1] Department of Applied Mathematics, The University of Western Ontario, London, ON, Canada
{wfan54,djeffrey}@uwo.ca
[2] Maplesoft, Waterloo, Canada

Abstract. We consider some integer sequences connected with combinatorial applications. Specifically, we consider Stirling partition and cycle numbers, associated Stirling partition and cycle numbers, and Eulerian numbers of the first and second kinds. We consider their evaluation in different contexts. One context is the calculation of a single value based on single input arguments. A more common context, however, is the calculation of a sequence of values. We compare strategies for both. Where possible, we compare with existing Maple implementations.

1 Introduction

For extended discussions of Stirling and Eulerian numbers, we refer to [1,2,7]. These and similar numbers arise frequently in combinatorial applications, and have therefore been implemented in several computer algebra systems. To date, the standard libraries of most systems have included Stirling numbers, but not *associated* Stirling numbers [3], even though they have found several applications in recent years. For example, they have appeared in series expansions for the Lambert W function [4], and also appeared in one form of Stirling's series for the Gamma function [2]. (Stirling did not define the associated numbers.)

Another feature of many implementations is that the functions expect a single argument, and return a single value. In practice, however, an application will usually require a sequence of values, for example, to provide successive coefficients in a series. The requirement of returning multiple values has already been recognized in some Maple functions, for example, in the implementation of Bernoulli numbers: they accept a `mode` parameter. To quote from Maple help:

> The mode parameter controls whether or not the `bernoulli` routine computes additional Bernoulli numbers in parallel with the requested one. For example, if your computer has 4 cores, then the command bernoulli (1000, singleton = false) will compute and store bernoulli (1002), bernoulli (1004), and bernoulli (1006). Since in practice nearly all computations which use Bernoulli numbers require many of them, and require them in sequence, this results in considerable efficiency gains.

V.P. Gerdt et al. (Eds.): CASC 2017, LNCS 10490, pp. 118–133, 2017.
DOI: 10.1007/978-3-319-66320-3_10

This paper addresses both the computation of single values and of the integer sequences associated with the combinatorial functions under consideration. As a matter of terminology, we shall call a function that accepts a unique argument and returns the corresponding unique result a *singleton* function, and the corresponding operation a singleton computation. In contrast, a function accepting a range (explicit or implicit) of arguments and returning the corresponding list of values will be a sequence function, and the calculation a sequence calculation.

1.1 Definitions of Numbers

We collect here the definitions of all numbers considered.

Definition 1. *The r-associated Stirling numbers of the first kind, more briefly Stirling r-cycle numbers, are defined by the generating function*

$$\left(\ln\frac{1}{1-z}-\sum_{j=1}^{r-1}\frac{z^j}{j}\right)^m = m!\sum_{n\geq 0}\left[{n \atop m}\right]_{\geq r}\frac{z^n}{n!}. \tag{1}$$

Remark 1. The number $\left[{n \atop m}\right]_{\geq r}$ gives the number of permutations of n distinct objects into m cycles, each cycle having a minimum cardinality r [2, p. 256].

Definition 2. *The r-associated Stirling numbers of the second kind, called more briefly here Stirling r-partition numbers, are defined, using Karamata–Knuth notation, by the generating function*

$$\left(e^z-\sum_{j=0}^{r-1}\frac{z^j}{j!}\right)^m = m!\sum_{n\geq 0}\left\{{n \atop m}\right\}_{\geq r}\frac{z^n}{n!}. \tag{2}$$

Remark 2. The number $\left\{{n \atop m}\right\}_{\geq r}$ gives the number of partitions of a set of size n into m subsets, each subset having a minimum cardinality of r [2,5,6].

Definition 3. *The Eulerian numbers of the first kind* $\left\langle{n \atop k}\right\rangle$ *are defined as the number of permutations* $\pi_1\pi_2\ldots\pi_n$ *of* $\{1,2,\ldots n\}$ *that have* k *ascents, i.e.* k *places where* $\pi_j < \pi_{j+1}$.

Definition 4. *The Eulerian numbers of the second kind* $\left\langle\!\left\langle{n \atop k}\right\rangle\!\right\rangle$ *are defined as the number of permutations of the multiset* $\{1,1,2,2,\ldots,n,n\}$ *for which all numbers between the two occurrences of every* m*, with* $1 \leq m \leq n$*, are greater than* m*, for each permutation having* k *ascents, i.e.* k *places where* $\pi_j < \pi_{j+1}$.

Remark 3. Note that m is not an argument. For example, given the multiset $\{112233\}$, permutations such as 122133 or 123321 are permitted, but 211233 is not. Amongst these permitted permutations, we count those with k ascents.

Nomenclature: In [1], the numbers $\left\langle{n \atop k}\right\rangle$ are called simply 'Eulerian numbers', while the numbers $\left\langle\!\left\langle{n \atop k}\right\rangle\!\right\rangle$ are called 'second-order Eulerian numbers'.

2 Stirling Partition Numbers

The Maple 2017 implementation is a singleton function, denoted `stirling2` in the `combinat` package. It uses the formula

$$\left\{ {n \atop m} \right\} = \frac{1}{m!} \sum_{k=0}^{m} (-1)^{m-k} \binom{m}{k} k^n. \tag{3}$$

For the singleton computation, Table 1 shows that the times[1] are much less using (3). In this table, we compared the Maple function `stirling2` with the method given below using the recurrence relation (7). Timings for a sequence calculation, however, given in Table 2, show the new method is more efficient.

Table 1. Timings (sec) for generating a singleton Stirling partition number. The time using (7) is compared with the Maple `stirling2` function.

n	m	Recurrence	`stirling2`
100	50	0.002	0.002
200	100	0.010	0.003
500	250	0.079	0.007
4000	200	2.700	0.009
5000	250	6.940	0.013

2.1 Sequence Calculation

Given n, m, we wish to compute all Stirling partition numbers $\left\{ {i \atop j} \right\}$ such that $i \le n$ and $j \le m$. We use the recurrence relation

$$\left\{ {i \atop j} \right\} = j \left\{ {i-1 \atop j} \right\} + \left\{ {i-1 \atop j-1} \right\}, \tag{4}$$

subject to the boundary conditions

$$\left\{ {j \atop j} \right\} = 1, \quad \text{and} \quad \left\{ {i \atop 1} \right\} = 1. \tag{5}$$

Since $\left\{ {i \atop j} \right\} = 0$ for $j > i$ (see Fig. 1), we define a matrix P which will not store these zeros.

$$P_{ij} = \left\{ {i+j-1 \atop j} \right\}. \tag{6}$$

Then the recurrence relation becomes

$$P(i,j) = j\,P(i-1,j) + P(i,j-1). \tag{7}$$

The boundary conditions then become, respectively, $P(1,j) = \left\{ {j \atop j} \right\} = 1$, and $P(i,1) = 1$.

[1] Product placement: times found using an Intel i7 in a Lenovo Ultrabook.

Timings: Table 2 shows the timings for filling matrices of various sizes with integer sequences of Stirling partition numbers. The recurrence relation (7) is compared with creating each entry through a call to Maple's `stirling2`. Filling the square matrix $P(n, n)$ actually calculates all partition numbers $\left\{ {i \atop j} \right\}$ with $i \leq 2n$ and $j \leq n$. This is done for timing convenience, and the matrix can be reshaped for other applications.

Table 2. Timings (sec) for generating sequences of Stirling partition numbers. The time using (7) compared with Maple `stirling2` function.

n	m	Recurrence	`stirling2`
100	100	0.031	7.87
200	200	0.093	69.2
300	300	0.265	259
400	400	0.437	667
500	500	0.843	1450

3 Stirling Cycle Numbers

We consider the computation of $\left[{n \atop m} \right]$, implemented in Maple 2017 as `stirling1` in the `combinat` package. The computational method used by `stirling1` is based on Stirling's original definition of his numbers:

$$x^{\underline{n}} = \sum_k \left[{n \atop k} \right] (-1)^{n-k} x^k. \tag{8}$$

For given n, `stirling1` constructs the product on the left, which is then collected in powers of x, so that by equating the coefficients of x^k, all numbers $\left[{n \atop k} \right]$ for $1 \leq k \leq n$ are determined and stored. Thus, a future call to $\left[{n \atop m} \right]$ with $1 \leq m \leq n$ will be returned by table lookup, but a future call with a different n will initiate a new computation. It is interesting that although the interface appears to offer the user only a singleton computation, in fact a particular integer sequence has been computed silently.

3.1 Singleton Computation

A singleton computation returns the value of a function for a single pair of input arguments. We implement the known recurrence relation

$$\left[{n \atop m} \right] = (n-1) \left[{n-1 \atop m} \right] + \left[{n-1 \atop m-1} \right], \tag{9}$$

subject to boundary conditions

$$\begin{bmatrix} m \\ m \end{bmatrix} = 1, \text{ for } m \geq 1, \tag{10}$$

$$\begin{bmatrix} n \\ 1 \end{bmatrix} = (n-1)!. \tag{11}$$

We define the vector

$$u_j^{(i)} = \begin{bmatrix} i+j-1 \\ j \end{bmatrix}.$$

In Fig. 1, we see that for fixed $i, u_j^{(i)}$ describes numbers along the ith diagonal line, counting from the left. The recurrence relation (9) can be written in terms of u as

$$u_j^{(i)} = (i+j-2)u_j^{(i-1)} + u_{j-1}^{(i)},$$

with $u_1^{(i)} = (i-1)!$. We note that once $u_j^{(i-1)}$ is used, it does not need to be stored further, so we can overwrite storage. Our iteration scheme is thus (Maple notation for the ith element of a vector is $u[i]$)

$$u[j] = (i+j-2)u[j] + u[j-1].$$

Therefore, we initialize $u[1] = u_1^{(i)} = \left[\begin{smallmatrix} i \\ 1 \end{smallmatrix}\right] = (i-1)!$ and fill in diagonal lines successively.

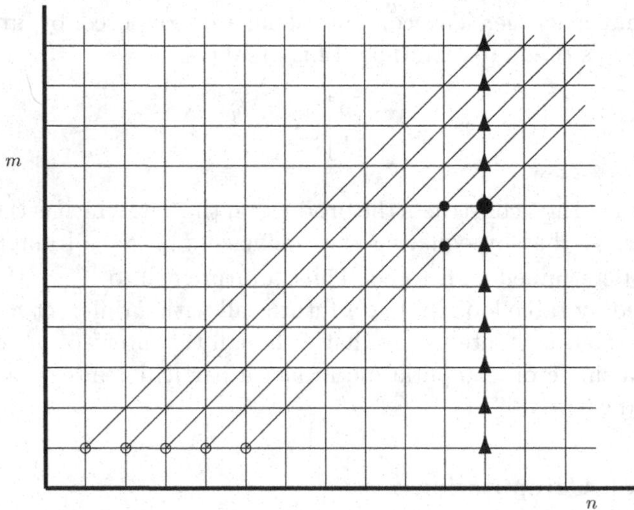

Fig. 1. Scheme for calculating singleton Stirling cycle $\left[\begin{smallmatrix} n \\ m \end{smallmatrix}\right]$ or partition numbers $\left\{\begin{smallmatrix} n \\ m \end{smallmatrix}\right\}$. The computation proceeds from left to right and bottom to top. At each stage only the numbers on one diagonal need to be stored in the vector $u_j^{(i)}$ which is progressively overwritten. The open circles show the base of each successive loop. The black filled circles show the recurrence relation used. The larger circle is calculated from the two smaller ones. The triangles line show the points computed by one call to `stirling1`.

Complexity. The aim of this subsection is to gain insight into the best ways to test the implementations, by identifying the worst cases for the methods. A full bit complexity is beyond the scope of this paper, and will require more work on estimates for the sizes of Stirling numbers. As pointed out by Wilf [9], the available estimates are for $\left[{n \atop k}\right]$ when k is fixed and $n \to \infty$, whereas the present algorithms require knowledge of the opposite case.

In order to calculate the number $\left[{n \atop m}\right]$, a vector of length m must be recomputed (overwritten) $n-m$ times. Each iteration requires one multiplication and 3 additions. Therefore the complexity is $m(n-m)$. We can therefore expect that the worst case for the method will be $m = n/2$.

Since Maple's approach and the present one calculate different sets of numbers, a direct comparison is not very meaningful, and so we simply make a brief comparison between one-time calculations. Notice that in Table 3, the times taken by `stirling1` are approximately independent of m as expected.

Table 3. Times for a single call to Maple's `stirling1` and the present singleton computation. Timings (sec) based on 10 trials, with memory being cleared before each call.

n	m	`stirling1`	Present scheme
300	150	0.023	0.035
400	200	0.063	0.052
400	20	0.062	0.011
1000	500	0.612	0.491
2000	500	4.92	3.00

3.2 A Finite Sum

For completeness, we mention that a singleton cycle number can be found from a finite sum, as was done for a singleton partition number. We have

$$\left[{n \atop m}\right] = \sum_{j=0}^{n-m} (-1)^{n-k+j} \binom{n-1+j}{n-k+j} \binom{2n-k}{n-k-j} \left\{{n-k+j \atop j}\right\}. \tag{12}$$

Combining this with (3), we can express a cycle number as a double sum. This, however, is too slow to warrant further consideration.

3.3 Sequence Calculation

The method used above for partition numbers can be readily adapted for cycle numbers. Given n, m, we compute all Stirling cycle numbers $\left[{i \atop j}\right]$ such that $i \le n$ and $j \le m$. We use the recurrence relation

$$\left[{i \atop j}\right] = (i-1)\left[{i-1 \atop j}\right] + \left[{i-1 \atop j-1}\right], \tag{13}$$

subject to the boundary conditions

$$\begin{bmatrix} j \\ j \end{bmatrix} = 1, \quad \text{and} \quad \begin{bmatrix} i \\ 1 \end{bmatrix} = (i-1)!. \tag{14}$$

Since $\left[{i \atop j}\right] = 0$ for $j > i$ (see Fig. 1), we define a matrix C which will not store these zeros.

$$C_{ij} = \begin{bmatrix} i+j-1 \\ j \end{bmatrix}. \tag{15}$$

Then the boundary conditions are $C(1,j) = \left[{j \atop j}\right] = 1$, and $C(i,1) = (i-1)!$. The recurrence relation becomes

$$C(i,j) = (i+j-2)\,C(i-1,j) + C(i,j-1). \tag{16}$$

Timings. Table 4 shows the timing for filling matrices of various sizes with integer sequences of Stirling cycle numbers. The recurrence relation (16) is compared with creating each entry through a call to Maple's `stirling1`. Filling the square matrix $C(n,n)$ actually calculated all cycle numbers $\left[{i \atop j}\right]$ with $i \le 2n$. This is done for timing purposes, and the matrix can be reshaped for other applications. The comparison is to compute the same numbers using the sequence calculation function `stirling1`. Larger values of (n,m) are not tabulated because a bug in Maple 2016 (and earlier) caused larger arguments to fail. This will be corrected in Maple 2017.

Table 4. Timings (sec) for generating sequences of Stirling cycle numbers. The time using (16) compared with Maple's `stirling1` function.

n	m	Recurrence	`stirling1`
40	40	0.000	0.842
60	60	0.000	4.446
80	80	0.015	14.414
100	100	0.015	35.037
120	120	0.015	193.004

4 Associated Stirling Numbers

There are no known analogues of (3) or (8) for the associated Stirling numbers for $r \ge 2$; hence we must use either the generating functions (1) and (2), or the following recurrence relations.

$$\left\{ {n+1 \atop k} \right\}_{\ge r} = k \left\{ {n \atop k} \right\}_{\ge r} + \binom{n}{r-1} \left\{ {n-r+1 \atop k-1} \right\}_{\ge r}, \tag{17}$$

$$\begin{bmatrix} n+1 \\ k \end{bmatrix}_{\geq r} = n \begin{bmatrix} n \\ k \end{bmatrix}_{\geq r} + n^{\underline{r-1}} \begin{bmatrix} n-r+1 \\ k-1 \end{bmatrix}_{\geq r}. \tag{18}$$

Note that $n^{\underline{0}} = 1$. The boundary cases are

$$\begin{Bmatrix} n \\ 1 \end{Bmatrix}_{\geq r} = 1, \quad n \geq r, \tag{19}$$

$$\begin{bmatrix} n \\ 1 \end{bmatrix}_{\geq r} = (n-1)!, \quad n \geq r, \tag{20}$$

$$\begin{Bmatrix} kr \\ k \end{Bmatrix}_{\geq r} = \frac{(rk)!}{(r!)^k \, k!}, \quad k \geq 1, \tag{21}$$

$$\begin{bmatrix} kr \\ k \end{bmatrix}_{\geq r} = \frac{(rk)!}{r^k k!}, \quad k \geq 1. \tag{22}$$

4.1 Singleton Stirling 2-Partition and 2-Cycle

The two computations have the same structure, and can be described in parallel. We choose to implement

$$\begin{Bmatrix} n \\ m \end{Bmatrix}_{\geq 2} = m \begin{Bmatrix} n-1 \\ m \end{Bmatrix}_{\geq 2} + (n-1) \begin{Bmatrix} n-2 \\ m-1 \end{Bmatrix}_{\geq 2}, \tag{23}$$

$$\begin{bmatrix} n \\ m \end{bmatrix}_{\geq 2} = (n-1) \begin{bmatrix} n-1 \\ m \end{bmatrix}_{\geq 2} + (n-1) \begin{bmatrix} n-2 \\ m-1 \end{bmatrix}_{\geq 2}. \tag{24}$$

We also have boundary conditions

$$\begin{bmatrix} 2n \\ n \end{bmatrix}_{\geq 2} = \begin{Bmatrix} 2n \\ n \end{Bmatrix}_{\geq 2} = \frac{(2n)!}{n! 2^n} = (2n-1)!!,$$

$$\begin{bmatrix} 2n+1 \\ n \end{bmatrix}_{\geq 2} = 2 \frac{(2n+1)!}{3(n-1)! 2^n} = 2 \begin{Bmatrix} 2n+1 \\ n \end{Bmatrix}_{\geq 2}.$$

We define the vector

$$u_j^{(i)} = \begin{bmatrix} i+2j-1 \\ j \end{bmatrix}_{\geq 2},$$

and similarly for 2-partition numbers. In Fig. 2, we see that if we fix i, then $u_j^{(i)}$ describes numbers along the ith diagonal line. Now $u_1^{(i)} = i!$ and

$$u_j^{(i)} = (i + 2j - 2)u_j^{(i-1)} + (i + 2j - 2)u_{j-1}^{(i)}.$$

We note that once $u_j^{(i-1)}$ is used, it does not need to be stored further, so we can overwrite storage. Our iteration scheme is thus

$$u[j] = (i + 2j - 2)(u[j] + u[j - 1]).$$

For initialization, we can use a special case of (24):

$$\begin{bmatrix} 2j+2 \\ j+1 \end{bmatrix}_{\geq 2} = u_{j+1}^{(1)} = (2j+1)\begin{bmatrix} 2j \\ j \end{bmatrix}_{\geq 2} = (2j+1)u_j^{(1)}.$$

Therefore, we initialize u to $i = 1$ using $u_j^{(1)} = u[j] = 1$ and fill in one line at a time by fixing i and looping over j. Each j loop starts setting $u_1^{(i)} = i! = iu_1^{(i-1)}$. We then loop over i.

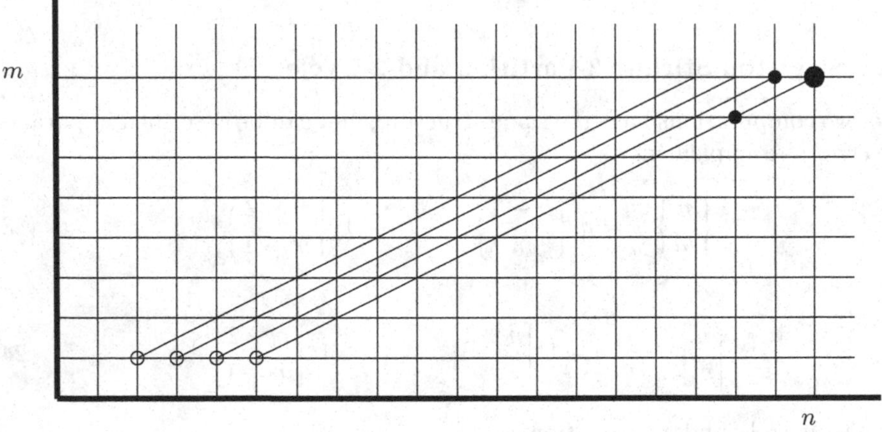

Fig. 2. Calculating 2-partition and 2-cycle numbers. As with the $r = 1$ case, only numbers on one sloping line need to be kept at any stage of the computation. The same convention for illustrating the recurrence relation is used.

4.2 Sequence Calculation of 2-Partition and 2-Cycle Numbers

Given n, m, we compute all Stirling 2-partition numbers $\left\{ {i \atop j} \right\}_{\geq 2}$ or 2-cycle numbers $\left[{i \atop j} \right]_{\geq 2}$ such that $i \leq n$ and $j \leq m$. We use the recurrence relations (23) or (24) as appropriate. Since ${i \brace j}_{\geq 2} = {i \brack j}_{\geq 2} = 0$ for $2j > i$ (see Fig. 2), we define a matrix C which will not store these zeros.

$$C_{ij} = {i+2j-1 \brack j}_{\geq 2}. \tag{25}$$

Then the recurrence relation for 2-partition becomes

$$C(i,j) = j\,C(i-1,j) + (i+2j-2)C(i-1,j-1). \tag{26}$$

The recurrence relation for 2-cycle becomes

$$C(i,j) = (i+2j-2)\,C(i-1,j) + (i+2j-2)C(i-1,j-1). \tag{27}$$

4.3 Singleton Stirling r-Partition and r-Cycle Numbers

From the above discussion of 1-associated and 2-associated numbers, the generalization is clear. We have to implement

$${n+1 \brack m}_{\geq r} = n{n \brack m}_{\geq r} + n(n-1)(n-2)\dots(n-r+2){n-r+1 \brack m-1}_{\geq r}. \tag{28}$$

We define the vector

$$u_j^{(i)} = {i+rj-1 \brack j}_{\geq r}.$$

The generalization of Fig. 2 to one containing lines of slope $1/r$ is not shown. For fixed i, $u_j^{(i)}$ describes numbers along one of the lines, with $u_1^{(i)} = (i+r-2)!$ and

$$u_j^{(i)} = (i+rj-2)u_j^{(i-1)} + (i+rj-2)(i+rj-3)\dots(i+rj-r)u_{j-1}^{(i)}.$$

We note that once $u_j^{(i-1)}$ is used, it does not need to be stored further, so we can overwrite storage. Our iteration scheme is thus

$$u[j] = (i+rj-2)u[j] + (i+rj-2)\dots(i+rj-r)u[j-1].$$

For initialization, we can use a special case of (28):

$$\begin{aligned}{rj+r \brack j+1}_{\geq r} = u_{j+1}^{(1)} &= (rj+r-1)(rj+r-2)\dots(rj+1){rj \brack j}_{\geq r} \\ &= (rj+r-1)(rj+r-2)\dots(rj+1)u_j^{(1)}.\end{aligned}$$

We initialize u using $u_j^{(1)} = u[j] = 1$ and fill in each line by fixing i and looping over j. We start each j loop by setting $u_1^{(i)} = (i+r-2)! = (i+r-2)u_1^{(i-1)}$. We then loop over i.

4.4 Sequence Calculation of r-Partition and r-Cycle Numbers

Given n, m, we compute all Stirling r-partition numbers $\left\{ {i \atop j} \right\}_{\geq r}$ or r-cycle numbers $\left[{i \atop j} \right]_{\geq r}$ such that $i \leq n$ and $j \leq m$. The recurrence relation applied here can refer to (17) and (18), which is subject to the boundary conditions

$$\left\{ {1 \atop 1} \right\}_{\geq r} = \left[{1 \atop 1} \right]_{\geq r} = 1. \tag{29}$$

Since $\left\{ {i \atop j} \right\}_{\geq r} = \left[{i \atop j} \right]_{\geq r} = 0$ for $rj > i$, we define a matrix C which will not store these zeros.

$$C_{ij} = \left[{i + rj - 1 \atop j} \right]_{\geq r}. \tag{30}$$

Then the recurrence relation for r-partition becomes

$$C(i,j) = j\, C(i-1,j) + \binom{i+rj-2}{r-1} C(i-r+1, j-1). \tag{31}$$

The recurrence relation for r-cycle becomes

$$C(i,j) = (i{+}rj{-}2)\, C(i{-}1,j){+}(i{+}rj{-}2)(i{+}rj{-}3)\ldots(i{+}rj{-}r)\, C(i{-}r{+}1,j{-}1). \tag{32}$$

4.5 Implementation in Maple

In our implementation of Stirling numbers, we provide procedures for users to compute either a singleton Stirling number or a sequence of Stirling numbers. The procedures are

1. StirlingRCycle: to calculate a singleton Stirling r-cycle number.
2. StirlingRCycleMatrix: to calculate a sequence of Stirling r-cycle numbers.
3. StirlingRPartition: to calculate a singleton Stirling r-partition number.
4. StirlingRPartitionMatrix: to calculate a sequence of Stirling r-partition numbers.

Neither Maple nor Mathematica has an implementation with which to compare our programs. Therefore we have programmed the recurrence relations, as well as the generating functions in Maple. In Table 5 below, we compared our new scheme for computing a singleton r-associated Stirling cycle number with using the generating function. The generating function for Stirling r-cycle numbers is:

```
StirRCycleGen := proc(n,k,r) local t, z, p;
      t:=series((ln(1/(1-z)) - add(z^p/p , p=1..r-1))^k, z=0,n+1);
      n!*coeff(t, z, n)/k!;
end proc;
```

Table 5. Timings in seconds of computations of single Stirling r-cycle number. Column headings give the functions used. The numbers tested were $\left[{1700 \atop m}\right]_{\geq 100}$.

m	Singleton scheme	Generating function
2	0.062	2.979
3	0.093	14.461
4	0.109	26.707
5	0.140	41.184
6	0.156	38.797
7	0.171	51.121
8	0.171	45.240
9	0.171	53.055
10	0.171	53.289

Table 5 shows that the singleton scheme is much faster than the generating function for the computation of single r-associated Stirling cycle number. For the computation of a sequence of r-associated Stirling cycle numbers, we compared three methods: (1) a loop calling the singleton function; (2) a loop calling the generating function; (3) the sequence procedure. The results are collected in Tables 6 and 7, and show that the sequence procedure is fastest.

Table 6. Timings in seconds of computations of a sequence of r-associated Stirling cycle numbers. Column headings give the functions used. The input argument is n, and the return is an $n \times n$ matrix.

n	Singleton scheme	Generating function
10	0.011	2.402
20	0.024	8.746
30	0.037	18.952
40	0.063	36.477
50	0.771	72.817

Similar tests were performed for r-associated Stirling partition numbers. The generating function for Stirling r-partition numbers is:

```
StirRPartGen := proc(n,k,r) local t, z, p;
      t:=series((exp(z) - add(z^p/p! , p=0..r-1))^k, z=0, n+1);
      n!*coeff(t, z, n)/k!;
end proc;
```

The test data are collected in Tables 8, 9 and 10. Since the pattern is similar to that for cycle numbers, the discussion and tables are abbreviated.

Table 7. Timings in seconds of computations of a sequence of r-associated Stirling cycle numbers. Column headings give the functions used. The input argument is n, and the return is an $n \times n$ matrix.

n	Singleton scheme	Sequence scheme
100	7.145	0.015
150	36.411	0.031
200	117.734	0.062
250	295.102	0.124
300	638.474	0.202

Table 8. Timings in seconds of computations of single r-associated Stirling partition number. Column headings give the functions being used. The numbers tested were $\left\{ {1000 \atop m} \right\}_{\geq 18}$.

m	Singleton scheme	Generating function
2	0.031	7.129
4	0.062	18.969
6	0.093	29.250
8	0.140	32.276
10	0.156	43.664

Table 9. Timings in seconds of computations of a sequence of r-associated Stirling partition number. Column headings give the functions being used. The input argument is n, and the return is an $n \times n$ matrix.

n	Singleton scheme	Generating function
10	0.020	1.864
20	0.035	8.345
30	0.070	22.047
40	0.144	44.074
50	0.201	79.233

Table 10. Timings in seconds of computations of a sequence of r-associated Stirling partition numbers. Column headings give the functions used. The input argument is n, and the return is an $n \times n$ matrix.

n	Singleton scheme	Sequence scheme
100	7.145	0.015
200	117.734	0.062
250	295.102	0.124
300	638.474	0.202

5 A Multiple Threads Approach to Sequence Calculations

The Maple help for Bernoulli numbers, quoted in the introduction, states that additional values of Bernoulli numbers are calculated in parallel. This section explored ways in which parallel computation could be applied to Stirling numbers. For this, we use the `Threads` package in Maple. When we generate the numbers inside a matrix, instead of filling the matrix row by row and column by column, we fill each diagonal from left to right. Here is the main part in the sequential code to fill Stirling r-cycle numbers in the matrix by diagonal with given input arguments (n, r) where n is the size of matrix.

```
for N from 3 to n do
  for k from 2 to N-1 do
    pd := mul(N-k+r-1-l, l = 1 .. r-1);
    A(N-k+r, k) := pd*A(N-k, k-1)+(N-k+r-2)*A(N-k+r-1, k);
  end do;
end do;
```

According to the recurrence relation, we know that we can divide such diagonal into a left half and right half. So we define two subroutines accordingly.

```
fileft := proc (N, r) local k, Nsplit, pd, l; global A;
Nsplit := floor((1/2)*N+1/2);
for k from 2 to Nsplit do
  pd := mul(N-k+r-1-l, l = 1 .. r-1);
  A(N-k+r, k) := pd*A(N-k, k-1)+(N-k+r-2)*A(N-k+r-1, k);
end do; end proc;
```

and

```
filrght := proc (N, r) local k, Nsplit, pd, l; global A;
Nsplit := floor((1/2)*N+1/2);
for k from Nsplit+1 to N-1 do
  pd := mul(N-k+r-1-l, l = 1 .. r-1);
  A(N-k+r, k) := pd*A(N-k, k-1)+(N-k+r-2)*A(N-k+r-1, k);
end do; end proc;
```

And for each half of the diagonal, we can establish an independent thread to fulfill the task. We implemented this approach in Maple.

```
Threaded := proc (n, r) local N, k, Nsplit; global A;
A := Matrix(n, n, fill = 0);
A(1, 1) := 1;
for N from 3 to n do
  Threads:-Task:-Start(null, Task = [fileft, N, r],
                             Task = [filrght, N, r])
end do;  end proc;
```

Table 11 compares the threaded scheme with the sequential scheme in the computation of an $n \times n$ matrix of Stirling cycle numbers. The table reflects the limitation that there is an overhead cost to setting up new threads, and the benefit of the threaded approach is felt only when the amount of work achieved within a thread outweighs the overhead. In this implementation, new threads are created for each loop. We are exploring new methods of calculation which will allow the threads to work more efficiently, with less overhead.

Table 11. Timings in seconds of comparison of threaded code with sequential code in generating sequences of Stirling Cycle numbers. The tests were made on an AMD 8-core processor.

n	Threaded scheme	Sequential scheme
500	0.856	0.329
1000	2.420	1.380
2000	9.240	9.610
2500	17.900	24.900
3000	29.880	39.870
4000	87.960	142.800

6 Implementation of Eulerian Numbers

The Eulerian numbers share many similarities with the Stirling numbers, and all the methods described above can be applied to their case. The numbers obey the following recurrence relations [1].

$$\left\langle {n \atop m} \right\rangle = (m+1)\left\langle {n-1 \atop m} \right\rangle + (n-m)\left\langle {n-1 \atop m-1} \right\rangle, \tag{33}$$

$$\left\langle\!\left\langle {n \atop m} \right\rangle\!\right\rangle = (m+1)\left\langle\!\left\langle {n-1 \atop m} \right\rangle\!\right\rangle + (2n-m-1)\left\langle\!\left\langle {n-1 \atop m-1} \right\rangle\!\right\rangle. \tag{34}$$

The present Maple functions `eulerian1` and `eulerian2` are recursively programmed implementations of these equations. As a consequence, they are very slow for large arguments. The new implementation of these numbers consists of 4 functions, which follow the patterns of the Stirling number implementations.

1. Eulerian1: calculates a singleton Eulerian number of the first kind. As with Stirling partition numbers, a finite sum is known which is distinctly the fastest method for a singleton computation [8]:

$$\left\langle {n \atop k} \right\rangle = \sum_{j=0}^{k+1} (-1)^j \binom{n+1}{j} (k-j+1)^n. \tag{35}$$

2. Eulerian1Matrix: calculates a sequence of Eulerian numbers of the first kind. This follows the sequence calculation of Stirling numbers, using (33).
3. Eulerian2: calculates singleton Eulerian numbers of the second kind. This follows the simpleton method used earlier for Stirling cycle numbers. There is no counterpart of (8) for Eulerian numbers.
4. Eulerian2Matrix: calculates a sequence of Eulerian numbers of the second kind.
This follows the sequence calculation of Stirling numbers.

6.1 Timings for Eulerian Number Calculations

In view of the similarities with Stirling numbers, we shall not labour the comparisons between methods, since they form the same procession of speeds seen before. Table 12 compares the implementations, following the patterns set above.

Table 12. Timings in seconds of computations of Eulerian numbers. Column headings give the functions used.

n	Eulerian1	Eulerian1Matrix	Eulerian2	Eulerian2Matrix
60	2.776	0.015	3.135	0.000
80	9.656	0.015	10.795	0.015
100	23.743	0.015	27.924	0.015
120	52.884	0.015	60.684	0.015
140	101.634	0.046	115.159	0.046
160	180.430	0.062	204.049	0.062
180	300.036	0.062	342.952	0.062

References

1. Graham, R.L., Knuth, D.E., Patashnik, O.: Concrete Mathematics. Addison-Wesley Publishing Co., Reading (1994)
2. Comtet, L.: Advanced Combinatorics. D. Reidel Publishing Co., Dordrecht (1974)
3. Howard, F.T.: Associated stirling numbers. Fibonacci Quart. **18**(4), 303–315 (1980)
4. Corless, R.M., Jeffrey, D.J., Knuth, D.E.: A sequence of series for the Lambert W Function. In: Kuechlin, W.W. (ed.) Proceedings of the ISSAC 1997. ACM Press (1997)
5. Karamata, J.: Theoreme sur la sommabilite exponentielle et d'autres sommabilites rattachant. Mathematica, Cluj **9**, 164–178 (1935). Romania
6. Knuth, D.E.: Two notes on notation. Am. Math. Mon. **99**, 403–422 (1992)
7. Stirling, J.: Methodus Differentialis, London (1730)
8. Lehmer, D.H.: Generalized Eulerian numbers. J. Combin. Theory Ser. A **32**, 195–215 (1982)
9. Wilf, H.S.: The asymptotic behavior of the stirling numbers of the first kind. J. Combin. Theory Ser. A **64**, 344–349 (1993)

Symbolic-Numerical Algorithm for Generating Interpolation Multivariate Hermite Polynomials of High-Accuracy Finite Element Method

A.A. Gusev[1(✉)], V.P. Gerdt[1,2], O. Chuluunbaatar[1,3], G. Chuluunbaatar[1], S.I. Vinitsky[1,2], V.L. Derbov[4], and A. Góźdź[5]

[1] Joint Institute for Nuclear Research, Dubna, Russia
gooseff@jinr.ru
[2] RUDN University, 6 Miklukho-Maklaya St., Moscow 117198, Russia
[3] Institute of Mathematics, National University of Mongolia, Ulaanbaatar, Mongolia
[4] N.G. Chernyshevsky Saratov National Research State University, Saratov, Russia
[5] Institute of Physics, University of M. Curie-Skłodowska, Lublin, Poland

Abstract. A symbolic-numerical algorithm implemented in Maple for constructing Hermitian finite elements is presented. The basis functions of finite elements are high-order polynomials, determined from a specially constructed set of values of the polynomials themselves, their partial derivatives, and their derivatives along the directions of the normals to the boundaries of finite elements. Such a choice of the polynomials allows us to construct a piecewise polynomial basis continuous across the boundaries of elements together with the derivatives up to a given order, which is used to solve elliptic boundary value problems using the high-accuracy finite element method. The efficiency and the accuracy order of the finite element scheme, algorithm and program are demonstrated by the example of the exactly solvable boundary-value problem for a triangular membrane, depending on the number of finite elements of the partition of the domain and the number of piecewise polynomial basis functions.

Keywords: Hermite interpolation polynomials · Boundary-value problem · High-accuracy finite element method

1 Introduction

In Refs. [9,10], the symbolic-numeric algorithms and programs for the solution of boundary-value problems for a system of second-order ordinary differential equations using the finite element method (FEM) of high accuracy order with Hermite interpolation polynomials (HIP) were developed, aimed at the calculation of spectral and optical characteristics of quantum systems.

It is known that the approximating function of the boundary-value problem solution in the entire domain can be expressed by means of its values and the values of its derivatives at the node points of the domain via the basis functions,

V.P. Gerdt et al. (Eds.): CASC 2017, LNCS 10490, pp. 134–150, 2017.
DOI: 10.1007/978-3-319-66320-3_11

referred to as Lagrange interpolation polynomials (LIP), which are nonzero only on a few elements, adjacent to the corresponding nodes. Generally, the approximating function for the entire domain is represented in terms of linear combinations of the basis functions. The coefficients of these linear combinations are the values of the approximating function and its directional derivatives on a given mesh of nodes. The basis functions themselves or their directional derivatives take a unit value at one of the nodes. In many cases, the schemes are restricted to the set of node values of the basis functions themselves. However, there are problems, in which the values of directional derivatives are also necessary. They are of particular importance when high smoothness between the elements is required, or when the gradient of solution is to be determined with increased accuracy. The construction of such basis functions, referred to as Hermite interpolation polynomials, is not possible on an arbitrary mesh of nodes. It is one of the most important and difficult problems in the finite element method and its applications in different fields, solved to date in the explicit form only for certain particular cases [1,2,4–8,11,13,14,17,19,21].

This motivation determines the aim of the present work, namely, the development of a symbolic-numerical algorithm implemented in any CAS for computing in analytical form the basis functions of Hermitian finite elements for a few variables and their application to constructing the FEM schemes with high order of accuracy.

In the paper, we present the symbolic-numeric algorithm implemented in the CAS Maple [15] for constructing the interpolation polynomials (basis functions) of Hermitian finite elements of a few variables based on a specially constructed set of values of the polynomials themselves, their partial derivatives, and derivatives along the normals to the boundaries of finite elements. The corresponding piecewise continuous basis of the high-order accuracy FEM provides the continuity not only of the approximate solution, but also of its derivatives to a given order depending on the smoothness of the variable coefficients of the equation and the domain boundary. This basis is used to construct the FEM scheme for high-accuracy solution of elliptic boundary-value problems in the bounded domain of multidimensional Euclidean space, specified as a polyhedral domain. We also used the symbolic algorithm to generate Fortran routines that allow the solution of the generalized algebraic eigenvalue problem with high-dimension matrices. The efficiency of the FEM scheme, the algorithm, and the program is demonstrated by constructing typical bases of Hermitian finite elements and their application to the benchmark exactly solvable boundary-value eigenvalue problem for a triangle membrane.

The paper is organized as follows. In Sect. 2, the setting of the boundary-value eigenvalue problem is given. In Sect. 3, we formulate the symbolic-numeric algorithm for generating the bases of Hermitian finite elements with multiple variables. In Sect. 4, we present the results of the calculations for the benchmark boundary-value problem, demonstrating the efficiency of the FEM scheme. In the Conclusion, we discuss the prospects of development of the proposed algorithm of constructing the Hermitian finite elements and its applications to high-order accuracy FEM schemes.

2 Setting of the Problem

Consider a self-adjoint boundary-value problem for the elliptic differential equation of the second order:

$$(D-E)\Phi(z) \equiv \left(-\frac{1}{g_0(z)}\sum_{ij=1}^{d}\frac{\partial}{\partial z_i}g_{ij}(z)\frac{\partial}{\partial z_j}+V(z)-E\right)\Phi(z)=0. \quad (1)$$

For the principal part coefficients of Eq. (1), the condition of uniform ellipticity holds in the bounded domain $z=(z_1,\ldots,z_d)\in\Omega$ of the Euclidean space $\mathcal{R}^d$, i.e., the constants $\mu>0$, $\nu>0$ exist such that $\mu\xi^2\leq\sum_{ij=1}^{d}g_{ij}(z)\xi_i\xi_j\leq\nu\xi^2$, $\xi^2=\sum_{i=1}^{d}\xi_i^2$ $\forall\xi\in\mathcal{R}^d$. The left-hand side of this inequality expresses the requirement of ellipticity, while the right-hand side expresses the boundedness of the coefficients $g_{ij}(z)$. It is also assumed that $g_0(z)>0$, $g_{ji}(z)=g_{ij}(z)$ and $V(z)$ are real-valued functions, continuous together with their generalized derivatives to a given order in the domain $z\in\bar{\Omega}=\Omega\cup\partial\Omega$ with the piecewise continuous boundary $S=\partial\Omega$, which provide the existence of nontrivial solutions obeying the boundary conditions [6,12] of the first kind

$$\Phi(z)|_S=0, \quad (2)$$

or the second kind

$$\left.\frac{\partial\Phi(z)}{\partial n_D}\right|_S=0,\quad \frac{\partial\Phi(z)}{\partial n_D}=\sum_{ij=1}^{d}(\hat{n},\hat{e}_i)g_{ij}(z)\frac{\partial\Phi(z)}{\partial z_j}, \quad (3)$$

where $\frac{\partial\Phi_m(z)}{\partial n_D}$ is the derivative along the conormal direction, $\hat{n}$ is the outer normal to the boundary of the domain $S=\partial\Omega$, $\hat{e}_i$ is the unit vector of $z=\sum_{i=1}^{d}\hat{e}_iz_i$, and $(\hat{n},\hat{e}_i)$ is the scalar product in $\mathcal{R}^d$.

For a discrete spectrum problem, the functions $\Phi_m(z)$ from the Sobolev space $H_2^{s\geq1}(\Omega)$, $\Phi_m(z)\in H_2^{s\geq1}(\Omega)$, corresponding to the real eigenvalues E: $E_1\leq E_2\leq\ldots\leq E_m\leq\ldots$ satisfy the conditions of normalization and orthogonality

$$\langle\Phi_m(z)|\Phi_{m'}(z)\rangle=\int_\Omega dz g_0(z)\Phi_m(z)\Phi_{m'}(z)=\delta_{mm'},\quad dz=dz_1\ldots dz_d. \quad (4)$$

The FEM solution of the boundary-value problems (1)–(4) is reduced to the determination of stationary points of the variational functional [3,6]

$$\Xi(\Phi_m,E_m,z)\equiv\int_\Omega dz g_0(z)\Phi_m(z)\left(D-E_m\right)\Phi(z)=\Pi(\Phi_m,E_m,z), \quad (5)$$

where $\Pi(\Phi_m,E_m,z)$ is the symmetric quadratic functional

$$\Pi(\Phi_m,E_m,z)=\int_\Omega dz\left[\sum_{ij=1}^{d}g_{ij}(z)\frac{\partial\Phi_m(z)}{\partial z_i}\frac{\partial\Phi_m(z)}{\partial z_j}+g_0(z)\Phi_m(z)(V(z)-E_m)\Phi_m(z)\right].$$

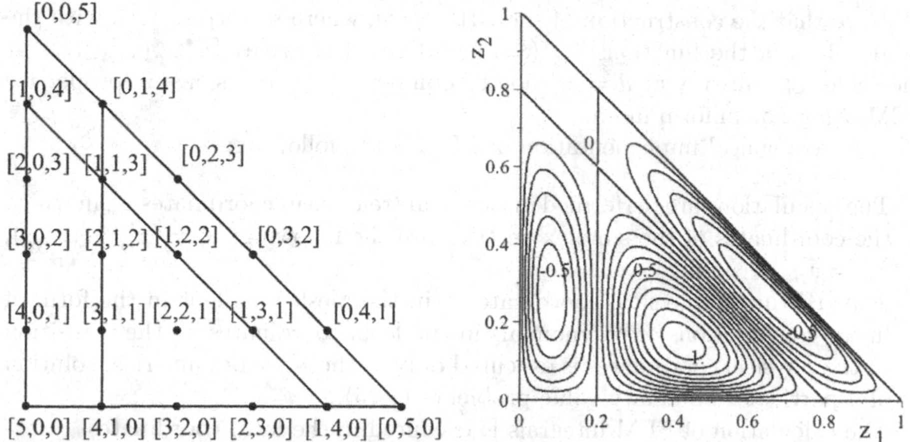

Fig. 1. (a) Enumeration of nodes A_r, $r = 1, \ldots, (p+1)(p+2)/2$ with sets of numbers $[n_0, n_1, n_2]$ for the standard (canonical) triangle element Δ in the scheme with the fifth-order LIP $p' = p = 5$ at $d = 2$. The lines (five crossing straight lines) are zeros of LIP $\varphi_{14}(z')$ from (12), equal to 1 at the point labeled with the number triple $[n_0, n_1, n_2] = [2, 2, 1]$. (b) LIP isolines of $\varphi_{14}(z')$

3 FEM Calculation Scheme

In FEM, the domain $\Omega = \Omega_h(z) = \bigcup_{q=1}^{Q} \Delta_q$, specified as a polyhedral domain, is covered with finite elements, in the present case, the simplexes Δ_q with $d+1$ vertices $\hat{z}_i = (\hat{z}_{i1}, \hat{z}_{i2}, \ldots, \hat{z}_{id})$ with $i = 0, \ldots, d$. Each edge of the simplex Δ_q is divided into p equal parts, and the families of parallel hyperplanes $H(i, k)$ are drawn, numbered with the integers $k = 0, \ldots, p$, starting from the corresponding face, e.g., as shown for $d = 2$ in Fig. 1 (see also [6]). The equation of the hyperplane is $H(i, k) : H(i; z) - k/p = 0$, where $H(i; z)$ is a linear function of z.

The node points of hyperplanes crossing A_r are enumerated with sets of integers $[n_0, \ldots, n_d]$, $n_i \geq 0$, $n_0 + \ldots + n_d = p$, where n_i, $i = 0, 1, \ldots, d$ are the numbers of hyperplanes, parallel to the simplex face, not containing the i-th vertex $\hat{z}_i = (\hat{z}_{i1}, \ldots \hat{z}_{id})$. The coordinates $\xi_r = (\xi_{r1}, \ldots, \xi_{rd})$ of the node point $A_r \in \Delta_q$ are calculated using the formula

$$(\xi_{r1}, \ldots, \xi_{rd}) = (\hat{z}_{01}, \ldots, \hat{z}_{0d})n_0/p + (\hat{z}_{11}, \ldots, \hat{z}_{1d})n_1/p + \ldots + (\hat{z}_{d1}, \ldots, \hat{z}_{dd})n_d/p \quad (6)$$

from the coordinates of the vertices $\hat{z}_j = (\hat{z}_{j1}, \ldots, \hat{z}_{jd})$. Then the LIP $\varphi_r(z)$ equal to one at the point A_r with the coordinates $\xi_r = (\xi_{r1}, \ldots, \xi_{rd})$, characterized by the numbers $[n_0, n_1, \ldots, n_d]$, and equal to zero at the remaining points $\xi_{r'}$, i.e., $\varphi_r(\xi_{r'}) = \delta_{rr'}$, has the form

$$\varphi_r(z) = \left(\prod_{i=0}^{d} \prod_{n_i'=0}^{n_i-1} \frac{H(i; z) - n_i'/p}{H(i; \xi_r) - n_i'/p} \right). \quad (7)$$

Note that the construction of the HIP $\varphi_r^\kappa(z)$, where $\kappa \equiv \kappa_1, \ldots, \kappa_d$, with the fixed values of the functions $\{\varphi_r^\kappa(\xi_{r'})\}$ and the derivatives $\{\partial_\bullet^\bullet \varphi_r^\kappa(z)|_{z=\xi_{r'}}\}$ at the nodes $\xi_{r'}$, already at $d = 2$ leads to cumbersome expressions, improper for FEM using nonuniform mesh.

The economical implementation of FEM is the following:

1. The calculations are performed in the local (reference) coordinates z', in which the coordinates of the simplex vertices are the following: $\hat{z}'_j = (\hat{z}'_{j1}, \ldots, \hat{z}'_{jd})$, $\hat{z}'_{jk} = \delta_{jk}$,
2. The HIP in the physical coordinates z in the mesh is sought in the form of linear combinations of polynomials in the local coordinates z', the transition to the physical coordinates is executed only at the stage of numerical solution of a particular boundary-value problem (1)–(5),
3. The calculation of FEM integrals is executed in the local coordinates.

Let us construct the HIP on an arbitrary d-dimensional simplex Δ_q with the $d+1$ vertices $\hat{z}_i = (\hat{z}_{i1}, \hat{z}_{i2}, \ldots, \hat{z}_{id})$, $i = 0, \ldots, d$. For this purpose, we introduce the local coordinate system $z' = (z'_1, z'_2, \ldots, z'_d) \in \mathcal{R}^d$, in which the coordinates of the simplex vertices are the following: $\hat{z}'_i = (\hat{z}'_{ik} = \delta_{ik}, k = 1, \ldots, d)$. The relation between the coordinates is given by the formula:

$$z_i = \hat{z}_{0i} + \sum_{j=1}^{d} \hat{J}_{ij} z'_j, \quad i = 1, \ldots, d, \quad \hat{J}_{ij} = \hat{z}_{ji} - \hat{z}_{0i}. \tag{8}$$

The inverse transformation and the relation between the differentiation operators are given by the formulas

$$z'_i = \sum_{j=1}^{d} (\hat{J}^{-1})_{ij} (z_j - \hat{z}_{0j}), \tag{9}$$

$$\frac{\partial}{\partial z'_i} = \sum_{j=1}^{d} \hat{J}_{ji} \frac{\partial}{\partial z_j}, \quad \frac{\partial}{\partial z_i} = \sum_{j=1}^{d} (\hat{J}^{-1})_{ji} \frac{\partial}{\partial z'_j}. \tag{10}$$

Equation (10) is used to calculate the HIP $\varphi_r^\kappa(z') = \{\check{\varphi}_r^\kappa(z'), Q_s(z')\}$ from (20) that satisfy the conditions (13), (17), and (18) of the next section, with the fixed derivatives to the given order at the nodes $\xi_{r'}$. In this case, the derivatives along the normal to the element boundary in the physical coordinate system are, generally, not those in the local coordinates z'. When constructing the HIP in the local coordinates z' one has to recalculate the fixed derivatives at the nodes $\xi_{r'}$ of the element Δ_q to the nodes $\xi'_{r'}$ of the element Δ, using the matrices $\hat{J}^{-1}$, given by cumbersome expressions. Therefore, the required recalculation is executed based on the relations (8)–(10) for each finite element Δ_q at the stage of the formation of the HIP basis $\{\varphi_r^{\bar{\kappa}'}(z')\}_{r=1}^{P}$ on the finite element Δ, implemented numerically using the analytical formulas, presented in the next section.

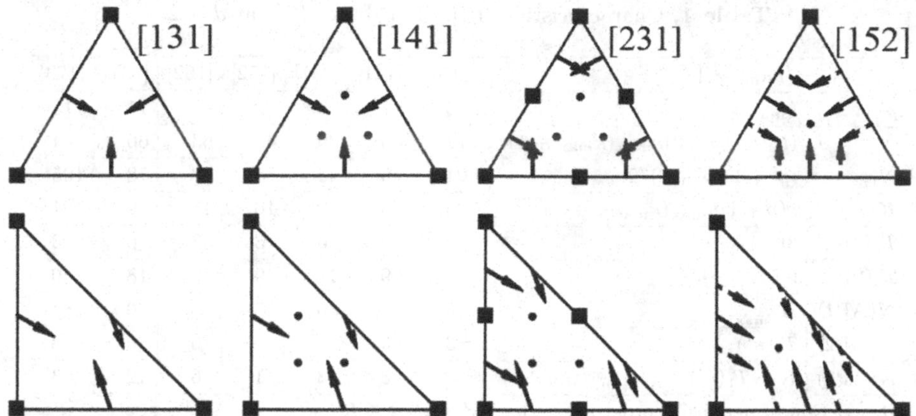

Fig. 2. Schematic diagram of the conditions on the element Δ_q (upper panel) and Δ (lower panel) for constructing the basis of HIP $[p\kappa_{\max}\kappa']$: [131], [141], [231], [152]. The squares are the points ξ'_r, where the values of the functions and their derivatives are fixed according to the conditions (13), (16); the solid (dashed) arrows begin at the points η'_s, where the values of the first (second) derivative in the direction of the normal in the physical coordinates are fixed, according to the condition (17), respectively; the circles are the points ζ'_s, where the values of the functions are fixed according to the condition (18)

The integrals that enter the variational functional (5) on the domain $\Omega_h(z) = \bigcup_{q=1}^{Q} \Delta_q$, are expressed via the integrals, calculated on the element Δ_q, and recalculated to the local coordinates z' on the element Δ,

$$\int_{\Delta_q} dz g_0(z)\varphi_r^{\kappa}(z)\varphi_{r'}^{\kappa''}(z)U(z) = J\int_{\Delta} dz' g_0(z(z'))\varphi_r^{\kappa}(z')\varphi_{r'}^{\kappa''}(z')U(z(z')), \tag{11}$$

$$\int_{\Delta_q} dz g_{s_1 s_2}(z)\frac{\partial\varphi_r^{\kappa}(z)}{\partial z_{s_1}}\frac{\partial\varphi_{r'}^{\kappa''}(z)}{\partial z_{s_2}} = J\sum_{t_1,t_2=1}^{d}\hat{J}^{-1}_{s_1 s_2;t_1 t_2}\int_{\Delta} dz' g_{s_1 s_2}(z(z'))\frac{\partial\varphi_r^{\kappa}(z')}{\partial z'_{t_1}}\frac{\partial\varphi_{r'}^{\kappa''}(z')}{\partial z'_{t_2}},$$

where $J = \det\hat{J} > 0$ is the determinant of the matrix $\hat{J}$ from Eq. (8), $\hat{J}^{-1}_{s_1 s_2;t_1 t_2} = (\hat{J}^{-1})_{t_1 s_1}(\hat{J}^{-1})_{t_2 s_2}$, $dz' = dz'_1 \ldots dz'_d$, and $\varphi_r^{\kappa}(z') = \{\check{\varphi}_r^{\kappa}(z'), Q_s(z')\}$ from Eq. (20).

3.1 Lagrange Interpolation Polynomials

In the local coordinates, the LIP $\varphi_r(z')$ is equal to one at the node point ξ'_r characterized by the numbers $[n_0, n_1, \ldots, n_d]$, and zero at the remaining node points $\xi'_{r'}$, i.e., $\varphi_r(\xi'_{r'}) = \delta_{rr'}$, are determined by Eq. (7) at $H(0; z') = 1 - z'_1 - \ldots - z'_d$, $H(i; z') = z'_i$, $i = 1, \ldots, d$:

Table 1. Characteristics of the HIP bases (20) at $d = 2$

	$[p\kappa_{\max}\kappa']$	[131]	[141]	[231]	[152]	[162]	[241]	[173]
p'	$\kappa_{\max}(p+1)-1$	5	7	8	9	11	11	13
$N_{\kappa_{\max}p'}$	$(p+1)(p+2)\kappa_{\max}(\kappa_{\max}+1)/4$	18	30	36	45	63	60	84
$N_{1p'}$	$(p'+1)(p'+2)/2$	21	36	45	55	78	78	105
K	$p(p+1)\kappa_{\max}(\kappa_{\max}-1)/4$	3	6	9	10	15	9	21
$T_1(1)$	$3p$	3	3	6	3	3	6	3
$T_1(2)$	$9p$	9	9	18	9	9	18	9
N(AP1)	$N_{\kappa_{\max}p'}$	18	30	36	45	63	60	84
N(AP2)	$T_1(\kappa')$	3	3	6	9	9	6	18
N(AP3)	$K - T_1(\kappa')$	0	3	3	1	6	12	3

Restriction of derivative order κ' : $3p\kappa'(\kappa'+1)/2 \le K$

$$\varphi_r(z') = \left(\prod_{i=1}^{d}\prod_{n'_i=0}^{n_i-1}\frac{z'_i - n'_i/p}{n_i/p - n'_i/p}\right)\left(\prod_{n'_0=0}^{n_0-1}\frac{1 - z'_1 - \ldots - z'_d - n'_0/p}{n_0/p - n'_0/p}\right). \quad (12)$$

Setting the numerators in Eq. (12) equal to zero yields the families of equations for the straight lines, directed "horizontally", "vertically", and "diagonally" in the local coordinate system of the element Δ, which is related by the affine transformation with the "oblique" family of straight lines of the element Δ_q. In Fig. 1, an example is presented that illustrates the construction of the LIP at $d = 2$, $r, r' = 1, \ldots, (p+1)(p+2)/2$, $p = 5$ on the element Δ in the form of a rectangular triangle with the vertices $\hat{z}'_0 = (\hat{z}'_{01}, \hat{z}'_{02}) = (0,0)$, $\hat{z}'_1 = (\hat{z}'_{11}, \hat{z}'_{12}) = (1,0)$, $\hat{z}'_2 = (\hat{z}'_{21}, \hat{z}'_{22}) = (0,1)$.

The piecewise polynomial functions $P_{\bar{l}}(z)$ forming the finite-element basis $\{P_{\bar{l}}(z)\}_{\bar{l}=1}^{P}$, which are constructed by joining the LIP $\varphi_r(z)$ of Eq. (7), obtained from Eq. (12) by means of the transformation (9), on the finite elements Δ_q:

$$P_l(z) = \{\varphi_l(z),\ A_l \in \Delta_q;\ 0, A_l \notin \Delta_q\},$$

are continuous, but their derivatives are discontinuous at the boundaries of the elements Δ_q.

3.2 Algorithm for Calculating the Basis of Hermite Interpolating Polynomials

Let us construct the HIP of the order p' by joining of which the piecewise polynomial functions (27) with the continuous derivatives up to the given order κ' can be obtained.

Step 1. Auxiliary Polynomials (AP1). To construct HIP in the local coordinates z', let us introduce the set of auxiliary polynomials (AP1)

$$\varphi_r^{\kappa_1\ldots\kappa_d}(\xi'_r) = \delta_{rr'}\delta_{\kappa_1 0}\ldots\delta_{\kappa_d 0}, \quad \left.\frac{\partial^{\mu_1\ldots\mu_d}\varphi_r^{\kappa_1\ldots\kappa_d}(z')}{\partial z_1'^{\mu_1}\ldots\partial z_d'^{\mu_d}}\right|_{z'=\xi'_{r'}} = \delta_{rr'}\delta_{\kappa_1\mu_1}\ldots\delta_{\kappa_d\mu_d}, \quad (13)$$
$$0 \le \kappa_1+\kappa_2+\ldots+\kappa_d \le \kappa_{\max}-1, \quad 0 \le \mu_1+\mu_2+\ldots+\mu_d \le \kappa_{\max}-1.$$

Here at the node points ξ'_r, defined according to (6), in contrast to LIP, the values of not only the functions themselves, but of their derivatives to the order $\kappa_{\max}-1$ are specified. AP1 are given by the expressions

$$\varphi_r^{\kappa_1\kappa_2\ldots\kappa_d}(z') = w_r(z')\sum_{\mu\in\Delta_\kappa} a_r^{\kappa_1\ldots\kappa_d,\mu_1\ldots\mu_d}(z'_1-\xi'_{r1})^{\mu_1}\times\ldots\times(z'_d-\xi'_{rd})^{\mu_d}, \quad (14)$$
$$w_r(z') = \left(\prod_{i=1}^{d}\prod_{n'_i=0}^{n_i-1}\frac{(z'_i-n'_i/p)^{\kappa^{\max}}}{(n_i/p-n'_i/p)^{\kappa^{\max}}}\right)\left(\prod_{n'_0=0}^{n_0-1}\frac{(1-z'_1-\ldots-z'_d-n'_0/p)^{\kappa^{\max}}}{(n_0/p-n'_0/p)^{\kappa^{\max}}}\right),$$
$$w_r(\xi'_r) = 1,$$

where the coefficients $a_r^{\kappa_1\ldots\kappa_d,\mu_1\ldots\mu_d}$ are calculated from recurrence relations obtained by substitution of Eq. (14) into conditions (13),

$$a_r^{\kappa_1\ldots\kappa_d,\mu_1\ldots\mu_d} = \begin{cases} 0, & \mu_1+\ldots+\mu_d \le \kappa_1+\ldots+\kappa_d, (\mu_1,\ldots,\mu_d)\ne(\kappa_1,\ldots,\kappa_d),\\ \prod_{i=1}^{d}\frac{1}{\mu_i!}, & (\mu_1,\ldots,\mu_d)=(\kappa_1,\ldots,\kappa_d);\\ -\sum_{\nu\in\Delta_\nu}\left(\prod_{i=1}^{d}\frac{1}{(\mu_i-\nu_i)!}\right) g_r^{\mu_1-\nu_1,\ldots,\mu_d-\nu_d}(\xi'_r)a_r^{\kappa_1\ldots\kappa_d,\nu_1\ldots\nu_d}, & \\ & \mu_1+\ldots+\mu_d > \kappa_1+\ldots+\kappa_d; \end{cases} \quad (15)$$
$$g^{\kappa_1\kappa_2\ldots\kappa_d}(z') = \frac{1}{w_r(z')}\frac{\partial^{\kappa_1\kappa_2\ldots\kappa_d}w_r(z')}{\partial z_1'^{\kappa_1}\partial z_2'^{\kappa_2}\ldots\partial z_d'^{\kappa_d}}.$$

For $d>1$ and $\kappa_{\max}>1$, the number $N_{\kappa_{\max}p'}$ of HIP of the order p' and the multiplicity of nodes $\kappa_{\max}$ are smaller than the number $N_{1p'}$ of the polynomials that form the basis in the space of polynomials of the order p' (e.g., the LIP from (12)), i.e., the polynomials satisfying Eq. (13) are determined not uniquely.

Table 2. The HIP $p=1$, $\kappa_{\max}=3$, $\kappa'=1$, $p'=5$ (the Argyris element [5,6,14])

AP1 : $\xi_1=(0,1),\ \xi_2=(1,0),\xi_3=(0,0)$		
$\varphi_1^{0,0}=z_2^3(6z_2^2-15z_2+10)$	$\varphi_2^{0,0}=z_1^3(6z_1^2-15z_1+10)$	$\varphi_3^{0,0}=z_0^3(6z_0^2-15z_0+10)$
$\varphi_1^{0,1}=-z_2^3(z_2-1)(3z_2-4)$	$\varphi_2^{0,1}=-z_1^3z_2(3z_1-4)$	$\varphi_3^{0,1}=-z_0^3z_2(3z_0-4)$
$\varphi_1^{1,0}=-z_1z_2^3(3z_2-4)$	$\varphi_2^{1,0}=-z_1^3(z_1-1)(3z_1-4)$	$\varphi_3^{1,0}=-z_0^3z_1(3z_0-4)$
$\varphi_1^{0,2}=z_2^3(z_2-1)^2/2$	$\varphi_2^{0,2}=z_1^3z_2^2/2$	$\varphi_3^{0,2}=z_0^3z_2^2/2$
$\varphi_1^{1,1}=z_1z_2^3(z_2-1)$	$\varphi_2^{1,1}=(z_1-1)z_1^3z_2$	$\varphi_3^{1,1}=z_0^3z_1z_2$
$\varphi_1^{2,0}=z_1^2z_2^3/2$	$\varphi_2^{2,0}=z_1^3(z_1-1)^2/2$	$\varphi_3^{2,0}=z_0^3z_1^2/2$
AP2 : $\eta_1=(0,1/2),\eta_2=(1/2,0),\eta_3=(1/2,1/2)$		
$Q_1=16z_0^2z_1z_2^2/f_{11}$	$Q_2=16z_0^2z_1^2z_2/f_{22}$	$Q_3=-8z_0z_1^2z_2^2/f_{01}$

Step 2. Auxiliary Polynomials (AP2 and AP3). For unique determination of the polynomial basis let us introduce $K = N_{1p'} - N_{\kappa_{\max}p'}$ auxiliary polynomials $Q_s(z)$ of two types: AP2 and AP3, linearly independent of AP1 from (14) and satisfying the following conditions at the node points $\xi'_{r'}$ of AP1:

$$Q_s(\xi'_{r'}) = 0, \quad \left.\frac{\partial^{\kappa'_1 \kappa'_2 \ldots \kappa'_d} Q_s(z')}{\partial {z'_1}^{\mu_1} \partial {z'_2}^{\mu_2} \ldots \partial {z'_d}^{\mu_d}}\right|_{z'=\xi'_{r'}} = 0, \quad s = 1, \ldots, K, \tag{16}$$

$$0 \le \kappa_1 + \kappa_2 + \ldots + \kappa_d \le \kappa_{\max} - 1, \quad 0 \le \mu_1 + \mu_2 + \ldots + \mu_d \le \kappa_{\max} - 1.$$

Note that to provide the continuity of derivatives the part of polynomials referred to as AP2 must satisfy the condition

$$\left.\frac{\partial^k Q_s(z')}{\partial n^k_{i(s)}}\right|_{z'=\eta'_{s'}} = \delta_{ss'}, \quad s, s' = 1, \ldots, T_1(\kappa'), \quad k = k(s'), \tag{17}$$

where $\eta'_{s'} = (\eta'_{s'1}, \ldots, \eta'_{s'd})$ are the chosen points lying on the faces of various dimensionalities (from 1 to $d-1$) of the d-dimensional simplex Δ and not coincident with the nodal points of HIP ξ'_r, where (13) is valid, $\partial/\partial n_{i(s)}$ is the directional derivative along the vector n_i, normal to the corresponding ith face of the d-dimensional simplex Δ_q at the point $\eta_{s'}$ in the physical coordinate system, which is recalculated to the point $\eta'_{s'}$ of the face of the simplex Δ in the local coordinate system using relations (8)–(10), e.g., for $d = 2$ see Eq. (25). Calculating the number $T_1(\kappa)$ of independent parameters required to provide the continuity of derivatives to the order κ, we determine its maximal value κ' that can be obtained for the schemes with given p and $\kappa_{\max}$ and, correspondingly, the additional conditions (17).

$T_2 = K - T_1(\kappa')$ parameters remain independent and, correspondingly, T_2 additional conditions are added, necessary for the unique determination of the polynomials referred to as AP3,

$$Q_s(\zeta'_{s'}) = \delta_{ss'}, \quad s, s' = T_1(\kappa') + 1, \ldots, K, \tag{18}$$

where $\zeta'_{s'} = (\zeta'_{s'1}, \ldots, \zeta'_{s'd}) \in \Delta$ are the chosen points belonging to the simplex without the boundary, but not coincident with the node points of AP1 ξ'_r.

The auxiliary polynomials AP2 are given by the expression

$$Q_s(z') = \left(\prod_{t=0}^{d} {z'_t}^{k_t}\right) \sum_{j_1,\ldots,j_d} b_{j_1,\ldots,j_d;s} {z'_1}^{j_1} \ldots {z'_d}^{j_d}, \quad z'_0 = 1 - z'_1 - \ldots - z'_d, \tag{19}$$

where $k_t = 1$, if the point η_s, in which the additional conditions (17) are specified, lies on the corresponding face of the simplex Δ, i.e., $H(t, \eta_s) = 0$, and $k_t = \kappa'$, if $H(t, \eta_s) \neq 0$. The auxiliary polynomials AP3 are given by the expression (19) at $k_t = \kappa'$. The coefficients $b_{j_1,\ldots,j_d;s}$ are determined from the uniquely solvable system of linear equations, obtained as a result of the substitution of the expression (19) into conditions (16)–(18).

Table 3. The HIP $p=1$, $\kappa_{\max}=4$, $\kappa'=1$, $p'=7$

AP1 : $\xi_1=(0,1)$, $\xi_2=(1,0)$, $\xi_3=(0,0)$		
$\varphi_1^{0,0}=-z_2^4P_0(z_3)$	$\varphi_2^{0,0}=-z_1^4P_0(z_1)$	$\varphi_3^{0,0}=-z_0^4P_0(z_0)$
$\varphi_1^{0,1}=z_2^4(z_2-1)P_1(z_2)$	$\varphi_2^{0,1}=z_1^4z_2P_1(z_1)$	$\varphi_3^{0,1}=z_0^4z_2P_1(z_0)$
$\varphi_1^{1,0}=z_1z_2^4P_1(z_2)$	$\varphi_2^{1,0}=z_1^4(z_1-1)P_1(z_1)$	$\varphi_3^{1,0}=z_0^4z_1P_1(z_0)$
$\varphi_1^{0,2}=-z_2^4(z_2-1)^2(4z_2-5)/2$	$\varphi_2^{0,2}=-(1/2)z_1^4z_2^2(4z_1-5)$	$\varphi_3^{0,2}=-z_0^4z_2^2(4z_0-5)/2$
$\varphi_1^{1,1}=-z_1z_2^4(z_2-1)(4z_2-5)$	$\varphi_2^{1,1}=-z_1^4z_2(z_1-1)(4z_1-5)$	$\varphi_3^{1,1}=-z_0^4z_1z_2(4z_0-5)$
$\varphi_1^{2,0}=-z_1^2z_2^4(4z_2-5)/2$	$\varphi_2^{2,0}=-z_1^4(z_1-1)^2(4z_1-5)/2$	$\varphi_3^{2,0}=-z_0^4z_1^2(4z_0-5)/2$
$\varphi_1^{0,3}=z_2^4(z_2-1)^3/6$	$\varphi_2^{0,3}=z_1^4z_2^3/6$	$\varphi_3^{0,3}=z_0^4z_2^3/6$
$\varphi_1^{1,2}=z_1z_2^4(z_2-1)^2/2$	$\varphi_2^{1,2}=z_1^4z_2^2(z_1-1)/2$	$\varphi_3^{1,2}=z_0^4z_1z_2^2/2$
$\varphi_1^{2,1}=z_1^2z_2^4(z_2-1)/2$	$\varphi_2^{2,1}=z_1^4z_2(z_1-1)^2/2$	$\varphi_3^{2,1}=z_0^4z_1^2z_2/2$
$\varphi_1^{3,0}=z_1^3z_2^4/6$	$\varphi_2^{3,0}=z_1^4(z_1-1)^3/6$	$\varphi_3^{3,0}=z_0^4z_1^3/6$
AP2 : $\eta_1=(0,1/2)$, $\eta_2=(1/2,0)$, $\eta_3=(1/2,1/2)$		
$Q_1=8z_1z_2^2z_0^2(12z_1^2-7z_1-8z_1z_2-8z_2^2+8z_2)/f_{11}$		
$Q_2=-8z_1^2z_2z_0^2(8z_1^2+8z_1z_2-8z_1+7z_2-12z_2^2)/f_{22}$		
$Q_3=4z_1^2z_2^2z_0(12z_2^2-17z_2+5-17z_1+32z_1z_2+12z_1^2)/f_{01}$		
AP3 : $\zeta_4=(1/4,1/2)$, $\zeta_5=(1/2,1/4)$, $\zeta_6=(1/4,1/4)$		
$Q_4=1024z_0^2z_1^2z_2^2(4z_2-1)$	$Q_5=1024z_0^2z_1^2z_2^2(4z_1-1)$	$Q_6=1024z_0^2z_1^2z_2^2(4z_0-1)$
$P_0(z_j)=(20z_j^3-70z_j^2+84z_j-35)$, $P_1(z_j)=(10z_j^2-24z_j+15)$		

Step 3. As a result, we get the required set of basis HIP

$$\varphi_r^{\kappa}(z')=\{\breve{\varphi}_r^{\kappa}(z'),Q_s(z')\},\quad \kappa=\kappa_1,\ldots,\kappa_d, \tag{20}$$

composed of the polynomials $Q_s(z')$ of the type AP2 and AP3, and the polynomials $\breve{\varphi}_r^{\kappa}(z')$ of the type AP1 that satisfy the conditions

$$\breve{\varphi}_r^{\kappa_1\ldots\kappa_d}(\xi_r')=\delta_{rr'}\delta_{\kappa_10}\ldots\delta_{\kappa_d0},\quad \left.\frac{\partial^{\mu_1\ldots\mu_d}\breve{\varphi}_r^{\kappa_1\ldots\kappa_d}(z')}{\partial {z'_1}^{\mu_1}\ldots\partial {z'_d}^{\mu_d}}\right|_{z'=\xi'_{r'}}=\delta_{rr'}\delta_{\kappa_1\mu_1}\ldots\delta_{\kappa_d\mu_d}, \tag{21}$$

$$0\le\kappa_1+\kappa_2+\ldots+\kappa_d\le\kappa_{\max}-1,\quad 0\le\mu_1+\mu_2+\ldots+\mu_d\le\kappa_{\max}-1;$$

$$\left.\frac{\partial^k\breve{\varphi}_r^{\kappa_1\ldots\kappa_d}(z')}{\partial n_{i(s)}^k}\right|_{z'=\eta'_{s'}}=0,\quad s'=1,\ldots,T_1(\kappa'),\quad k=k(s'), \tag{22}$$

$$\breve{\varphi}_r^{\kappa_1\ldots\kappa_d}(\zeta'_{s'})=0,\quad s'=T_1(\kappa')+1,\ldots,K, \tag{23}$$

and are calculated using the formulas

$$\breve{\varphi}_r^{\kappa}(z')=\varphi_r^{\kappa}(z')-\sum_{s=1}^{K}c_{\kappa;r;s}Q_s(z'),\ c_{\kappa;r;s}=\begin{cases}\left.\dfrac{\partial^k\varphi_r^{\kappa}(z')}{\partial n_{i(s)}^k}\right|_{z'=\eta'_s}, & Q_s(z')\in\mathrm{AP2},\\ \varphi_r^{\kappa}(\zeta_s), & Q_s(z')\in\mathrm{AP3}.\end{cases} \tag{24}$$

Step 4. The AP1 $\breve{\varphi}_r^{\kappa}(z')$ from (20), where κ denotes the directional derivatives along the local coordinate axes, are recalculated using formulas (10) into $\breve{\varphi}_r^{\kappa}(z')$, specified in the local coordinates, but now κ denotes already the directional derivatives along the physical coordinate axes.

Table 4. The HIP $p = 2$, $\kappa_{\max} = 3$, $\kappa' = 1$, $p' = 8$

AP1 : $\xi_1 = (0,1), \xi_2 = (1/2,1/2), \xi_3 = (1,0), \xi_4 = (0,1/2), \xi_5 = (1/2,0), \xi_6 = (0,0)$		
$\varphi_1^{0,0} = z_2^3(2z_2-1)^3 S_0(z_2)$	$\varphi_3^{0,0} = z_1^3(2z_1-1)^3 S_0(z_1)$	$\varphi_6^{0,0} = z_0^3(2z_0-1)^3 S_0(z_0)$
$\varphi_1^{0,1} = -z_2^3(z_2-1)S_1(z_2)$	$\varphi_3^{0,1} = -z_1^3 z_2 S_1(z_1)$	$\varphi_6^{0,1} = -z_0^3 z_2 S_1(z_0)$
$\varphi_1^{1,0} = -z_1 z_2^3 S_1(z_2)$	$\varphi_3^{1,0} = -z_1^3(z_1-1)S_1(z_1)$	$\varphi_6^{1,0} = -z_0^3 z_1 S_1(z_0)$
$\varphi_1^{0,2} = z_2^3(z_2-1)^2(2z_2-1)^3/2$	$\varphi_3^{0,2} = z_1^3(2z_1-1)^3 z_2^2/2$	$\varphi_6^{0,2} = z_0^3 z_2^2(2z_0-1)^3/2$
$\varphi_1^{1,1} = z_2^3(2z_2-1)^3 z_1(z_2-1)$	$\varphi_3^{1,1} = z_1^3 z_2(z_1-1)(2z_1-1)^3$	$\varphi_6^{1,1} = z_0^3 z_1 z_2(2z_0-1)^3$
$\varphi_1^{2,0} = z_2^3(2z_2-1)^3 z_1^2/2$	$\varphi_3^{2,0} = z_1^3(z_1-1)^2(2z_1-1)^3/2$	$\varphi_6^{2,0} = z_0^3 z_1^2(2z_0-1)^3/2$
$\varphi_2^{0,0} = 64z_1^3 z_2^3 S_2(z_0)$	$\varphi_4^{0,0} = 64z_0^3 z_2^3 S_2(z_1)$	$\varphi_5^{0,0} = 64z_0^3 z_1^3 S_2(z_2)$
$\varphi_2^{0,1} = 32z_1^3 z_2^3 S_3(z_2,z_0)$	$\varphi_4^{0,1} = 32z_0^3 z_2^3 S_3(z_2,z_1)$	$\varphi_5^{0,1} = 64z_0^3 z_1^3 z_2(6z_2+1)$
$\varphi_2^{1,0} = 32z_1^3 z_2^3 S_3(z_1,z_0)$	$\varphi_4^{1,0} = 64z_0^3 z_1 z_2^3(6z_1+1)$	$\varphi_5^{1,0} = 32z_0^3 z_1^3 S_3(z_1,z_2)$
$\varphi_2^{0,2} = 8z_1^3 z_2^3(2z_2-1)^2$	$\varphi_4^{0,2} = 8z_0^3 z_2^3(2z_2-1)^2$	$\varphi_5^{0,2} = 32z_0^3 z_1^3 z_2^2$
$\varphi_2^{1,1} = 16z_1^3 z_2^3(2z_1-1)(2z_2-1)$	$\varphi_4^{1,1} = 32z_0^3 z_1 z_2^3(2z_2-1)$	$\varphi_5^{1,1} = 32z_0^3 z_1^3 z_2(2z_1-1)$
$\varphi_2^{2,0} = 8z_1^3 z_2^3(2z_1-1)^2$	$\varphi_4^{2,0} = 32z_0^3 z_1^2 z_2^3$	$\varphi_5^{2,0} = 8z_0^3 z_1^3(2z_1-1)^2$
AP2 : $\eta_1 = (0,1/4), \eta_2 = (0,3/4), \eta_3 = (1/4,0), \eta_4 = (3/4,0), \eta_5 = (1/4,3/4), \eta_6 = (3/4,1/4)$		
$Q_1 = (512/9)z_0^2 z_1 z_2^2(2z_0-1)(2z_2-1)(4z_0-1)/f_{11}$		
$Q_2 = -(512/9)z_0^2 z_1 z_2^2(2z_0-1)(2z_2-1)(4z_2-1)/f_{11}$		
$Q_3 = -(512/9)z_0^2 z_1^2 z_2(2z_0-1)(2z_1-1)(4z_0-1)/f_{22}$		
$Q_4 = -(512/9)z_0^2 z_1^2 z_2(2z_0-1)(2z_1-1)(4z_1-1)/f_{22}$		
$Q_5 = (256/9)z_0 z_1^2 z_2^2(2z_1-1)(2z_2-1)(4z_2-1)/f_{01}$		
$Q_6 = (256/9)z_0 z_1^2 z_2^2(2z_1-1)(2z_2-1)(4z_1-1)/f_{01}$		
AP3 : $\zeta_7 = (1/4,1/2), \zeta_8 = (1/2,1/4), \zeta_9 = (1/4,1/4)$		
$Q_7 = 4096z_0^2 z_1^2 z_2^2(2z_0-1)(2z_1-1)$		
$Q_8 = 4096z_0^2 z_1^2 z_2^2(2z_0-1)(2z_2-1)$		
$Q_9 = 4096z_0^2 z_1^2 z_2^2(2z_1-1)(2z_2-1)$		
$S_0(z_j) = (48z_2^2 - 105z_2 + 58)$, $S_1(z_j) = (2z_j-1)^3(9z_j-10)$,		
$S_2(z_j) = (24z_j^2 - 12z_0 z_1 z_2/z_j + 4)$, $S_3(z_i,z_j) = (2z_i-1)(6z_j+1)$		

Step 5. The final transition to the physical coordinates is implemented by means of transformation (9).

3.3 Example: HIP for $d = 2$

For $d = 2$, the order p' of the polynomial with respect to the tangential variable t at the boundary of the triangle $\frac{\partial^{\kappa'+1}}{\partial n^{\kappa'}\partial t}, \ldots, \frac{\partial^{\kappa_{\max}}}{\partial n^{\kappa'}\partial t^{\kappa_{\max}-\kappa'-1}}$. Thus, since the triangle has three sides, the unique determination of the derivatives to the order of κ' at the boundary requires $T_1(\kappa') = 3p + \ldots + 3\kappa' p = 3p\kappa'(\kappa'+1)/2$ parameters and, correspondingly, the additional conditions (17).

For example, if $p = 1$ and $\kappa_{\max} = 4$, then there are $K = 6$ additional conditions for the determination of AP2 and AP3. The order $p' = 7$ of the polynomial in the tangential variable t at the boundary of the triangle coincides with the order of the polynomial of two variables, and its unique determination requires $p' + 1 = 8$ parameters. The first-order derivative $\kappa' = 1$ in the variable

Table 5. The HIP $p = 1$, $\kappa_{\max} = 5$, $\kappa' = 2$, $p' = 9$

AP1 : $\xi_1 = (0,1),\ \xi_2 = (1,0), \xi_3 = (0,0)$		
$\varphi_1^{0,0} = z_2^5 T_0(z_2)$	$\varphi_2^{0,0} = z_1^5 T_0(z_1)$	$\varphi_3^{0,0} = z_0^5 T_0(z_0)$
$\varphi_1^{0,1} = -z_2^5(z_2-1)T_1(z_2)$	$\varphi_2^{0,1} = -z_1^5 z_2 T_1(z_1)$	$\varphi_3^{0,1} = -z_0^5 z_2 T_1(z_0)$
$\varphi_1^{1,0} = -z_1 z_2^5 T_1(z_2)$	$\varphi_2^{1,0} = -z_1^5(z_1-1)T_1(z_1)$	$\varphi_3^{1,0} = -z_0^5 z_1 T_1(z_0)$
$\varphi_1^{0,2} = z_2^5(z_2-1)^2 T_2(z_2)/2$	$\varphi_2^{0,2} = z_1^5 z_2^2 T_2(z_1)/2$	$\varphi_3^{0,2} = z_0^5 z_2^2 T_2(z_0)/2$
$\varphi_1^{1,1} = z_1 z_2^5(z_2-1)T_2(z_2)$	$\varphi_2^{1,1} = z_1^5 z_2(z_1-1)T_2(z_1)$	$\varphi_3^{1,1} = z_0^5 z_1 z_2 T_2(z_0)$
$\varphi_1^{2,0} = z_1^2 z_2^5 T_2(z_2)/2$	$\varphi_2^{2,0} = z_1^5(z_1-1)^2 T_2(z_1)/2$	$\varphi_3^{2,0} = z_0^5 z_1^2 T_2(z_0)/2$
$\varphi_1^{0,3} = -z_2^5(z_2-1)^3(5z_2-6)/6$	$\varphi_2^{0,3} = -z_1^5 z_2^3(5z_1-6)/6$	$\varphi_3^{0,3} = -z_0^5 z_2^3(5z_0-6)/6$
$\varphi_1^{1,2} = -z_1 z_2^5(z_2-1)^2(5z_2-6)/2$	$\varphi_2^{1,2} = -z_1^5 z_2^2(z_1-1)(5z_1-6)/2$	$\varphi_3^{1,2} = -z_0^5 z_1 z_2^2(5z_0-6)/2$
$\varphi_1^{2,1} = -z_1^2 z_2^5(z_2-1)(5z_2-6)/2$	$\varphi_2^{2,1} = -z_1^5 z_2(z_1-1)^2(5z_1-6)/2$	$\varphi_3^{2,1} = -z_0^5 z_1^2 z_2(5z_0-6)/2$
$\varphi_1^{3,0} = -z_1^3 z_2^5(5z_2-6)/6$	$\varphi_2^{3,0} = -z_1^5(z_1-1)^3(5z_1-6)/6$	$\varphi_3^{3,0} = -z_0^5 z_1^3(5z_0-6)/6$
$\varphi_1^{0,4} = z_2^5(z_2-1)^4/24$	$\varphi_2^{0,4} = z_1^5 z_2^4/24$	$\varphi_3^{0,4} = z_0^5 z_2^4/24$
$\varphi_1^{1,3} = z_1 z_2^5(z_2-1)^3/6$	$\varphi_2^{1,3} = z_1^5 z_2^3(z_1-1)/6$	$\varphi_3^{1,3} = z_0^5 z_1 z_2^3/6$
$\varphi_1^{2,2} = z_1^2 z_2^5(z_2-1)^2/4$	$\varphi_2^{2,2} = z_1^5 z_2^2(z_1-1)^2/4$	$\varphi_3^{2,2} = z_0^5 z_1^2 z_2^2/4$
$\varphi_1^{3,1} = z_1^3 z_2^5(z_2-1)/6$	$\varphi_2^{3,1} = z_1^5 z_2(z_1-1)^3/6$	$\varphi_3^{3,1} = z_0^5 z_1^3 z_2/6$
$\varphi_1^{4,0} = z_1^4 z_2^5/24$	$\varphi_2^{4,0} = z_1^5(z_1-1)^4/24$	$\varphi_3^{4,0} = z_0^5 z_1^4/24$
AP2 : $\eta_1 = (0,1/2), \eta_2 = (1/2,0), \eta_3 = (1/2,1/2), \eta_4 = (0,1/3), \eta_5 = (0,2/3)$,		
$\eta_6 = (1/3,0), \eta_7 = (2/3,0), \eta_8 = (1/3,2/3), \eta_9 = (2/3,1/3)$		
$Q_1 = 256 z_0^3 z_1 z_2^3((3z_1z_2 - 5z_1^2 - z_2^2 + z_2)f_{11} - 4z_1(z_2 - z_0)f_{12})/f_{11}^2$		
$Q_2 = 256 z_0^3 z_1^3 z_2((3z_1z_2 - 5z_2^2 - z_1^2 + z_1)f_{22} + 4z_2(z_1 - z_0)f_{21})/f_{22}^2$		
$Q_3 = 128 z_0 z_1^3 z_2^3((7z_0^2 - 2z_0 - z_1z_2)f_{01} + 2z_0(z_1 - z_2)f_{02})/f_{01}^2$		
$Q_4 = (729/16)z_0^3 z_1^2 z_2^2(3z_0 - 1)/f_{11}^2$		
$Q_5 = (729/16)z_0^3 z_1^2 z_2^3(3z_2 - 1)/f_{11}^2$		
$Q_6 = (729/16)z_0^3 z_1^3 z_2^2(3z_0 - 1)/f_{22}^2$		
$Q_7 = (729/16)z_0^3 z_1^3 z_2^2(3z_1 - 1)/f_{22}^2$		
$Q_8 = (729/64)z_0^2 z_1^3 z_2^3(3z_2 - 1)/f_{01}^2$		
$Q_9 = (729/64)z_0^2 z_1^3 z_2^3(3z_1 - 1)/f_{01}^2$		
AP3 : $\zeta_{10} = (1/3,1/3)$	$Q_{10} = 19683 z_0^3 z_1^3 z_2^3$	
$T_0(z_j) = (70z_j^4 - 315z_j^3 + 540z_j^2 - 420z_j + 126)$		
$T_1(z_j) = (35z_2^3 - 120z_2^2 + 140z_2 - 56), T_2(z_j) = (15z_j^2 - 35z_j + 21)$		

normal to the boundary will be a polynomial of the order $p' - \kappa' = 6$, and its unique determination will require $p' - \kappa' + 1 = 7$ parameters. However, it is determined by only $p' - \kappa'(p+1) = 6$ parameters: the mixed derivatives $\frac{\partial}{\partial n}$, $\frac{\partial^2}{\partial n \partial t}$ and $\frac{\partial^3}{\partial n \partial t^2}$, specified at two vertices. The missing parameter can be determined by specifying the directional derivative along the direction, non-parallel to the triangle boundary, at one of the points on its side (e.g., in the middle of the side). Thus, for $p = 1$ and $\kappa_{\max} = 4$, one can construct HIP with the fixed values of the first derivative on the boundary of the triangle, and $6 - 3 = 3$ parameters remain free.

The second-order derivative $\kappa' = 2$ in the variable normal to the boundary is a polynomial of the order $p' - \kappa' = 5$, and its unique determination requires $p' - \kappa' + 1 = 6$ parameters. However, it is determined by only $p' - \kappa'(p+1) = 4$ parameters: the mixed derivatives $\frac{\partial^2}{\partial n^2}$ and $\frac{\partial^3}{\partial n^2 \partial t}$ specified at two vertices of the triangle. Thus, the unique determination of the second derivative will require 6 parameters. This fact means that using this algorithm for $p = 1$ and $\kappa_{\max} = 4$, it is impossible to construct the FEM scheme with continuous second derivative. In this case, one should use the scheme with $\kappa_{\max} > 4$, e.g., denoted as [152] in Table 1 and Fig. 2. Then the three remaining free parameters are used to construct AP3. Note that it is possible to construct the schemes providing the continuity of the second derivatives at some boundaries of the finite elements. This case is not considered in the present paper.

For $d = 2$, the derivatives $\partial/\partial n_i$ along the direction n_i, perpendicular to the appropriate face $i = 0, 1, 2$ in the physical coordinate system are given in terms of the partial derivatives $\partial/\partial z'_j$, $j = 1, 2$ in the local coordinate system Δ, using (8)–(10), by the expressions

$$\frac{\partial}{\partial n_i} = f_{i1}\frac{\partial}{\partial z'_1} + f_{i2}\frac{\partial}{\partial z'_2}, \quad i = 1, 2, \quad \frac{\partial}{\partial n_0} = (f_{01} + f_{02})\frac{\partial}{\partial z'_1} + (f_{01} - f_{02})\frac{\partial}{\partial z'_2}, \tag{25}$$

where $f_{ij} = f_{ij}(\hat{z}_0, \hat{z}_1, \hat{z}_2)$ are the functions of the coordinates of vertices $\hat{z}_0, \hat{z}_1, \hat{z}_2$ of the triangle Δ_q in the physical coordinate system

$$\begin{aligned}
&f_{11} = J^{-1}R(\hat{z}_2, \hat{z}_0), \quad f_{12} = -\frac{(\hat{z}_{12} - \hat{z}_{02})(\hat{z}_{22} - \hat{z}_{02}) + (\hat{z}_{21} - \hat{z}_{01})(\hat{z}_{11} - \hat{z}_{01})}{JR(\hat{z}_2, \hat{z}_0)},\\
&f_{22} = J^{-1}R(\hat{z}_1, \hat{z}_0), \quad f_{21} = -\frac{(\hat{z}_{12} - \hat{z}_{02})(\hat{z}_{22} - \hat{z}_{02}) + (\hat{z}_{21} - \hat{z}_{01})(\hat{z}_{11} - \hat{z}_{01})}{JR(\hat{z}_1, \hat{z}_0)},\\
&f_{01} = -(2J)^{-1}R(\hat{z}_2, \hat{z}_1), \; f_{02} = \frac{(\hat{z}_{11} - \hat{z}_{01})^2 + (\hat{z}_{12} - \hat{z}_{02})^2 - (\hat{z}_{22} - \hat{z}_{02})^2 - (\hat{z}_{21} - \hat{z}_{01})^2}{2JR(\hat{z}_2, \hat{z}_1)},\\
&J = (\hat{z}_{11} - \hat{z}_{01})(\hat{z}_{22} - \hat{z}_{02}) - (\hat{z}_{12} - \hat{z}_{02})(\hat{z}_{21} - \hat{z}_{01}),\\
&R(\hat{z}_j, \hat{z}_{j'}) = ((\hat{z}_{1j} - \hat{z}_{1j'})^2 + (\hat{z}_{2j} - \hat{z}_{2j'})^2)^{1/2}.
\end{aligned} \tag{26}$$

The implementation of conditions (13), (16), (17), and (18), using which the basis HIP were constructed, is schematically shown for $d = 2$ in Fig. 2. The characteristics of the polynomial basis of HIP on the element Δ at $d = 2$ are presented in Table 1.

Tables 2, 3, 4 and 5 present the results of executing the Algorithm from Sect. 3.2 for the HIP ($p = 1$, $\kappa_{\max} = 3$, $\kappa' = 1$, $p' = 5$), ($p = 1$, $\kappa_{\max} = 4$, $\kappa' = 1$, $p' = 7$), ($p = 2$, $\kappa_{\max} = 3$, $\kappa' = 1$, $p' = 8$) and ($p = 1$, $\kappa_{\max} = 5$, $\kappa' = 2$, $p' = 9$): AP1 $\varphi_r^k(z')$, AP2 and AP3 $Q_s^k(z')$, and the corresponding coefficients $c_{\kappa;r;s}$ are calculated using Eq. (24). The notations are as follows: ξ_r, η_s, ζ_s are the coordinates of the nodes, in which the right-hand side of Eqs. (21), (17) or (18) equals one, $z_0 = 1 - z_1 - z_2$, f_{ij} is found from formulas (26), the arguments of functions and the primes at the notations of independent variables are omitted. The explicit expressions for the HIPs ($p = 1$, $\kappa_{\max} = 6$, $\kappa' = 2$, $p' = 11$), ($p = 2$, $\kappa_{\max} = 4$, $\kappa' = 1$, $p' = 11$), and ($p = 1$, $\kappa_{\max} = 7$, $\kappa' = 3$, $p' = 13$) were calculated too, but are not presented here because of the paper size limitations (one can receive it with request to authors or using program TRIAHP implemented in Maple which will be published in the library JINRLIB). The calculations were carried out using the computer Intel Pentium CPU 987, ×64, 4 GB RAM, the Maple 16. The computing time for the considered examples did not exceed 6 s.

Remark 1. At $\kappa' = 1$ on uniform grids, one can make use of the basis with continuous first derivative consisting of the reduced HIP $\check{\varphi}_r^k(z')$ and $Q_s(z')$ for $f_{01} = f_{11} = f_{22} = 1$. In this case, the derivatives of such polynomials along the direction normal to the boundary generally do not satisfy conditions (17).

Fig. 3. (a) The mesh on the domain $\Omega_h(z) = \bigcup_{q=1}^{Q} \Delta_q$ of the triangle membrane composed of triangle elements Δ_q (b) the profiles of the fourth eigenfunction $\Phi_4^h(z)$ with $E_4^h = 3 + 1.90 \cdot 10^{-17}$ obtained using the LIP of the order $p' = p = 8$

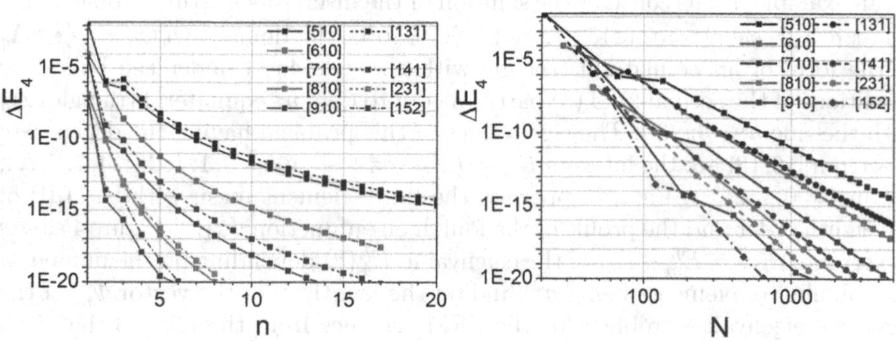

Fig. 4. The error ΔE_4^h of the eigenvalue E_4^h versus the number of elements n and the length of the vector N

3.4 Piecewise Polynomial Functions

The piecewise polynomial functions $P_l(z)$ with continuous derivatives to the order κ' are constructed by joining the polynomials $\varphi_r^\kappa(z) = \{\check{\varphi}_r^\kappa(z), Q_s(z)\}$ from (20), obtained using the Algorithm on the finite elements $\Delta_q \in \Omega_h(z) = \bigcup_{q=1}^{Q} \Delta_q$:

$$P_{l'}(z) = \left\{ \pm\varphi^{\kappa}_{l(l')}(z), A_{l(l')} \in \Delta_q; 0, A_{l(l')} \notin \Delta_q \right\}, \tag{27}$$

where the sign "–" can appear only for AP2, when it is necessary to join the normal derivatives of the odd order.

The expansion of the sought solution $\Phi_m(z)$ in the basis of piecewise polynomial functions $P_{l'}(z)$, $\Phi^h_m(z) = \sum_{l'=1}^{N} P_{l'}(z)\Phi^h_{l'm}$ and its substitution into the variational functional (5) leads to the generalized algebraic eigenvalue problem, $(A - BE^h_m)\Phi^h_m = 0$, solved using the standard method (see, e.g., [3]). The elements of the symmetric matrices of stiffness A and mass B comprise the integrals like Eq. (5), which are calculated on the elements in the domain $\Delta_q \in \Omega_h(z) = \bigcup_{q=1}^{Q} \Delta_q$, recalculated into the local coordinates on the element Δ.

The deviation of the approximate solution $E^h_m, \Phi^h_m(z) \in \mathcal{H}_2^{\kappa'+1\geq 1}(\Omega_h)$ from the exact one $E_m, \Phi_m(z) \in \mathcal{H}^2_2(\Omega)$ is theoretically estimated as [6,20]

$$\left|E_m - E^h_m\right| \leq c_1 h^{2p'}, \quad \left\|\Phi_m(z) - \Phi^h_m(z)\right\|_0 \leq c_2 h^{p'+1}, \tag{28}$$

where $\|\Phi_i(z)\|_0^2 = \int_\Omega g_0(z)dz\overline{\Phi_i}(z)\Phi_i(z)$, h is the maximal size of the finite element Δ_q, p' is the order of the FEM scheme, m is the number of the eigenvalue, c_1 and c_2 are coefficients independent of h.

4 Results and Discussion

As an example, let us consider the solution of the discrete-spectrum problem (1)–(4) at $d = 2$, $g_0(z) = g_{ij}(z) = 1$, and $V(z) = 0$ in the domain $\Omega_h(z) = \cup_{q=1}^{Q}\Delta_q$ in the form of an equilateral triangle with the side $4\pi/3$ under the boundary conditions of the second kind (3) partitioned into $Q = n^2$ equilateral triangles Δ_q with the side $h = 4\pi/3n$. The eigenvalues of this problem having the degenerate spectrum [16,18] are the integers $E_m = m_1^2 + m_2^2 + m_1 m_2 = 0, 1, 1, 3, 4, 4, 7, 7, \ldots$, $m_1, m_2 = 0, 1, 2, \ldots$. Figure 3 presents the finite-element mesh with the LIP of the eighth order and the profile of the fourth eigenfunction $\Phi^h_4(z)$. Figure 4 shows the errors $\Delta E_m = E^h_m - E_m$ of the eigenvalue $E^h_4(z)$ depending on the number n (the number of elements being n^2) and on the length N of the vector Φ^h_m of the algebraic eigenvalue problem for the FEM schemes from the fifth to the ninth order of accuracy: using LIP with the labels $[p\kappa_{\max}\kappa'] = [510], \ldots, [910]$, and using HIP with the labels [131], [141], [231] and [152] from Table 1, conserving the continuity of the first and the second derivative of the approximate solution, respectively.

As seen from Fig. 4, the errors of the eigenvalue $\Delta E^h_4(z)$ of the FEM schemes of the same order are nearly similar and correspond to the theoretical estimates (28), but in the FEM schemes conserving the continuity of the first and the second derivatives of the approximate solution, the matrices of smaller dimension are used that correspond to the length of the vector N smaller by 1.5–2 times than in the schemes with LIP that conserve only the continuity of the functions themselves at the boundaries of the finite elements. The calculations were carried out using the computer 2× Xeon 3.2 GHz, 4 GB RAM, the Intel Fortran 77 with quadruple precision real*16, with 32 significant digits. The computing time for the considered examples did not exceed 3 min.

5 Conclusion

We presented a symbolic-numeric algorithm, implemented in the Maple system for analytical calculation of the basis of Hermite interpolation polynomials of several variables, which is used to construct a FEM computational scheme of high-order accuracy. The scheme is intended for solving the eigenvalue problem for the elliptic partial differential equation in a bounded domain of multidimensional Euclidean space. The procedure provides the continuity not only of the approximate solution itself, but also of its derivatives to a given order. By the example of the exactly solvable problem for the triangle membrane it is shown that the errors for the eigenvalue are nearly the same for the FEM schemes of the same order and correspond to the theoretical estimates. To achieve the given accuracy of the approximate solution the FEM schemes with HIP, providing the continuity of the first and the second derivatives of the approximate solutions the required matrices have smaller dimension, corresponding to the length of the vector N smaller by 1.5–2 times than for the schemes with LIP, providing only the continuity of the approximate solution itself at the boundaries of the finite elements.

The FEM computational schemes are oriented at the calculations of the spectral and optical characteristics of quantum dots and other quantum mechanical systems. The implementation of FEM with HIP in the space with $d \geq 2$ and the domains different from a polyhedral domain will be presented elsewhere.

The work was partially supported by the Russian Foundation for Basic Research (grants Nos. 16-01-00080 and 17-51-44003 Mong_a) and the Bogoliubov-Infeld program. The reported study was partially funded within the Agreement N 02.03.21.0008 dated 24.04.2016 between the MES RF and RUDN University.

References

1. Ames, W.F.: Numerical Methods for Partial Differential Equations. Academic Press, London (1992)
2. Argyris, J.H., Buck, K.E., Scharpf, D.W., Hilber, H.M., Mareczek, G.: Some new elements for the matrix displacement method. In: Proceedings of the Conference on Matrix Methods in Structural Mechanics (2nd), Wright-Patterson Air Force Base, Ohio, 15–17 October 1968
3. Bathe, K.J.: Finite Element Procedures in Engineering Analysis. Prentice Hall, Englewood Cliffs/New York (1982)
4. Bell, K.: A refined triangular plate bending element. Int. J. Numer. Methods Eng. **1**, 101–122 (1969)

5. Brenner, S.C., Scott, L.R.: The Mathematical Theory of Finite Element Methods. Texts in Applied Mathematics, vol. 15, 3rd edn. Springer, New York (2008). doi:10.1007/978-0-387-75934-0
6. Ciarlet, P.: The Finite Element Method for Elliptic Problems. North-Holland Publishing Company, Amsterdam (1978)
7. Dhatt, G., Touzot, G., Lefrançois, E.: Finite Element Method. Wiley, Hoboken (2012)
8. Gasca, M., Sauer, T.: On the history of multivariate polynomial interpolation. J. Comp. Appl. Math. **122**, 23–35 (2000)
9. Gusev, A.A., Chuluunbaatar, O., Vinitsky, S.I., Derbov, V.L., Góźdź, A., Le Hai, L., Rostovtsev, V.A.: Symbolic-numerical solution of boundary-value problems with self-adjoint second-order differential equation using the finite element method with interpolation hermite polynomials. In: Gerdt, V.P., Koepf, W., Seiler, W.M., Vorozhtsov, E.V. (eds.) CASC 2014. LNCS, vol. 8660, pp. 138–154. Springer, Cham (2014). doi:10.1007/978-3-319-10515-4_11
10. Gusev, A.A., Hai, L.L., Chuluunbaatar, O., Vinitsky, S.I.: KANTBP 4M: Program for Solving Boundary Problems of the System of Ordinary Second Order Differential Equations. http://wwwinfo.jinr.ru/programs/jinrlib/indexe.html
11. Habib, A.W., Goldman, R.N., Lyche, T.: A recursive algorithm for Hermite interpolation over a triangular grid. J. Comput. Appl. Math. **73**, 95–118 (1996)
12. Ladyzhenskaya, O.A.: The Boundary Value Problems of Mathematical Physics. Applied Mathematical Sciences, vol. 49. Springer, New York (1985). doi:10.1007/978-1-4757-4317-3
13. Lekien, F., Marsden, J.: Tricubic interpolation in three dimensions. Int. J. Numer. Meth. Eng. **63**, 455–471 (2005)
14. Logg, A., Mardal, K.-A., Wells, G.N. (eds.): Automated Solution of Differential Equations by the Finite Element Method (The FEniCS Book). Springer, Heidelberg (2012). doi:10.1007/978-3-642-23099-8
15. www.maplesoft.com
16. McCartin, B.J.: Laplacian Eigenstructure of the Equilateral Triangle. Hikari Ltd., Ruse, Bulgaria (2011)
17. Mitchell, A.R., Wait, R.: The Finite Element Method in Partial Differential Equations. Wiley, Chichester (1977)
18. Pockels, F.: Über die Partielle Differential-Gleichung $\Delta u + k^2 u = 0$ und deren Auftreten in der Mathematischen Physik. B.G. Teubner, Leipzig (1891)
19. Ramdas Ram-Mohan, L.: Finite Element and Boundary Element Aplications in Quantum Mechanics. Oxford University Press, New York (2002)
20. Strang, G., Fix, G.J.: An Analysis of the Finite Element Method. Prentice-Hall, Englewood Cliffs/New York (1973)
21. Zienkiewicz, O.C.: Finite elements. The background story. In: Whiteman, J.R. (ed.) The Mathematics of Finite Elements and Applications, p. 1. Academic Press, London (1973)

Symbolic-Numerical Algorithms for Solving the Parametric Self-adjoint 2D Elliptic Boundary-Value Problem Using High-Accuracy Finite Element Method

A.A. Gusev[1], V.P. Gerdt[1,2], O. Chuluunbaatar[1,3], G. Chuluunbaatar[1,2], S.I. Vinitsky[1,2(✉)], V.L. Derbov[4], and A. Góźdź[5]

[1] Joint Institute for Nuclear Research, Dubna, Russia
gooseff@jinr.ru, vinitsky@theor.jinr.ru
[2] RUDN University, 6 Miklukho-Maklaya St., Moscow 117198, Russia
[3] Institute of Mathematics, National University of Mongolia, Ulaanbaatar, Mongolia
[4] N.G. Chernyshevsky Saratov National Research State University, Saratov, Russia
[5] Institute of Physics, University of Maria Curie-Skłodowska, Lublin, Poland

Abstract. We propose new symbolic-numerical algorithms implemented in Maple-Fortran environment for solving the parametric self-adjoint elliptic boundary-value problem (BVP) in a 2D finite domain, using high-accuracy finite element method (FEM) with triangular elements and high-order fully symmetric Gaussian quadratures with positive weights, and no points are outside the triangle (PI type). The algorithms and the programs calculate with the given accuracy the eigenvalues, the surface eigenfunctions and their first derivatives with respect to the parameter of the BVP for parametric self-adjoint elliptic differential equation with the Dirichlet and/or Neumann type boundary conditions on the 2D finite domain, and the potential matrix elements, expressed as integrals of the products of surface eigenfunctions and/or their first derivatives with respect to the parameter. We demonstrated an efficiency of algorithms and program by benchmark calculations of helium atom ground state.

Keywords: Parametric elliptic boundary-value problem · Finite element method · High-order fully symmetric high-order Gaussian quadratures · Kantorovich method · Systems of second-order ordinary differential equations

1 Introduction

The adiabatic representation is widely applied for solving multichannel scattering and bound-state problems for systems of several quantum particles in molecular, atomic and nuclear physics [6,7,11,14].

V.P. Gerdt et al. (Eds.): CASC 2017, LNCS 10490, pp. 151–166, 2017.
DOI: 10.1007/978-3-319-66320-3_12

Such problems are described by elliptic boundary value problems (BVPs) in a multidimensional domain of the configuration space, solved using the Kantorovich method, i.e., the reduction to a system of self-adjoint ordinary differential equations (SODEs) using the basis of surface functions of an auxiliary BVP depending on the independent variable of the SODEs parametrically [10, 16]. The elements of matrices of variable coefficients of these SODEs including the matrix of the first derivatives are determined by the integrals of products of surface eigenfunctions and/or their first derivatives with respect to the parameter [4].

Thus, the key problem of such a method is to develop effective algorithms and programs for calculating with given accuracy the surface eigenfunctions and the corresponding eigenvalues of the auxiliary BVP, together with their derivatives with respect to the parameter, and the corresponding integrals that present the matrix elements of the effective potentials in the SODEs [9].

In this paper, we propose new calculation schemes and symbolic-numerical algorithms implemented in Maple-Fortran environment for the solution of the parametric 2D elliptic boundary-value problem using high-accuracy finite element method (FEM) with triangular elements. For the integration, the new high-order fully symmetric high-order Gaussian quadratures on a triangle are performed. We used the symbolic algorithms to generate Fortran routines that allow the solution of the algebraic eigenvalue problem with high-dimension matrices. The algorithms were implemented in a package of programs that calculate with the given accuracy eigenvalues, eigenfunctions, and their first derivatives with respect to the parameter of the parametric self-adjoint elliptic differential equations with the boundary conditions of the Dirichlet and/or Neumann type in the 2D finite domain and the integrals of products of the surface eigenfunctions and their first derivatives with respect to the parameters that express the matrix elements of the effective potentials in the SODEs. Efficiency of the FEM scheme is demonstrated by benchmark calculations of Helium atom ground state.

The structure of the paper is the following. In Sect. 2, the 2D FEM schemes and algorithms for solving the parametric 2D BVP are presented. In Sect. 3, fully symmetric high-order Gaussian quadratures are constructed. In Sect. 4, the algorithm for calculating the parametric derivatives of eigenfunctions and effective potentials is presented. In Sect. 5, the benchmark calculations of 2D FEM algorithms and programs are analyzed. In the Conclusion we discuss the results and perspectives.

2 FEM Algorithm for Solving the Parametric 2D BVP

Let us consider a BVP for the parametric self-adjoint 2D PDE in the domain Ω_x, $x = (x_1, x_2)$ with the piecewise continuous boundary $S = \partial\Omega_x$,

$$(D(x;z)-\varepsilon_i(z))\,\Phi_i(x;z) = 0, \tag{1}$$

$$D\equiv D(x;z) = -\frac{1}{g_0(x)}\left(\sum_{ij=1}^{2}\frac{\partial}{\partial x_i}g_{ij}(x)\frac{\partial}{\partial x_j}\right) + U(x;z), \tag{2}$$

with the mixed Dirichlet/Neumann boundary conditions

$$(I): \ \Phi(x;z)|_S = 0, \tag{3}$$

$$(II): \ \frac{\partial \Phi(x;z)}{\partial n_D}\Big|_S = 0, \quad \frac{\partial \Phi(x;z)}{\partial n_D} = \sum_{ij=1}^{2} (\hat{n}, \hat{e}_i) g_{ij}(x) \frac{\partial \Phi(x;z)}{\partial x_j}. \tag{4}$$

Here $z \in \Omega_z = [z_{\min}, z_{\max}]$ is a parameter, the functions $g_0(x) > 0$, $g_{ij}(x) > 0$, and $\partial_{x_k} g_{ij}(x)$, $U(x;z)$, $\partial_z U(x;z)$ and $\partial_z \Phi_i(x;z)$ are continuous and bounded for $x \in \Omega_x$; $g_{12}(x) = g_{21}(x)$, $g_{11}(x)g_{22}(x) - g_{12}^2(x) > 0$. Also assume that the BVP (1), (3) has only the discrete spectrum, so that $\varepsilon(z) : \varepsilon_1(z) < \ldots < \varepsilon_{j_{\max}}(z) < \ldots$ is the desired set of real eigenvalues. The eigenfunctions satisfy the orthonormality conditions

$$\langle \Phi_i | \Phi_j \rangle = \int_{\Omega} g_0(x) \Phi_i(x;z) \Phi_j(x;z) dx = \delta_{ij}, \quad dx = dx_1 dx_2. \tag{5}$$

The FEM solution of the boundary-value problems (1), (3) is reduced to the determination of stationary points of the variational functional [1,2]

$$\Xi(\Phi_m, \varepsilon_m(z)) \equiv \int_{\Omega} dx g_0(x) \Phi_m(x;z) \left(D - \varepsilon_m(z)\right) \Phi(x;z) = \Pi(\Phi_m, \varepsilon_m(z)), \tag{6}$$

where $\Pi \equiv \Pi(\Phi_m, \varepsilon_m(z))$, $\Phi_m \equiv \Phi_m(x;z)$ is the symmetric quadratic functional

$$\Pi = \int_{\Omega} dx \left[\sum_{ij=1}^{2} g_{ij}(x) \frac{\partial \Phi_m}{\partial x_i} \frac{\partial \Phi_m}{\partial x_j} + g_0(x) \Phi_m (U(x;z) - \varepsilon_m(z)) \Phi_m \right].$$

The domain $\Omega(x,y) = \bigcup_{q=1}^{Q} \Delta_q$, specified as a polygon in the plane $(x_1, x_2) \in \mathcal{R}^2$, is covered with finite elements, the triangles Δ_q with the vertices (x_{11}, x_{21}), (x_{12}, x_{22}), (x_{13}, x_{23}) (here $x_{ik} \equiv x_{ik;q}$, $i = \overline{1,2}$, $k = \overline{1,3}$, $q = \overline{1,Q}$). On each of the triangles Δ_q (the boundary is considered to belong to the triangle), the shape functions $\varphi_l^p(x_1, x_2)$ are introduced. For this purpose we divide the sides of the triangle into p equal parts and draw three families of parallel straight lines through the partition points. The straight lines of each family are numbered from 0 to p, so that the line passing through the side of the triangle has the number 0, and the line passing through the opposite vertex of the triangle has the number p.

Three straight lines from different families intersect at one point $A_l \in \Delta_q$, which will be numbered by the triplet (n_1, n_2, n_3), $n_i \geq 0$, $n_1 + n_2 + n_3 = p$, where n_1, n_2, and n_3 are the numbers of the straight lines passing parallel to the side of the triangle that does not contain the vertex (x_{11}, x_{21}), (x_{12}, x_{22}), and (x_{13}, x_{23}), respectively. The coordinates of this point $x_l = (x_{1l}, x_{2l})$ are determined by the expression $(x_{1l}, x_{2l}) = (x_{11}, x_{21}) n_1/p + (x_{12}, x_{22}) n_2/p + (x_{13}, x_{23}) n_3/p$.

As shape functions we use the Lagrange triangular polynomials $\varphi_l^p(x)$ of the order p that satisfy the condition $\varphi_l^p(x_{1l'}, x_{2l'}) = \delta_{ll'}$, i.e., equal 1 at one of the points A_l and zero at the other points.

In this method, the piecewise polynomial functions $N_l^p(x)$ in the domain Ω are constructed by joining the shape functions $\varphi_l^p(x)$ in the triangle Δ_q:

$$N_l^p(x) = \{\varphi_l^p(x), A_l \in \Delta_q; 0, A_l \notin \Delta_q\}$$

and possess the following properties: the functions $N_l^p(x)$ are continuous in the domain Ω; the functions $N_l^p(x)$ equal 1 at one of the points A_l and zero at the rest points; $N_l^p(x_{1l'}, x_{2l'}) = \delta_{ll'}$ in the entire domain Ω. Here l takes the values $l = \overline{1, N}$.

The functions $N_l^p(x)$ form a basis in the space of polynomials of the pth order. Now, the function $\Phi(x; z) \in \mathcal{F}_z^h \sim \mathcal{H}^1(\Omega_x)$ is approximated by a finite sum of piecewise basis functions $N_l^p(x)$

$$\Phi^h(x; z) = \sum_{l=1}^{N} \Phi_l^h(z) N_l^p(x). \tag{7}$$

The vector function $\boldsymbol{\Phi}^h = \{\Phi_l^h(z)\}_{l=1}^N$ has a generalized first-order partial derivative and belongs to the Sobolev space $\mathcal{H}^1(\Omega_x)$ [13]. After substituting expansion (7) into the variational functional and minimizing it [1,13], we obtain the generalized eigenvalue problem

$$\mathbf{A}^p \boldsymbol{\Phi}^h = \varepsilon^h \mathbf{B}^p \boldsymbol{\Phi}^h. \tag{8}$$

Here $\mathbf{A}^p$ is the stiffness matrix; $\mathbf{B}^p$ is the positive definite mass matrix; $\boldsymbol{\Phi}^h$ is the vector approximating the solution on the finite-element grid; and $\varepsilon^h \equiv \varepsilon^h(z)$ is the corresponding eigenvalue. The matrices $\mathbf{A}^p$ and $\mathbf{B}^p$ have the form:

$$\mathbf{A}^p = \{a_{ll'}^p\}_{ll'=1}^N, \ \mathbf{B}^p = \{b_{ll'}^p\}_{ll'=1}^N, \tag{9}$$

where the matrix elements $a_{ll'}^p$ and $b_{ll'}^p$ are calculated for triangular elements as

$$a_{ll'}^p = \int_{\Delta_q} g_0(x) \varphi_l^p(x; z) \varphi_{l'}^p(x; z) U(x; z)\, dx + \sum_{i,j=1}^{2} \int_{\Delta_q} g_{ij} \frac{\partial \varphi_l^p(x; z)}{\partial x_i} \frac{\partial \varphi_{l'}^p(x; z)}{\partial x_j}\, dx,$$

$$b_{ll'}^p = \int_{\Delta_q} g_0(x) \varphi_l^p(x; z) \varphi_{l'}^p(x; z) dx.$$

Let us construct the LIP on a triangle Δ_q with the vertices $\hat{x}_i = (x_{i1}, x_{i2}, x_{3d})$. For this purpose we introduce the local coordinate system $x' = (x_1', x_2') \in \mathcal{R}^2$, in which the coordinates of the simplex vertices are the following: $\hat{x}_i' = (x_{ik}' = \delta_{ik}, k = 1, 2)$. The relation between the coordinates and derivatives is given by the formula:

$$x_i = x_{0i} + \sum_{j=1}^{2} \hat{J}_{ij} x_j', \quad x_i' = \sum_{j=1}^{2} (\hat{J}^{-1})_{ij} (x_j - x_{0j}), \quad i = 1, 2, \tag{10}$$

$$\frac{\partial}{\partial x_i'} = \sum_{j=1}^{2} \hat{J}_{ji} \frac{\partial}{\partial x_j}, \quad \frac{\partial}{\partial x_i} = \sum_{j=1}^{2} (\hat{J}^{-1})_{ji} \frac{\partial}{\partial x_j'}, \tag{11}$$

where $\hat{J}_{ij} = \hat{x}_{ji} - \hat{x}_{0i}$. When constructing the LIP in the local coordinates x' one has to recalculate the fixed derivatives at the nodes $\Phi_{r'}$ of the element Δ_q to the nodes $\Phi'_{r'}$ of the element Δ, using the matrices $\hat{J}^{-1}$, given by cumbersome expressions. Therefore, the required recalculation is executed based on the relations (10) and (11) for each finite element Δ_q at the stage of the formation of the LIP basis $\{\varphi^p_r(x')\}_{r=1}^N$ on the finite element Δ, implemented numerically using the analytical formulas

$$\int_{\Delta_q} dx g_0(x)\varphi^p_r\varphi^p_{r'}U(x;z) = J\int_{\Delta} dx' g_0(x')\varphi^p_r(x';z)\varphi^p_{r'}(x';z)U(x';z),$$

$$\int_{\Delta_q} dx g_{s_1s_2}(x)\frac{\partial\varphi^p_r}{\partial x_{s_1}}\frac{\partial\varphi^p_{r'}}{\partial x_{s_2}} = J\sum_{t_1,t_2=1}^{2}\hat{J}^{-1}_{s_1s_2;t_1t_2}\int_{\Delta} dx' g_{s_1s_2}(x')\frac{\partial\varphi^p_r(x';z)}{\partial x'_{t_1}}\frac{\partial\varphi^p_{r'}(x';z)}{\partial x'_{t_2}},$$

where $\varphi^p_r \equiv \varphi_r(x;z)$, $J = \det\hat{J} > 0$ is the determinant of the matrix $\hat{J}$ from Eq. (10), $\hat{J}^{-1}_{s_1s_2;t_1t_2} = (\hat{J}^{-1})_{t_1s_1}(\hat{J}^{-1})_{t_2s_2}$, $dx' = dx'_1dx'_2$. In this case, we have explicit expression for shape functions $\varphi^p_l(z'_1, z'_2)$:

$$\varphi^p_l(x') = \prod_{n'_1=0}^{n_1-1}\frac{1 - x'_1 - x'_2 - n'_1/p}{n_1/p - n'_1/p}\prod_{n'_2=0}^{n_2-1}\frac{x'_1 - n'_2/p}{n_2/p - n'_2/p}\prod_{n'_3=0}^{n_3-1}\frac{x'_2 - n'_3/p}{n_3/p - n'_3/p}. \quad (12)$$

The integrals (10) are evaluated using the $2p$-order 2D Gaussian quadrature.

In order to solve the generalized eigenvalue problem (8), the subspace iteration method [1,13] elaborated by Bathe [1] for the solution of large symmetric banded-matrix eigenvalue problems has been chosen. This method uses the skyline storage mode which stores the components of the matrix column vectors within the banded region of the matrix, and is ideally suited for banded finite-element matrices. The procedure chooses a vector subspace of the full solution space and iterates upon the successive solutions in the subspace (for details, see [1]). The iterations continue until the desired set of solutions in the iteration subspace converges to within the specified tolerance on the Rayleigh quotients for the eigenpairs. If the matrix $\mathbf{A}^p$ in Eq. (8) is not positively defined, the problem (8) is replaced with the following problem:

$$\tilde{\mathbf{A}}^p\,\boldsymbol{\Phi}^h = \tilde{\varepsilon}^h\,\mathbf{B}^p\,\boldsymbol{\Phi}^h, \quad \tilde{\mathbf{A}}^p = \mathbf{A}^p - \alpha\mathbf{B}^p. \quad (13)$$

The number α (the shift of the energy spectrum) is chosen such that the matrix $\tilde{\mathbf{A}}^p$ is positive defined. The eigenvector of problem (13) is the same, and $\varepsilon^h = \tilde{\varepsilon}^h + \alpha$.

3 Fully Symmetric High-Order Gaussian Quadratures

Let consider the two-dimensional integral on triangular domain $\triangle_{xy}$ with vertices (x_1, y_1), (x_2, y_2), (x_3, y_3):

$$I = \frac{1}{S_{\triangle_{xy}}}\int_{\triangle_{xy}} f(x,y)dydx \quad (14)$$

Table 1. The quadrature rule for $p = 15$ with $n_p = 52$, $[n_0, n_1, n_2] = [1, 5, 6]$, N_i^w is the number of different permutations of the areal coordinates (α_i, β_i, γ_i).

N_i^w	w_i	α_i, β_i, γ_i		
1	0.033266408301048	0.333333333333333	0.333333333333333	0.333333333333333
3	0.045542949984995	0.202687173029433	0.398656413485283	0.398656413485283
3	0.018936193317852	0.075705168935176	0.462147415532411	0.462147415532411
3	0.046595625404608	0.555517449279976	0.222241275360011	0.222241275360011
3	0.014390824709404	0.878972401688571	0.060513799155714	0.060513799155714
3	0.000733389561154	0.822518347845233	0.088740826077383	0.088740826077383
6	0.011157489727398	0.016416695030487	0.426971506367034	0.556611798602478
6	0.031443815585368	0.096704376730713	0.328778565825110	0.574517057444176
6	0.014551780499648	0.019017867773827	0.282103601487049	0.698878530739123
6	0.010312560870261	0.015907369998417	0.141176714757054	0.842915915244527
6	0.027717303713350	0.089942179570517	0.180738614626992	0.729319205802489
6	0.002839823398123	0.004434769410597	0.037262719444011	0.958302511145391

where $S_{\triangle_{xy}}$ is a square of triangular domain $\triangle_{xy}$. Using change of variables

$$x = x_1\gamma + x_2\alpha + x_3\beta, \quad y = y_1\gamma + y_2\alpha + y_3\beta, \quad \gamma = 1 - \alpha - \beta, \tag{15}$$

we obtain

$$I = \frac{|J|}{S_{\triangle_{xy}}} \int_{\triangle_{\alpha\beta}} f(\alpha, \beta) d\beta d\alpha = 2 \int_0^1 \int_0^{1-\alpha} f(\alpha, \beta) d\beta d\alpha, \tag{16}$$

where J is a Jacobian and $|J| = 2S_{\triangle_{xy}}$, and domain $\triangle_{\alpha\beta}$ is an isosceles right triangle with vertices (0,0), (0,1), (1,0). The pth ordered fully symmetrical Gaussian quadrature rules for this integral may be written as

$$I \approx \sum_{i=1}^{n_p} w_i f(\alpha_i, \beta_i). \tag{17}$$

We consider fully symmetric rules, where if a point with areal coordinates (α, β, γ) is used in the quadrature, then all points resulting from the N_i^w permutation of the areal coordinates are also used, with the same weight w_i. Integration points in a fully symmetric rule can thus belong to one of three different types of point sets, or orbits, depending on the number of areal coordinates which are equal. The number of points for such a rule is $n_p = n_0 + 3n_1 + 6n_2$. Here n_0 is the number of points which three areal coordinates are equal, i.e., $n_0 = 0$ or 1. n_1 is the number of points which two areal coordinates are equal, i.e., we get three points which lie on the medians of the triangle. n_2 is the number of points which three areal coordinates are different, i.e., we get six points.

In paper [5], the weights and coordinates of the fully symmetric rules were presented up to order $p = 20$ with minimal number of points using the moment equations. Calculation was performed with double precision accuracy. However,

Table 2. The quadrature rule for $p = 16$, $n_p = 58$, type $[n_0, n_1, n_2] = [1, 7, 6]$

N_i^w	w_i	α_i, β_i, γ_i		
1	0.0415207350648329	0.3333333333333333	0.3333333333333333	0.3333333333333333
3	0.0101046137864021	0.0121739816884923	0.4939130091557539	0.4939130091557539
3	0.0363778998629740	0.1778835483267153	0.4110582258366423	0.4110582258366423
3	0.0253955775082257	0.0671491113178838	0.4664254443410581	0.4664254443410581
3	0.0359208834794810	0.5001385533336064	0.2499307233331968	0.2499307233331968
3	0.0267742614985530	0.6719362487011838	0.1640318756494081	0.1640318756494081
3	0.0136749716214666	0.8476751119345034	0.0761624440327483	0.0761624440327483
3	0.0031626040488014	0.9688994524978406	0.0155502737510797	0.0155502737510797
6	0.0266514412829383	0.1235525166817187	0.3005378086834664	0.5759096746348149
6	0.0089313378511684	0.0119532031311031	0.3372065794794446	0.6508402173894523
6	0.0152078872638436	0.0523853085701298	0.3143393035872713	0.6332753878425989
6	0.0183760532268712	0.0658032190776827	0.1786829962718098	0.7555137846505075
6	0.0080645623746130	0.0117710730623248	0.1921850841541305	0.7960438427835448
6	0.0068098562534747	0.0149594704947242	0.0806342445495042	0.9044062849557716

Table 3. The quadrature rule for $p = 18$, $n_p = 76$, type $[n_0, n_1, n_2] = [1, 9, 8]$

N_i^w	w_i	α_i, β_i, γ_i		
1	0.0223535614716711	0.3333333333333333	0.3333333333333333	0.3333333333333333
3	0.0059334988479546	0.0460021789844010	0.4769989105077995	0.4769989105077995
3	0.0165585324593954	0.0730729604309092	0.4634635197845454	0.4634635197845454
3	0.0195910892704527	0.1551748557050338	0.4224125721474831	0.4224125721474831
3	0.0074160344816382	0.1550933080132821	0.4224533459933590	0.4224533459933590
3	0.0174049699198115	0.2365578681901632	0.3817210659049184	0.3817210659049184
3	0.0296996298680842	0.4863851422108091	0.2568074288945954	0.2568074288945954
3	0.0222906281899201	0.6736478731957263	0.1631760634021368	0.1631760634021368
3	0.0134460768460945	0.8559888247875595	0.0720055876062202	0.0720055876062202
3	0.0005486878691143	0.9921639450656871	0.0039180274671564	0.0039180274671564
6	0.0098384904247447	0.0120605799230755	0.4119329978824294	0.5760064221944951
6	0.0262821659985039	0.1420581973687457	0.2846674905460437	0.5732743120852107
6	0.0161450882618767	0.0645759925263757	0.3322842391902052	0.6031397682834191
6	0.0078521623046175	0.0411153725698427	0.2629574865443483	0.6959271408858090
6	0.0066043565050862	0.0091463267009754	0.2594416877532075	0.7314119855458171
6	0.0174843686058097	0.0725930398678583	0.1734580428423163	0.7539489172898254
6	0.0080232785271782	0.0147258776438553	0.1349402463458236	0.8503338760103211
6	0.0042665885840052	0.0124575576578779	0.0477763926862289	0.9397660496558932

some rules have the points outside the triangle and/or negative weights. We need to use Gaussian quadrature rules with positive weights, and no points are outside the triangle (so-called PI type).

The above Gaussian quadrature rules are constructed with Algorithm:

Step 1. Transfer the isosceles right triangular domain $\Delta_{\alpha\beta}$ to the equilateral triangular domain with vertices $(-1,0)$, $(1/2,-\sqrt{3}/2)$, $(1/2,\sqrt{3}/2)$, which centroid of triangle located at the origin of the coordinate system.
Step 2. Write the moment equations in polar coordinates [5].
Step 3. Minimize nonlinear moment equation for solving $n_0+2n_1+3n_2$ unknowns using the Levenberg–Marquardt algorithm.
Step 4. Transformation of the calculated areal coordinates to the isosceles right triangular domain $\Delta_{\alpha\beta}$.

A new high ordered PI type rules that are not listed in the Encyclopedia of Quadrature Formulas [3,12] are presented in Tables 1, 2, 3 and 4 calculated by the above algorithm implemented in Maple. In the considered problems, the maximal number of the nonlinear moment equations equals 44, and the number of unknowns equals 47 at $p=20$. The explicit expressions for Gauss quadratures weights and areal coordinates with 32 significant digits were calculated, but are not presented here because of the paper size limitations. Note, the alternative equilateral triangle quadrature formulas were calculated in [17].

Table 4. The quadrature rule for $p=20$, $n_p=85$, type $[n_0,n_1,n_2]=[1,8,10]$

N_i^w	w_i	α_i, β_i, γ_i		
1	0.0284956488386955	0.3333333333333333	0.3333333333333333	0.3333333333333333
3	0.0142039534279209	0.0474234283023599	0.4762882858488200	0.4762882858488200
3	0.0194408133550425	0.1095872167894353	0.4452063916052824	0.4452063916052824
3	0.0273065929935536	0.4916898571477065	0.2541550714261467	0.2541550714261467
3	0.0190593173083705	0.6282404953903102	0.1858797523048449	0.1858797523048449
3	0.0153240833856847	0.7827490888114787	0.1086254555942607	0.1086254555942607
3	0.0003407707226317	0.8487205009418537	0.0756397495290731	0.0756397495290731
3	0.0046354964939763	0.9218908161548015	0.0390545919225992	0.0390545919225992
3	0.0016717238812827	0.9775115344410667	0.0112442327794667	0.0112442327794667
6	0.0146283618671282	0.2120524546203612	0.3758687560757836	0.4120787893038552
6	0.0172080000328995	0.0546435084561301	0.3335452223628692	0.6118112691810008
6	0.0073409966477119	0.0097859886040601	0.4202306323332298	0.5699833790627102
6	0.0232450825127741	0.1383472868057439	0.3152308903849581	0.5464218228092980
6	0.0070480826238744	0.0106040218922527	0.2811743979692607	0.7082215801384866
6	0.0153834272762777	0.1032538874333241	0.2130007906781420	0.6837453218885339
6	0.0041951209853354	0.0070915889018085	0.1595497908201870	0.8333586202780045
6	0.0114288995104660	0.0449113089652980	0.1997044919178251	0.7553841991168769
6	0.0081140164445318	0.0377602618140266	0.1028511090917952	0.8593886290941782
6	0.0023340281749869	0.0051697211528337	0.0641281242816143	0.9307021545655520

4 The Algorithm for Calculating the Parametric Derivatives of Eigenfunctions and Effective Potentials

Taking a derivative of the boundary-value problem (1)–(5) with respect to the parameter z, we find that $\partial_z\Phi_i(x;z)$ is a solution of the following boundary-value problem with the mixed boundary conditions

$$(D(x;z)-\varepsilon_i(z))\frac{\partial\Phi_i(x;z)}{\partial z}=-\left[\frac{\partial}{\partial z}(U(x;z)-\varepsilon_i(z))\right]\Phi_i(x;z),$$
$$\left.\frac{\partial\Phi(x;z)}{\partial z}\right|_S=0 \text{ or } \left.\frac{\partial^2\Phi(x;z)}{\partial n_D\partial z}\right|_S=0. \tag{18}$$

The parametric BVP (18) has a unique solution, if and only if it satisfies the conditions

$$\frac{\partial\varepsilon_i(z)}{\partial z}=\int_\Omega dx g_0(x)\,(\Phi_i(x;z))\frac{\partial U(x;z)}{\partial z}\Phi_i(x;z), \tag{19}$$
$$\int_\Omega dx g_0(x)\Phi_i(x;z)\frac{\partial\Phi_i(x;z)}{\partial z}=0. \tag{20}$$

Below we present an efficient numerical method that allows the calculation of $\partial_z\Phi_i(x;z)$ with the same accuracy as achieved for the eigenfunctions of the BVP (1)–(5) and the use of it for computing the matrices of the effective potentials defined as

$$H_{ij}(z)=H_{ji}(z)=\int_\Omega dx g_0(x)\frac{\partial\Phi_i(x;z)}{\partial z}\frac{\partial\Phi_j(x;z)}{\partial z}, \tag{21}$$
$$Q_{ij}(z)=-Q_{ji}(z)=-\int_\Omega dx g_0(x)\Phi_i(x;z)\frac{\partial\Phi_j(x;z)}{\partial z}. \tag{22}$$

The boundary-value problem (18)–(20) is reduced to the linear system of inhomogeneous algebraic equations with respect to the unknown $\partial\boldsymbol{\Phi}^h/\partial z$:

$$\mathbf{L}\frac{\partial\boldsymbol{\Phi}^h}{\partial z}\equiv(\mathbf{A}^p-\varepsilon^h\mathbf{B}^p)\frac{\partial\boldsymbol{\Phi}^h}{\partial z}=b,\quad b=-\left(\frac{\partial\mathbf{A}^p}{\partial z}-\frac{\partial\varepsilon^h}{\partial z}\mathbf{B}^p\right)\boldsymbol{\Phi}^h. \tag{23}$$

The normalization condition (5), the condition of orthogonality between the function and its parametric derivative (20), and the additional conditions (19) for the solution of (23) read as

$$\left(\boldsymbol{\Phi}^h\right)^T\mathbf{B}^p\boldsymbol{\Phi}^h=1,\quad \left(\frac{\partial\boldsymbol{\Phi}^h}{\partial z}\right)^T\mathbf{B}^p\boldsymbol{\Phi}^h=0,\quad \frac{\partial\varepsilon^h}{\partial z}=\left(\boldsymbol{\Phi}^h\right)^T\frac{\partial\mathbf{A}^p}{\partial z}\boldsymbol{\Phi}^h. \tag{24}$$

Then the potential matrix elements $H_{ij}^h(z)$ and $Q_{ij}^h(z)$ (21) can be calculated using the formulas

$$H_{ij}^h(z)=\left(\frac{\partial\boldsymbol{\Phi}_i^h}{\partial z}\right)^T\mathbf{B}^p\frac{\partial\boldsymbol{\Phi}_j^h}{\partial z},\quad Q_{ij}^h(z)=-\left(\boldsymbol{\Phi}_i^h\right)^T\mathbf{B}^p\frac{\partial\boldsymbol{\Phi}_j^h}{\partial z}. \tag{25}$$

Since ε^h is an eigenvalue of (8), the matrix $\mathbf{L}$ in Eq. (23) is degenerate. In this case, the algorithm for solving Eq. (23) can be written in three steps as follows:

Step k1. Calculate the solutions $\mathbf{v}$ and $\mathbf{w}$ of the auxiliary inhomogeneous systems of algebraic equations

$$\bar{\mathbf{L}}\mathbf{v} = \bar{\mathbf{b}}, \quad \bar{\mathbf{L}}\mathbf{w} = \mathbf{d}, \tag{26}$$

with the non-degenerate matrix $\bar{\mathbf{L}}$ and the right-hand sides $\bar{\mathbf{b}}$ and $\mathbf{d}$

$$\bar{L}_{ss'} = \begin{cases} L_{ss'}, \ (s-S)(s'-S) \neq 0, \\ \delta_{ss'}, \ (s-S)(s'-S) = 0, \end{cases} \tag{27}$$

$$\bar{b}_s = \begin{cases} b_s, \ s \neq S, \\ 0, \ \ s = S, \end{cases} \quad d_s = \begin{cases} L_{sS}, \ s \neq S, \\ 0, \quad\ s = S, \end{cases} \tag{28}$$

where S is the number of the element of the vector $\mathbf{B}^p\boldsymbol{\Phi}^h$ having the greatest absolute value.

Step k2. Evaluate the coefficient γ

$$\gamma = -\frac{\gamma_1}{(\mathbf{D}_S - \gamma_2)}, \quad \gamma_1 = \mathbf{v}^T\mathbf{B}^p\boldsymbol{\Phi}^h, \quad \gamma_2 = \mathbf{w}^T\mathbf{B}^p\boldsymbol{\Phi}^h, \quad \mathbf{D}_S = (\mathbf{B}^p\boldsymbol{\Phi}^h)_S. \tag{29}$$

Step k3. Evaluate the vector $\partial_z\boldsymbol{\Phi}^h$

$$\frac{\partial \Phi_s^h}{\partial z} = \begin{cases} v_s - \gamma w_s, \ s \neq S, \\ \gamma, \qquad\qquad s = S. \end{cases} \tag{30}$$

From the above consideration, it is evident that the computed derivative has the same accuracy as the calculated eigenfunction.

Let $D(x;z)$ in Eq. (1) be a continuous and bounded positive definite operator on the space $\mathcal{H}^1$ with the energy norm, $\varepsilon_i(z)$, $\Phi_i(x,z) \in \mathcal{H}^2$ being the exact solutions of Eqs. (1)–(5), and $\varepsilon_i^h(z)$, $\Phi_i^h(x;z) \in \mathcal{H}^1$ being the corresponding numerical solutions. Then the following estimates are valid [13]

$$\left|\varepsilon_i(z) - \varepsilon_i^h(z)\right| \leq c_1 h^{2p}, \quad \left\|\Phi_i(x;z) - \Phi_i^h(x;z)\right\|_0 \leq c_2 h^{p+1}, \tag{31}$$

$$\|\Phi_i(x;z)\|_0^2 = \int_{\Omega_x} dx g_0(x)\Phi_i(x;z)\Phi_i(x;z), \tag{32}$$

where h is the largest distance between any two points in Δ_q, p is the order of the finite elements, i is the number of the corresponding solutions, and the constants c_1 and c_2 are independent of the step h.

The following theorem can be formulated.

Theorem. Let $D(x;z)$ in Eq. (1) be a continuous and bounded positive definite operator on the space $\mathcal{H}^1$ with the energy norm. Also let $\partial_z U(x;z)$ be continuous and bounded for each value of the parameter z. Then for the exact values of the solutions $\partial_z\varepsilon_i(z)$, $\partial_z\Phi_i(x;z) \in \mathcal{H}^2$, $H_{ij}(z)$, $Q_{ij}(z)$ from (18)–(21) and the corresponding numerical values $\partial_z\varepsilon_i^h(z)$, $\partial_z\Phi_i^h(x;z) \in \mathcal{H}^1$, $H_{ij}^h(z)$, $Q_{ij}^h(z)$ from (23)–(25), the following estimates are valid:

$$\left|\frac{\partial\varepsilon_i(z)}{\partial z} - \frac{\partial\varepsilon_i^h(z)}{\partial z}\right| \le c_3h^{2p}, \qquad \left\|\frac{\partial\Phi_i(x;z)}{\partial z} - \frac{\partial\Phi_i^h(x;z)}{\partial z}\right\|_0 \le c_4h^{p+1},$$
$$\left|Q_{ij}(z) - Q_{ij}^h(z)\right| \le c_5h^{2p}, \quad \left|H_{ij}(z) - H_{ij}^h(z)\right| \le c_6h^{2p}, \tag{33}$$

where h is the largest distance between any two points of the finite element Δ_q, p is the order of finite elements, i, j are the numbers of the corresponding solutions, and the constants c_3, c_4, c_5, and c_6 are independent of the step h.

The proof is straightforward following the scheme in accordance with [13].

5 Benchmark Calculations of Helium Atom Ground State

In the hyperspheroidal coordinates $0 \le R < \infty$, $1 \le \xi < \infty$, $-1 \le \eta \le 1$

$$r_{12} = \frac{\sqrt{2}R}{\sqrt{\xi^2+\eta^2}}, \quad r_1 = \frac{R(\xi+\eta)}{\sqrt{2}\sqrt{\xi^2+\eta^2}}, \quad r_2 = \frac{R(\xi-\eta)}{\sqrt{2}\sqrt{\xi^2+\eta^2}} \tag{34}$$

the equation for the wave functions $\Psi(R,\xi,\eta) = \sqrt{\xi^2+\eta^2}\Phi(R,\xi,\eta)$ for S-states of the Helium atom reads as [15]

$$\left[-\frac{1}{R^5}\frac{\partial}{\partial R}R^5\frac{\partial}{\partial R} - \frac{3}{R^2} - \frac{1}{R^2}\frac{(\xi^2+\eta^2)^2}{\xi^2-\eta^2}\left(\frac{\partial}{\partial\xi}(\xi^2-1)\frac{\partial}{\partial\xi} + \frac{\partial}{\partial\eta}(1-\eta^2)\frac{\partial}{\partial\eta}\right)\right.$$
$$\left.+\sqrt{2}\frac{\sqrt{\xi^2+\eta^2}}{R}\left(1-\frac{8\xi}{\xi^2-\eta^2}\right) - 2E\right]\Phi(R,\xi,\eta) = 0. \tag{35}$$

The function $\Phi(R,\xi,\eta)$ satisfies the boundary conditions

$$\lim_{R\to 0} R^5\frac{\partial\Phi(R,\xi,\eta)}{\partial R} = 0, \quad \lim_{R\to\infty} R^5\Phi(R,\xi,\eta) = 0,$$
$$\lim_{\xi\to 1}(\xi^2-1)\frac{\partial\Phi(R,\xi,\eta)}{\partial\xi} = 0, \quad \lim_{\xi\to\infty}\Phi(R,\xi,\eta) = 0,$$
$$\lim_{\eta\to\pm 1}(1-\eta^2)\frac{\partial\Phi(R,\xi,\eta)}{\partial\eta} = 0, \tag{36}$$

and is normalized by the condition

$$8\pi^2\int_0^\infty dRR^5\int_1^\infty d\xi\int_{-1}^1 d\eta\frac{\xi^2-\eta^2}{(\xi^2+\eta^2)^2}\Phi^2(R,\xi,\eta) = 1. \tag{37}$$

The parametric function $\phi_i \equiv \phi_i(\xi,\eta;R)$ and the corresponding potential curves $\varepsilon_i(R)$ are eigensolutions of the 2D BVP having a purely discrete spectrum

$$\left[-\frac{\partial}{\partial\xi}(\xi^2-1)\frac{\partial}{\partial\xi} - \frac{\partial}{\partial\eta}(1-\eta^2)\frac{\partial}{\partial\eta} + \frac{\sqrt{2}R\,(\xi^2-\eta^2-8\xi)}{\sqrt{\xi^2+\eta^2}^3} - \varepsilon_i(R)\frac{\xi^2-\eta^2}{(\xi^2+\eta^2)^2}\right]\phi_i = 0 \tag{38}$$

subject to the following boundary conditions

$$\lim_{\xi\to1}(\xi^2-1)\frac{\partial\phi_i(\xi,\eta;R)}{\partial\xi}=0,\quad \lim_{\xi\to\infty}\phi_i(\xi,\eta;R)=0,\quad \lim_{\eta\to\pm1}(1-\eta^2)\frac{\partial\phi_i(\xi,\eta;R)}{\partial\eta}=0,$$

and the normalization condition

$$\int_1^\infty d\xi\int_{-1}^1 d\eta\frac{\xi^2-\eta^2}{(\xi^2+\eta^2)^2}\phi_i^2(\xi,\eta;R)=1. \tag{39}$$

In terms of scaled variable and parametric surface functions

$$\xi=\frac{1+\lambda}{1-\lambda},\quad 0\le\lambda<1,\quad \phi_i(\xi,\eta;R)=\frac{p_i(\xi,\eta;R)}{\xi+1}\equiv\frac{p_i(\lambda,\eta;R)}{\xi+1}, \tag{40}$$

we rewrite the 2D BVP (38)–(39) in the form

$$\left[-\frac{\partial}{\partial\lambda}\lambda(1-\lambda)^2\frac{\partial}{\partial\lambda}-\frac{\partial}{\partial\eta}(1-\eta^2)\frac{\partial}{\partial\eta}+\sqrt{2}R(1-\lambda)\frac{(1+\lambda)^2-(1-\lambda)^2\eta^2-8(1-\lambda^2)}{\sqrt{(1+\lambda)^2+(1-\lambda)^2\eta^2}^3}\right.$$
$$\left.+1-\lambda-\varepsilon_i(R)(1-\lambda)^2\frac{(1+\lambda)^2-(1-\lambda)^2\eta^2}{((1+\lambda)^2+(1-\lambda)^2\eta^2)^2}\right]p_i(\lambda,\eta;R)=0. \tag{41}$$

The surface functions $p_i(\lambda,\eta;R)$ satisfy the following boundary and normalization conditions

$$\lim_{\lambda\to0,1}\lambda(1-\lambda)\frac{\partial p_i(\lambda,\eta;R)}{\partial\lambda}=0,\quad \lim_{\eta\to\pm1}(1-\eta^2)\frac{\partial p_i(\lambda,\eta;R)}{\partial\eta}=0, \tag{42}$$

$$\frac{1}{2}\int_0^1 d\lambda\int_{-1}^1 d\eta(1-\lambda)^2\frac{(1+\lambda)^2-(1-\lambda)^2\eta^2}{((1+\lambda)^2+(1-\lambda)^2\eta^2)^2}p_i^2(\lambda,\eta;R)=1. \tag{43}$$

The numerical experiments in the finite-element grids have shown a strict correspondence with the theoretical estimations (31) and (33) for the eigenvalues, eigenfunctions, and the matrix elements. In particular, we calculated the values of the Runge coefficients

$$\beta_l=\log_2\left|(\sigma_l^h-\sigma_l^{h/2})/(\sigma_l^{h/2}-\sigma_l^{h/4})\right|,\quad l=1,2, \tag{44}$$

with absolute errors on three twice condensed grids for their eigenvalues and eigenfunctions, respectively

$$\sigma_1^h=|E_m^{2h}(z)-E_m^h(z)|,\quad \sigma_2^h=\|\Phi_m^{2h}(x;z)-\Phi_m^h(x;z)\|_0. \tag{45}$$

The Runge coefficients for six eigenvalues presented in Table 5 equal 7.52 ÷ 8.19 and for their parametric derivatives equal 7.46 ÷ 7.76 are nearly similar and correspond to the theoretical estimates (31) and (33) for the fourth-order scheme ($2p\approx8$).

The calculations were carried out using the server 2 × 4 kernels i7k (i7-3770K 4.5 GHz, 32 GB RAM, GPU GTX680), and the Intel Fortran compiler 17.0. The

Table 5. Comparison of the transformed potential curves $E_j(R) = (\varepsilon_j(R) - 3)/4$ and their first derivative with respect to parameter R with results [9] at $j_{\max} = 12$. The mesh points are $\lambda = \{0(L)1\}$ and $\eta = \{0(L)1\}$, and $R = 7.65$ a.u.

j	$E_j(R)$ $(L = 40)$	$\partial_R E_j(R)$ $(L = 40)$	$E_j(R)$ [9]	$\partial_R E_j(R)$ [9]
1	−63.499 153 248	−15.796 136 178	−63.499 153 256	−15.796 136 189
2	−21.451 891 391	−3.997 429 168	−21.451 886 907	−3.997 431 891
3	−19.082 406 592	−4.142 660 217	−19.082 325 834	−4.142 711 985
4	−13.371 481 961	−3.897 822 460	−13.371 480 623	−3.897 824 374
5	−11.876 679 683	−3.314 363 652	−11.876 677 566	−3.314 347 679
6	−8.898 981 042	−2.705 445 931	−8.897 839 854	−2.705 544 197
j	$E_j(R)$ $(L = 20)$	$\partial_R E_j(R)$ $(L = 20)$	$E_j(R)$ $(L = 10)$	$\partial_R E_j(R)$ $(L = 10)$
1	−63.499 151 482	−15.796 133 881	−63.498 825 358	−15.795 727 590
2	−21.451 891 369	−3.997 429 139	−21.451 886 770	−3.997 423 220
3	−19.082 406 568	−4.142 660 186	−19.082 401 572	−4.142 653 692
4	−13.371 481 948	−3.897 822 446	−13.371 479 034	−3.897 819 472
5	−11.876 679 657	−3.314 363 641	−11.876 674 062	−3.314 361 245
6	−8.898 980 996	−2.705 445 914	−8.898 971 861	−2.705 442 515

Table 6. Matrix elements $H_{ji}(R)$, $i, j = 1, ..., 6$ at $R = 7.65$.

.1291804E-1	−.1264117E-1	.7293917E-2	.3763094E-2	−.1051774E-1	−.6007265E-2
−.1264117E-1	.3871021E-1	−.4493495E-2	−.1899806E-1	.2378084E-1	.5400750E-2
.7293917E-2	−.4493495E-2	.3270711E-1	.2565576E-1	.2270581E-1	−.1199926E-1
.3763094E-2	−.1899806E-1	.2565576E-1	.8136326E-1	.9664928E-2	−.2314799E-1
−.1051774E-1	.2378084E-1	.2270581E-1	.9664928E-2	.8335278E-1	.1949047E-1
−.6007265E-2	.5400750E-2	−.1199926E-1	−.2314799E-1	.1949047E-1	.2743837E-1

Table 7. Matrix elements $Q_{ji}(R)$, $i, j = 1, ..., 6$ at $R = 7.65$.

.37E-15	−.5859058E-1	.2863643E-1	.4422091E-1	.3362249E-1	.1621148E-1
.5859058E-1	.43E-16	.2502732E-1	−.1657796E+0	−.6079201E-1	−.1728211E-1
−.2863643E-1	−.2502732E-1	.36E-15	−.4584596E-1	.1345970E+0	.8980072E-1
−.4422091E-1	.1657796E+0	.4584596E-1	−.12E-15	.2029277E+0	.1556143E-1
−.3362249E-1	.6079201E-1	−.1345970E+0	−.2029277E+0	.92E-16	.1142082E+0
−.1621148E-1	.1728211E-1	−.8980072E-1	−.1556143E-1	−.1142082E+0	.13E-15

computing time for the considered examples with 10^{-12} accuracy on the uniform grids $\lambda = \{0(L)1\}$, $\eta = \{0(L)1\}$ at $L = 10, 20, 40$ is 0.38, 5.08, and 41.21 s, respectively. The matrix elements $Q_{ij}(R)$ and $H_{ij}(R)$ are presented in Tables 6 and 7. As an example eigenfunctions and their parametric derivatives are shown in Figs. 1 and 2.

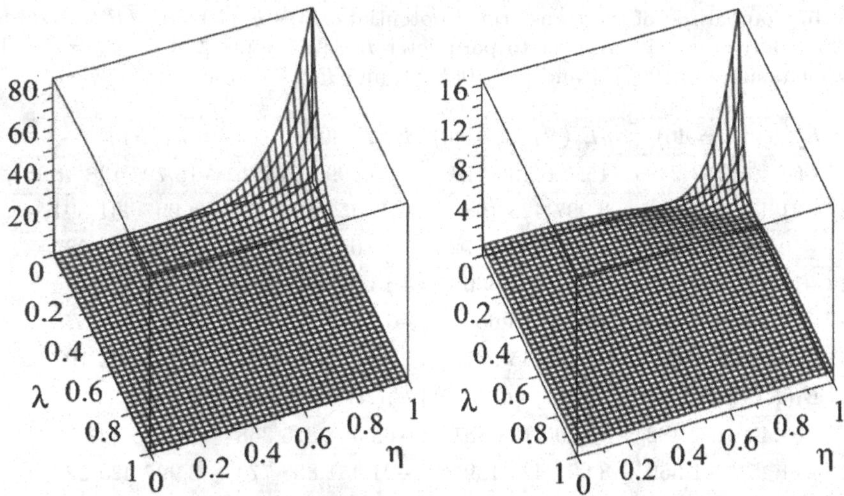

Fig. 1. The eigenfunction $p_1(\lambda, \eta; R)$ and its parametric derivative $\partial_R p_1(\lambda, \eta; R)$ at $R = 7.65$.

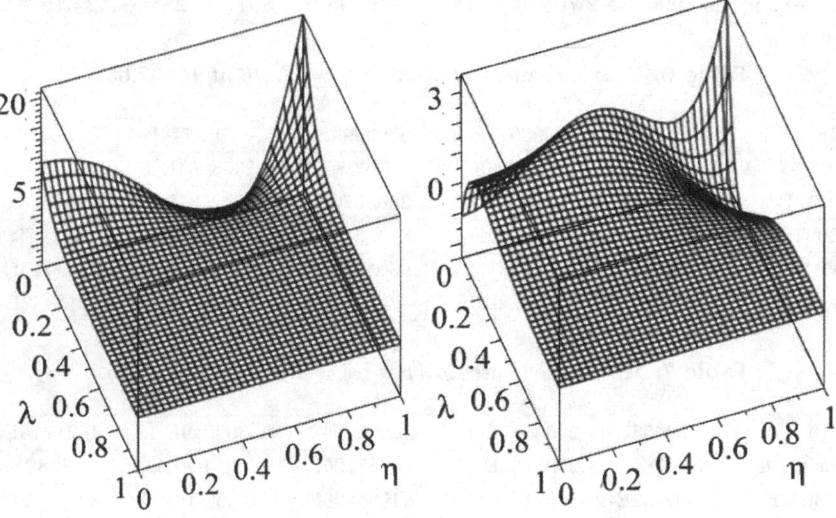

Fig. 2. The eigenfunction $p_4(\lambda, \eta; R)$ and its parametric derivative $\partial_R p_4(\lambda, \eta; R)$ at $R = 7.65$.

We seek for the solution of the BVP (35)–(37) by Kantorovich expansion

$$\Phi(R, \xi, \eta) = \sum_{j=1}^{j_{\max}} \phi_j(\xi, \eta; R)\chi_j(R) \tag{46}$$

over the eigenfunctions $\phi_j(\xi,\eta;R)$ of the parametric 2D BVP having a purely discrete spectrum $E_j(R) = (\varepsilon_j(R) - 3)/R^2, j = 1, 2, \ldots$. Substituting expansion (46) into the 3D BVP Eqs. (35)–(37), we get the 1D BVP for a finite set of $j_{\max}$ coupled SOODEs for $\boldsymbol{\chi}(R) = \{\chi_1(R), \ldots, \chi_N(R)\}^T$

$$\left(-\frac{1}{R^5}\mathbf{I}\frac{d}{dR}R^5\frac{d}{dR}+\mathbf{V}(R)+\mathbf{Q}(R)\frac{d}{dR}+\frac{1}{R^5}\frac{dR^5\mathbf{Q}(R)}{dR}-2E\mathbf{I}\right)\boldsymbol{\chi}(R)=0,$$

with the boundary and normalization conditions

$$\lim_{R\to 0} R^5\frac{d\boldsymbol{\chi}(R)}{dR}=0,\quad \lim_{R\to\infty} R^5\boldsymbol{\chi}(R)=0,\quad 8\pi^2\int_0^\infty dRR^5(\boldsymbol{\chi}(R))^T\boldsymbol{\chi}(R)=1.$$

The solution of this BVP with the help of KANTBP program [8] on the non-uniform grids $R = \{0(50), 5, (75), 20\}$ using calculated $E_j(R)$, $V_{ij}(R) = H_{ij}(R)$, $V_{jj}(R) = H_{jj}(R) + E_j(R)$, $Q_{ij}(R)$, $i, j = 1, \ldots, 12$ gives us the energy of Helium atom ground state $E_1 = -2.90372430$ a.u. with 8 significant digits.

6 Conclusion

We have elaborated new calculation schemes, algorithms, and the program for solving the parametric 2D elliptic BVP using the high-accuracy FEM with triangular elements. The program calculates the potential matrix elements, the integrals of the eigenfunctions multiplied by their first derivatives with respect to the parameter. The parametric eigenvalues (potential curves) and the matrix elements computed by the program can be used for solving the bound-state and multi-channel scattering problems for a system of the coupled second-order ODES with using the Kantorovich method. We demonstrated the efficiency of the proposed finite element schemes, algorithms, and codes by benchmark calculations of BVPs of helium atom ground state.

The work was partially supported by the Russian Foundation for Basic Research (grants Nos. 16-01-00080 and 17-51-44003 Mong_a) and the Bogoliubov-Infeld program. The reported study was funded by the Agreement N 02.03.21.0008 dated 24.04.2016 between the MES of the RF and RUDN University.

References

1. Bathe, K.J.: Finite Element Procedures in Engineering Analysis. Prentice Hall, Englewood Cliffs (1982)
2. Ciarlet, P.: The Finite Element Method for Elliptic Problems. North-Holland Publ. Comp., Amsterdam (1978)
3. Cools, R.: An encyclopaedia of quadrature Formulas. J. Complex. **19**, 445 (2003). http://nines.cs.kuleuven.be/ecf/
4. Chuluunbaatar, O., Gusev, A.A., Abrashkevich, A.G., Amaya-Tapia, A., Kaschiev, M.S., Larsen, S.Y., Vinitsky, S.I.: KANTBP: a program for computing energy levels, reaction matrix and radial wave functions in the coupled-channel hyperspherical adiabatic approach. Comput. Phys. Commun. **177**, 649–675 (2007)

5. Dunavant, D.A.: High degree efficient symmetrical Gaussian quadrature rules for the triangle. Int. J. Numer. Methods Eng. **21**, 1129–1148 (1985)
6. Esry, B.D., Lin, C.D., Greene, C.H.: Adiabatic hyperspherical study of the helium trimer. Phys. Rev. A **54**, 394–401 (1996)
7. Fano, U., Rau, A.R.P.: Atomic Collisions and Spectra. Academic Press, Florida (1986)
8. Gusev, A.A., Chuluunbaatar, O., Vinitsky, S.I., Abrashkevich, A.G.: KANTBP 3.0: new version of a program for computing energy levels, reflection and transmission matrices, and corresponding wave functions in the coupled-channel adiabatic approach. Comput. Phys. Commun. **185**, 3341–3343 (2014)
9. Gusev, A.A., Chuluunbaatar, O., Vinitsky, S.I., Abrashkevich, A.G.: POTHEA: a program for computing eigenvalues and eigenfunctions and their first derivatives with respect to the parameter of the parametric self-adjoined 2D elliptic partial differential equation. Comput. Phys. Commun. **185**, 2636–2654 (2014)
10. Kantorovich, L.V., Krylov, V.I.: Approximate Methods of Higher Analysis. Wiley, New York (1964)
11. Kress, J.D., Parker, G.A., Pack, R.T., Archer, B.J., Cook, W.A.: Comparison of Lanczos and subspace iterations for hyperspherical reaction path calculations. Comput. Phys. Commun. **53**, 91–108 (1989)
12. Papanicolopulos, S.-A.: Analytical computation of moderate-degree fully-symmetric quadrature rules on the triangle. arXiv:1111.3827v1 [math.NA]
13. Strang, G., Fix, G.J.: An Analysis of the Finite Element Method. Prentice-Hall, Englewood Cliffs (1973)
14. Vinitskii, S.I., Ponomarev, L.I.: Adiabatic representation in the three-body problem with Coulomb interaction. Sov. J. Part. Nucl. **13**, 557–587 (1982)
15. Vinitsky, S.I., Gusev, A.A., Chuluunbaatar, O., Derbov, V.L., Zotkina, A.S.: On calculations of two-electron atoms in spheroidal coordinates mapping on hypersphere. In: Proceedings of SPIE, vol. 9917, p. 99172Z (2016)
16. Vlasova, Z.A.: On the method of reduction to ordinary differential equations. Trudy Mat. Inst. Steklov. **53**, 16–36 (1959)
17. Zhang, L., Cui, T., Liu, H.: A set of symmetric quadrature rules on triangles and tetrahedra. J. Comput. Math. **27**, 89–96 (2009)

A Symbolic Study of the Satellite Dynamics Subject to Damping Torques

Sergey A. Gutnik[1(✉)] and Vasily A. Sarychev[2]

[1] Moscow State Institute of International Relations (University), 76, Prospekt Vernadskogo, Moscow 119454, Russia
s.gutnik@inno.mgimo.ru

[2] Keldysh Institute of Applied Mathematics (Russian Academy of Sciences), 4, Miusskaya Square, Moscow 125047, Russia
vas31@rambler.ru

Abstract. The dynamics of the rotational motion of a satellite moving in the central Newtonian force field in a circular orbit under the influence of gravitational and active damping torques is investigated with the help of computer algebra methods. The properties of a nonlinear algebraic system that determines equilibrium orientations of a satellite under the action of gravitational and active damping torques were studied. An algorithm for the construction of a Gröbner basis is proposed for determining the equilibrium orientations of a satellite with given central moments of inertia and given damping torques. The conditions of the equilibria's existence were obtained by the analysis of real roots of algebraic equations from the constructed Gröbner basis. The domains with an equal number of equilibria were specified by using algebraic methods for the construction of discriminant hypersurfaces. The conditions of asymptotic stability of the satellite's equilibria were determined as a result of the analysis of linearized equations of motion using Routh–Hurwitz criterion.

1 Introduction

In this paper, a symbolic investigation of a satellite dynamics under the influence of gravitational and active damping torques is presented. The gravity orientation systems are based on the fact that a satellite with different moments of inertia in the central Newtonian force field in a circular orbit has 24 equilibrium orientations and four of them are stable [1]. An important property of gravity orientation systems is that these systems can operate for a long time without spending energy. The problem to be analyzed in the present work is related to the behavior of the satellite acted upon by the gravity gradient and active damping torques. We assume that active damping torques depend on the projections of the angular velocity of the satellite. Such active damping torques can be provided by using the angular velocity sensor. The action of damping torques both leads to new equilibrium orientations and can provide the asymptotic stability of the well known equilibria of the gravity oriented satellites. Therefore, it is necessary to study the joint action of gravitational and active damping torques

V.P. Gerdt et al. (Eds.): CASC 2017, LNCS 10490, pp. 167–182, 2017.
DOI: 10.1007/978-3-319-66320-3_13

and, in particular, to analyze the necessary and sufficient conditions for asymptotic stability of the satellite's equilibria in a circular orbit. Such solutions can be used in practical space technology in the design of control orientation systems of the satellites.

In the present work, the problem of determination of the classes of equilibrium orientations and the conditions for asymptotic stability of defined equilibria for the general values of damping torques is considered. The equilibrium orientations are determined by real roots of the system of algebraic equations. The investigation of equilibria was performed by using the computer algebra Gröbner basis methods. The evolution of domains with a fixed number of equilibria is investigated by the analysis of the singular points of the discriminant hypersurface depending on three dimensionless damping parameters.

The conditions of equilibria stability are determined as a result of an analysis of the linearized equations of motion using the Routh–Hurwitz criterion. The detailed investigation of the regions of the necessary and sufficient conditions of stability is carried out by a numerical-analytical method in the plane of two dimensionless inertia parameters at different values of damping coefficients. The types of transition decay processes of spatial oscillations of a satellite at different damping parameters have been investigated numerically.

The computer algebra methods for determination of the equilibrium orientation of a satellite had been successfully used earlier to analyze the equilibrium orientations of a satellite under the influence of gravitational and constant torques [3]. The study of the equilibria of polynomial dynamical systems by means of symbolic computation is a very popular application of computer algebra. The detailed analysis of typical problems on parametric dynamical systems and computer algebra algorithms for solving this problem was presented at the CASC 2011 Workshop [4]. The symbolic methods for analyzing the stability of the equilibria of polynomial dynamical systems were presented at the CASC 2002 [5] and CASC 2007 Workshops [6].

2 Equations of Motion

Consider the attitude motion of a satellite subjected to gravitational and active damping torques in a circular orbit. We assume that the satellite is a triaxial rigid body, and active damping torques depend on the projections of the angular velocity of the satellite. To write the equations of motion we introduce two right-handed Cartesian coordinate systems with origin at the satellite's center of mass O. The orbital coordinate system is $OXYZ$, where the OZ axis is directed along the radius-vector connecting the centers of mass of the Earth and the satellite, the OX axis is in the direction of a satellite orbital motion. Then, the OY axis is directed along the normal to the orbital plane. The satellite body coordinate system is $Oxyz$, where Ox, Oy, and Oz are the principal central axes of inertia of the satellite. The orientation of the satellite body coordinate system $Oxyz$ with respect to the orbital coordinate system is determined by means of the aircraft angles of pitch (α), yaw (β), and roll (γ), and the direction cosines in the transformation matrix between the orbital coordinate system $OXYZ$ and $Oxyz$ are represented by the following expressions [2]:

$$\begin{aligned}
a_{11} &= \cos(x, X) = \cos\alpha\cos\beta,\\
a_{12} &= \cos(y, X) = \sin\alpha\sin\gamma - \cos\alpha\sin\beta\cos\gamma,\\
a_{13} &= \cos(z, X) = \sin\alpha\cos\gamma + \cos\alpha\sin\beta\sin\gamma,\\
a_{21} &= \cos(x, Y) = \sin\beta,\\
a_{22} &= \cos(y, Y) = \cos\beta\cos\gamma,\\
a_{23} &= \cos(z, Y) = -\cos\beta\sin\gamma,\\
a_{31} &= \cos(x, Z) = -\sin\alpha\cos\beta,\\
a_{32} &= \cos(y, Z) = \cos\alpha\sin\gamma + \sin\alpha\sin\beta\cos\gamma,\\
a_{33} &= \cos(z, Z) = \cos\alpha\cos\beta - \sin\alpha\sin\beta\sin\gamma. \qquad (1)
\end{aligned}$$

For small oscillations of the satellite, the angles of pitch, yaw, and roll correspond to the rotations around the OY, OZ, and OX axes, respectively.

Let the satellite be acted upon by the moments of active damping, their integral vector projections on the axes Ox, Oy, and Oz are equal to the following values: $M_x = \bar{k}_1 p_1, M_y = \bar{k}_2(q_1 - \omega_0)$, and $M_z = \bar{k}_3 r_1$. Here $\bar{k}_1, \bar{k}_2$, and $\bar{k}_3$ are the damping coefficients, p_1, q_1, and r_1 are the projections of the satellite's angular velocity onto the axes Ox, Oy, and Oz; ω_0 is the angular velocity of the orbital motion of the satellite's center of mass. The equations of satellite attitude motion can then be written in the Euler form:

$$\begin{aligned}
&Ap_1' + (C - B)q_1 r_1 - 3\omega_0^2(C - B)a_{32}a_{33} + \bar{k}_1 p_1 = 0,\\
&Bq_1' + (A - C)r_1 p_1 - 3\omega_0^2(A - C)a_{31}a_{33} + \bar{k}_2(q_1 - \omega_0) = 0,\\
&Cr_1' + (B - A)p_1 q_1 - 3\omega_0^2(B - A)a_{31}a_{32} + \bar{k}_3 r_1 = 0, \qquad (2)
\end{aligned}$$

$$\begin{aligned}
p_1 &= (\alpha' + \omega_0)a_{21} + \gamma',\\
q_1 &= (\alpha' + \omega_0)a_{22} + \beta'\sin\gamma,\\
r_1 &= (\alpha' + \omega_0)a_{23} + \beta'\cos\gamma. \qquad (3)
\end{aligned}$$

Here A, B, and C are the principal central moments of inertia of the satellite. The prime denotes the differentiation with respect to time t.

After the introduction of dimensionless parameters $\theta_A = A/B, \theta_C = C/B$, $p = p_1/\omega_0, q = q_1/\omega_0, r = r_1/\omega_0, \tilde{k}_1 = \bar{k}_1/B\omega_0, \tilde{k}_2 = \bar{k}_2/B\omega_0, \tilde{k}_3 = \bar{k}_3/B\omega_0$, and $\tau = \omega_0 t$ one can rewrite system (2)–(3) in the form

$$\begin{aligned}
&\theta_A\dot{p} + (\theta_C - 1)qr - 3(\theta_C - 1)a_{32}a_{33} + \tilde{k}_1 p = 0,\\
&\dot{q} + (\theta_A - \theta_C)rp - 3(\theta_A - \theta_C)a_{31}a_{33} + \tilde{k}_2(q - 1) = 0,\\
&\theta_C\dot{r} + (1 - \theta_A)pq - 3(1 - \theta_A)a_{31}a_{32} + \tilde{k}_3 r = 0, \qquad (4)
\end{aligned}$$

$$\begin{aligned}
p &= (\dot{\alpha} + 1)a_{21} + \dot{\gamma},\\
q &= (\dot{\alpha} + 1)a_{22} + \dot{\beta}\sin\gamma,\\
r &= (\dot{\alpha} + 1)a_{23} + \dot{\beta}\cos\gamma. \qquad (5)
\end{aligned}$$

The dot denotes the differentiation with respect to τ.

3 Equilibrium Orientations of Satellite

Setting in (2) and (3) $\alpha = \alpha_0 = \text{const}$, $\beta = \beta_0 = \text{const}, \gamma = \gamma_0 = \text{const}$, we obtain at $A \neq B \neq C$ the equations

$$\begin{aligned} a_{22}a_{23} - 3a_{32}a_{33} + k_1 a_{21} &= 0, \\ a_{21}a_{23} - 3a_{31}a_{33} + k_2(a_{22} - 1) &= 0, \\ a_{21}a_{22} - 3a_{31}a_{32} + k_3 a_{23} &= 0, \end{aligned} \tag{6}$$

which allow us to determine the satellite equilibria in the orbital coordinate system. Here $k_1 = \tilde{k}_1/(C-B), k_2 = \tilde{k}_2/(A-C)$, and $k_3 = \tilde{k}_3/(B-A)$. We will consider the case when damping coefficients k_1, k_2, and k_3 are positive. Substituting the expressions for the direction cosines from (1) in terms of the aircraft angles into Eq. (6), we obtain three equations with three unknowns α, β, and γ. Another way of closing Eq. (6) is to add the following three conditions for the orthogonality of direction cosines:

$$\begin{aligned} a_{21}^2 + a_{22}^2 + a_{23}^2 - 1 &= 0, \\ a_{31}^2 + a_{32}^2 + a_{33}^2 - 1 &= 0, \\ a_{21}a_{31} + a_{22}a_{32} + a_{23}a_{33} &= 0. \end{aligned} \tag{7}$$

Equations (6) and (7) form a closed system of equations with respect to the six direction cosines identifying the satellite equilibrium orientations. For this system of equations, we formulate the following problem: for given values of k_1, k_2, and k_3, it is required to determine all the nine directional cosines, i.e., all satellite equilibrium orientations in the orbital coordinate system. After $a_{21}, a_{22}, a_{23}, a_{31}, a_{32}$, and a_{33} are found, the direction cosines a_{11}, a_{12}, and a_{13} can be determined from the conditions of orthogonality.

It should be noted that to solve system (6), (7) it is sufficient to find the values of two unknowns a_{21} and a_{22}. Indeed, for each value a_{21} and a_{22}, one can find two values of a_{23} from the first equation of system (7) and then uniquely determine their corresponding values a_{31}, a_{32}, and a_{33} from system (6), (7).

To find solutions of the algebraic system (6), (7) we used the algorithm for constructing the Gröbner bases [7]. The method for constructing a Gröbner basis is an algorithmic procedure that reduces the problem in the case of polynomials of several variables to a problem with a polynomial of a single variable.

In our study, for Gröbner bases construction, we applied the command `Groebner[Basis]` from the package `Groebner` implemented in the computer algebra system Maple 15 [8]. We constructed the Gröbner basis of the system of six second-order polynomials (6), (7) with six variables a_{ij} $(i = 2, 3; j = 1, 2, 3)$, with respect to the lexicographic ordering of variables by using option `plex`. In the list of polynomials F:=$[f_i(i = 1, 2, \ldots 6)]$, f_i are the left–hand sides of the algebraic equations (6), (7):

```
G:=map(factor,Groebner[Basis]([F, plex(a31, ... a22))).
```

Here, calculating the Gröbner basis over the field of rational functions in k_1, k_2, and k_3, we compute the generic solutions of our problem only. In our task from the area of satellite dynamics, the main goal of the study is to estimate a wide range of system parameters for which the satellite's equilibria exist, and the task is to determine the regions in the space of parameters for which these equilibria are asymptotically stable.

Taking into account the errors of the angular velocity sensors and the errors of the signals, which generate damping torques, the exact bifurcation values of the coefficients are very difficult to obtain in practice. We are interested in estimating the size of regions in the space of damping parameters where equilibria exist. In the case of parametric dynamical system solving, when the parameters reach non-generic solutions, the symbolic application based on comprehensive Gröbner bases [9], discriminant varieties [10] and comprehensive triangular decomposition [4] methods are used.

Here we write down the polynomial in the Gröbner basis that depends only on one variable $x = a_{22}$. This polynomial has the form

$$P(a_{22}, k_1, k_2, k_3) = (a_{22}^2 - 1)[(k_1k_2 + k_2k_3 + k_1k_3 - 4)a_{22} - k_2(k_1 + k_3)] = 0. \quad (8)$$

To determine the equilibria it is required to consider separately the following three cases: $a_{22} = 1, a_{22} = -1$ and $(k_1k_2 + k_2k_3 + k_1k_3 - 4)a_{22} - k_2(k_1 + k_3) = 0$.

In the first case, when $a_{22} = 1$ $(a_{21} = a_{23} = 0)$, we will get the following eight equilibrium solutions from system (6) and (7):

$$a_{31}^2 = 1, a_{32} = a_{33} = 0; a_{32}^2 = 1, a_{31} = a_{33} = 0; a_{33}^2 = 1, a_{32} = a_{33} = 0. \quad (9)$$

In the second case, when $a_{22} = -1$, system (6), (7) takes the form

$$\begin{aligned} a_{32}a_{33} = 0, a_{31}a_{32} = 0, \\ a_{31}a_{33} + 2k_2 = 0, \\ a_{31}^2 + a_{33}^2 = 1. \end{aligned} \quad (10)$$

From (10) we obtain the following solutions:

$$\begin{aligned} a_{32} = 0, a_{31} = -2k_2/3a_{33}; \\ 9a_{33}^4 - 9a_{33}^2 + 4k_2^2 = 0, \\ a_{33}^2 = \frac{3 - \sqrt{9 - 16k_2^2}}{2}. \end{aligned} \quad (11)$$

Solutions (11) exist in the case when the discriminant of the biquadratic equation $9a_{33}^4 - 9a_{33}^2 + 4k_2^2 = 0$ is non-negative and $a_{33}^2 \le 1$. These conditions are satisfied when $k_2^2 \le 1/2$.

Now let us consider the third case, where the satellite equilibrium solutions are determined by the linear equation $(k_1k_2+k_2k_3+k_1k_3-4)a_{22}-k_2(k_1+k_3) = 0$, from which we can obtain:

$$a_{22} = \frac{k_2(k_1 + k_3)}{k_1k_2 + k_2k_3 + k_1k_3 - 4}. \quad (12)$$

From the condition for the existence of a solution for the direction cosine $a_{22} \le 1$, we obtain the inequality $k_1k_3 \ge 4$. From the condition $a_{22} \ge -1$, we obtain the inequality $2k_1k_2+2k_2k_3+k_1k_3 \geq 4$. Consequently, solution (12) is possible when the inequality $k_1k_3 \geq 4$ holds.

Thus, from Eq. (8), we obtain all possible values of the direction cosine a_{22} satisfying the initial system (6), (7).

To find the a_{21} values, we have recalculated the Gröbner basis with respect to the variable a_{21}. The polynomial depending on only one variable a_{21} in the Gröbner basis obtained is given by

$$P(a_{21}) = p_0a_{21}^8 + p_1a_{21}^6 + p_2a_{21}^4 + p_3a_{21}^2 + p_4 = 0, \tag{13}$$

where

$$\begin{aligned}
p_0 &= p_{01}^8, \quad p_{01} = k_1k_2 + k_2k_3 + k_1k_3 - 4,\\
p_1 &= -2p_{01}^6 p_{11},\\
p_{11} &= (k_1k_3 - 4)^2 + 2k_2(k_1 + k_3)(k_1k_3 - 4) + k_2^2(k_1^2 - k_3^2),\\
p_2 &= p_{01}^4 p_{21},\\
p_{21} &= (k_1k_3 - 4)^4 + 4k_2(k_1 + k_3)(k_1k_3 - 4)^3\\
&\quad + k_2^2(6k_1^2 + 8k_1k_3 + 17)(k_1k_3 - 4)^2\\
&\quad + 2k_2^3(k_1 + k_3)(2k_1^2 - 4k_3^2 + 17)(k_1k_3 - 4)\\
&\quad + k_2^4(k_1 + k_3)^2((k_1 - k_3)^2 + 25),\\
p_3 &= p_{01}^2 p_{31} p_{32},\\
p_{31} &= k_2^2(k_1k_3 - 4)(2k_1k_2 + 2k_2k_3 + k_1k_3 - 4),\\
p_{32} &= (2k_3^2 - 17)(k_1k_3 - 4)^2 + 2k_2(2k_3^2 - 17)(k_1 + k_3)(k_1k_3 - 4)\\
&\quad + k_2^2(k_1 + k_3)(2k_3^2(k_1 - k_3) - 17k_1 - 33k_3),\\
p_4 &= (k_3^2 + 4)^2 p_{31}^2.
\end{aligned}$$

Equation (13) together with (12), (6), and (7) can be used to determine all the equilibrium orientations of the satellite under the influence of gravitational and active damping torques.

The number of real roots of the algebraic equation (13) is even and does not exceed 8. Let us show that each real root a_{21} of Eq. (13) corresponds to two equilibrium solutions of the original system (6), (7). Indeed, for each solution a_{21} of Eq. (13) and a_{22} of Eq. (12), one can find two values of a_{23} from the first equation of system (7) and then uniquely determine their corresponding values a_{31}, a_{32}, and a_{33} from system (6), (7). Once the set of six values a_{21}, a_{22}, a_{23}, a_{31}, a_{32}, and a_{33} is found, the remaining three values a_{11}, a_{12}, and a_{13} can be uniquely determined from the conditions of the orthogonality of the directional cosines. Since the number of real roots of Eq. (13) does not exceed eight, the number of the satellite equilibria in this case does not exceed sixteen.

4 Conditions for the Existence of Equilibrium Orientations of the Satellite

Equations (6)–(8) and (12), (13) make it possible to determine all the equilibrium orientations of the satellite due to gravity and active damping torques for the given values of dimensionless damping parameters k_1, k_2, and k_3 of the problem.

In studying the satellite equilibrium orientations, we determine the domains with an equal number of real roots of Eq. (13) in the space of parameters. To identify these domains, we use the Meiman theorem [11], which yields that the decomposition of the space of parameters into domains with an equal number of real roots is determined by the discriminant hypersurface. It is also possible to calculate the number of real roots of a polynomial by means of ith subdiscriminants using Jacobi theorem [12,13].

In our case, the discriminant hypersurface is given by the discriminant of polynomial (13). This hypersurface contains a component of codimension 1, which is the boundary of domains with an equal number of real roots. The set of singular points of the discriminant hypersurface in the space of parameters k_1, k_2, and k_3 is given by the following system of algebraic equations:

$$P(x) = 0, \quad P'(x) = 0. \tag{14}$$

Here $x = a_{21}^2$, and the prime denotes the differentiation with respect to x.

We eliminate the variable x from system (14) by calculating the determinant of the resultant matrix of Eq. (14) with the help of symbolic matrix functions in Maple and obtain an algebraic equation of the discriminant hypersurface as

$$P_1(k_1, k_2, k_3)P_2(k_1, k_2, k_3) = 0. \tag{15}$$

Here $P_1(k_1, k_2, k_3)$ and $P_2(k_1, k_2, k_3)$ are 14th and 8th degree polynomials, respectively, in terms of k_2. The polynomial $P_1(k_1, k_2, k_3)$ has the form

$$P_1(k_1, k_2, k_3) = 625k_2^8 p_{01}^4 (k_1k_3 - 4)^2 (2k_1k_2 + 2k_2k_3 + k_1k_3 - 4)^2. \tag{16}$$

Here $p_{01} = k_1k_2 + k_2k_3 + k_1k_3 - 4$. $P_2(k_1, k_2, k_3)$ has the form

$$\begin{aligned} P_2(k_1, k_2, k_3) &= p_{2,0}k_2^8 + p_{2,1}k_2^7 + p_{2,2}k_2^6 + p_{2,3}k_2^5 + p_{2,4}k_2^4 \\ &\quad + p_{2,5}k_2^3 + p_{2,6}k_2^2 + p_{2,7}k_2 + p_{2,8} = 0, \end{aligned} \tag{17}$$

where

$$\begin{aligned} p_{2,0} &= 4(k_1 + k_3)^4[4(k_1k_3 - 4)^2 - 9((k_1 + k_3)^2][((k_1 - k_3)^2 + 25]^2, \\ p_{2,1} &= 8(k_1k_3 - 4)(k_1 + k_3)^3[4(4k_3^2 - 9)k_1^6 - 2k_3(16k_3^2 + 39)k_1^5 \\ &\quad + (32k_3^4 + 220k_3^2 - 7)k_1^4 - 2k_3(16k_3^4 + 506k_3^2 - 1557)k_1^3 \\ &\quad + (16k_3^6 + 220k_3^4 - 14186k_3^2 - 15827)k_1^2 \\ &\quad - 6k_3(13k_3^4 - 519k_3^2 + 9041)k_1 - 36k_3^6 - 7k_3^4 - 15827k_3^2 + 28800], \\ p_{2,2} &= 4(k_1k_3 - 4)^2(k_1 + k_3)^2[28(4k_3^2 - 9)k_1^6 - 2k_3(16k_3^2 + 339)k_1^5 \end{aligned}$$

$$
\begin{aligned}
&+ 2(48k_3^4 - 1158k_3^2 + 2325)k_1^4 - 2k_3(16k_3^4 + 3106k_3^2 - 15007)k_1^3 \\
&+ (112k_3^6 - 2316k_3^4 + 54848k_3^2 - 58559)k_1^2 - 2k_3(339k_3^4 \\
&- 15007k_3^2 + 65823)k_1 - 252k_3^6 + 4650k_3^4 - 58559k_3^2 + 49536], \\
p_{2,3} &= 4(k_1k_3 - 4)^3(k_1 + k_3)[56(4k_3^2 - 9)k_1^6 + 8k_3(32k_3^2 - 197)k_1^5 \\
&+ (320k_3^4 - 7176k_3^2 + 11101)k_1^4 + 4k_3(64k_3^4 - 3276k_3^2 + 10397)k_1^3 \\
&+ (224k_3^6 - 7176k_3^4 + 57150k_3^2 - 53748)k_1^2 - 4k_3(394k_3^4 \\
&- 10397k_3^2 + 24858)k_1 - 504k_3^6 + 11101k_3^4 - 53748k_3^2 + 20736], \\
p_{2,4} &= 4(k_1k_3 - 4)^4[280(4k_3^2 - 9)k_1^6 + 40k_3(64k_3^2 - 219)k_1^5 \\
&+ (3136k_3^4 - 34408k_3^2 + 44617)k_1^4 \\
&+ 4k_3(640k_3^4 - 14308k_3^2 + 30053)k_1^3 \\
&+ (1120k_3^6 - 34408k_3^4 + 147366k_3^2 - 108828)k_1^2 \\
&- 4k_3(2190k_3^4 - 30053k_3^2 + 52398)k_1 - 2520k_3^6 \\
&+ 44617k_3^4 - 108828k_3^2 + 20736], \\
p_{2,5} &= 4(k_1k_3 - 4)^5(k_1 + k_3)[56(4k_3^2 - 9)k_1^4 \\
&+ 4k_3(64k_3^2 - 219)k_1^3 + (224k_3^4 - 3240k_3^2 + 5481)k_1^2 \\
&- 6k_3(146k_3^2 - 441)k_1 - 504k_3^4 + 5481k_3^2 - 8262], \\
p_{2,6} &= 2(k_1k_3 - 4)^6[56(4k_3^2 - 9)k_1^4 + 4k_3(96k_3^2 - 241)k_1^3 \\
&+ (224k_3^4 - 1752k_3^2 + 2583)k_1^2 - 2k_3(482k_3^2 - 1197)k_1 \\
&- 504k_3^4 + 2583k_3^2 - 2754], \\
p_{2,7} &= 8(4k_1^2 - 9)(4k_3^2 - 9)(k_1 + k_3)(k_1k_3 - 4)^7, \\
p_{2,8} &= (4k_1^2 - 9)(4k_3^2 - 9)(k_1k_3 - 4)^8.
\end{aligned}
$$

Now we should check the change in the number of equilibria when one of the surfaces (15) is intersected. This can be done numerically by determining the number of equilibria at a point of each domain $P_1(k_1, k_2, k_3) = 0$ and $P_2(k_1, k_2, k_3) = 0$ in the space of parameters k_1, k_2 and k_3.

It should be noted that when the boundaries of the surface $P_1(k_1, k_2, k_3) = 0$ are intersected no change in the equilibria occurs due to the condition (12). From (12) it follows that the factor $k_1k_2 + k_2k_3 + k_1k_3 - 4$ from (16) is not equal to zero. When $k_1k_3 - 4 = 0$, then $a_{22} = 1$ and $a_{21} = 0$; when $2k_1k_2 + 2k_2k_3 + k_1k_3 - 4 = 0$, then $a_{22} = -1$ and $a_{21} = 0$. Thus, in these cases, we have only zero solutions.

To study the evolution of the domains of the existence of a different number of equilibrium orientations depending on the magnitude of the damping torque vector in the space of dimensionless parameters k_1, k_2, and k_3, we perform a detailed analysis of the surface $P_2(k_1, k_2, k_3) = 0$. The satellite equilibrium orientations exist when Eqs. (12) and (13) have real solutions. Equation (12) has a solution if the condition $k_1k_3 \geq 4$ is satisfied.

Below we present the results of the numerical and analytical analysis of the properties and form of the discriminant hypersurface $P_2(k_1, k_2, k_3) = 0$, which are two-dimensional cross sections of the surface in the plane (k_1, k_3) at a fixed value of parameter k_2 (Figs. 1, 2 and 3).

Figures 1, 2 and 3 show the distributions of domains with an equal number of real roots of Eq. (13) for the cases of significantly changed characteristics. The distributions are classified for the values of k_2 in the range $0.1 \leq k_2 \leq 5$. The figures demonstrate the domains with a fixed number of real solutions in the plane (k_1, k_3) (here, k_1 is the vertical axis, and k_3 is the horizontal axis), and the domain boundaries are cross sections of the surface $P_2(k_1, k_2, k_3) = 0$ with the plane $k_2 = \text{const}$.

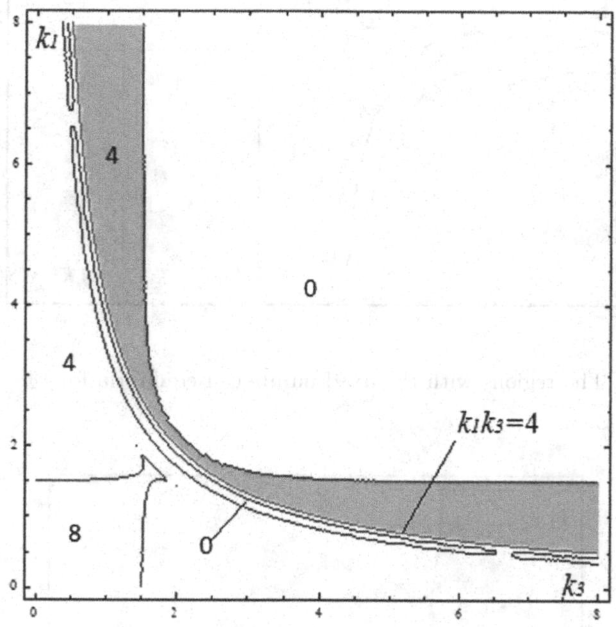

Fig. 1. The regions with the fixed number of equilibria for $k_2 = 0.1$

The figures indicate the domains where eight and four real solutions exist as well as the domains where no real solutions exist (marked by 0). It can be seen from Fig. 1 that for small values of k_2 ($k_2 < 0.5$), there are eight real roots of Eq. (13) in the region near the origin of the coordinate system. In these cases, there is only one region located above the positive branch of the hyperbola $k_1 k_3 = 4$, where eight equilibria of the satellite exist (four real roots of Eq. (13)). Grey shaded regions correspond to the existence of equilibria.

For $k_2 = 0.5$, the regions with the number of real roots of Eq. (13) equal to 8 disappear in the positive quadrant $k_1 \geq 0, k_3 \geq 0$ (Fig. 2) and, with further increase of parameter k_2, there are regions with the number of real roots equal to 4, and the regions with no real roots (Fig. 3). There is only one region (marked by

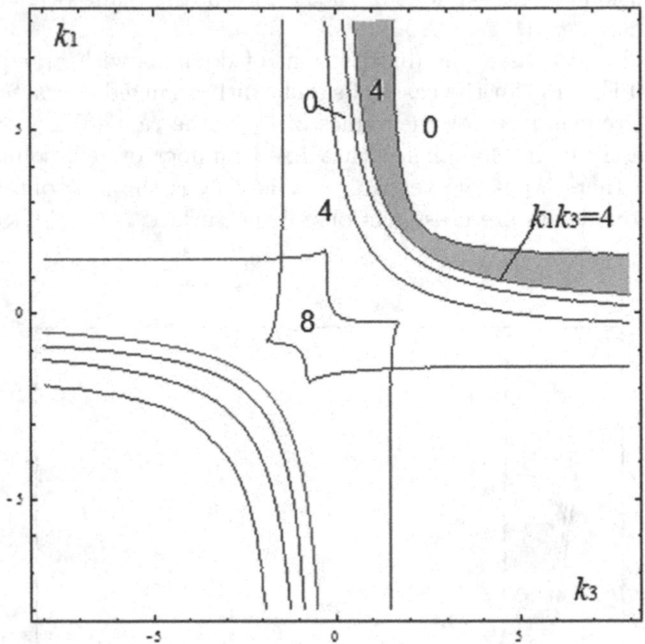

Fig. 2. The regions with the fixed number of equilibria for $k_2 = 0.5$

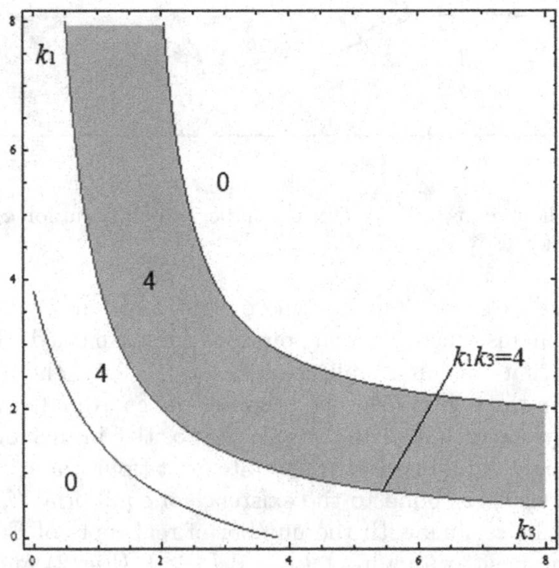

Fig. 3. The regions with the fixed number of equilibria for $k_2 = 5.0$

grey color), which is located above the positive branch of the hyperbola $k_1k_3 = 4$, with 8 equilibrium orientations (four real roots of Eq. (13)).

The results of the analysis of the equilibria total number in the third case can be summarized as follows. The curves $P_2(k_1, k_2, k_3) = 0$ and $k_1k_3 = 4$ decompose the plane (k_1, k_3) into three domains where no equilibria (8 or 4 real roots exist), 8 equilibria (4 real roots exist), and no equilibria (no real roots) exist.

The final decomposition of the plane (k_1, k_3) for $k_2 = 0.1$, $k_2 = 0.5$, and $k_2 = 5.0$ is presented in Figs. 1, 2 and 3.

5 Necessary and Sufficient Conditions of Asymptotic Stability of the Equilibrium Orientations of Satellite

In order to study the necessary and sufficient conditions of asymptotic stability of the above-determined equilibrium orientations of system (6)–(7) let us linearize the system of Eqs. (4) and (5) in the vicinity of the equilibrium solution $\alpha = \alpha_0$, $\beta = \beta_0, \gamma = \gamma_0$. We represent α, β, and γ in the form $\alpha = \alpha_0 + \bar{\alpha}, \beta = \beta_0 + \bar{\beta}$, $\gamma = \gamma_0 + \bar{\gamma}$, where $\bar{\alpha}, \bar{\beta}$ and $\bar{\gamma}$ are small deviations from the equilibrium orientation of the satellite $\alpha = \alpha_0$, $\beta = \beta_0, \gamma = \gamma_0$.

After rather exhausting symbolic transformations, the linearized system of equations of motion takes the following form:

$$\begin{aligned}
&\theta_A \ddot{\bar{\alpha}} \sin\beta_0 + [2(\theta_C - 1)a_{22}a_{23} + k_1 a_{21}]\dot{\bar{\alpha}} + 3(\theta_C - 1)(a_{12}a_{33} + a_{13}a_{32})\bar{\alpha} \\
&\quad + \cos\beta_0[(\theta_A + \theta_C - 1) - 2(\theta_C - 1)\sin^2\gamma_0]\dot{\bar{\beta}} + \cos\beta_0[(\theta_C - 1) \\
&\quad [(1 + 3\sin^2\alpha_0)\sin\beta_0\sin 2\gamma_0 - \frac{3}{2}\sin 2\alpha_0\cos 2\gamma_0] + k_1]\bar{\beta} + \theta_A\ddot{\bar{\gamma}} + k_1\dot{\bar{\gamma}} \\
&\quad + (\theta_C - 1)[(a_{23}^2 - a_{22}^2) - 3((a_{33}^2 - a_{32}^2)]\bar{\gamma} = 0, \\
&\ddot{\bar{\alpha}} a_{22} + [2(\theta_A - \theta_C)a_{21}a_{23} + k_2 a_{22}]\dot{\bar{\alpha}} + 3(\theta_A - \theta_C)(a_{13}a_{31} + a_{11}a_{33})\bar{\alpha} \\
&\quad + \ddot{\bar{\beta}}\sin\gamma_0 + [(\theta_A + \theta_C - 1)\sin\beta_0\cos\gamma_0 + k_2\sin\gamma_0]\dot{\bar{\beta}} - [(\theta_A - \theta_C) \\
&\quad [(1 + 3\sin^2\alpha_0)\cos 2\beta_0\sin\gamma_0 + \frac{3}{2}\sin 2\alpha_0\sin\beta_0\cos\gamma_0] + k_2\sin\beta_0\cos\gamma_0]\bar{\beta} \\
&\quad + (\theta_A + \theta_C - 1)a_{23}\dot{\bar{\gamma}} + [(\theta_C - \theta_A)(a_{21}a_{22} - 3a_{31}a_{32}) + k_2 a_{23}]\bar{\gamma} = 0, \\
&\theta_C\ddot{\bar{\alpha}} a_{23} + \cos 2\beta_0[2(1 - \theta_A)\sin\beta_0\cos\gamma_0 - k_3\sin\gamma_0]\dot{\bar{\alpha}} + \theta_C\ddot{\bar{\beta}}\cos\gamma_0 \\
&\quad + [(\theta_C - \theta_A + 1)\sin\beta_0\sin\gamma_0 + k_3\cos\gamma_0]\dot{\bar{\beta}} \\
&\quad + [(1 - \theta_A)[(1 + 3\sin^2\alpha_0)\cos 2\beta_0\cos\gamma_0 - \frac{3}{2}\sin 2\alpha_0\sin\beta_0\sin\gamma_0] \\
&\quad + k_3\sin\beta_0\sin\gamma_0]\bar{\beta} + 3(1 - \theta_A)(a_{11}a_{32} + a_{12}a_{31})\bar{\alpha} \\
&\quad - (\theta_A + \theta_C - 1)a_{22}\dot{\bar{\gamma}} + [(1 - \theta_A)(a_{21}a_{23} - 3a_{31}a_{33}) - k_3 a_{22}]\bar{\gamma} = 0.
\end{aligned} \tag{18}$$

Now let us consider small oscillations of the satellite in the vicinity of the specific equilibrium orientation, when the principal axes of inertia of the satellite coincide with the orbital coordinate system:

$$\alpha_0 = \beta_0 = \gamma_0 = 0. \tag{19}$$

This is one of the equilibrium solutions from (9), when $a_{22} = 1, a_{11} = 1$, and $a_{33} = 1$. Taking into account expressions (1) for solution (19), we get $\sin\alpha_0 = 0$, $\sin\beta_0 = 0$, $\sin\gamma_0 = 0$, and linearized equations (18) take the form

$$\begin{aligned}
\ddot{\bar{\alpha}} + k_2\dot{\bar{\alpha}} + 3(\theta_A - \theta_C)\bar{\alpha} = 0,\\
\theta_C\ddot{\bar{\beta}} + k_3\dot{\bar{\beta}} - (\theta_A + \theta_C - 1)\dot{\bar{\gamma}} + (1 - \theta_A)\bar{\beta} - k_3\bar{\gamma} = 0,\\
\theta_A\ddot{\bar{\gamma}} + k_1\dot{\bar{\gamma}} + (\theta_A + \theta_C - 1)\dot{\bar{\beta}} + 4(1 - \theta_C)\bar{\gamma} + k_1\bar{\beta} = 0.
\end{aligned} \tag{20}$$

The characteristic equation of system (20)

$$[\lambda^2 + k_2\lambda + 3(\theta_A - \theta_C)](A_0\lambda^4 + A_1\lambda^3 + A_2\lambda^2 + A_3\lambda + A_4) = 0 \tag{21}$$

decomposes into quadratic and 4th degree equations. Here the following notations are introduced:

$$\begin{aligned}
A_0 &= \theta_A\theta_C, \quad A_1 = k_1\theta_C + k_3\theta_A,\\
A_2 &= k_1k_3 + (\theta_A + \theta_C - 1)^2 + \theta_A(1 - \theta_A) + 4\theta_C(1 - \theta_C),\\
A_3 &= k_1\theta_C + k_3(\theta_A - 3\theta_C + 3), \quad A_4 = k_1k_3 + 4(1 - \theta_A)(1 - \theta_C).
\end{aligned}$$

The necessary and sufficient conditions for asymptotic stability (Routh–Hurwitz criterion) of the equilibrium solution (19) take the following form:

$$\begin{aligned}
&k_2 > 0, \quad \theta_A - \theta_C > 0,\\
&\Delta_1 = A_1 = k_1\theta_C + k_3\theta_A > 0,\\
&\Delta_2 = A_1A_2 - A_0A_3 = k_1^2k_3\theta_C + k_1k_3^2\theta_A\\
&\quad + (1 - \theta_C)[k_1\theta_C(3\theta_C - \theta_A + 1) + k_3\theta_A(1 - \theta_A)] > 0,\\
&\Delta_3 = A_1A_2A_3 - A_0A_3^2 - A_1^2A_4 = 3(1 - \theta_C)[k_1^2k_3^2\theta_C + k_1k_3^3\theta_A\\
&\quad + k_1^2\theta_C^2(\theta_A + \theta_C - 1) + k_1k_3\theta_C[(\theta_A + \theta_C - 1)(2\theta_A - 1)\\
&\quad + 3\theta_C(1 - \theta_C)] - k_3^2\theta_A(1 - \theta_A)(\theta_A + \theta_C - 1)] > 0,\\
&\Delta_4 = \Delta_3A_4 > 0, \quad A_4 = k_1k_3 + 4(1 - \theta_A)(1 - \theta_C) > 0.
\end{aligned} \tag{22}$$

Let us consider the special case when $k_1 = k_2 = k_3 = k$. In this case, conditions (22) take a simpler form

$$\begin{aligned}
&k > 0, \quad \theta_A - \theta_C > 0,\\
&\Delta_1 = k(\theta_C + \theta_A) > 0,\\
&\Delta_2 = k[k^2 + (1 - \theta_C)^2]\theta_A\\
&\quad + k[k^2 + (1 - \theta_C)(1 + 3\theta_C)]\theta_C - k(1 - \theta_C)\theta_A^2 > 0,\\
&\Delta_3 = 3k^2(1 - \theta_C)[\theta_A^3 + (3\theta_C - 2)\theta_A^2\\
&\quad + [k^2 + (1 - \theta_C)(1 - 3\theta_C)]\theta_A\\
&\quad + \theta_C[k^2 + (1 - \theta_C)(1 + 2\theta_C)]] > 0,\\
&\Delta_4 = \Delta_3A_4 > 0, \quad A_4 = k^2 + 4(1 - \theta_A)(1 - \theta_C) > 0.
\end{aligned} \tag{23}$$

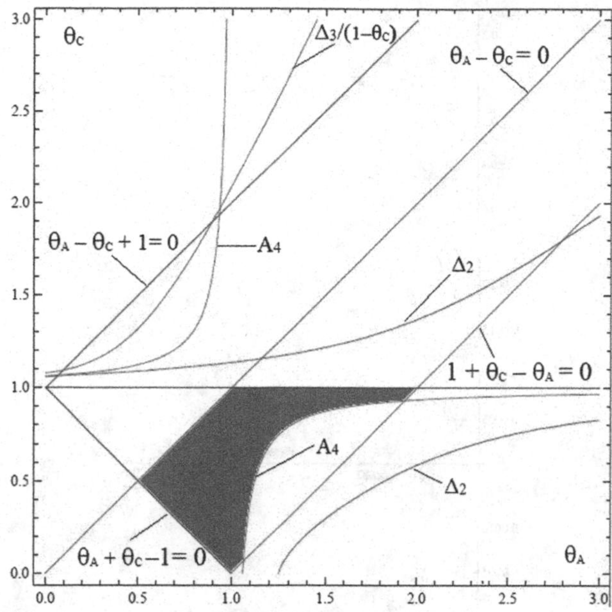

Fig. 4. The region of fulfillment of the asymptotic stability conditions for $k = 0.5$

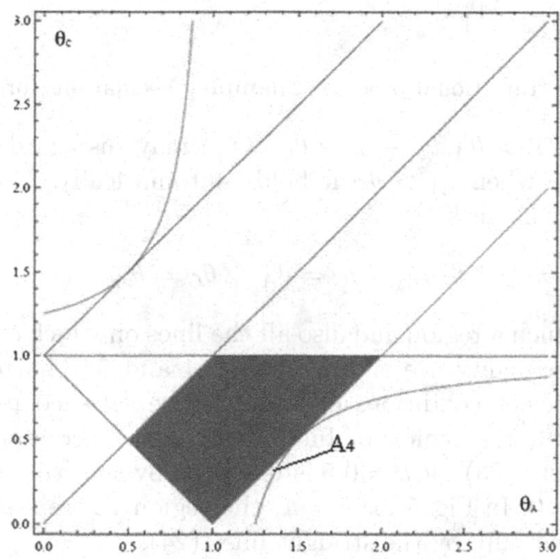

Fig. 5. The region of fulfillment of the asymptotic stability conditions for $k = 1.0$

The detailed analysis of the regions where necessary and sufficient conditions of stability (23) hold is studied in the plane of two dimensionless inertia parameters (θ_A, θ_C) at different values of damping coefficient k. It is evident that along with (23), the triangle inequalities should also be satisfied:

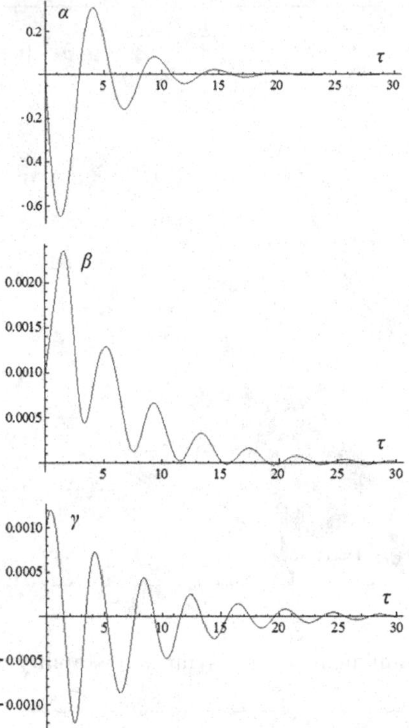

Fig. 6. The transitional process of damping oscillations for $k = 0.5$

$\theta_A + \theta_C > 1, \theta_C + 1 > \theta_A, \theta_A + 1 > \theta_C$. One may disregard the third triangle inequality, since when $\theta_A > \theta_C$ it holds automatically. Thus, the region is limited by the straight lines

$$\theta_C = 1 - \theta_A, \quad \theta_C = \theta_A, \quad \theta_C = \theta_A - 1. \tag{24}$$

An example of such a region and also all the lines on which one of inequalities (23) converts into equality are shown in Figs. 4 and 5. The region where the necessary and sufficient conditions of stability are satisfied is marked out by gray color. In Fig. 4, the region of fulfillment of the necessary and sufficient conditions of stability (23) for $k = 0.5$ is bounded by the straight lines (24) and by hyperbola $A_4 = 0$. In Fig. 5 for $k = 1$, the region where stability conditions (23) hold is bounded only by the straight lines (24).

The numerical integration of system (4) and (5) has been done in the special case when $k_1 = k_2 = k_3 = k$. The different types of transition decay processes of spatial oscillations of the satellite at different damping parameters have been investigated numerically. Figure 6 shows an example of transition decay processes of spatial oscillations for $k = 0.5$ and for inertia parameters $\theta_A = 1, \theta_C = 0.5$ where conditions of asymptotic stability (23) hold. The system in this case reaches the equilibrium position (19) at all three angles at the τ value, equal to 25.

6 Conclusion

In this paper, we have analyzed the rotational motion of the satellite relative to the center of mass in a circular orbit due to gravity and active damping torques. The main focus is the study of satellite equilibrium orientations and the conditions for their stability. A computer algebra method (based on the construction of Gröbner bases) has been proposed to determine all the equilibria of the satellite in the orbital coordinate system for the given values of the active damping torque vector in the general case; the conditions of their existence have been obtained.

The two-dimensional cross sections of domains with equal number of equilibrium orientations using algebraic methods for the construction of discriminant hypersurfaces have been classified. We have made a detailed analysis of the evolution of different domains of existence of equilibrium orientations in the plane of parameters k_1 and k_3 for the fixed values of parameter k_2.

Necessary and sufficient conditions for asymptotic stability of the equilibrium orientations were obtained with the help of the Routh–Hurwitz criterion. The transition decay processes of spatial oscillations of the satellite have been investigated numerically. The results of this study can be used for a preliminary design of gravitational systems to control the satellite's orientation and make it possible to simulate the influence of the damping torque on its orientation.

Acknowledgements. The authors thank the reviewers for very useful remarks and suggestions and Professor V. Gerdt for the advice on the effectiveness of methods and algorithms of Gröbner basis construction.

References

1. Beletsky, V.V.: Attitude Motion of Satellite in Gravitational Field. MGU Press, Moscow (1975)
2. Sarychev, V.A.: Problems of orientation of satellites. Itogi Nauki i Tekhniki. Ser. "Space Research", **11** (1978). VINITI, Moscow
3. Gutnik, S.A., Guerman, A., Sarychev, V.A.: Application of computer algebra methods to investigation of influence of constant torque on stationary motions of satellite. In: Gerdt, V.P., Koepf, W., Seiler, W.M., Vorozhtsov, E.V. (eds.) CASC 2015. LNCS, vol. 9301, pp. 198–209. Springer, Cham (2015). doi:10.1007/978-3-319-24021-3_15
4. Chen, C., Maza, M.M.: Semi-algebraic description of the equilibria of dynamical systems. In: Gerdt, V.P., Koepf, W., Mayr, E.W., Vorozhtsov, E.V. (eds.) CASC 2011. LNCS, vol. 6885, pp. 101–125. Springer, Heidelberg (2011). doi:10.1007/978-3-642-23568-9_9
5. El Kahoui, M., Weber, A.: Symbolic equilibrium point analysis of parameterized polynomial vector fields. In: Ganzha, V., Mayr, E.W., Vorozhtsov, E.V. (eds.) Computer Algebra in Scientific Computing (CASC 2002), pp. 71–83. Institut für Informatik. Technische Universität München, Garching (2002)

6. Chen, C., Golubitsky, O., Lemaire, F., Maza, M.M., Pan, W.: Comprehensive triangular decomposition. In: Ganzha, V.G., Mayr, E.W., Vorozhtsov, E.V. (eds.) CASC 2007. LNCS, vol. 4770, pp. 73–101. Springer, Heidelberg (2007). doi:10.1007/978-3-540-75187-8_7
7. Buchberger, B.: A theoretical basis for the reduction of polynomials to canonical forms. SIGSAM Bull. **10**(3), 19–29 (1976)
8. Char, B.W., Geddes, K.O., Gonnet, G.H., Monagan, M.B., Watt, S.M.: Maple Reference Manual. Watcom Publications Limited, Waterloo (1992)
9. Weispfenning, V.: Comprehensive Gröbner bases. J. Symb. Comput. **14**(1), 1–30 (1992)
10. Lazard, D., Rouillier, F.: Solving parametric polynomial systems. J. Symb. Comput. **42**(6), 636–667 (1992)
11. Meiman, N.N.: Some problems on the distribution of the zeros of polynomials. Uspekhi Mat. Nauk **34**, 154–188 (1949)
12. Gantmacher, F.R.: The Theory of Matrices. Chelsea Publishing Company, New York (1959)
13. Batkhin, A.B.: Parameterization of the discriminant set of a polynomial. Program. Comput. Softw. **42**(2), 65–76 (2016)

Characteristic Set Method for Laurent Differential Polynomial Systems

Youren Hu(✉) and Xiao-Shan Gao

KLMM, UCAS, Academy of Mathematics and Systems Science, Chinese Academy of Sciences, Beijing 100190, China
huyouren14@mails.ucas.ac.cn

Abstract. In this paper, a characteristic set method for Laurent (differential) polynomial systems is given. In the Laurent polynomial case, the concept of Laurent regular chain is introduced and a characteristic set algorithm for Laurent polynomial system is given. In the Laurent differential polynomial case, we give a partial method to decide whether a Laurent differential chain $\mathscr{A}$ is Laurent regular.

Keywords: Characteristic set · Gröbner basis · Laurent differential polynomial · Laurent regular

1 Introduction

The characteristic set method can be used to decompose the zero set of a general polynomial set into the union of zero sets of polynomials in triangular form. This method has applications in automated reasoning, robotics, computer vision, computer-aided design, and analysis of cryptosystems, etc [20]. The characteristic set method was proposed by Ritt and was extensively studied in the past thirty years for polynomial systems [7,18,19,21], semi-algebraic sets [6], polynomial systems over finite fields [9,15], differential polynomial systems [2,3,8,13,17,23], and difference polynomial systems [11].

In this paper, we consider the characteristic set method for Laurent polynomial systems and Laurent differential polynomial systems. This is motivated by the work on difference binomial ideals [10], where the characteristic set method in the Laurent case plays a key role.

In the Laurent polynomial case, we introduce the concept of Laurent regular chain (or triangular set) and prove that it has similar properties with regular chains in the non-Laurent case. Then, a characteristic set algorithm is given to decompose the zero set of a Laurent polynomial system into the union of zero sets of Laurent regular chains. We also introduce the concept of Laurent Gröbner basis for a Laurent polynomial ideal and use it to give a minimal triangular set decomposition for the zero set of a Laurent polynomial system. Laurent

Partially supported by an NSFC grant No. 11688101.

V.P. Gerdt et al. (Eds.): CASC 2017, LNCS 10490, pp. 183–195, 2017.
DOI: 10.1007/978-3-319-66320-3_14

Gröbner bases were defined in [16,22]. The method given in [16] is direct but quite complicated. The method given in [22] is similar to our definition, but our treatment is simpler and more direct.

In the Laurent differential polynomial case, we introduce the concept of Laurent regular differential chain. But, the problem of deciding whether a Laurent differential chain is Laurent regular is quite difficult and is still open. We first show that deciding whether a Laurent differential chain is Laurent regular can be reduced to deciding whether a univariate differential polynomial $f(z)$ is Laurent regular, then give a partial method to decide whether $f(z)$ is Laurent regular, which is complete when $f(z)$ is of the first order. The method is based on the work of Cano [4,5] and Grigoriev-Singer [12], where a special type of Newton polygon is introduced to describe the minimal monomial of the series solution to a differential equation.

It should be noticed that the extension of the characteristic set method [11] to Laurent difference polynomial systems is straightforward. The reason is that if a difference polynomial f is invertible w.r.t a proper irreducible regular and coherent difference chain $\mathscr{A}$, then $\sigma^k f$ is also invertible w.r.t $\mathscr{A}$ for the difference operator σ and any $k \in \mathbb{N}$ [11], which is not true in the differential case.

The paper is organized as follows. In Sect. 2, we present the characteristic set method for Laurent polynomial systems. In Sect. 3, we present the characteristic set method for Laurent differential polynomial systems.

2 Laurent Polynomial Systems

2.1 Laurent Regular Chain

In this section, we will define and prove the basic properties of Laurent regular chains. For the basic concepts about the characteristic set, please refer to [2,3,18,19].

Let k be a field and $\mathbb{Y} = \{y_1, \ldots, y_n\}$ a set of indeterminates. We denote $k[\mathbb{Y}] = k[y_1, \ldots, y_n]$ to be the polynomial ring in $\mathbb{Y}$ and $k[\mathbb{Y}^{\pm}]$ the Laurent polynomial ring in $\mathbb{Y}$. For $F \subseteq k[\mathbb{Y}]$ and $G \subseteq k[\mathbb{Y}^{\pm}]$, we denote by (F) or $(G)_{k[\mathbb{Y}^{\pm}]}$ to be the ideals generated by F or G in $k[\mathbb{Y}]$ or $k[\mathbb{Y}^{\pm}]$, respectively.

A polynomial $f \in k[\mathbb{Y}]$ is called *monomial-primitive* if no y_i divides f. The *normal form* of a Laurent polynomial $f = \sum_{i=1}^{s} a_i \frac{m_i}{n_i} \in k[\mathbb{Y}^{\pm}]$, where m_i, n_i are monomials in $k[\mathbb{Y}]$ and $\gcd(m_i, n_i) = 1$, is the monomial-primitive polynomial $\frac{f \operatorname{lcm}(n_1,\ldots,n_s)}{\gcd(m_1,\ldots,m_s)}$, denoted by $\tilde{f}$.

As in the characteristic set theory, we fix a variable order $y_1 < \cdots < y_n$. For $f \in k[\mathbb{Y}]$, denote $\operatorname{lv}(f)$ to be the largest variable occurring in f.

Definition 1. *A Laurent ascending chain, or simply a Laurent chain $\mathscr{A} = A_1, \ldots, A_p$ in $k[\mathbb{Y}^{\pm}]$ is a chain in $k[\mathbb{Y}]$ such that $A_1, \ldots, A_p \in k[\mathbb{Y}]$ are monomial-primitive. A characteristic set of a Laurent polynomial set $\mathbb{P} \subseteq k[\mathbb{Y}^{\pm}]$ is defined to be the characteristic set of $\widetilde{\mathbb{P}}$.*

A Laurent polynomial f is said to be *reduced* w.r.t a Laurent chain $\mathscr{A}$ if $\widetilde{f}$ is reduced w.r.t $\mathscr{A}$. We define the pseudo-remainder of a Laurent polynomial f w.r.t a Laurent chain $\mathscr{A}$ as $\text{lprem}(f, \mathscr{A}) = \text{prem}(\tilde{f}, \mathscr{A})$, where $\text{prem}(\tilde{f}, \mathscr{A})$ is the usual pseudo-remainder.

Assume $\mathscr{A} = A_1, \ldots, A_p$ in $k[\mathbb{Y}]$ is a chain and let x_i be the leading variable of A_i, $\mathbb{X} = \{x_1, \ldots, x_p\}$ and $\mathbb{U} = \mathbb{Y} - \mathbb{X}$. $\mathbb{U}$ and $\mathbb{X}$ are called the *parameter set* and the *leading variable set* of $\mathscr{A}$ respectively. We denote $k[\mathbb{Y}]$ as $k[\mathbb{U}, \mathbb{X}]$. A polynomial $f \in k[\mathbb{U}, \mathbb{X}]$ is said to be *invertible* w.r.t $\mathscr{A}$ if either $f \in k[\mathbb{U}]$ or $(f, A_1, \ldots, A_s) \bigcap k[\mathbb{U}] \neq \{0\}$ where $\text{lv}(f) = \text{lv}(A_s)$. $\mathscr{A}$ is called *regular* if the initial of A_i is invertible w.r.t $\mathscr{A}$, for $i = 1, \ldots, n$. Regular chains have the following nice properties [1,3].

Lemma 1. *The following statements are equivalent.*

1. *$\mathscr{A}$ is a regular chain.*
2. *$\mathscr{A}$ is a characteristic set of* $\text{sat}(\mathscr{A})$*, where* $\text{sat}(\mathscr{A})$ *is the saturation ideal of $\mathscr{A}$.*
3. *If a polynomial N is invertible w.r.t $\mathscr{A}$ such that $Nf \in \text{sat}(\mathscr{A})$, then $f \in \text{sat}(\mathscr{A})$.*

Lemma 2. *A polynomial f is not invertible w.r.t a regular chain $\mathscr{A}$ if and only if there exists a nonzero polynomial N reduced w.r.t $\mathscr{A}$, such that $Nf \in (\mathscr{A})$.*

Let $\mathscr{A}$ be a Laurent chain and $I_{\mathscr{A}}$ the product of the initials of $\mathscr{A}$. The *Laurent saturation ideal* of $\mathscr{A}$ in $k[\mathbb{Y}^{\pm}]$ is defined as follows

$$\text{lsat}(\mathscr{A}) = \{f \in k[\mathbb{Y}^{\pm}] \mid \exists s \in \mathbb{N}, I_{\mathscr{A}}^s f \in (\mathscr{A})_{k[\mathbb{Y}^{\pm}]}\}.$$

Definition 2. *A Laurent chain $\mathscr{A}$ is called a* Laurent regular chain *if it is a regular chain in $k[\mathbb{Y}]$ and y_i is invertible w.r.t $\mathscr{A}$ for any $i = 1, \ldots, n$. Let $\mathscr{A}$ be a Laurent regular chain, $H \in k[\mathbb{Y}^{\pm}]$ is said to be* invertible *w.r.t $\mathscr{A}$ if $\widetilde{H}$ is invertible w.r.t $\mathscr{A}$ in $k[\mathbb{Y}]$.*

A Laurent regular chain has similar properties to that of a regular chain.

Lemma 3. *If $\mathscr{A}$ is a Laurent regular chain, then $f \in \text{lsat}(\mathscr{A})$ if and only if* $\text{lprem}(f, \mathscr{A}) = 0$.

Proof. It is obvious that if $\text{lprem}(f, \mathscr{A}) = 0$, then $f \in \text{lsat}(\mathscr{A})$. We need only to prove the converse implication. Let $f \in \text{lsat}(\mathscr{A})$. Then there exists a monomial M and $m \in \mathbb{N}$ such that $I_{\mathscr{A}}^m M\tilde{f} \in (\mathscr{A})$, or equivalently $M\tilde{f} \in \text{sat}(\mathscr{A})$. Because $\mathscr{A}$ is Laurent regular and M is invertible w.r.t $\mathscr{A}$, by Lemma 1, $\tilde{f} \in \text{sat}(\mathscr{A})$. By Lemma 1, $\text{prem}(\tilde{f}, \mathscr{A}) = 0$, that is $\text{lprem}(f, \mathscr{A}) = 0$. □

Lemma 4. *Assume $\mathscr{A}$ is a Laurent regular chain and let $\mathbb{U}$ be the parameter set of $\mathscr{A}$ and $D \in k[\mathbb{U}^{\pm}]$, then $PD \in \text{lsat}(\mathscr{A})$ implies $P \in \text{lsat}(\mathscr{A})$.*

Proof. Suppose $H = \text{lprem}(P, \mathscr{A})$, then there exists $m \in \mathbb{N}$ such that $I_{\mathscr{A}}^m \widetilde{P} - H \in (\mathscr{A})$. So $DP \in \text{lsat}(\mathscr{A})$ implies $DH \in \text{lsat}(\mathscr{A})$. Since $\mathscr{A}$ is Laurent regular and DH is reduced w.r.t $\mathscr{A}$, by Lemma 3, $DH = 0$. So $H = 0$, which means that $P \in \text{lsat}(\mathscr{A})$. □

Lemma 5. *Let $\mathscr{A}$ be a Laurent regular chain. Then $P \in k[\mathbb{Y}^{\pm}]$ is not invertible w.r.t $\mathscr{A}$ if and only if there exists a nonzero Laurent polynomial Q reduced w.r.t $\mathscr{A}$, such that $QP \in (\mathscr{A})_{k[\mathbb{Y}^{\pm}]}$.*

Proof. First, suppose that $P \in k[\mathbb{Y}^{\pm}]$ is not invertible w.r.t $\mathscr{A}$. By Lemma 2, there exists a nonzero polynomial $Q \in k[\mathbb{Y}]$ reduced w.r.t $\mathscr{A}$, such that $\tilde{P}Q \in (\mathscr{A})$, so $PQ \in (\mathscr{A})_{k[\mathbb{Y}^{\pm}]}$. To prove the inverse implication, consider some P invertible w.r.t $\mathscr{A}$. Then there exist $U \in k[\mathbb{U}]$ and $A \in k[\mathbb{Y}]$ such that $U - A\tilde{P} \in (\mathscr{A})$. So for nonzero Laurent polynomial Q reduced w.r.t $\mathscr{A}$, we have $UQ - A\tilde{P}Q \in (\mathscr{A})_{k[\mathbb{Y}^{\pm}]} \subseteq \text{lsat}(\mathscr{A})$. By Lemmas 3 and 4, UQ cannot be in $\text{lsat}(\mathscr{A})$, which implies $\tilde{P}Q$ cannot be in $\text{lsat}(\mathscr{A})$. So PQ cannot be in $(\mathscr{A})_{k[\mathbb{Y}^{\pm}]}$, which is a contradiction. □

Theorem 1. *A Laurent chain $\mathscr{A}$ is Laurent regular if and only if $\mathscr{A}$ is the characteristic set of* $\text{lsat}(\mathscr{A})$.

Proof. Assume that $\mathscr{A}$ is Laurent regular. By Lemma 3, for $P \in \text{lsat}(\mathscr{A})$, $\text{lprem}(P, \mathscr{A}) = 0$. Thus $\mathscr{A}$ is the characteristic set of $\text{lsat}(\mathscr{A})$. Now suppose that $\mathscr{A}$ is the characteristic set of $\text{lsat}(\mathscr{A})$. Then $\mathscr{A}$ is the characteristic set of $\text{sat}(\mathscr{A})$. By Lemma 1, $\mathscr{A}$ is regular. If some monomial M is not invertible w.r.t $\mathscr{A}$, by Lemma 5, there exists a nonzero Laurent polynomial Q reduced w.r.t $\mathscr{A}$ such that $MQ \in (\mathscr{A})_{k[\mathbb{Y}^{\pm}]} \subseteq \text{lsat}(\mathscr{A})$. Thus $Q \in \text{lsat}(\mathscr{A})$, which is a contradiction. □

Let $\mathscr{A} = \{A_1, \ldots, A_p\}$ be a regular chain in $k[\mathbb{Y}] = k[\mathbb{U}, \mathbb{X}]$, where $\mathbb{U}$ is the parameter set of $\mathscr{A}$. A polynomial P is called *reducible* modulo $\mathscr{A}$ if there exist $0 \neq M \in k[\mathbb{U}]$, $P_1, P_2 \in k[\mathbb{Y}]$ with the same leading variable as P and the initials of $P_i, i = 1, 2$ are invertible w.r.t $\mathscr{A}$ such that $MP = P_1P_2 \bmod(\mathscr{A})$. If such M, P_1, P_2 do not exist, P is called *irreducible* modulo $\mathscr{A}$. $\mathscr{A}$ is called *irreducible* if A_1 is irreducible as a polynomial in $k[\mathbb{U}][x_1]$ and A_i is irreducible modulo $\mathscr{A}_{i-1} = \{A_1, \ldots, A_{i-1}\}$, $i = 2, \ldots, p$.

Let K be an algebraically closed extension of k and $K^* = K \setminus \{0\}$. For a Laurent polynomial set $S \subset k[\mathbb{Y}^{\pm}]$, we use $\text{LZero}(S)$ to denote the elements $e \in (K^*)^n$, which are zeros of the Laurent polynomials in S.

2.2 Characteristic Set Method

In this subsection, we present the characteristic set method for Laurent polynomial systems, which is basically the same as that given in [7,19], and the correctness can be similarly proved. We first show how to check whether a polynomial is invertible w.r.t a regular chain.

Algorithm 1. Invert$(f, \mathscr{A})$
Input: a polynomial f and a regular chain $\mathscr{A}$.
Output: a pair (test,g) such that
test = *true*, if f is invertible w.r.t $\mathscr{A}$, and $g = 0$.

test = *false*, if f is not invertible w.r.t $\mathscr{A}$, and g is a nonzero polynomial reduced w.r.t $\mathscr{A}$ such that $fg \in (\mathscr{A})$.
Begin
  $w :=$ a new indeterminate;
  $P(U,w) = \text{Res}(w - f, \mathscr{A})$, and $\text{Res}(g, \mathscr{A})$: the resultant of g w.r.t $\mathscr{A}$ [21].
  If $P(U,0) \neq 0$, then $test := true$ and $g = 0$
  else $test := false$ and $g = \text{prem}(\overline{P}(f), \mathscr{A})$, where $\overline{P}(U,w) = P(U,w)/w$.
End.

The above algorithm is based on methods in [3,21], where the details could be found. We now show how to check whether a chain is Laurent regular.

Algorithm 2. LRegular($\mathscr{A}$)
Input: A chain $\mathscr{A} = f_1(U,x_1),\ldots,f_p(U,x_1,\ldots,x_p) \subseteq k[U,x_1,\ldots,x_p]$.
Output: If $\mathscr{A}$ is Laurent regular, output $(\varnothing,\varnothing)$. Otherwise let i be the largest number such that $\mathscr{A}_{i-1} = f_1,\ldots,f_{i-1}$ is Laurent regular. If $I_i = \text{init}(f_i)$ is not invertible w.r.t $\mathscr{A}_{i-1}$, output (I_i, g) such that $I_i g \in (A_{i-1})$ and g is reduced w.r.t $\mathscr{A}_{i-1}$. If x_i is not invertible w.r.t $\mathscr{A}_{i-1}$, then output (x_i, g) such that $x_i g \in (\mathscr{A}_{i-1})$, where g is reduced w.r.t $\mathscr{A}_{i-1}$.
  Begin
  Let s be the largest integer such that $x_1^s | f_1$.
  If $s > 0$, then return $(x_1, f_1/x_1^s)$.
  i=2;
  while $i \leq p$
    $(test_1, g_1) =$ **Invert**$(I_i, \mathscr{A}_{i-1})$. If $test_1 = false$ return $(I,g) = (I_i, g_1)$
    $(test_2, g_2) =$ **Invert**$(x_i, \mathscr{A}_{i-1})$ If $test_2 = false$ return $(I,g) = (x_i, g_2)$
    $i = i + 1$
  end while
  if $i = p + 1$ then return $(\varnothing, \varnothing)$
End.

We now give the main algorithm.

Algorithm 3. ZDec(P)
Input: a finite set P of Laurent polynomials in $k[\mathbb{Y}^{\pm}]$.
Output: $W = \{T_1,\ldots,T_k\}$ such that each T_i is a Laurent regular chain and

$$\text{LZero}(P) = \bigcup_{i=1}^{k} \text{LZero}(\text{lsat}(T_i)).$$

Begin
  $P =$ **Normalize**(P): returns normal forms of the polynomials in P.
  $C =$ **Charset**(P): C is a Wu-characteristic set of P [19].
  $C =$ **Normalize**(C): Remove the monomial factors from elements in C.
  $(I,g) =$ **LRegular**(C).
    If $I = y_i$ for some i then $W =$ **ZDec**$(P \bigcup \{g\} \bigcup C)$.
    If $I \neq \varnothing$ then $W =$ **ZDec**$(P \bigcup \{I\} \bigcup C) \bigcup$ **ZDec**$(P \bigcup \{g\} \bigcup C)$.
    If $I = \varnothing$ then $W = \{C\} \bigcup \bigcup_{I \in I_C}$ **ZDec** $(P \bigcup \{I\} \bigcup C)$,
      where I_c is the set of initials of C.
End.

2.3 Laurent Gröbner Basis and Minimal Decomposition

To obtain a minimal zero decomposition, we need the concept of Laurent Gröbner basis.

Definition 3. *A finite set G of monomial-primitive polynomials is said to be a* Laurent Gröbner basis *of the Laurent polynomial ideal $I \subseteq k[\mathbb{Y}^{\pm}]$ if $G \subseteq I$ and for any $f \in I$,* $\mathrm{grem}(\widetilde{f}, G) = 0$, *where* $\mathrm{grem}(\widetilde{f}, G)$ *is the normal form of $\widetilde{f}$ w.r.t G as defined in the Gröbner basis theory in the polynomial ring.*

Assuming that $F = \{f_1, \ldots, f_s\} \subseteq k[\mathbb{Y}^{\pm}]$, we can compute the Laurent Gröbner basis of $(F)_{k[\mathbb{Y}^{\pm}]}$ as follows.

Theorem 2. *Let G_0 be the reduced Gröbner basis of $(\widetilde{f}_1, \ldots, \widetilde{f}_s, y_i z_i - 1, i = 1, \ldots, n) \subseteq k[y_1, \ldots, y_n, z_1, \ldots, z_n]$ w.r.t some monomial order satisfying $y_i \prec z_j$ for $i, j = 1, \ldots, n$ and $G = G_0 \bigcap k[y_1, \ldots, y_n]$. Then G is a Laurent Gröbner basis of $(F)_{k[\mathbb{Y}^{\pm}]}$.*

Proof. We claim $G \subseteq (F)_{k[\mathbb{Y}^{\pm}]}$. Since $G \subseteq (\widetilde{f}_1, \ldots, \widetilde{f}_s, y_i z_i - 1, i = 1, \ldots, n)$, any $g \in G$ can be written as $g = \sum_{i=1}^{s} a_i \widetilde{f}_i + \sum_{i=1}^{n} b_i (y_i z_i - 1)$, where $a_i, b_i \in k[y_1, \ldots, y_n, z_1, \ldots, z_n]$. Substituting $1/y_i$ for z_i, we get $Mg = \sum_{i=1}^{s} \widetilde{a}_i \widetilde{f}_i$, where M is a monomial in $k[y_1, \ldots, y_n]$ and $\widetilde{a}_i \in k[y_1, \ldots, y_n]$. So the claim is proved. For any $f \in (F)_{k[\mathbb{Y}^{\pm}]}$, there exists a monomial N in $k[y_1, \ldots, y_n]$ such that $Nf = \sum_{i=0}^{s} c_i \widetilde{f}_i$, where $c_i \in k[y_1, \ldots, y_n]$. By the definition of $\widetilde{f}$, we have $H\widetilde{f} = Nf$, for some monomial H in $k[y_1, \ldots, y_n]$. Assume that $H = y_{i_0} H_1$, for some i_0. Then $y_{i_0} H_1 \widetilde{f} = \sum_{i=0}^{s} c_i \widetilde{f}_i$, we have $z_{i_0} y_{i_0} H_1 \widetilde{f} = \sum_{i=0}^{s} c_i z_{i_0} \widetilde{f}_i$. Then $H_1 \widetilde{f} = \sum_{i=0}^{s} c_i z_{i_0} \widetilde{f}_i - (y_{i_0} z_{i_0} - 1) H_1 \widetilde{f} \in (\widetilde{f}_1, \ldots, \widetilde{f}_s, y_i z_i - 1, i = 1, \ldots, n) \bigcap k[y_1, \ldots, y_n]$. Repeating the above process, we have $\widetilde{f} \in (\widetilde{f}_1, \ldots, \widetilde{f}_s, y_i z_i - 1, i = 1, \ldots, n) \bigcap k[y_1, \ldots, y_n]$. Since G is a Gröbner basis of $(\widetilde{f}_1, \ldots, \widetilde{f}_s, y_i z_i - 1, i = 1, \ldots, n)$ under the order $y_i \prec z_j$, for $i, j = 1, \ldots, n$, $\mathrm{grem}(\widetilde{f}, G_0) = 0$ which implies $\mathrm{grem}(\widetilde{f}, G) = 0$.

We prove that G_0 is monomial-primitive. If $g \in G_0$ is not monomial-primitive, then there exists an $i \in \{1, \ldots, n\}$ and a $g' \in k[\mathbb{Y}]$ such that $g = y_i g'$. So we have $g' = z_i g + (1 - y_i z_i) g' \in (\widetilde{f}_1, \ldots, \widetilde{f}_s, y_i z_i - 1, i = 1, \ldots, n) \subseteq k[y_1, \ldots, y_n, z_1, \ldots, z_n]$, which implies that there exists a $q \in G_0$ such that $lt(q) | lt(g') | lt(g)$. It is a contradiction to the fact that G_0 is a reduced Gröbner basis. Thus G_0 is monomial-primitive. □

Example 1. Let $F = \{y_1 y_3 - y_2, y_2 y_4 - y_1\}$. Then F is already a Gröbner basis, but not a Laurent Gröbner basis. With the method given in Theorem 2, we can compute the Laurent Gröbner basis of $(F)_{k[\mathbb{Y}^{\pm}]}$: $\{y_3 y_4 - 1, y_1 y_3 - y_2, y_2 y_4 - y_1\}$.

Remark: To obtain a minimal decomposition, we first compute a decomposition $\mathrm{LZero}(P) = \bigcup_{i=1}^{k} \mathrm{LZero}(\mathrm{lsat}(T_i))$, where T_i are irreducible [19]. Then, compute the Laurent Gröbner basis B_i of $\mathrm{lsat}(T_i)$ with Theorem 2. We have $\mathrm{LZero}(P) = \bigcup \mathrm{LZero}(B_i)$ and a minimal decomposition can be obtained easily.

3 Differential Polynomial Systems

3.1 Laurent Regular Differential Chains

In this section, we will extend the characteristic set method to the Laurent differential case. For details of differential characteristic set method, please refer to [2,8,13,17].

Let $\mathcal{F}$ be a differential field with the differential operator δ, $\mathbb{Y} = \{y_1, y_2, \ldots, y_n\}$ differential indeterminates, and $\mathcal{F}\{\mathbb{Y}\}$ the differential polynomial ring in $\mathbb{Y}$ over $\mathcal{F}$. Let $f \in \mathcal{F}\{\mathbb{Y}\}$, denote $\mathrm{ld}(f)$ to be the leader of f and $\mathrm{ord}(f, y_i)$ the order of f in y_i.

Let $\mathscr{A} = f_1, \ldots, f_p$ be a differential chain, $c_i = \mathrm{ord}(f_i, \mathrm{ld}(f_i))$, and $o \in \mathbb{N}$. Then denote

$$\mathscr{A}^{(o)} = f_1, f_1^{(1)}, \ldots, f_1^{(\widehat{o}-c_1)}, \ldots, f_p, f_p^{(1)}, \ldots, f_p^{(\widehat{o}-c_p)}$$

where $\widehat{o} = \max\{o, c_1, \ldots, c_p\}$. Note that $\mathscr{A}^{(o)}$ is a chain in the polynomial ring $\mathcal{F}[\mathbb{Y}, \ldots, \mathbb{Y}^{(\widehat{o})}]$. Let $f \in \mathcal{F}\{\mathbb{Y}\}$. Then f is called *invertible* w.r.t $\mathscr{A}$ if f is invertible w.r.t $\mathscr{A}^{(\mathrm{ord}(f))}$ in the polynomial ring $\mathcal{F}[\Theta(\mathbb{Y})]$.

Definition 4. *If the initials and separants of a differential chain $\mathscr{A}$ are invertible w.r.t $\mathscr{A}$, then $\mathscr{A}$ is called* differential regular. *Besides, if $y_i^{(j)}$ is invertible w.r.t $\mathscr{A}$, for any $i = 1, \ldots, n, j \in \mathbb{N}$, $\mathscr{A}$ is called* Laurent regular. *If $\mathscr{A}$ consists of only one element, then this polynomial is called a Laurent regular polynomial.*

Let $\mathscr{A}$ be a differential chain and $H_{\mathscr{A}}$ be the product of the initials and separants of $\mathscr{A}$. We define the Laurent saturation ideal of $\mathscr{A}$ to be

$$\mathrm{ldsat}(\mathscr{A}) = [\mathscr{A}] : H_{\mathscr{A}}^{\infty} = \{f \in \mathcal{F}\{\mathbb{Y}^{\pm}\} \mid \exists m \in \mathbb{N}, s.t.\ H_{\mathscr{A}}^{m} f \in [\mathscr{A}]_{\mathcal{F}\{\mathbb{Y}^{\pm}\}}\}.$$

Theorem 1 can be easily extended to the following differential version.

Theorem 3. *$\mathscr{A}$ is Laurent differential regular if and only if $\mathscr{A}$ is the characteristic set of* $\mathrm{ldsat}(\mathscr{A})$.

If we can solve the following problem, then we can extend the characteristic set method to the Laurent differential case.

Problem LR. For an irreducible and differential regular chain $\mathscr{A}$, either decide $\mathscr{A}$ is Laurent regular or find some $y_c^{(e)}$ which is not invertible w.r.t $\mathscr{A}$.

The decision of whether a differential chain $\mathscr{A}$ is Laurent differential regular is still open and we will give some partial answers to this problem in the rest of this paper. We first give an example to show why the problem is difficult.

Example 2. Let $f = xy_1' - ky_1, k \in \mathbb{N}$. Then $f^{(i)} = xy_1^{(i+1)} + (i-k)y^{(i)}$ for $i = 1, \ldots, k$. Hence $f^{(k)} = xy_1^{(k+1)}$. That is, $y_1^{(k+1)}$ is not invertible w.r.t f and f is not a Laurent regular differential polynomial.

In this section, a differential chain $\mathscr{A}$ is always irreducible and regular.

Lemma 6. *A differential regular chain $\mathcal{A}$ is not Laurent differential regular if and only if* $\mathrm{prem}(y_i^{(m)}, \mathcal{A}) = 0$ *for some* $i \in \{1, \ldots, n\}$ *and* $m \in \mathbb{N}$.

Proof. If $\mathcal{A}$ is not Laurent differential regular, then $y_i^{(m)}$ is not invertible w.r.t $\mathcal{A}$ for some $i \in [1, n]$ and $m \in \mathbb{N}$. By Lemma 2, there exists an N reduced w.r.t $\mathcal{A}$ and $Ny_i^{(m)} \in [\mathcal{A}]$. Since $\mathcal{A}$ is regular and irreducible, we have $y_i^{(m)} \in \mathrm{dsat}(\mathcal{A})$ and thus $\mathrm{prem}(y_i^{(m)}, \mathcal{A}) = 0$. On the other hand, assume $\mathrm{prem}(y_i^{(m)}, \mathcal{A}) = 0$. Then we have $Hy_i^{(m)} = 0 \mod [\mathcal{A}]$, where H is a power of the product of the initials and separants of $\mathcal{A}$. Since $\mathcal{A}$ is regular, H is invertible w.r.t $\mathcal{A}$. By Lemma 1, $y_i^{(m)}$ is not invertible w.r.t $\mathcal{A}$ and hence $\mathcal{A}$ is not Laurent regular. □

As a consequence, we have the following results.

Corollary 1. *If* $y_p^{(m)}$ *is not invertible w.r.t* $\mathcal{A}$, *then* $\mathrm{dsat}(\mathcal{A}) \bigcap \mathcal{F}\{y_p\} \neq \{0\}$.

Corollary 2. *If* $f \in \mathcal{F}\{\mathbb{Y}\}$ *is irreducible and there are more than one differential indeterminate in* f, *then* f *is Laurent regular.*

Theorem 4. *Let $\mathcal{A}$ be a differential regular and irreducible differential chain. Then deciding whether $\mathcal{A}$ is Laurent regular can be reduced to deciding whether a univariate differential polynomial is Laurent regular.*

Proof. Let $U = \{u_1, \ldots, u_q\}$ be the parameter set of $\mathcal{A}$ and $X = \mathbb{Y} \setminus U = \{x_1, \ldots, x_p\} (p + q = n)$. Then $\mathcal{A}$ can be written as $\mathcal{A} = f_1(U, x_1), f_2(U, x_1, x_2), \ldots, f_p(U, x_1, \ldots, x_p)$ with $\mathrm{ld}(f_i) = x_i^{(o_i)}$. By Lemma 6 and Corollary 1, $u_i^{(j)}, i = 1, \ldots, q,\ j \in \mathbb{N}$ is invertible w.r.t $\mathcal{A}$. We now consider whether $x_c^{(e)}, c = 1, \ldots, p, e \in \mathbb{N}$ is invertible w.r.t $\mathcal{A}$. For each c, we can use the change of order algorithm given in [2] to compute a regular and irreducible differential chain $\mathcal{A}_c$ under the variable order $U < x_c < x_1 < \cdots x_{c-1} < x_{c+1} < \cdots < x_p$ such that $\mathrm{dsat}(\mathcal{A}) = \mathrm{dsat}(\mathcal{A}_c)$. Since $\mathcal{A}$ is irreducible, it is clear that

$$\mathcal{A}_c = g_1(U, x_c), g_2(U, x_c, x_1), \ldots, f_p(U, x_c, x_1, \ldots, x_{c-1}, x_{c+1}, \ldots, x_p).$$

Since $\mathcal{A}_c$ is differential regular, we have $\mathrm{dsat}(\mathcal{A}) \cap \mathcal{F}\{x_c\} = \mathrm{dsat}(\mathcal{A}_c) \cap \mathcal{F}\{x_c\} = \mathrm{dsat}(g_1) \cap \mathcal{F}\{x_c\}$. If g_1 contains some u_i, then $\mathrm{dsat}(\mathcal{A}_c) \cap \mathcal{F}\{x_c\} = \mathrm{dsat}(g_1) \cap \mathcal{F}\{x_c\} = \{0\}$. By Corollary 1, $x_c^{(e)}$ is invertible w.r.t $\mathcal{A}$ for any $e \in \mathbb{N}$. If $g_1 \in \mathcal{F}\{x_c\}$, then $\mathrm{dsat}(\mathcal{A}_c) \cap \mathcal{F}\{x_c\} = \mathrm{dsat}(g_1) \cap \mathcal{F}\{x_c\} = \mathrm{dsat}(g_1)$. By Lemma 6, $x_c^{(e)}$ is invertible w.r.t $\mathcal{A}$ if and only if it is invertible w.r.t g_1, that is, g_1 is Laurent regular. The theorem is proved. □

3.2 Decision of Univariate Laurent Regular Differential Polynomial

Let z be a differential indeterminate and $\mathcal{F} = \mathbb{Q}(x)$ with $\delta = \frac{\mathrm{d}}{\mathrm{d}x}$, and $f(z) \in \mathcal{F}\{z\}$ a univariate irreducible differential polynomial. In this section, we will give a partial solution to decide whether $f(z)$ is Laurent regular.

Let $n = \text{ord}(f)$ and $z_i = z^{(i)}$ for $i \in \mathbb{N}$ with $z_0 = z$. Then, we can write f as a polynomial in $\mathbb{Q}[x, z, z_1, \ldots, z_n]$:

$$f(z) = \sum_{(\alpha,\beta)\in\mathbb{N}\times\mathbb{N}^{n+1}} c_{\alpha,\beta} x^{\alpha} z_0^{\beta_0} \ldots z_n^{\beta_n} \quad (1)$$

where $\beta = (\beta_0, \ldots, \beta_n)$ and $c_{\alpha,\beta} \in \mathbb{Q}$ are not zero for a finite number of terms.

Note that $z_m \in \text{sat}(f)$ if and only if $f(z) = 0$ has a *generic polynomial solution*. Therefore, we need to give a method to find polynomial solutions of $f(z) = 0$. Following [4,5], we define a special Newton polygon.

Definition 5. *For $f(z)$ given in (1), we set*

$$\varepsilon(f(z)) = \{P(\alpha, \beta) = (\alpha - \beta_1 - 2\beta_2 - \ldots - n\beta_n, \beta_0 + \ldots + \beta_n) : c_{(\alpha,\beta)} \neq 0\}.$$

Define $NP(f(z))$ to be the convex hull of $\varepsilon(f(z))$ in $\mathbb{R}^2$. In the rest of this paper, we use u and v to represent horizontal and vertical axes of $\mathbb{R}^2$, respectively.

Example 3. Assume $f(z) = z'^2 - 2z' - 4z + 4x$. Then $\varepsilon(f) = \{(-2, 2), (-1, 1), (0, 1), (1, 0)\}$ and $NP(f)$ is given in Fig. 1.

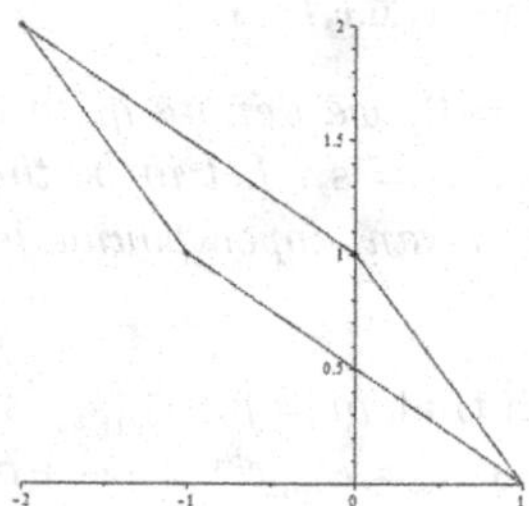

Fig. 1. $NP(f(z))$ in Example 3

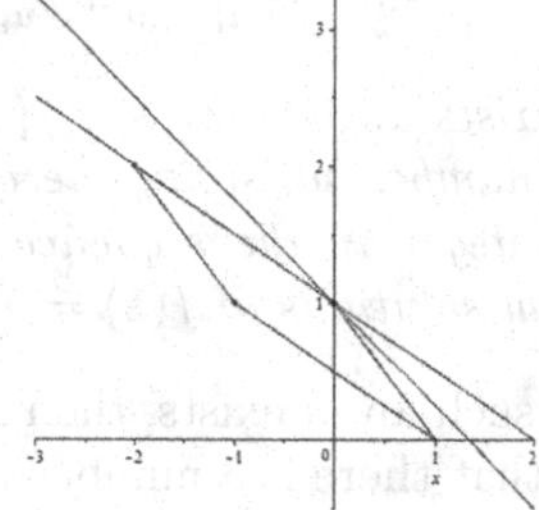

Fig. 2. $NP(f(z))$ in Example 4

Definition 6. *Given $f(z)$ as before and $\mu \in \mathbb{R}$, a straight line $L(f, \mu)$ in $\mathbb{R}^2$ is called* feasible *for $f(z)$ if $L(f, \mu)$ is a line with slope $-1/\mu$, which has the defining equation $u + \mu v = w$, for some $w \in \mathbb{R}$, and $L(f, \mu)$ left-supports $NP(f(z))$, that is, $NP(f(z)) \bigcap L(f, \mu) \neq \varnothing$ and $NP(f(z))$ lies in the left side of $L(f, \mu)$.*

Example 4. Assume that f is defined in Example 3. The μ corresponding to the feasible lines in Fig. 2 are 2 and 4/3.

Definition 7. *Let $f(z)$ be defined as in (1). We denote the vertices of $NP(f)$ from the top-right to the right-most as $P_1, \ldots, P_t$ clockwise, where $P_i = (u_i, v_i)$. Then the line P_iP_{i+1}, $i = 1, \ldots, t-1$, is denoted l_i and its slope is $\frac{v_{i+1}-v_i}{u_{i+1}-u_i} < 0$. We set $s_i = -\frac{u_{i+1}-u_i}{v_{i+1}-v_i} > 0$, for $i = 1, \ldots, t-1$, $s_0 = +\infty$ and $s_t = 0$, for $t \geq 2$. If $t = 1$, set $s_1 = 0$.*

It is easy to see that

Lemma 7. *We have* $s_0 > s_1 > \ldots > s_{t-1} > s_t$ *and the lines*

$$L(f, \delta_i) : u + \delta_i v = u_i + \delta_i v_i, \delta_i \in [s_i, s_{i-1}), i = 1, \ldots, t-1$$

are all the feasible lines whose slopes are negative.

Associated to $P_i = (u_i, v_i)$, we define the polynomials

$$\Phi_i(\mu) = \sum_{P(\alpha,\beta)=(u_i,v_i)} c_{\alpha,\beta}(\mu)_1^{\beta_1} \ldots (\mu)_n^{\beta_n} \in \mathbb{Q}[\mu], \tag{2}$$

where $i = 1, \ldots, t$ and $(\mu)_k = \mu(\mu-1)\ldots(\mu-k+1)$.

Example 5. Let f be defined in Example 3. Then $P_1 = (-2, 2), P_2 = (0, 1), P_3 = (1, 0), s_0 = +\infty, s_1 = 2, s_2 = 1, s_3 = 0$ and $\Phi_1 = \mu^2, \Phi_2 = -4, \Phi_3 = 4$.

We first consider a differential polynomial of first order, for which there exists a complete method to decide whether $f(z)$ has a polynomial solution.

Theorem 5. *Assuming* $f(z) = \sum_{(\alpha,\beta)\in\mathbb{N}\times\mathbb{N}^2} c_{\alpha,\beta} x^\alpha z_0^{\beta_0} z_1^{\beta_1}$ *and* $P_1, \ldots, P_t$ *be defined as in Definition 7, then associated to* P_i, Φ_i *becomes*

$$\Phi_i = c_{u_i,v_i,0} + c_{u_i+1,v_i-1,1}\mu + \cdots + c_{u_i+v_i,0,v_i}\mu^{v_i}. \tag{3}$$

If there exists an $n \in (s_i, s_{i-1}) \bigcap \mathbb{N}$ *such that* $\Phi_i(n) = 0$, *we denote* η_i *to be the maximal number satisfying the condition. Otherwise,* $\eta_i = s_i$. *Let* m *be the first positive integer in the sequence* $\eta_1, \ldots, \eta_t$. *Then* m *is an upper bound for the polynomial solutions of* $f(z) = 0$.

Proof. If such an m exists, then $\exists i_0 \in \{1, \ldots, t\}$ such that $m = \eta_{i_0} \in [s_{i_0}, s_{i_0-1})$. Suppose that there is a number $m < \mu_0 \in \mathbb{N}$ such that $\tilde{z} = a_{\mu_0} x^{\mu_0} + \cdots + a_0$ is a polynomial solution of $f(z) = 0$. Then μ_0 must be in $[s_i, s_{i-1})$ for some $i \leq i_0$. Substitute $\tilde{z}$ into $f(z)$ and we get

$$f(\tilde{z}) = \sum c_{\alpha,\beta_0,\beta_1} x^\alpha (a_{\mu_0} x^{\mu_0} + \ldots + a_0)^{\beta_0} (\mu_0 a_{\mu_0} x^{\mu_0-1} + \ldots + a_1)^{\beta_1}.$$

The leading term in the expansion of $c_{\alpha,\beta_0,\beta_1} x^\alpha z^{\beta_0} z_1^{\beta_1}$ in $f(\tilde{z})$ is

$$c_{\alpha,\beta_0,\beta_1} a_{\mu_0}^{\beta_0+\beta_1} \mu_0^{\beta_1} x^{\alpha-\beta_1+\mu_0(\beta_0+\beta_1)}.$$

If we choose the maximal number in $\{\alpha - \beta_1 + \mu_0(\beta_0 + \beta_1) : c_{\alpha,\beta_0,\beta_1} \neq 0\}$, denoted by $u_0 + \mu_0 v_0$, where $u_0 = \alpha - \beta_1$, $v_0 = \beta_0 + \beta_1$ for some $c_{\alpha,\beta_0,\beta_1} \neq 0$. Then $u + \mu_0 v \leq u_0 + \mu_0 v_0$, for any $(u, v) \in \varepsilon(f)$. So we can define the line $L(f, \mu_0) : u + \mu_0 v = u_0 + \mu_0 v_0$, which is feasible for $f(z)$. Because $\mu_0 \in [s_i, s_{i-1})$, $(u_0, v_0) = (u_i, v_i)$ or (u_{i+1}, v_{i+1}). Then we consider the coefficient of $x^{u_0+\mu_0 v_0}$ in the expansion of $f(\tilde{z})$: $H = \sum_{\alpha-\beta_1+\mu_0(\beta_0+\beta_1)=u_0+\mu_0 v_0} c_{\alpha,\beta_0,\beta_1} \mu_0^{\beta_1} a_{\mu_0}^{\beta_0+\beta_1} = \sum_{(\alpha-\beta_1,\beta_0+\beta_1)\in L(f,\mu_0)} c_{\alpha,\beta_0,\beta_1} \mu_0^{\beta_1} a_{\mu_0}^{\beta_0+\beta_1}$, and it must be zero. Now we consider two cases.

1. If there exists only one vertex $(u_0, v_0) \in L(f, \mu_0) \bigcap NP(f)$, then H can be written as $H = \sum_{k=0}^{v_0} c_{u_0+k,v_0-k,k}\mu_0^k a_{\mu_0}^{v_0} = a_{\mu_0}^{v_0}\Phi_i(\mu_0)$. So μ_0 must be a zero of Φ_i. Thus if $i < i_0$, we have $m < \mu_0 \leq \eta_i \in \mathbb{N}$, a contradiction to the choice of m. Otherwise, $i = i_0$ implies $m \geq \mu_0$, a contradiction to the choice of μ_0.
2. If $L(f, \mu_0) \bigcap NP(f)$ is an edge of $NP(f)$, $\mu_0 = s_i$. If $i = i_0$, $\mu_0 = s_{i_0} \leq m < \mu_0$, a contradiction. If $i < i_0$, $m < \mu_0 \leq \eta_i \in \mathbb{N}$, a contradiction to the choice of m. □

Example 6. Let f be defined in Example 3. We have $\eta_1 = 2, \eta_2 = 1, \eta_3 = 0$. By Theorem 5, an upper bound for the degrees of the polynomial solutions of $f(z) = 0$ is 2.

We now consider the general case and a partial solution is given.

Theorem 6. *For $f(z) \in \mathcal{F}\{z\}$ of order n as defined in* (1),

$$\Phi_1(\mu) = \sum_{P(\alpha,\beta)=(u_1,v_1)} c_{\alpha,\beta}(\mu)_1^{\beta_1} \dots (\mu)_n^{\beta_n}.$$

Use the notations in Definition 7. Then

1. *If Φ_1 is not the zero polynomial and $t \geq 2$, let m be the maximal integer root of $\Phi_1(\mu) = 0$ in $[s_1, +\infty)$. If such a number exists, then it is an upper bound of the polynomial solutions of $f(z) = 0$. Otherwise, $m = \lfloor s_1 \rfloor$ is an upper bound of the polynomial solutions of $f(z) = 0$.*
2. *If Φ_1 is not the zero polynomial and $t = 1$, let m be the maximal integer root of $\Phi_1(\mu) = 0$ in $(0, +\infty)$. If such a number exists, then it is an upper bound of the polynomial solutions of $f(z) = 0$. Otherwise, $f = 0$ has no polynomial solutions.*
3. *If $\Phi_1 \equiv 0$ and $t \geq 2$, let $m = \lfloor s_1 \rfloor$. If* $\text{prem}(z_{m+1}, f) = 0$*, then m is an upper bound for the polynomial solutions of $f(z) = 0$.*

Proof. 1. If $M > s_1$ and $\tilde{z} = \sum_{i=0}^{M} a_i x^i$ is a polynomial solution of $f(z) = 0$ with $a_M \neq 0$. Then $u_1 + Mv_1 > u + Mv$ for any $(u_1, v_1) \neq (u, v) \in \varepsilon(f)$. Indeed, from the line $L(f, s_1) : u + s_1 v = u_1 + s_1 v_1$, we have $u + s_1 v \leq u_1 + s_1 v_1$, for any $(u_1, v_1) \neq (u, v) \in \varepsilon(f)$. Then $u_1 + Mv_1 = u_1 + s_1 v_1 + (M - s_1)v_1 \geq u + s_1 v + (M - s_1)v_1 > u + s_1 v + (M - s_1)v = u + Mv$, for any $(u_1, v_1) \neq (u, v) \in \varepsilon(f)$. So as in the proof of Theorem 5, the leading term of $f(\tilde{z})$ is $\sum_{P(\alpha,\beta)=(u_1,v_1)} c_{\alpha,\beta}(M)_1^{\beta_1} \dots (M)_n^{\beta_n} a_M^{v_1} x^{u_1+Mv_1}$. Because $f(\tilde{z}) = 0$ and $a_M \neq 0$, we have

$$\Phi_1(M) = \sum_{P(\alpha,\beta)=(u_1,v_1)} c_{\alpha,\beta}(M)_1^{\beta_1} \dots (M)_n^{\beta_n} = 0.$$

So if $\Phi_1(\mu)$ has no positive integer roots larger than s_1, the degree of the polynomial solutions of $f(z) = 0$ is not larger than s_1. Otherwise, the number m exists and we have $M < m$ by the choice of m.

2. $t = 1$ implies that $L(f,\delta) : u + \delta v = u_1 + \delta v_1, 0 < \delta < +\infty$ are all the feasible lines and $L(f,\delta) \bigcap NP(f) = \{(u_1, v_1)\}$, which means that for any $(u', v') \in NP(f), (u', v') \neq (u_1, v_1)$ we have $u' + \delta v' < u_1 + \delta v_1$. Then if $z' = \sum_{i=0}^{N} a_i x^i$ is a solution of $f(z) = 0$, where $a_N \neq 0$, substitute z' into $f(z)$ and we have the leading term of $f(z'(x))$ is $a_N^{v_1} \Phi_1 x^{u_1+Nv_1}$. So $\Phi_1(N) = 0$. Because Φ_1 is not the zero polynomial, we have $N \leq m$. Note that case 3 is trivial.

Remark: An upper bound for the polynomial solutions is not given if $\Phi_1 \equiv 0$, $t \geq 2$ and $\text{prem}(z_{m+1}, f) \neq 0$ or $\Phi_1 \equiv 0$ and $t = 1$.

Example 7. For $f(y) = x^2y'^4 - 2xyy'^3 - x^2y^2y''^2 + y^2y'^2$, $\varepsilon(f(y))$ has one point $(-2, 4)$ and $\Phi_1 \equiv 0$ and $t = 1$. For $g(y) = x^2y'^4 - 2xyy'^3 - x^2y^2y''^2 + y^2y'^2 + x^6$, $\varepsilon(g(y))$ has two points $P_1 = (-2, 4)$ and $P_2 = (6, 0)$ and $\Phi_1 \equiv 0$. We have m which is 2 and $\text{prem}(y''', g) \neq 0$. Then, we can not bound the degrees of the polynomial solutions of $f = 0$ and $g = 0$ in these two cases.

After an upper bound d for the degrees of the polynomial solutions of $f(z) = 0$ is given, we can check whether f is Laurent regular with the following theorem.

Theorem 7. *Let $f(z)$ be an irreducible differential polynomial in $\mathcal{F}\{z\}$, $n = \text{ord}(f)$, $d = \deg(f, \{z, z_1, \ldots, z_n\})$, and $m \in \mathbb{N}_{>0}$. Then $f(z) = 0$ has a generic polynomial solution of degree no more than m if and only if* $\text{prem}(z^{(m+1)}, f(z)) = 0$, *which can be done with $2^{2.4}[2(d+1)^{m+3}]^{2.4(n+m+2)}$ $\mathcal{F}$-arithmetic operations.*

Proof. If $\tilde{z}$ is a generic polynomial solution of $f(z) = 0$ with degree not larger than m, then $\tilde{z}^{(m+1)} = 0$. Suppose $g = \text{prem}(z^{(m+1)}, f)$. Then $g = \sum a_i f^{(i)} + Hz^{(m+1)}$, where $a_i \in \mathcal{F}\{z\}$ and H is a power of the product of the initial and separant of f. So we have $g(\tilde{z}) = 0$. But g is reduced w.r.t f and f is irreducible, thus $g = 0$. Assume $\text{prem}(z^{(m+1)}, f(z)) = 0$. Then we have $\sum b_i f^{(i)} = H'z^{(m+1)}$, for some $b_i \in \mathcal{F}\{z\}$ and H' a power of the product of the initial and separant of f. So if $\tilde{z}$ is a generic zero of $f(z) = 0$, then $H'(\tilde{z}) \neq 0$, because f is irreducible. Thus we have $\tilde{z}^{(m+1)} = 0$, which implies that $\tilde{z}$ is a polynomial of order less than $m + 1$. The complexity follows from [14, Theorem 3.15]. □

References

1. Aubry, P., Lazard, D., Maza, M.M.: On the theories of triangular sets. J. Symb. Comput. **28**, 105–124 (1999)
2. Boulier, F., Lemaire, F., Maza, M.M.: Computing differential characteristic sets by change of ordering. J. Symb. Comput. **45**, 124–149 (2010)
3. Bouziane, D., Kandri Rody, A., Maârouf, H.: Unmixed-dimensional decomposition of a finitely generated perfect differential ideal. J. Symb. Comput. **31**, 631–649 (2010)
4. Cano, J.: An extension of the Newton-Puiseux polygon construction to give solutions of Pfaffian forms. Ann. Inst. Fourier **43**, 125–142 (1993)
5. Cano, J.: On the series defined by differential equations, with an extension of the Puiseux polygon construction to these equations. Analysis **13**, 103–120 (1993)

6. Chen, C., Davenport, J.H., May, J.P., Maza, M.M., Xia, B., Xiao, R.: Triangular decomposition of semi-algebraic systems. J. Symb. Comput. **49**, 3–26 (2013)
7. Chou, S.-C., Gao, X.-S.: Ritt-Wu's decomposition algorithm and geometry theorem proving. In: Stickel, M.E. (ed.) CADE 1990. LNCS, vol. 449, pp. 207–220. Springer, Heidelberg (1990). doi:10.1007/3-540-52885-7_89
8. Chou, S.C., Gao, X.S.: Automated reasoning in differential geometry and mechanics: part I. An improved version of Ritt-Wu's decomposition algorithm. J. Autom. Reason. **10**, 161–172 (1993)
9. Gao, X.S., Huang, Z.: Characteristic set algorithms for equation solving in finite fields. J. Symb. Comput. **47**, 655–679 (2012)
10. Gao, X.S., Huang, Z., Yuan, C.M.: Binomial difference ideals. J. Symb. Comput. **80**, 665–706 (2017)
11. Gao, X.S., Luo, Y., Yuan, C.: A characteristic set method for difference polynomial systems. J. Symb. Comput. **44**, 242–260 (2009)
12. Grigoriev, D.Y., Singer, M.: Solving ordinary differential equations in terms of series with real exponents. Trans. AMS **327**, 329–351 (1991)
13. Hubert, E.: Factorization-free decomposition algorithms in differential algebra. J. Symb. Comput. **129**, 641–662 (2000)
14. Li, W., Li, Y.H.: Computation of differential chow forms for ordinary prime differential ideals. Adv. Appl. Math. **72**, 77–112 (2016)
15. Li, X., Mou, C., Wang, D.: Decomposing polynomial sets into simple sets over finite fields. Comput. Math. Appl. **60**, 2983–2997 (2010)
16. Pauer, F., Unterkircher, A.: Gröbner bases for ideals in Laurent polynomial rings and their application to systems of difference equations. AAECC **9**, 271–291 (1999)
17. Sit, W.: The Ritt-Kolchin theory for diffferential polynomials. In: Differential Algebra and Related Topics, pp. 1–70 (2002)
18. Wang, D.: Elimination Methods. Springer Science & Business Media, Heidelberg (2012)
19. Wu, W.T.: Mathematics Mechanization. Science Press/Kluwer, Beijing (2001)
20. Wu, W.T., Gao, X.S.: Mathematics mechanization and applications after thirty years. Front. Comput. Sci. **1**, 1–8 (2007)
21. Yang, L., Zhang, J.: Searching dependency between algebraic equations. ICTP, IC/91/6 (1991)
22. Zampieri, S.: A solution of the Cauchy problem for multidimensional discrete linear shift-invariant systems. Linear Algebra Appl. **202**, 143–162 (1994)
23. Zhu, W., Gao, X.S.: A triangular decomposition algorithm for differential polynomial systems with elemenray complexity. J. Syst. Sci. Complex. **30**, 464–483 (2017)

Sparse Polynomial Interpolation with Finitely Many Values for the Coefficients

Qiao-Long Huang(✉) and Xiao-Shan Gao

KLMM, UCAS, Academy of Mathematics and Systems Science,
Chinese Academy of Sciences, Beijing 100190, China
huangqiaolong13@mails.ucas.ac.cn

Abstract. In this paper, we give new sparse interpolation algorithms for black box polynomial f whose coefficients are from a finite set. In the univariate case, we recover f from one evaluation $f(\beta)$ for a sufficiently large number β. In the multivariate case, we introduce the modified Kronecker substitution to reduce the interpolation of a multivariate polynomial to that of the univariate case. Both algorithms have polynomial bit-size complexity.

Keywords: Sparse polynomial interpolation · Modified Kronecker substitution · Polynomial time algorithms

1 Introduction

The interpolation for a sparse multivariate polynomial $f(x_1, \ldots, x_n)$ given as a black box is a basic computational problem. Interpolation algorithms were given when we know an upper bound for the terms of f [3] and upper bounds for the terms and the degrees of f [12]. These algorithms were significantly improved and these works can be found in the references of [1].

In this paper, we consider the sparse interpolation for f whose coefficients are taken from a known finite set. For example, f could be in $\mathbb{Z}[x_1, \ldots, x_n]$ with an upper bound on the absolute values of coefficients of f, or f is in $\mathbb{Q}[x_1, \ldots, x_n]$ with upper bounds both on the absolute values of coefficients and their denominators. This kind of interpolation is motivated by the following applications. The interpolation of sparse rational functions leads to interpolation of sparse polynomials whose coefficients have bounded denominators [6, p. 6]. In [7], a new method is introduced to reduce the interpolation of a multivariate polynomial f to the interpolation of univariate polynomials, where we need to obtain the terms of f from a larger set of terms and the method given in this paper is needed to solve this problem.

In the univariate case, we show that if β is larger than a given bound depending on the coefficients of f, then f can be recovered from $f(\beta)$. Based on this idea, we give a sparse interpolation algorithm for univariate polynomials with rational numbers as coefficients, whose bit complexity is $\mathcal{O}((td \log H(\log C + \log H))$

Partially supported by an NSFC grant No. 11688101.

V.P. Gerdt et al. (Eds.): CASC 2017, LNCS 10490, pp. 196–209, 2017.
DOI: 10.1007/978-3-319-66320-3_15

or $\widetilde{\mathcal{O}}(td)$, where t is the number of terms of f, d is the degree of f, C and H are upper bounds for the coefficients and the denominators of the coefficients of f. It seems that the algorithm has the optimal bit complexity $\widetilde{\mathcal{O}}(td)$ in all known deterministic and exact interpolation algorithms for black box univariate polynomials as discussed in Remark 2.

In the multivariate case, we show that by choosing a good prime, the interpolation of a multivariate polynomial can be reduced to that of the univariate case in polynomial-time. As a consequence, a new sparse interpolation algorithm for multivariate polynomials is given, which has polynomial bit-size complexity. We also give its probabilistic version.

There exist many methods for reducing the interpolation of a multivariate polynomial into that of univariate polynomials, like the classical Kronecker substitution, randomize Kronecker substitutions [2], Zipple's algorithm [12], Klivans-Spielman's algorithm [9], Garg-Schost's algorithm [4], and Giesbrecht-Roche's algorithm [5]. Using the original Kronecker substitution [10], interpolation for multivariate polynomials can be easily reduced to the univariate case. The main problem with this approach is that the highest degree of the univariate polynomial and the height of the data in the algorithm are exponential. In this paper, we give the following modified Kronecker substitution

$$x_i = x^{\mathbf{mod}((D+1)^{i-1},p)}, i = 1, 2, \ldots, n$$

to reduce multivariate interpolations to univariate interpolations. Our approach simplifies and builds on previous work by Garg-Schost [4], Giesbrecht-Roche [5], and Klivans-Spielman [9]. The first two are for straight-line programs. Our interpolation algorithm works for the more general setting of black box sampling.

The rest of this paper is organized as follows. In Sect. 2, we give interpolation algorithms about univariate polynomials. In Sect. 3, we give interpolation algorithms about multivariate polynomials. In Sect. 4, experimental results are presented.

2 Univariate Polynomial Interpolation

2.1 Sparse Interpolation with Finitely Many Coefficients

In this section, we always assume

$$f(x) = c_1x^{d_1} + c_2x^{d_2} + \cdots + c_tx^{d_t} \tag{1}$$

where $d_1, d_2, \ldots, d_t \in \mathbb{N}, d_1 < d_2 < \cdots < d_t$, and $c_1, c_2, \cdots, c_t \in A$, where $A \subset \mathbb{C}$ is a finite set. Introduce the following notations

$$C := \max_{a\in A}(|a|), \quad \varepsilon := \min(\varepsilon_1, \varepsilon_2) \tag{2}$$

where $\varepsilon_1 := \min_{a,b\in A, a\neq b} |a - b|$ and $\varepsilon_2 := \min_{a\in A, a\neq 0} |a|$.

Theorem 1. *If $\beta \geq \frac{2C}{\varepsilon} + 1$, then $f(x)$ can be uniquely determined by $f(\beta)$.*

Proof. Firstly, for $k = 1, 2, \cdots$, we have $\beta \geq \frac{2C}{\varepsilon} + 1 \Longrightarrow \beta - 1 \geq \frac{2C}{\varepsilon} \Longrightarrow \beta - 1 > \frac{2C}{\varepsilon}\frac{\beta^k - 1}{\beta^k} \Longrightarrow \varepsilon\beta^k > 2C\frac{\beta^k - 1}{\beta - 1} \Longrightarrow \varepsilon\beta^k > 2C(\beta^{k-1} + \beta^{k-2} + \cdots + \beta + 1)$.

From (1), we have $f(\beta) = c_1\beta^{d_1} + c_2\beta^{d_2} + \cdots + c_t\beta^{d_t}$. Assume that there is another form $f(\beta) = a_1\beta^{k_1} + a_2\beta^{k_2} + \cdots + a_s\beta^{k_s}$, where $a_1, a_2, \ldots, a_s \in A$ and $k_1 < k_2 < \cdots < k_s$. It suffices to show that $c_t\beta^{d_t} = a_s\beta^{k_s}$. The rest can be proved by induction. First assume that $d_t \neq k_s$. Without loss of generality, let $d_t > k_s$. Then we have $0 = |(c_1\beta^{d_1} + c_2\beta^{d_2} + \cdots + c_t\beta^{d_t}) - (a_1\beta^{k_1} + a_2\beta^{k_2} + \cdots + a_s\beta^{k_s})| \geq |c_t|\beta^{d_t} - C(\beta^{d_t - 1} + \cdots + \beta + 1) - C(\beta^{k_s} + \cdots + \beta + 1) \geq |c_t|\beta^{d_t} - 2C(\beta^{d_t-1} + \cdots + \beta + 1) > |c_t|\beta^{d_t} - \varepsilon\beta^{d_t} \geq 0$. It is a contradiction, so $d_t = k_s$.

Assume $c_t \neq a_s$, then $0 = |(c_1\beta^{d_1} + c_2\beta^{d_2} + \cdots + c_t\beta^{d_t}) - (a_1\beta^{k_1} + a_2\beta^{k_2} + \cdots + a_s\beta^{k_s}) \geq |c_t - a_s|\beta^{d_t} - 2C(\beta^{d_t-1} + \cdots + \beta + 1) > |c_t - a_s|\beta^{d_t} - \varepsilon\beta^{d_t} \geq 0$. It is a contradiction, so $c_t = a_s$. The theorem has been proved. □

2.2 The Sparse Interpolation Algorithm

The idea of the algorithm is first to obtain the maximum term m of f, then subtract $m(\beta)$ from $f(\beta)$ and repeat the procedure until $f(\beta)$ becomes 0.

We first show how to compute the leading degree d_t.

Lemma 1. *Assume* $\beta \geq \frac{2C}{\varepsilon} + 1$. *If* $k \leq d_t$, *then* $|\frac{f(\beta)}{\beta^k}| > \frac{\varepsilon}{2}$; *if* $k > d_t$, *then* $|\frac{f(\beta)}{\beta^k}| < \frac{\varepsilon}{2}$.

Proof. From $|f(\beta)| = |c_1\beta^{d_1} + c_2\beta^{d_2} + \cdots + c_t\beta^{d_t}| \leq C(\beta^{d_t} + \cdots + \beta + 1) = C(\frac{\beta^{d_t+1} - 1}{\beta - 1})$ and $|f(\beta)| = |c_1\beta^{d_1} + c_2\beta^{d_2} + \cdots + c_t\beta^{d_t}| \geq |c_t|\beta^{d_t} - C(\beta^{d_t-1} + \cdots + \beta + 1) = |c_t|\beta^{d_t} - C\frac{\beta^{d_t} - 1}{\beta - 1}$, we have $|c_t|\beta^{d_t} - C\frac{\beta^{d_t}-1}{\beta-1} \leq |f(\beta)| \leq C(\frac{\beta^{d_t+1}-1}{\beta-1})$.

When $k \leq d_t$, $|\frac{f(\beta)}{\beta^k}| \geq |c_t|\beta^{d_t-k} - \frac{C}{\beta-1}(\beta^{d_t-k} - \frac{1}{\beta^k}) \geq \varepsilon\beta^{d_t-k} - \frac{\varepsilon}{2}(\beta^{d_t-k} - \frac{1}{\beta^k}) \geq \frac{\varepsilon}{2}\beta^{d_t-k} + \frac{\varepsilon}{2}\frac{1}{\beta^k} > \frac{\varepsilon}{2}$. When $k > d_t$, $|\frac{f(\beta)}{\beta^k}| \leq \frac{C}{\beta-1}(\beta^{d_t+1-k} - \frac{1}{\beta^k}) \leq \frac{\varepsilon}{2}(\beta^{d_t+1-k} - \frac{1}{\beta^k}) \leq \frac{\varepsilon}{2}\beta^{d_t+1-k} - \frac{\varepsilon}{2}\frac{1}{\beta^k} < \frac{\varepsilon}{2}$. □

If we can use logarithm, we can change the above lemma into the following form.

Lemma 2. *If* $\beta \geq \frac{2C}{\varepsilon} + 1$, *then* $d_t = \lfloor \log_\beta \frac{2|f(\beta)|}{\varepsilon} \rfloor$.

Proof. By Lemma 1, we know $\frac{|f(\beta)|}{\beta^{d_t}} > \frac{\varepsilon}{2}$ and $\frac{|f(\beta)|}{\beta^{d_t+1}} < \frac{\varepsilon}{2}$. Then we have $\log_\beta \frac{|f(\beta)|}{\beta^{d_t}} > \log_\beta \frac{\varepsilon}{2}$ and $\log_\beta \frac{|f(\beta)|}{\beta^{d_t+1}} < \log_\beta \frac{\varepsilon}{2}$, this can be reduced to $\log_\beta \frac{2|f(\beta)|}{\varepsilon} - 1 < d_t < \log_\beta \frac{2|f(\beta)|}{\varepsilon}$. As d_t is an integer, then we have $d_t = \lfloor \log_\beta \frac{2|f(\beta)|}{\varepsilon} \rfloor$. □

Based on Lemma 2, we have the following algorithm which will be used in several places.

Algorithm 2 (UDeg)

Input: $f(\beta), \varepsilon$, where $\beta \geq \frac{2C}{\varepsilon} + 1$.
Output: the degree of $f(x)$.

Step 1: return $\lfloor \log_\beta(\frac{2|f(\beta)|}{\varepsilon}) \rfloor$.

Remark 1. If we cannot use logarithm operation, then it is easy to show that we need $\mathcal{O}(\log^2 D)$ arithmetic operations to obtain the degree based on Lemma 1. In the following section, we will regard logarithm as a basic step.

Now we will show how to compute the leading coefficient c_t.

Lemma 3. *If $\beta \geq \frac{2C}{\varepsilon} + 1$, then c_t is the only element in A that satisfies $|\frac{f(\beta)}{\beta^{d_t}} - c_t| < \frac{\varepsilon}{2}$.*

Proof. First we show that c_t satisfies $|\frac{f(\beta)}{\beta^{d_t}} - c_t| < \frac{\varepsilon}{2}$. We rewrite $f(\beta)$ as $f(\beta) = c_t\beta^{d_t} + g(\beta)$, where $g(x) := c_{t-1}x^{d_{t-1}} + c_{t-2}x^{d_{t-2}} + \cdots + c_1x^{d_1}$. So $\frac{f(\beta)}{\beta^{d_t}} = c_t + \frac{g(\beta)}{\beta^{d_t}}$. As $\deg(g) < d_t$, by Lemma 1, we have $|\frac{g(\beta)}{\beta^{d_t}}| < \frac{\varepsilon}{2}$. So $|\frac{f(\beta)}{\beta^{d_t}} - c_t| < \frac{\varepsilon}{2}$. Assume there is another $c \in A$ also have $|\frac{f(\beta)}{\beta^{d_t}} - c| < \frac{\varepsilon}{2}$, then $|c_t - c| \leq |\frac{f(\beta)}{\beta^{d_t}} - c| + |\frac{f(\beta)}{\beta^{d_t}} - c_t| < \varepsilon$. This can only happen when $c_t = c$, so we have proved the uniqueness. □

Based on Lemma 3, we give the algorithm to obtain the leading coefficient.

Algorithm 3 (ULCoef)

Input: $f(\beta), \beta, \varepsilon, d_t$
Output: the leading coefficient of $f(x)$

Step 1: Find the element c in A such that $|\frac{f(\beta)}{\beta^{d_t}} - c| < \frac{\varepsilon}{2}$.
Step 2: Return c.

Now we can give the complete algorithm.

Algorithm 4 (UPolySI)

Input: A black box univariate polynomial $f(x)$, whose coefficients are in A.
Output: The exact form of $f(x)$.

Step 1: Find the bounds C and ε of A, as defined in (2).
Step 2: Let $\beta := \frac{2C}{\varepsilon} + 1$.
Step 3: Let $g := 0, u := f(\beta)$.
Step 4: **while** $u \neq 0$ **do**
$d :=$**UDeg**(u, ε, β); $c :=$**ULCoef**$(u, \beta, \varepsilon, d)$; $u := u - c\beta^d$; $g := g + cx^d$;
end do.
Step 5: Return g.

Note that the complexity of Algorithm 3 depends on A, which is denoted by O_A. Note that $O_A \leq |A|$. We have the following theorem.

Theorem 5. *The arithmetic complexity of the Algorithm 4 is $\mathcal{O}(tO_A) \leq \mathcal{O}(t|A|)$, where t is the number of terms in f.*

Proof. Since finding the maximum degree needs one operation and finding the coefficient of the maximum term needs O_A operations, and finding the maximum term needs $\mathcal{O}(O_A)$ operations, we have proved the theorem. □

2.3 The Rational Number Coefficients Case

In this section, we assume that the coefficients of $f(x)$ are rational numbers in

$$A = \{\frac{b}{a} \mid 0 < a \le H, |\frac{b}{a}| \le C, a, b \in \mathbb{Z}\} \tag{3}$$

and we have $\varepsilon = \frac{1}{H(H-1)}$. Notice that in Algorithm 4, only Algorithm 3 (**ULCoef**) needs refinement. We first consider the following general problem about rational numbers.

Lemma 4. *Let $0 < r_1 < r_2$ be rational numbers. Then we can find the smallest $d > 0$ such that a rational number with denominator d is in (r_1, r_2) with computational complexity $\mathcal{O}(\log(r_2 - r_1))$.*

Proof. We consider three cases.

1. If one of the r_1 and r_2 is an integer and the other one is not, then the smallest positive integer d such that $(r_2 - r_1)d > 1$ is the smallest denominator, and $d = \lceil \frac{1}{r_2 - r_1} \rceil$.
2. Both of r_1, r_2 are integers. If $r_2 - r_1 > 1$, then 1 is the smallest denominator. If $r_2 - r_1 = 1$, then 2 is the smallest denominator.
3. Both of r_1, r_2 are not integers. This is the most complicated case.

First, we check if there exists an integer in (r_1, r_2). If $\lceil r_1 \rceil < r_2$, then $\lceil r_1 \rceil$ is in the interval which has the smallest denominator 1.

Now we consider the case that (r_1, r_2) does not contain an integer. Assume $r_1 < \frac{d_1}{d} < r_2$, where $d > 1$ is the smallest denominator. Denote $w :=$trunc(r_1), $\epsilon_1 := r_1 - w, \epsilon_2 := r_2 - w$. Then $\epsilon_1 < \epsilon_2 < 1$ and d is smallest positive integer such that (dr_1, dr_2) contains an integer. Since $dr_1 = d(w + \epsilon_1), dr_2 = d(w + \epsilon_2)$, d is the smallest positive integer such that interval $(d\epsilon_1, d\epsilon_2)$ contains an integer. We still denote it d_1. Then $d\epsilon_1 < d_1 < d\epsilon_2$, so $\frac{d_1}{\epsilon_2} < d < \frac{d_1}{\epsilon_1}$, and we can see that d_1 is the the smallest integer such that $(\frac{d_1}{\epsilon_2}, \frac{d_1}{\epsilon_1})$ contains an integer. Suppose we know how to compute the number d_1. Then $d = \lceil \frac{d_1}{\epsilon_2} \rceil$ when $\frac{d_1}{\epsilon_2}$ is not an integer, and $d = \frac{d_1}{\epsilon_2} + 1$ when $\frac{d_1}{\epsilon_2}$ is an integer.

Note that d_1 is the smallest denominator such that some rational number $\frac{d}{d_1}$ is in $(\frac{1}{\epsilon_2}, \frac{1}{\epsilon_1})$. To find d_1, we need to repeat the above procedure to $(\frac{1}{\epsilon_2}, \frac{1}{\epsilon_1})$ and obtain a sequence of intervals $(r_1, r_2) \to (\frac{1}{\epsilon_2}, \frac{1}{\epsilon_1}) \to \cdots$. The denominators of end points of the intervals becomes smaller after each repetition. So the algorithm will terminate.

Now we prove that the number of operations of the procedure is $\mathcal{O}(\log(r_2 - r_1))$. First, we know the length of the interval (r_1, r_2) is $r_2 - r_1$. Now we prove that every time we run one or two recursive steps, the length of the new interval will be 2 times bigger. Let $(\frac{b_1}{a_1}, \frac{b_2}{a_2})$ be the first interval. If it contains an integer, then we finish the algorithm. We assume that case does not happen, so we can assume $|\frac{b_1}{a_1}| \le 1, |\frac{b_2}{a_2}| \le 1$. Then the second interval is $(\frac{a_2}{b_2}, \frac{a_1}{b_1})$. Now the new interval length is $\frac{a_1}{b_1} - \frac{a_2}{b_2}$. If $\frac{b_1}{a_1} \le \frac{1}{2}$, then we have $\frac{\frac{a_1}{b_1} - \frac{a_2}{b_2}}{\frac{b_2}{a_2} - \frac{b_1}{a_1}} = \frac{\frac{a_1 b_2 - a_2 b_1}{b_1 b_2}}{\frac{a_1 b_2 - a_2 b_1}{a_1 a_2}} = \frac{a_1 a_2}{b_1 b_2} \ge 2$.

If $\frac{b_1}{a_1} > \frac{1}{2}$, then we let $a_1 = b_1 + c_1, a_2 = b_2 + c_2$ and the third interval is $(\frac{b_1}{c_1}, \frac{b_2}{c_2})$.

Then we have $\frac{\frac{b_2}{c_2}-\frac{b_1}{c_1}}{\frac{b_2}{a_2}-\frac{b_1}{a_1}} = \frac{\frac{c_1b_2-c_2b_1}{c_1c_2}}{\frac{a_1b_2-a_2b_1}{a_1a_2}} = \frac{a_1a_2}{c_1c_2} > 2$. In this case, if we have an interval whose length is bigger than 1, then the recursion will terminate. So if $(r_2 - r_1)2^k \geq 1$, then $2k$ is the upper bound of the number of recursions. So the complexity is $\mathcal{O}(\log(r_2 - r_1))$. We proved the lemma. □

Based on Lemma 4, we present a recursive algorithm to compute the rational number in an interval (r_1, r_2) with the smallest denominator.

Algorithm 6 (MiniDenom)

Input: r_1, r_2 are positive rational numbers.
Output: the minimum denominator of rational numbers in (r_1, r_2)

Step 1: **if** one of r_1, r_2 is an integer and the other one is not an integer **then** return $\lceil \frac{1}{r_2-r_1} \rceil$.
Step 2: **if** both of r_1 and r_2 are integers and $r_2 - r_1 > 1$ **then** return 1.
if both of r_1 and r_2 are integers and $r_2 - r_1 = 1$ **then** return 2.
Step 3: **if** $\lceil r_1 \rceil < r_2$, **then** return 1.
Step 4: let w :=trunc(r_1), $\epsilon_1 := r_1 - w$,$\epsilon_2 := r_2 - w$;
$d_1 :=$ **MiniDenom**$(\frac{1}{\epsilon_2}, \frac{1}{\epsilon_1})$;
if $\frac{d_1}{\epsilon_2}$ is a integer **then** return $\frac{d_1}{\epsilon_2} + 1$ **else** return $\lceil \frac{d_1}{\epsilon_2} \rceil$.

We now show how to compute the leading coefficient of $f(x)$.

Lemma 5. *Suppose* $c_t = \frac{b}{a}$, *where* $\gcd(a, b) = 1, a > 0$, *and* $I_i = (\frac{f(\beta)}{\beta^{d_t}}i - \frac{\varepsilon}{2}i, \frac{f(\beta)}{\beta^{d_t}}i + \frac{\varepsilon}{2}i), i = 1, 2, \ldots, H$. *Then* $I_a \cap \mathbb{Z} = \{b\}$ *and if* $I_{a_0} \cap \mathbb{Z} = \{b_0\}$ *then* $\frac{b}{a} = \frac{b_0}{a_0}$.

Proof. By Lemma 3, we have $\frac{f(\beta)}{\beta^{d_t}} - \frac{\varepsilon}{2} < \frac{b}{a} < \frac{f(\beta)}{\beta^{d_t}} + \frac{\varepsilon}{2}$, so $\frac{f(\beta)}{\beta^{d_t}}a - \frac{\varepsilon}{2}a < b < \frac{f(\beta)}{\beta^{d_t}}a + \frac{\varepsilon}{2}a$, and the existence is proved. The length of $(\frac{f(\beta)}{\beta^{d_t}}a - \frac{\varepsilon}{2}a, \frac{f(\beta)}{\beta^{d_t}}a + \frac{\varepsilon}{2}a)$ is $< 2\frac{\varepsilon}{2}a \leq \varepsilon H \leq \frac{1}{H-1} \leq 1$, so b is the unique integer in the interval.

Assume that there is another $a_0 \in \{1, 2, \ldots, H\}$, such that $(\frac{f(\beta)}{\beta^{d_t}}a_0 - \frac{\varepsilon}{2}a_0, \frac{f(\beta)}{\beta^{d_t}}a_0 + \frac{\varepsilon}{2}a_0)$ contains the integer b_0. Then $\frac{f(\beta)}{\beta^{d_t}}a_0 - \frac{\varepsilon}{2}a_0 < b_0 < \frac{f(\beta)}{\beta^{d_t}}a_0 + \frac{\varepsilon}{2}a_0$, so $\frac{f(\beta)}{\beta^{d_t}} - \frac{\varepsilon}{2} < \frac{b_0}{a_0} < \frac{f(\beta)}{\beta^{d_t}} + \frac{\varepsilon}{2}$. If $\frac{a}{b} \neq \frac{a_0}{b_0}$, then $|\frac{a}{b} - \frac{a_0}{b_0}| = |\frac{ab_0-a_0b}{bb_0}| \geq \frac{1}{H(H-1)} = \varepsilon$, which contradicts that the length of the interval is less than ε. □

Let $r_1 := \frac{f(\beta)}{\beta^{d_t}} - \frac{\varepsilon}{2}, r_2 := \frac{f(\beta)}{\beta^{d_t}} + \frac{\varepsilon}{2}$. By Lemma 5, if a_0 is the smallest positive integer such that (a_0r_1, a_0r_2) contains the unique integer b_0, then we have $c_t = \frac{b_0}{a_0}$. Note that a_0 is the smallest integer such that (a_0r_1, a_0r_2) contains the unique integer b_0 if and only if a_0 is the smallest integer such that b_0/a_0 is in (r_1, r_2), and such an a_0 can be found with Algorithm 6. This observation leads to the following algorithm to find the leading coefficient of $f(x)$.

Algorithm 7 (ULCoefRat)

Input: $\frac{f(\beta)}{\beta^{d_t}}, \varepsilon, d_t$
Output: the leading coefficient of $f(x)$.
Step 1: **if** $\frac{f(\beta)}{\beta^{d_t}} > 0$, **then** $r_1 := \frac{f(\beta)}{\beta^{d_t}} - \frac{\varepsilon}{2}$, $r_2 := \frac{f(\beta)}{\beta^{d_t}} + \frac{\varepsilon}{2}$; **else** $r_1 := -\frac{f(\beta)}{\beta^{d_t}} - \frac{\varepsilon}{2}$, $r_2 := -\frac{f(\beta)}{\beta^{d_t}} + \frac{\varepsilon}{2}$;
Step 2: Let $a :=$ MiniDenom (r_1, r_2);
Step 3: **Return** $\frac{\lceil a(\frac{f(\beta)}{\beta^{d_t}} - \frac{\varepsilon}{2})\rceil}{a}$

Replacing Algorithm **ULCoef** with Algorithm **ULCoefRat** in Algorithm **UPolySI**, we obtain the following interpolation algorithm for sparse polynomials with rational coefficients.

Algorithm 8 (UPolySIRat)

Input: A black box polynomial $f(x) \in \mathbb{Q}[x]$ whose coefficients are in A given in (3).
Output: The exact form of $f(x)$.

Theorem 9. *The arithmetic operations of Algorithm 8 are $\mathcal{O}(t \log H)$ and the bit complexity is $\mathcal{O}(td \log H(\log C + \log H))$, where d is the degree of $f(x)$.*

Proof. In order to obtain the degree, we need one log arithmetic operation in field $\mathbb{Q}$, while in order to obtain the coefficient, we need $\mathcal{O}(\log H)$ arithmetic operations, so the total complexity is $\mathcal{O}(t \log H)$. Assume $f(\beta) = \frac{a_1}{h_1}\beta^{d_1} + \frac{a_2}{h_2}\beta^{d_2} + \cdots + \frac{a_t}{h_t}\beta^{d_t}$ and let $H_i := h_1 \cdots h_{i-1}h_{i+1} \cdots h_t$. Then we have

$$f(\beta) = \frac{a_1H_1\beta^{d_1} + a_2H_2\beta^{d_2} + \cdots + a_tH_t\beta^{d_t}}{h_1h_2 \cdots h_t}$$

Then $|a_1H_1\beta^{d_1} + a_2H_2\beta^{d_2} + \cdots + a_tH_t\beta^{d_t}| \leq H^{t-1}C(\beta^{d_t} + \cdots + \beta + 1) = H^{t-1}\frac{C}{\beta-1}(\beta^{d_t+1} - 1)$, so its bit length is $\mathcal{O}(t \log H + d \log C + d \log H)$. It is easy to see that the bit length of $h_1h_2 \cdots h_t$ is $\mathcal{O}(t \log H)$. So the total bit complexity is $\mathcal{O}((t \log H)(t \log H + d \log C + d \log H))$. As $t \leq d$, the bit complexity is $\mathcal{O}(td \log H(\log C + \log H))$. □

Corollary 1. *If the coefficients of $f(x)$ are integers in $[-C, C]$, then Algorithm 8 computes $f(x)$ with arithmetic complexity $\mathcal{O}(t)$ and with bit complexity $\mathcal{O}(td \log C)$.*

Remark 2. The bit complexity of Algorithm 8 is $\widetilde{\mathcal{O}}(td)$, which seems to be the optimal bit complexity for deterministic and exact interpolation algorithms for a black box polynomial $f(x) \in Q[x]$. For a t-sparse polynomial, t terms are needed and the arithmetic complexity is at least $\mathcal{O}(t)$. For $\beta \in \mathbb{C}$, we have $|f(\beta)| \leq C\frac{\beta^{d+1}-1}{\beta-1}$, where C is defined in (2). If $|\beta| \neq 1$, then the height of $f(\beta)$ is $d|\log\beta| + \log C$ or $\widetilde{\mathcal{O}}(d)$. For a deterministic and exact algorithm, β satisfying $|\beta| = 1$ seems not usable. So the bit complexity is at least $\widetilde{\mathcal{O}}(td)$. For instance, the height of the data in Ben-or and Tiwari's algorithm is already $\widetilde{\mathcal{O}}(td)$ [3,8].

3 Multivariate Polynomial Sparse Interpolation with Modified Kronecker Substitution

In this section, we give a deterministic and a probabilistic polynomial-time reduction of multivariate polynomial interpolation to univariate polynomial interpolation.

3.1 Find a Good Prime

We will show how to find a prime number which can be used in the reduction.

We assume $f(x_1, x_2, \dots, x_n)$ is a multivariate polynomial in $\mathbb{Q}[x_1, x_2, \dots, x_n]$ with a degree bound D, a term bound T, and p is a prime. We use the substitution

$$x_i = x^{\mathbf{mod}((D+1)^{i-1},p)}, i = 1, 2, \dots, n. \tag{4}$$

For convenience of description, we denote

$$f_{x,p} := f(x, x^{\mathbf{mod}((D+1),p)}, \dots, x^{\mathbf{mod}((D+1)^{n-1},p)}). \tag{5}$$

Then the degree of $f_{x,p}$ is no more than $D(p-1)$ and the number of terms of $f_{x,p}$ is no more than T.

If the number of terms of $f_{x,p}$ is the same as that of $f(x_1, x_2, \dots, x_n)$, there is no collision in different monomials and we call such prime as a *good prime* for $f(x_1, x_2, \dots, x_n)$.

If p is a good prime, then we can consider a new substitution:

$$x_i = q_i x^{\mathbf{mod}((D+1)^{i-1},p)}, i = 1, 2, \dots, n, \tag{6}$$

where $q_i, i = 1, 2, \dots, n$ is the i-th prime. In this case, each coefficient will change according to monomials of f. Note that in [4], the substitution is

$f(x, x^{(D+1)}, \dots, x^{(D+1)^{n-1}})\mathbf{mod}(x^p - 1)$. With this substitution, the substitution (6) cannot be used.

We show how to find a good prime p. We first give a lemma.

Lemma 6. *Suppose p is a prime. If* $\mathbf{mod}(a_1 + a_2(D+1) + \cdots + a_n(D+1)^{n-1}, p) \neq 0$, *then* $a_1 + a_2\mathbf{mod}(D+1, p) + \cdots + a_n\mathbf{mod}((D+1)^{n-1}, p) \neq 0$.

Proof. If $a_1 + a_2\mathbf{mod}(D+1, p) + \cdots + a_n\mathbf{mod}((D+1)^{n-1}, p) = 0$, then $\mathbf{mod}(a_1 + a_2(D+1) + \cdots + a_n(D+1)^{n-1}, p) = 0$, which contradicts to the assumption. □

Now, we have the following theorem to find the good prime.

Theorem 10. *Let $f(x_1, \dots, x_n)$ be polynomial with degree at most D and $t \leq T$ terms. If*

$$N > \frac{T(T-1)}{2}\log_2[(D+1)^n - 1] - \frac{1}{4}T^2 + \frac{1}{2}T$$

then there at least one of N distinct odd primes $p_1, p_2, \dots, p_N$ is a good prime for f.

Proof. Assume $m_1, m_2, \ldots, m_t$ are all the monomials in f, and $m_i = x_1^{e_{i,1}} x_2^{e_{i,2}} \cdots x_n^{e_{i,n}}$. In order for p to be a good prime, we need $e_{i,1} + e_{i,2}(\mathbf{mod}(D+1,p)) + \cdots + e_{i,n}(\mathbf{mod}((D+1)^{n-1},p)) \neq e_{j,1} + e_{j,2}(\mathbf{mod}(D+1,p)) + \cdots + e_{j,n}(\mathbf{mod}((D+1)^{n-1},p))$, for all $i \neq j$. This can be changed into $(e_{i,1} - e_{j,1}) + (e_{i,2} - e_{j,2})(\mathbf{mod}(D+1,p)) + \cdots + (e_{i,n} - e_{j,n})(\mathbf{mod}((D+1)^{n-1},p)) \neq 0$. By Lemma 6, it is enough to show

$$\mathbf{mod}((e_{i,1} - e_{j,1}) + (e_{i,2} - e_{j,2})(D+1) + \cdots + (e_{i,n} - e_{j,n})(D+1)^{n-1}, p) \neq 0, i \neq j$$

Firstly, $|(e_{i,1} - e_{j,1}) + (e_{i,2} - e_{j,2})(D+1) + \cdots + (e_{i,n} - e_{j,n})(D+1)^{n-1}| \leq D(1 + (D+1) + \cdots + (D+1)^{n-1}) = (D+1)^n - 1$.

We assume that $\overline{f}(x) = a_1 x^{k_1} + a_2 x^{k_2} + \cdots + a_t x^{k_t}$ is the polynomial after the Kronecker substitution, where $k_i = e_{i,1} + e_{i,2}(D+1) + \cdots + e_{i,n}(D+1)^{n-1}$. If $t = 2$, it is trivial. So now we assume $t > 2$ and we analyse how many kinds of primes the number $\prod_{i>j}(k_i - k_j)$ has. Without loss of generality, assume $k_1, k_2 \ldots, k_w$ are even, $k_{w+1}, k_{w+2} \ldots, k_t$ are odd, denote $v := t - w$. It is easy to see that $k_i - k_j$ has factor 2 if $1 \leq i \neq j \leq w$ or $w+1 \leq i \neq j \leq t$. If one of w and v is zero, then $\prod_{i>j}(k_i - k_j)$ has a factor $2^{\frac{t(t-1)}{2}}$. If both w, v are not zero, then $\prod_{i>j}(k_i - k_j)$ has a factor $2^{\frac{w(w-1)}{2} + \frac{v(v-1)}{2}}$. We give a lower bound of $\frac{w(w-1)}{2} + \frac{v(v-1)}{2}$.

As $\frac{w(w-1)}{2} + \frac{v(v-1)}{2} = \frac{w^2+v^2-t}{2} \geq \frac{1/2(w+v)^2-t}{2} = \frac{1}{4}t^2 - \frac{1}{2}t$, $\prod_{i>j}(k_i - k_j)$ at least has a factor $2^{\frac{1}{4}t^2 - \frac{1}{2}t}$. Since $|k_i - k_j| \leq (D+1)^n - 1$, we have $\prod_{i>j}(k_i - k_j) \leq [(D+1)^n - 1]^{\frac{t(t-1)}{2}}$. If $p_1, p_2, \ldots, p_N$ are distinct primes satisfying $p_1 p_2 \ldots p_N > \frac{[(D+1)^n - 1]^{\frac{t(t-1)}{2}}}{2^{\frac{1}{4}t^2 - \frac{1}{2}t}}$, then at least one of the primes is a good prime. Since $p_i \geq 2$, $N > \frac{t(t-1)}{2} \log_2[(D+1)^n - 1] - \frac{1}{4}t^2 + \frac{1}{2}t$. As we just know the upper bound T of t, we can choose $T - t$ different positive integer $k_{t+1}, k_{t+2}, \ldots, k_T$ which are different from $k_1, k_2, \ldots, k_t$. So we still can use T as the number of the terms. We have proved the lemma. □

3.2 A Deterministic Algorithm

Lemma 7. *Assume* $f = \frac{c_1}{H_1}x^{d_1} + \frac{c_2}{H_2}x^{d_1} + \cdots + \frac{c_t}{H_t}x^{d_t}$, *where* $c_1, c_2, \ldots, c_t \in \mathbb{Z}, H_1, H_2, \ldots, H_t \in \mathbb{Z}_+, d_1, d_2, \ldots, d_t \in \mathbb{N}, d_1 < d_2 < \cdots < d_t, |\frac{c_i}{H_i}| \leq C$, $H_1, H_2, \ldots, H_t, d_1, d_2, \ldots, d_t$ *are known. Let* $H_{\max} := \max\{H_1, H_2, \ldots, H_t\}$. *If* $\beta \geq 2CH_{\max} + 1$, *then we can recover* $c_1, c_2, \ldots, c_t$ *from* $f(\beta)$.

Proof. It suffices to show that c_t can be recovered from $f(\beta)$. As $\beta - 1 \geq 2CH_{\max} \geq 2CH_t$, then $\frac{1}{2} \geq \frac{CH_t}{\beta-1}$. So $|f(\beta)H_t - c_t\beta^{d_t}| = |\frac{c_1 H_t}{H_1}\beta^{d_1} + \frac{c_2 H_t}{H_2}\beta^{d_2} + \cdots + \frac{c_{t-1}H_t}{H_{t-1}}\beta^{d_{t-1}}| \leq CH_t(\frac{\beta^{d_t}-1}{\beta-1}) \leq \frac{1}{2}(\beta^{d_t} - 1)$. So $|\frac{f(\beta)H_t}{\beta^{d_t}} - c_t| < \frac{1}{2}$. That is $\frac{f(\beta)H_t}{\beta^{d_t}} - \frac{1}{2} < c_t < \frac{f(\beta)H_t}{\beta^{d_t}} + \frac{1}{2}$. Since c_t is an integer, $c_t = \lceil \frac{f(\beta)H_t}{\beta^{d_t}} - \frac{1}{2} \rceil$. The rest can be proved by induction. □

Algorithm 11 (MPolySIMK)

Input: A black box polynomial $f(x_1, x_2, \ldots, x_n) \in A[x_1, x_2, \ldots, x_n]$, whose coefficients are in A given in (3), an upper bound D for the degree, an upper bound T of the number of terms, a list of n different primes $q_1, q_2, \ldots, q_n (q_1 < \cdots < q_n)$.
Output: The exact form of $f(x_1, x_2, \ldots, x_n)$.

Step 1: Randomly choose N different odd primes $p_1, p_2, \ldots, p_N$, where $N = \lfloor \frac{T(T-1)}{2} \log_2[(D+1)^n - 1] - \frac{1}{4}T^2 + \frac{1}{2}T \rfloor + 1$.
Step 2: **for** $i = 1, 2, \ldots, N$ let $f_i :=$ **UPolySIRat**(f_{x,p_i}, A, T) via Algorithm 8, where f_{x,p_i} is defined in (5).
Step 3: Let $S := \{\}$;
for $i = 1, 2, \ldots, N$ **do** if $f_i \neq failure$, then $S := S \bigcup \{f_i\}$. **end do**;
Step 4: **Repeat**:
Choose one integer i such that f_i has the most number of the terms in S.
if $f_i(j) = f_{x,p_i}(j)$ for $j = 1, 2, \ldots, D(p_i - 1) + 1$ **then** break **Repeat**;
$S := S \backslash \{f_i\}$
end Repeat
Let i_0 be the integer found and $f_{i_0} = \frac{c_1}{H_1} x^{d_1} + \frac{c_2}{H_2} x^{d_2} + \cdots + \frac{c_t}{H_t} x^{d_t}, d_1 < d_2 < \cdots < d_t$
Step 5: Let $\beta := 2Cq_n^D \max\{H_1, H_2, \ldots, H_t\} + 1$. [Lemma 7]
Denote $g = f(q_1 x, q_2 x^{\mathbf{mod}(D+1, p_{i_0})}, \ldots, q_n x^{\mathbf{mod}((D+1)^{n-1}, p_{i_0})})$ and let $u := g(\beta)$;
Step 6: Let $h := 0$;
for $i = t, t-1, \ldots, 1$ **do**
Let $b := \lceil \frac{u}{\beta^{d_i}} H_i - \frac{1}{2} \rceil$. Factor $\frac{b}{c_i}$ into $q_1^{e_1} q_2^{e_2} \cdots q_n^{e_n}$.
$h := h + \frac{c_i}{H_i} x_1^{e_1} x_2^{e_2} \cdots x_n^{e_n}$. $u := u - \frac{b}{H_i} \beta^{d_i}$.
Step 7: return h.

Remark 3. If p_i is not a good prime for f, then the substitution f_{x,p_i} of f has collisions. f_{x,p_i} may have some coefficients not in A. So we need to modify step 4 of Algorithm 8 as follows, with T as an extra input. For $c = \frac{a}{b}$, if $|c| > C$, $|b| > H$, or the number of the terms of f_i are more than T, then we let $f_i = failure$.

Theorem 12. *Algorithm 11 is correct and its bit complexity is* $\widetilde{\mathcal{O}}(n^2 T^5 D \log H \log C + n^2 T^5 D \log^2 H + n^3 T^6 D^2)$.

Proof. First, we show the correctness. If p_i is a good prime for f, then all the coefficients of f_{x,p_i} are in A. So in step 2, Algorithm 8 can be used to find $f_i = f_{x,p_i}$. It is sufficient to show that the prime p_{i_0} corresponding to i_0 obtained in step 4 is a good prime. In step 4, if there exists a j_0 such that $f_i(j_0) \neq f_{x,p_i}(j_0)$, then $f_i \neq f_{x,p_i}$. This only happens when some of the coefficients of f_{x,p_i} are not in A. That is, p_i is not a good prime for f. So we throw it away. If $f_{i_0}(j) = f_{x,p_{i_0}}(j)$ for $j = 1, 2, \ldots, D(p_{i_0} - 1) + 1$ for some i_0. Since $\deg f_{i_0} \leq D(p_{i_0} - 1)$, we have $f_{i_0} = f_{x,p_{i_0}}$.

Assume by contradiction that p_{i_0} is not a good prime for f, then the number of terms of f_{i_0} is less than that of f. Since S includes at least one f_{i_1} such that p_{i_1} is a good prime for f, the number of terms in f_{i_1} is more than f_{i_0}.

It contradicts that f_{i_0} has the most number of the terms in S. So p_{i_0} is a good prime for f.

As $f_{i_0} = \frac{c_1}{H_1}x^{d_1} + \frac{c_2}{H_2}x^{d_2} + \cdots + \frac{c_t}{H_t}x^{d_t}, d_1 < d_2 < \cdots < d_t$, we can assume $f = \frac{c_1}{H_1}m_1 + \frac{c_2}{H_2}m_2 + \cdots + \frac{c_t}{H_t}m_t$, where $m_i = x_1^{e_{i,1}}x_2^{e_{i,2}}\cdots x_n^{e_{i,n}}$. We can write g as $g = f(q_1x, q_2x^{\mathbf{mod}(D+1,p_{i_0})}, \ldots, q_nx^{\mathbf{mod}((D+1)^{n-1},p_{i_0})}) = \frac{c_1q_1^{e_{1,1}}q_2^{e_{1,2}}\cdots q_n^{e_{1,n}}}{H_1}x^{d_1} + \frac{c_2q_1^{e_{2,1}}q_2^{e_{2,2}}\cdots q_n^{e_{2,n}}}{H_2}x^{d_2} + \cdots + \frac{c_tq_1^{e_{t,1}}q_2^{e_{t,2}}\cdots q_n^{e_{t,n}}}{H_t}x^{d_t}$. Since $|\frac{c_iq_1^{e_{i,1}}q_2^{e_{i,2}}\cdots q_n^{e_{i,n}}}{H_i}| \leq Cq_n^D$, by Lemma 7, in step 6, $b = c_iq_1^{e_{i,1}}q_2^{e_{i,2}}\cdots q_n^{e_{i,n}}$. By Factoring $\frac{b}{c_i} = q_1^{e_{i,1}}q_2^{e_{i,2}}\cdots q_n^{e_{i,n}}$, we obtain the degrees of m_i. We have proved the correctness.

We now analyse the complexity. In step 2, we call Algorithm **UPolySIRat** $\mathcal{O}(nT^2\log D)$ times. The degree of f_{x,p_i} is bounded by $D(p_i - 1)$. Since the i-th prime is $\mathcal{O}(i\log i)$ and we use at most $\mathcal{O}(nT^2\log D)$ primes, the degree bound is $\widetilde{\mathcal{O}}(nT^2D)$. So by Theorem 9, the bit complexity of getting all f_i is $\widetilde{\mathcal{O}}((nT^3D\log H)(\log C + \log H)(nT^2\log D))$, that is $\widetilde{\mathcal{O}}(n^2T^5D\log H\log C + n^2T^5D\log^2 H)$.

In step 4, since $\deg f_i$ is $\widetilde{\mathcal{O}}(nT^2D)$, by fast multipoint evaluation [11, p. 299], it needs $\widetilde{\mathcal{O}}(nT^2D)$ operations. The number of the f_i that we need to check is at most $\widetilde{\mathcal{O}}(nT^2\log D)$, so the total arithmetic operations for evaluations is $\widetilde{\mathcal{O}}(n^2T^4D)$. As the coefficients of f_i are in A and the number of terms is less than T, the data is $\widetilde{\mathcal{O}}(TC(nT^2D)^{nT^2D}H^T)$. So the height of the data is $\widetilde{\mathcal{O}}(nT^2D + \log C + T\log H)$. The total bit complexity of step 4 is $\widetilde{\mathcal{O}}(n^3T^6D^2 + n^2T^4D\log C + n^2T^5D\log H)$.

In step 6, we need to obtain t terms of g. We analyse the bit complexity of one step of the cycle. To obtain b, we need $\mathcal{O}(1)$ arithmetic operations. The height of the data is $\widetilde{\mathcal{O}}(nT^2D(\log C + D\log n + \log H))$, so the bit complexity is $\widetilde{\mathcal{O}}(nT^2D\log C + nT^2D^2 + nT^2D\log H)$. To factor $\frac{b}{c_i}$, we need $n\log^2 D$ operations. The data of b and c_i is $\widetilde{\mathcal{O}}(Cq_n^DH)$, so the bit complexity is $\widetilde{\mathcal{O}}(n\log^2 D\log C + nD + n\log^2 D\log H)$. So the total bit complexity of step 6 is $\widetilde{\mathcal{O}}(nT^3D\log C + nT^3D^2 + nT^3D\log H)$.

Therefore, the bit complexity is $\widetilde{\mathcal{O}}(n^2T^5D\log H\log C + n^2T^5D\log^2 H + n^3T^6D^2)$. □

Remark 4. If $A = \{a|C \geq |a|, a \in \mathbb{Z}\}$, we can modify Algorithm 11. Assume $A_T = \{a|TC \geq |a|, a \in \mathbb{Z}\}$. In step 2, we let $f_i :=$ **UPolySIRat**(f_{x,p_i}, A_T). Note that f_{x,p_i} is an integer polynomial with coefficients bounded by TC, $f_i = f_{x,p_i}$. So in step 4, we just find the smallest integer i_0 that f_{i_0} has the most number of the terms in S. In this case, p_{i_0} is a good prime for f. The bit complexity of the algorithm will be $\widetilde{\mathcal{O}}(n^2T^5D\log C + nT^3D^2)$.

3.3 Probabilistic Algorithm

Giesbrecht and Roche [5, Lemma 2.1] proved that if $\lambda = \max\{21, \frac{5}{3}nT(T-1)\ln D\}$, then a prime p chosen at random in $[\lambda, 2\lambda]$ is a good prime for $f(x_1, \ldots, x_n)$ with probability at least $\frac{1}{2}$. Based on this result, we give a probabilistic algorithm.

Algorithm 13 (ProMPolySIMK)

Input: A black box polynomial $f(x_1,\ldots,x_n) \in A[x_1,\ldots,x_n]$, whose coefficients are in A given in (3), an upper bound D for the degree, an upper bound T of the number of terms, a list of n different primes $q_1, q_2, \ldots, q_n (q_1 < \cdots < q_n)$.
Output: The exact form of $f(x_1,\ldots,x_n)$ with probability $\geq \frac{1}{2}$.

Step 1: Let $\lambda := \max\{21, \frac{5}{3}nT(T-1)\ln D\}$, randomly choose a prime p in $[\lambda, 2\lambda]$.
Step 2: Let $f_p :=$ **UPolySIRat**$(f_{x,p}, A, T)$ via Algorithm 8.
  if $f_p = failure$ **then** return failure;
  Assume $f_p = \frac{c_1}{H_1}x^{d_1} + \frac{c_2}{H_2}x^{d_2} + \cdots + \frac{c_t}{H_t}x^{d_t}, d_1 < d_2 < \cdots < d_t$
Step 3: Let $\beta := 2Cq_n^D \max\{H_1, H_2, \ldots, H_t\} + 1$. [Lemma 7]
  Denote $g(x) = f(q_1 x, q_2 x^{\mathbf{mod}(D+1,p)}, \ldots, q_n x^{\mathbf{mod}((D+1)^{n-1},p)})$. Let $u := g(\beta)$;
Step 4: Let $s := 0$;
  for $i = t, t-1, \ldots, 1$ **do**
  Let $b := \lceil \frac{u}{\beta^{d_i}} H_i - \frac{1}{2} \rceil$
  Factor $\frac{b}{c_i} = kq_1^{e_1}q_2^{e_2}\cdots q_n^{e_n}$, where $q_i \nmid k, i = 1, 2, \ldots, n$
  if $k \neq 1$ or $e_1 + e_2 + \cdots + e_n > D$ **then** return failure;
  $s := s + \frac{c_i}{H_i}x_1^{e_1}x_2^{e_2}\cdots x_n^{e_n}$.
  $u := u - \frac{b}{H_i}x^{d_i}$.
  end do;
  if $u = 0$ **then** return s **else** return failure;

Theorem 14. *The bit complexity of Algorithm 13 is* $\widetilde{\mathcal{O}}(nT^3 D \log H \log C + nT^3 D \log^2 H + nT^3 D^2)$.

Proof. In step 2, the degree of $f_{x,p}$ is bounded by $D(p-1)$. As p is $\mathcal{O}(nT^2 \log D)$, the degree bound is $\widetilde{\mathcal{O}}(nT^2 D)$. By Theorem 9, the complexity is $\widetilde{\mathcal{O}}((nT^3 D \log H)(\log C + \log H))$, or $\widetilde{\mathcal{O}}(nT^3 D \log H \log C + nT^3 D \log^2 H)$.

In step 4, we need to obtain t terms of g. We analyse the bit complexity of one step of the cycle. To obtain b, we need $\mathcal{O}(1)$ arithmetic operations. The height of the data is $\widetilde{\mathcal{O}}(nT^2 D(\log C + D \log n + \log H))$, so the bit complexity is $\widetilde{\mathcal{O}}(nT^2 D \log C + nT^2 D^2 + nT^2 D \log H)$. To factor $\frac{b}{c_i}$, we need $n \log^2 D$ operations. The height of b and c_i is $\widetilde{\mathcal{O}}(Cq_n^D H)$, so the bit complexity is $\widetilde{\mathcal{O}}(n \log^2 D \log C + nD + n \log^2 D \log H)$. So the total bit complexity of step 4 is $\widetilde{\mathcal{O}}(nT^3 D \log C + nT^3 D^2 + nT^3 D \log H)$.

Therefore, the total bit complexity of the algorithm is $\widetilde{\mathcal{O}}(nT^3 D \log H \log C + nT^3 D^2 + nT^3 D \log^2 H)$. □

Remark 5. In Algorithm 13, we also modify step 4 of Algorithm 8 as in Remark 4.

4 Experimental Results

In this section, practical performances of the algorithms will be presented. The data are collected on a desktop with Windows system, 3.60 GHz Core $i7-4790$

CPU, and 8 GB RAM memory. The implementations in Maple can be found in http://www.mmrc.iss.ac.cn/~xgao/software/sicoeff.zip

We randomly construct five polynomials, then regard them as black box polynomials and reconstruct them with the algorithms. The average times are collected. The results for univariate interpolation are shown in Figs. 1, 2, 3 and 4. In each figure, three of the parameters C, H, D, T are fixed and one of them is variable. From these figures, we can see that Algorithm **UPolySIRat** is linear in T, approximately linear in D, logarithmic in C and H. The results in the multivariate case are shown in Figs. 5 and 6. We just test the probabilistic algorithm. From these figures, we can see that the Algorithm **ProMPolySIMK** is polynomial in T and D.

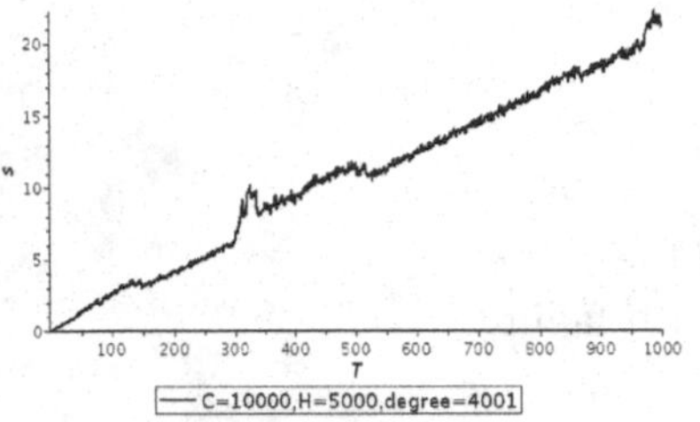

Fig. 1. UPolySIRat: average running times with varying T

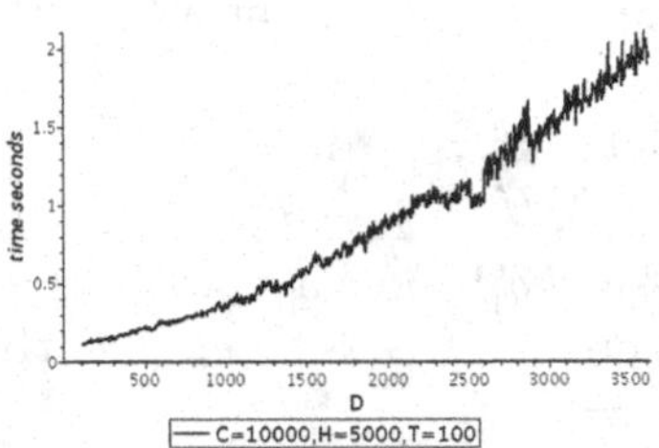

Fig. 2. UPolySIRat: average running times with varying D

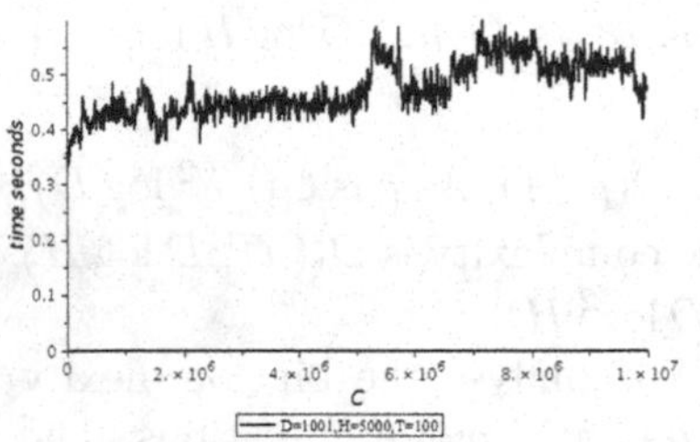

Fig. 3. UPolySIRat: average running times with varying C

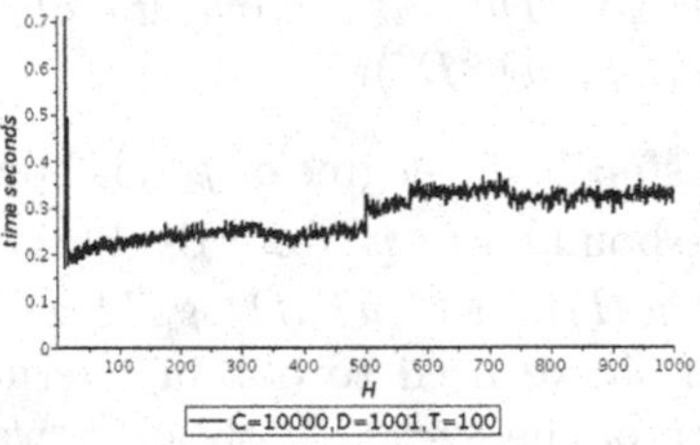

Fig. 4. UPolySIRat: average running times with varying H

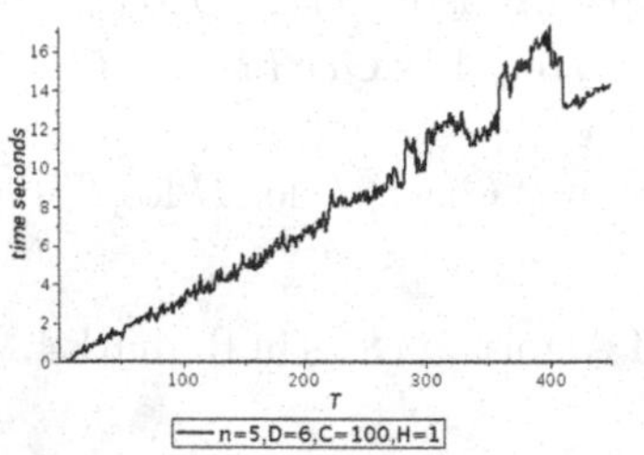

Fig. 5. ProMPolySIMK: average running times with varying T

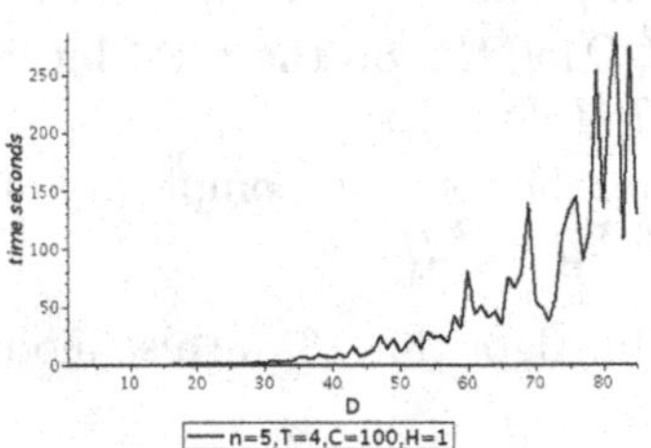

Fig. 6. ProMPolySIMK: average running times with varying D

5 Conclusion

In this paper, a new type of sparse interpolation is considered, that is, the coefficients of the black box polynomial f are from a finite set. Specifically, we assume that the coefficients are rational numbers such that the upper bounds of the absolute values of these numbers and their denominators are given, respectively. We first give an interpolation algorithm for a univariate polynomial f, where f is obtained from one evaluation $f(\beta)$ for a sufficiently large number β. Then, we introduce the modified Kronecker substitution to reduce the interpolation of a multivariate polynomial into the univariate case. Both algorithms have polynomial bit-size complexity and the algorithms can be used to recover quite large polynomials.

References

1. Arnold, A.: Sparse polynomial interpolation and testing. Ph.D. thesis, Waterloo Unversity, Canada (2016)
2. Arnold, A., Roche, D.S.: Multivariate sparse interpolation using randomized Kronecker substitutions. In: ISSAC 2014, 23–25 July, Kobe, Japan (2014)
3. Ben-Or, M., Tiwari, P.: A deterministic algorithm for sparse multivariate polynomial interpolation. In: 20th Annual ACM Symposium on Theory of Computing, pp. 301–309 (1988)
4. Garg, S., Schost, É.: Interpolation of polynomials given by straight-line programs. Theoret. Comput. Sci. **410**(27–29), 2659–2662 (2009)
5. Giesbrecht, M., Roche, D.S.: Diversification improves interpolation. In: Proceedings of the ISSAC 2011, pp. 123–130. ACM Press (2011)
6. Huang, Q.L., Gao, X.S.: Sparse rational function interpolation with finitely many values for the coefficients arXiv:1706.00914 (2017)
7. Huang, Q.L., Gao, X.S.: New algorithms for sparse interpolation and identity testing of multivariate polynomials. Preprint (2017)
8. Kaltofen, E., Yagati, L.: Improved sparse multivariate polynomial interpolation algorithms. In: Gianni, P. (ed.) ISSAC 1988. LNCS, vol. 358, pp. 467–474. Springer, Heidelberg (1989). doi:10.1007/3-540-51084-2_44
9. Klivans, A.R., Spielman, D.: Randomness efficient identity testing of multivariate polynomials. In: Proceedings of the STOC 2001, pp. 216–223. ACM Press (2001)
10. Kronecker, L.: Grundzüge einer arithmetischen Theorie der algebraischen Grössen. J. Reine Angew. Math. **92**, 1–122 (1882)
11. von zur Gathen, J., Gerhard, J.: Modern Computer Algebra. Cambridge University Press, Cambridge (1999)
12. Zippel, R.: Interpolating polynomials from their values. J. Symbolic Comput. **9**(3), 375–403 (1990)

On Stationary Motions of the Generalized Kowalewski Gyrostat and Their Stability

Valentin Irtegov and Tatyana Titorenko(✉)

Institute for System Dynamics and Control Theory SB RAS,
134, Lermontov str., Irkutsk 664033, Russia
{irteg,titor}@icc.ru

Abstract. The stationary motions of the Kowalewski gyrostat in two constant force fields are studied. It is revealed that the equations of motion of the gyrostat have the families of permanent rotations when the force fields are parallel, and the families of equilibria when these fields have special directions. It is shown that all the found solutions belong to an intersection of two invariant manifolds of codimension 2. The analysis of stability in the Lyapunov sense for these solutions is conducted.

1 Introduction

The problem of the rotational motion of a gyrostat (a rigid body with a symmetrical rotor placed inside it) in two constant force fields is considered. Mass distribution in the body is subject to the Kowalewski conditions [1]. Similar problems arise, e.g., in space dynamics [2], quantum mechanics [3]. The equations of motion of the gyrostat represent a completely Liouville integrable system: an additional first integral of the problem has been found in [4]. It should be noted that so far this system is poorly studied because of computational difficulties arising in the process of its investigation. There exists a series of works devoted to the topological analysis of the system (see, e.g., [5,6]). We conduct the qualitative analysis for the equations of motion of the gyrostat. Our approach to the study of similar problems is based on solving an extremum problem for the elements of the algebra of the problem's first integrals that enables us to reduce the problem of the qualitative analysis of differential equations to an algebraic one and to apply computer algebra tools in our study. In the present work, within the framework of the qualitative analysis of the system under consideration, we study the stationary motions of the gyrostat and their stability. By stationary motions, we mean solutions of the equations of motion on which the first integrals (or their combinations) in the problem under study take stationary values. The latter allows one to use these integrals for obtaining a Lyapunov function to investigate the stability of such solutions. All computations represented in this paper have been performed with "Mathematica" computer algebra system (CAS) as well as the software package [7].

V.P. Gerdt et al. (Eds.): CASC 2017, LNCS 10490, pp. 210–224, 2017.
DOI: 10.1007/978-3-319-66320-3_16

2 Formulation of the Problem

For describing the motion of the gyrostat, an inertial coordinate system with its origin at a fixed point O of the body is introduced. The $Oxyz$ frame is rigidly attached to the body. The axes of the frame are directed along the principal axes of inertia of the body. The rotor axis coincides with the Oz axis. The inertia moments of the gyrostat are related as follows: $A = B = 2C$.

The equations of motion of the gyrostat in the $Oxyz$ frame can be written as:

$$\begin{array}{lll} 2\dot{p} = q\,(r-\lambda) + b\,\delta_3, & \dot{\gamma}_1 = \gamma_2 r - \gamma_3 q, & \dot{\delta}_1 = \delta_2 r - \delta_3 q, \\ 2\dot{q} = x_0\gamma_3 - p\,(r-\lambda), & \dot{\gamma}_2 = \gamma_3 p - \gamma_1 r, & \dot{\delta}_2 = \delta_3 p - \delta_1 r, \\ \dot{r} = -b\delta_1 - x_0\gamma_2, & \dot{\gamma}_3 = \gamma_1 q - \gamma_2 p, & \dot{\delta}_3 = \delta_1 q - \delta_2 p. \end{array} \tag{1}$$

Here p, q, r are the projections of the angular velocity vector onto the axes of the $Oxyz$ frame; $\gamma_i\,(i = 1, 2, 3)$ are the components of the direction vector of the 1st force field; $\delta_i\,(i = 1, 2, 3)$ are the components of the direction vector of the 2nd force field; $(x_0, 0, 0)$, $(0, b, 0)$ are the radius vectors of the 1st and 2nd force centers, respectively; $\lambda = const$ is the gyrostatic parameter.

Equation (1) admit the following first integrals:

$$\begin{array}{l} 2H = 2(p^2+q^2) + r^2 + 2(x_0\gamma_1 - b\,\delta_2) = 2h, \\ V_1 = (p^2 - q^2 - x_0\gamma_1 - b\,\delta_2)^2 + (2p\,q - x_0\gamma_2 + b\,\delta_1)^2 \\ \qquad +2\lambda[(p^2+q^2)(r-\lambda) + 2(b\,q\delta_3 - p\,x_0\gamma_3)] = c_1, \\ V_2 = \gamma_1^2 + \gamma_2^2 + \gamma_3^2 = 1,\ V_3 = \delta_1^2 + \delta_2^2 + \delta_3^2 = 1, \\ V_4 = \gamma_1\delta_1 + \gamma_2\delta_2 + \gamma_3\delta_3 = c_2, \\ V_5 = x_0^2\,[p\gamma_1 + q\gamma_2 + \frac{1}{2}(r+\lambda)\gamma_3]^2 + b^2\,[p\delta_1 + q\delta_2 + \frac{1}{2}(r+\lambda)\delta_3]^2 \\ \qquad -x_0 b\,(r-\lambda)[(\gamma_2\delta_3 - \gamma_3\delta_2)p + (\gamma_3\delta_1 - \gamma_1\delta_3)q + \frac{1}{2}(r+\lambda)(\gamma_1\delta_2 - \gamma_2\delta_1)] \\ \qquad +x_0 b^2\gamma_1(\delta_1^2 + \delta_2^2 + \delta_3^2) - x_0^2 b\,\delta_2(\gamma_1^2 + \gamma_2^2 + \gamma_3^2) \\ \qquad -b x_0(b\delta_1 - \gamma_2 x_0)(\delta_1\gamma_1 + \delta_2\gamma_2 + \delta_3\gamma_3) = c_3, \end{array} \tag{2}$$

where V_5 is the additional first integral found in [4], h, c_1, c_2, c_3 are some constants.

Note, as c_2 is an arbitrary constant, Eq. (1) together with the integral $V_4 = c_2$ describe the motion of the gyrostat in the constant force fields having special directions (according to the value of the parameter c_2). Thus, system (1), (2) can be considered as a family of the systems parameterized by c_2.

When $b = 0$, system (1), (2) corresponds to an integrable case in the problem of motion of the Kowalewski gyrostat in a gravitational force field [8].

The purpose of this work is to find the stationary solutions of Eq. (1) and to investigate their qualitative properties.

3 Finding the Stationary Solutions

Traditionally, stationary solutions can be obtained from the conditions of stationarity for the first integrals of a problem (see, e.g., [9]). In the case under consideration, following this technique, we should solve a system of 9 nonhomogeneous cubic equations with respect to the phase variables $p, q, r, \gamma_i, \delta_i\,(i = 1, 2, 3)$. In the given work, we apply another technique represented below.

3.1 Permanent Rotations

For finding the desired solutions, we equate the right-hand sides of differential Eq. (1) to zero, and add relations $V_2 = 1, V_3 = 1$ (2) to them. For the polynomials of a resulting system (the system of quadratic equations), we construct a lexicographical Gröbner basis with respect to some part of the phase variables, e.g., $\delta_1, \delta_2, \delta_3, \gamma_1, \gamma_2, \gamma_3, p, q$ with the "Mathematica" procedure *GroebnerBasis*. A result will be a system which is decomposed into two subsystems:

$$
\begin{aligned}
\text{(I)}\quad & b^4r^2 - q^2(\lambda - r)^2\,[b^2(q^2 + r^2) + x_0^2 q^2] = 0,\ bp - x_0 q = 0,\\
& b\,\gamma_3 + q\,(\lambda - r) = 0,\ b\,r\gamma_2 + q^2\,(\lambda - r) = 0,\ b^2\,r\gamma_1 + x_0 q^2\,(\lambda - r) = 0,\\
& b\,\delta_3 - q\,(\lambda - r) = 0,\ b\,r\delta_2 - q^2\,(\lambda - r) = 0,\ -b^2\,r\delta_1 + x_0 q^2\,(\lambda - r) = 0.
\end{aligned} \tag{3}
$$

$$
\begin{aligned}
\text{(II)}\quad & b^4r^2 - q^2(\lambda - r)^2\,[b^2(q^2 + r^2) + x_0^2 q^2] = 0,\ bp + x_0 q = 0,\\
& b\,\gamma_3 - q\,(\lambda - r) = 0,\ b\,r\gamma_2 - q^2\,(\lambda - r) = 0,\ b^2\,r\gamma_1 - x_0 q^2\,(\lambda - r) = 0,\\
& b\,\delta_3 - q\,(\lambda - r) = 0,\ b\,r\delta_2 - q^2\,(\lambda - r) = 0,\ -b^2\,r\delta_1 - x_0 q^2\,(\lambda - r) = 0.
\end{aligned} \tag{4}
$$

As can easily be verified by invariant manifold (IM) definition, Eqs. (3), (4) define 2 one-dimensional IMs of Eq. (1).

The vector field on each IM is given by the equation $\dot{r} = 0$. It has the following family of solutions:

$$r = r_0 = \text{const}. \tag{5}$$

Hence, geometrically, IMs (3), (4) in space R^9 correspond to 2 curves, over each point of which the family of solutions (5) is defined.

Equations (3) with (5) represent 4 families of solutions for the equations of motion (1):

$$
\begin{aligned}
& p = \mp x_0\,\alpha_1\,r_0^{1/2}\,(\lambda - r_0)^{-1/2},\ q = \mp b\,\alpha_1\,r_0^{1/2}\,(\lambda - r_0)^{-1/2},\ r = r_0,\\
& \gamma_1 = -x_0\,\alpha_1^2,\ \gamma_2 = -b\,\alpha_1^2,\ \gamma_3 = \pm[r_0\,(\lambda - r_0)]^{1/2}\alpha_1, \delta_1 = x_0\,\alpha_1^2,\\
& \delta_2 = b\,\alpha_1^2,\ \delta_3 = \mp[r_0\,(\lambda - r_0)]^{1/2}\alpha_1;
\end{aligned} \tag{6}
$$

$$
\begin{aligned}
& p = \pm x_0\,\alpha_2\,r_0^{1/2}(r_0 - \lambda)^{-1/2},\ q = \pm b\,\alpha_2\,r_0^{1/2}(r_0 - \lambda)^{-1/2},\ r = r_0,\\
& \gamma_1 = x_0\,\alpha_2^2,\ \gamma_2 = b\,\alpha_2^2,\ \gamma_3 = \pm[r_0\,(r_0 - \lambda)]^{1/2}\alpha_2,\ \delta_1 = -x_0\,\alpha_2^2,\\
& \delta_2 = -b\,\alpha_2^2,\ \delta_3 = \mp[r_0\,(r_0 - \lambda)]^{1/2}\alpha_2.
\end{aligned} \tag{7}
$$

Here r_0 is the parameter of the families, $\alpha_1 = \rho_1\,\beta,\ \alpha_2 = \rho_2\,\beta,\ \beta = [2\,(b^2 + x_0^2)]^{-1/2},\ \rho_1 = [r_0\,(r_0 - \lambda) + (4(b^2 + x_0^2) + (\lambda - r_0)^2 r_0^2)^{1/2}]^{1/2}$,
$\rho_2 = [r_0\,(\lambda - r_0) + (4(b^2 + x_0^2) + (\lambda - r_0)^2 r_0^2)^{1/2}]^{1/2}$.

Equations (4), (5) determine the following 4 families of solutions for system (1):

$$
\begin{aligned}
& p = \pm x_0\,\alpha_1 r_0^{1/2}(\lambda - r_0)^{-1/2},\ q = \mp b\,\alpha_1 r_0^{1/2}(\lambda - r_0)^{-1/2},\ r = r_0,\\
& \gamma_1 = -x_0\,\alpha_1^2,\ \gamma_2 = b\,\alpha_1^2,\ \gamma_3 = \mp[r_0\,(\lambda - r_0)]^{1/2}\alpha_1,\ \delta_1 = -x_0\,\alpha_1^2,\\
& \delta_2 = b\,\alpha_1^2,\ \delta_3 = \mp[r_0\,(\lambda - r_0)]^{1/2}\alpha_1;
\end{aligned} \tag{8}
$$

$$p = \mp x_0\,\alpha_2\,r_0^{1/2}(r_0-\lambda)^{-1/2},\; q = \pm b\,\alpha_2\,r_0^{1/2}(r_0-\lambda)^{-1/2},\; r = r_0,$$
$$\gamma_1 = x_0\,\alpha_2^2,\; \gamma_2 = -b\,\alpha_2^2, \gamma_3 = \mp[r_0\,(r_0-\lambda)]^{1/2}\alpha_2,$$
$$\delta_1 = x_0\,\alpha_2^2,\; \delta_2 = -b\,\alpha_2^2,\; \delta_3 = \mp[r_0\,(r_0-\lambda)]^{1/2}\alpha_2. \tag{9}$$

On substituting solutions (6), (7) into integral V_4 (2), the latter is identically equal to -1. On solutions (8), (9), this integral becomes identically 1. So, with mechanical viewpoint, the elements of the families of solutions (6)–(9) correspond to the permanent rotations of the gyrostat in the parallel (or opposite in direction) force fields. The gyrostat rotates around the coinciding (or opposite) directions of the force fields with the angular velocity $\omega^2 = r_0\,(\rho_1^2/(2(\lambda-r_0))+r_0)$ $(\omega^2 = r_0\,(\rho_2^2/(2(r_0-\lambda))+r_0))$. The axis position of rotation in the body depends on the parameter r_0 and does not coincide with the principal axes of inertia of the body.

Now, we show that all the solutions found above are stationary. First, we consider the families of solutions (6).

Let

$$2K = 2\lambda_0 H - \lambda_1 V_1 - \lambda_2 V_2 - \lambda_3 V_3 - 2\lambda_4 V_4 - 4\lambda_5 V_5 \tag{10}$$

be the family of the problem's first integrals, where $\lambda_i\,(i = 0,\ldots,5)$ are the parameters of the family.

We write down the necessary conditions for the integral K to have an extremum with respect to the phase variables

$$\frac{\partial K}{\partial p} = 0,\; \frac{\partial K}{\partial q} = 0,\; \frac{\partial K}{\partial r} = 0,\; \frac{\partial K}{\partial \gamma_i} = 0,\; \frac{\partial K}{\partial \delta_i} = 0,\; (i = 1, 2, 3) \tag{11}$$

and find the values of $\lambda_2, \lambda_3, \lambda_5$

$$\lambda_2 = x_0^2\,\chi + \lambda_4,\; \lambda_3 = b^2\,\chi + \lambda_4,\; \lambda_5 = \frac{1}{(\lambda^2 - r_0^2 + \rho_1^2)}\Big[\frac{2\lambda_0}{\rho_1^2} - \frac{\lambda\lambda_1}{(\lambda - r_0)}\Big],$$

under which solutions (6) satisfy equation (11). Here $\chi = [2(r_0^2-\lambda^2)\lambda_0 - \lambda\rho_1^2(\lambda - 3r_0)\lambda_1]\,[\rho_1\,(\lambda-r_0)]^{-2}$.

Having substituted the above expressions into (10), we have the following family of integrals:

$$2K_1 = 2\tilde{K}_1 - (V_2 + V_3 + 2V_4)\,\lambda_4,$$
$$\text{where}\;\; 2\tilde{K}_1 = 2\Big[H + \frac{(\lambda + r_0)(x_0^2 V_2 + b^2 V_3)}{\rho_1^2(\lambda - r_0)} - \frac{4V_5}{\rho_1^2(\lambda^2 - r_0^2 + \rho_1^2)}\Big]\lambda_0$$
$$+\Big[\frac{\lambda\,(\lambda - 3r_0)(x_0^2 V_2 + b^2 V_3)}{(\lambda - r_0)^2} - V_1 + \frac{4\lambda V_5}{(\lambda - r_0)(\lambda^2 - r_0^2 + \rho_1^2)}\Big]\lambda_1. \tag{12}$$

It is split up into 3 subfamilies of the integrals which correspond to the coefficients of λ_0, λ_1, λ_4, respectively. The elements of both the family of the integrals

K_1 and its subfamilies take stationary values on the elements of families (6). The latter is verified by direct computation. So, the solutions under consideration are stationary. The family K_1 and its subfamilies – both individually and in combination – can be used for obtaining a Lyapunov function to analyze stability of solutions (6).

In a similar manner, we prove the stationarity of solutions (7)–(9). Note that IM (3) and IM (4), which the families of solutions under consideration belong to, are stationary. The integrals $V_2 + V_3 + 2V_4$ and $V_2 + V_3 - 2V_4$ take stationary values on these IMs, respectively.

3.2 Equilibria

Using the technique chosen, we have found another group of the stationary solutions of differential Eq. (1) in the case when $p = q = r = 0$.

The equations

$$p = 0,\ q = 0,\ r = 0,\ \gamma_3 = 0,\ \delta_3 = 0,$$
$$x_0\gamma_2 + b\delta_1 = 0,\ b^2\delta_1^2 + x_0^2(\gamma_1^2 - 1) = 0,\ \delta_1^2 + \delta_2^2 = 1 \tag{13}$$

define one-dimensional IM of Eq. (1). Likewise as above, it can be verified by IM definition. The vector field on this IM is described by the equation $\dot{\delta}_1 = 0$ which has the following family of solutions:

$$\delta_1 = \delta_1^0 = \text{const.} \tag{14}$$

Equations (13) with (14) determine 4 families of solutions for the equations of motion (1):

$$p = 0,\ q = 0,\ r = 0, \gamma_1 = \mp(x_0^2 - b^2\delta_1^{0^2})^{1/2}x_0{}^{-1},\ \gamma_2 = -b\,\delta_1^0 x_0^{-1},\ \gamma_3 = 0,$$
$$\delta_1 = \delta_1^0,\ \delta_2 = (1 - \delta_1^{0^2})^{1/2},\ \delta_3 = 0; \tag{15}$$
$$p = 0,\ q = 0,\ r = 0, \gamma_1 = \mp(x_0^2 - b^2\delta_1^{0^2})^{1/2}x_0^{-1},\ \gamma_2 = -b\,\delta_1^0 x_0^{-1},\ \gamma_3 = 0,$$
$$\delta_1 = \delta_1^0,\ \delta_2 = -(1 - \delta_1^{0^2})^{1/2},\ \delta_3 = 0. \tag{16}$$

Here δ_1^0 is the family parameter, $|\delta_1^0| \le 1$ and $|\delta_1^0| \le |x_0 b^{-1}|$ are the conditions for the solutions to be real.

With mechanical viewpoint, the elements of the families of solutions (15), (16) correspond to the equilibria of the gyrostat.

The integral V_4 takes the values $\pm\delta_1^0 x_0^{-1} b\,(1 - \delta_1^{0^2})^{1/2} \pm (x_0^2 - b^2\delta_1^{0^2})^{1/2}$ on the corresponding elements of the families of solutions (15), (16). So, each equilibrium corresponds to a definite angle between the directions of the force fields. In [5], four equilibria of the above families for the case of the orthogonal force fields are presented.

With the aid of the technique described above, we have found the families of the integrals the elements of which take stationary values on the corresponding elements of families (15), (16). Below, one of these families is represented.

$$\begin{aligned}
2K_2 = &\Big[\frac{2\,[4V_5 - ((\lambda^2+4z)z - 2(b^2+x_0^2))\,H]}{\lambda^2+2z} - V_1\Big]\,\lambda_1 \\
&+\Big[\frac{4V_5 - 2[(\lambda^2+3z)z - (b^2+x_0^2)]\,H}{x_0^2\,(\lambda^2+2z)} - V_2 - \frac{b^2V_3}{x_0^2}\Big]\,\lambda_2 + \Big[\frac{\bar{z}V_3}{\delta_1^0 x_0} \\
&+2\Big(\frac{1}{\delta_1^0 x_0} - \frac{2\,[\delta_1^{0^2} z - (x_0^2 - b^2\delta_1^{0^2})^{1/2}]}{\delta_1^0 x_0(\lambda^2+2z)}\Big)\,H - 2V_4 - \frac{4(1-\delta_1^{0^2})^{1/2}V_5}{b\,\delta_1^0 x_0(\lambda^2+2z)}\Big]\,\lambda_4.
\end{aligned} \tag{17}$$

Here $z = b\,(1-\delta_1^{0^2})^{1/2} + (x_0^2 - b^2\delta_1^{0^2})^{1/2}$, $\bar{z} = b\,(1-\delta_1^{0^2})^{1/2} - (x_0^2 - b^2\delta_1^{0^2})^{1/2}$.

The elements of the family K_2 and its subfamilies (the coefficients of $\lambda_1, \lambda_2, \lambda_4$) assume stationary values on the elements of the 1st family of solutions (15). The corresponding families of the integrals for other solutions (15), (16) are similar to the above family.

4 On Invariant Manifolds of Codimension 2

Let us show that all the solutions found above belong to two IMs of codimension 2 of the equations of motion (1). We shall obtain these IMs from the stationary conditions for integral K (10) by resolving them with respect to part of the phase variables and the parameters λ_i. The given technique, based on Gröbner basis method, was already applied by the authors repeatedly (see, e.g., [10]). In the problem under study, its direct application reduces to cumbersome computations. In order to avoid these difficulties, we replace the initial problem's variables with the following ones:

$$\begin{aligned}
&x_1 = -(x_0\gamma_1 + b\delta_2) - i(x_0\gamma_2 - b\delta_1),\ x_2 = -(x_0\gamma_1 + b\delta_2) + i(x_0\gamma_2 - b\delta_1),\\
&y_1 = -(x_0\gamma_1 - b\delta_2) - i(x_0\gamma_2 + b\delta_1),\ y_2 = -(x_0\gamma_1 - b\delta_2) + i(x_0\gamma_2 + b\delta_1),\\
&z_1 = -x_0\gamma_3 + ib\,\delta_3,\ z_2 = -x_0\gamma_3 - ib\,\delta_3,\ w_1 = p + iq,\ w_2 = p - iq,\ w_3 = r.
\end{aligned} \tag{18}$$

These are similar to (3.8) [5].

In the above variables, the equations of motion and the problem's first integrals can be written as

$$\begin{aligned}
&2\dot{w}_1 = i(w_1(\lambda - w_3) - z_1),\ \dot{x}_1 = i(w_1 z_1 - w_3 x_1),\ \ \dot{y}_1 = i(w_1 z_2 - w_3 y_1),\\
&2\dot{w}_2 = i(w_2(w_3 - \lambda) + z_2),\ \dot{x}_2 = i(w_3 x_2 - w_2 z_2),\ \ \dot{y}_2 = i(w_3 y_2 - w_2 z_1),\\
&2\dot{w}_3 = i(y_2 - y_1), 2\dot{z}_1 = i(w_2 x_1 - w_1 y_2),\ 2\dot{z}_2 = i(w_2 y_1 - w_1 x_2)
\end{aligned} \tag{19}$$

and

$$\begin{aligned}
&2\tilde{H} = 2w_1w_2 + w_3^2 - y_1 - y_2 = 3\tilde{h},\\
&\tilde{V}_1 = (w_1^2 + x_1)(w_2^2 + x_2) + 2\lambda\,[w_2 z_1 + w_1(w_2 w_3 + z_2)] - 2\lambda^2 w_1 w_2 = \tilde{c}_1,\\
&\tilde{V}_2 = (x_1 + y_1)(x_2 + y_2) + (z_1 + z_2)^2 = 1,\\
&\tilde{V}_3 = (x_1 - y_1)(x_2 - y_2) - (z_1 - z_2)^2 = 1,\\
&\tilde{V}_4 = x_1 y_2 - x_2 y_1 + z_1^2 - z_2^2 = \tilde{c}_2,\\
&\tilde{V}_5 = x_1 y_1(2w_2^2 + x_2) + x_2 y_2(2w_1^2 + x_1) - y_1 y_2(y_1 + y_2)\\
&\qquad + 2w_1(w_2 x_1 x_2 + w_2 y_1 y_2 + 2w_3 x_2 z_1) + 4w_2 w_3 x_1 z_2 - 2(y_1 + y_2) z_1 z_2
\end{aligned} \tag{20}$$

$$+2(x_2z_1^2 + x_1z_2^2) + (\lambda^2 - w_3^2)(x_1x_2 - y_1y_2) + 2(\lambda^2 + w_3^2)z_1z_2$$
$$+4\lambda(w_2y_1z_1 + w_1y_2z_2 + w_3z_1z_2) = \tilde{c}_3,$$

respectively.

Let $\hat{K}$ be integral K (10) in variables (18). We write down the stationary conditions for the integral $\hat{K}$

$$\frac{\partial \hat{K}}{\partial w_1} = 0, \quad \frac{\partial \hat{K}}{\partial w_2} = 0, \quad \frac{\partial \hat{K}}{\partial w_3} = 0, \quad \frac{\partial \hat{K}}{\partial x_i} = 0, \quad \frac{\partial \hat{K}}{\partial y_i} = 0, \quad \frac{\partial \hat{K}}{\partial z_i} = 0 \quad (i = 1, 2)$$

which represent a system of nonhomogeneous cubic equations with respect to the variables $w_1, w_2, w_3, x_1, x_2, y_1, y_2, z_1, z_2$ with the parameters λ, λ_i $(i = 0, \ldots, 5)$, and then, for the polynomials of this system, construct a lexicographical Gröbner basis with respect to $\lambda_0, \lambda_1, \lambda_2, \lambda_3, \lambda_4, y_1, y_2$. A result will be a system which is split up into two subsystems.

The 1st subsystem:

$$f_i(w_1, w_2, w_3, x_1, x_2, z_1, z_2, \lambda_0, \lambda_1, \lambda_2, \lambda_3, \lambda_4, \lambda) = 0 (i = 1, \ldots, 5) \tag{21}$$

$$\begin{aligned}
&w_2\,(\lambda - w_3)(w_2x_1 + w_1y_2) + (\lambda + w_3)(\lambda w_2 - w_2w_3 - z_2)z_1 - w_1x_2z_1 \\
&-w_2x_1z_2 = 0, \\
&w_1\,(\lambda - w_3)(w_1x_2 + w_2y_1) + (\lambda + w_3)(\lambda w_1 - w_1w_3 - z_1)z_2 - w_1x_2z_1 \\
&-w_2x_1z_2 = 0.
\end{aligned} \tag{22}$$

The 2nd subsystem:

$$g_i(w_1, w_2, w_3, x_1, x_2, z_1, z_2, \lambda_0, \lambda_1, \lambda_2, \lambda_3, \lambda_4, \lambda) = 0 (i = 1, \ldots, 5) \tag{23}$$

$$\begin{aligned}
&\lambda y_2(w_1w_2 + \lambda w_3)(w_1x_2 + \lambda z_2) - w_1(w_2^2 + x_2)\,[(w_2x_1 + \lambda z_1)z_2 \\
&+(w_1x_2 + \lambda z_2)z_1 - w_3x_1x_2] - \lambda x_1(w_2w_3 + z_2)(w_2z_2 - w_3x_2) \\
&+\lambda^2 z_1(w_3^2x_2 - 2w_2w_3z_2 - z_2^2) = 0, \\
&\lambda y_1(w_1w_2 + \lambda w_3)(w_2x_1 + \lambda z_1) - w_2(w_1^2 + x_1)\,[(w_2x_1 + \lambda z_1)z_2 \\
&+z_1(w_1x_2 + \lambda z_2) - w_3x_1x_2] - \lambda x_2(w_1w_3 + z_1)(w_1z_1 - w_3x_1) \\
&+\lambda^2 z_2(w_3^2x_1 - 2w_1w_3z_1 - z_1^2) = 0.
\end{aligned} \tag{24}$$

Here $f_i = 0$, $g_i = 0$ $(i = 1, \ldots, 5)$ are the linear equations with respect to $\lambda_0, \lambda_1, \lambda_2, \lambda_3, \lambda_4$.

Equations (22), (24) define two IMs of codimension 2 of differential Eq. (19). Equations (21), (23) enable us to obtain the first integrals of vector fields on these IMs. For this purpose, it is necessary to resolve the equations with respect to $\lambda_0, \lambda_1, \lambda_2, \lambda_3, \lambda_4$.

In the initial variables, Eq. (22) can be written as:

$$\begin{aligned}
&F + i\,G = 0,\ F - i\,G = 0, \text{ where} \\
&F = -b\,(b\,\delta_3 + qr)\sigma_2 + x_0[2b\,(p\,(\gamma_2\delta_3 - \gamma_3\delta_2) + q\,(\gamma_3\delta_1 - \gamma_1\delta_3)) + p\,r\sigma_1] \\
&\quad -x_0^2\gamma_3\sigma_1 + \lambda^2(bq\delta_3 - x_0p\gamma_3) + \lambda\,[b\,(2q\,(p\delta_1 + q\delta_2) - b\delta_3^2) \\
&\quad -x_0\,(x_0\gamma_3^2 + 2p\,(p\gamma_1 + q\gamma_2))], \\
&G = -r\,(bp\,\sigma_2 + x_0q\,\sigma_1) + 2\lambda\,[x_0q(p\gamma_1 + q\gamma_2) + bp\,(p\delta_1 + q\delta_2)] \\
&\quad +\lambda^2\,(bp\delta_3 + x_0q\gamma_3).
\end{aligned}$$

Here $\sigma_1 = 2p\gamma_1 + 2q\gamma_2 + r\gamma_3, \sigma_2 = 2p\delta_1 + 2q\delta_2 + r\delta_3$.

It is not difficult to verify that the equations $F = 0, G = 0$ define IM of codimension 2 of the equations of motion (1). In order to represent the IM equations in more compact form, we have constructed a lexicographical basis for the polynomials of left-hand sides of these equations, e.g., with respect to the variables γ_2, δ_1. As a result, we have:

$$\begin{aligned} &2b\,p\,(\gamma_2\delta_3 - \gamma_3\delta_2) - (p\,(\lambda - r) + x_0\gamma_3)\,[2\,(p\gamma_1 + q\gamma_2) + (\lambda + r)\gamma_3] = 0,\\ &2x_0\,q\,(\gamma_1\delta_3 - \gamma_3\delta_1) + (q\,(r - \lambda) + b\delta_3)\,[2\,(p\,\delta_1 + q\delta_2) + \delta_3\,(\lambda + r)] = 0. \end{aligned} \tag{25}$$

After analogous transformations of Eq. (24), we obtain:

$$\begin{aligned} &(p\,(p^2 + q^2) + b\,(\delta_1 q - \delta_2 p) - x_0(\gamma_1 p + \gamma_2 q))\,\sigma + \lambda\,(\varrho_0\lambda^2\\ &+\varrho_1\lambda + \varrho_2) = 0,\\ &(q\,(p^2 + q^2) + b\,(\delta_1 p + \delta_2 q) - x_0\,(p\gamma_2 - q\gamma_1))\,\sigma - \lambda\,(\bar{\varrho}_0\lambda^2\\ &+\bar{\varrho}_1\lambda + \bar{\varrho}_2) = 0. \end{aligned} \tag{26}$$

Here σ, ϱ_j, $\bar{\varrho}_j$ are the expressions of $p, q, r, \gamma_i, \delta_i$. Their full form is given in the Appendix.

Equation (26) define IM of codimension 2 of the equations of motion (1) that is verified by IM definition.

Integral V_4 (2) does not turn into any constant on IM (25). Hence, Eq. (25) together with this integral can be considered as a family of IMs. Each element of the given family corresponds to some element of the family of systems (1), (2) (according to the value of the parameter c_2). The above statement is also true for IM (26).

Now, we resolve Eqs. (3) and (4) with respect to the variables p, q, γ_i, δ_i $(i = 1, 2, 3)$, and then substitute the obtained expressions into (25) and (26). The latter equations turn into identities. Hence, IMs (3), (4) are submanifolds of both IM (25) and IM (26), i.e., they belong to their intersection. As the integral V_4 is identically equal to 1 (or -1) on IMs (3), (4), then IMs (25), (26) have to also satisfy this condition. In other words, the result under discussion corresponds to the cases $c_2 = \pm 1$ of the initial problem.

Analogously, one can show that IM (13) belongs to an intersection of IM (25) and IM (26). As the integral V_4 takes the form $\pm[b\,(1 - \delta_1^2)^{1/2} \pm (x_0^2 - b^2\delta_1^2)^{1/2}]\,\delta_1\,x_0^{-1} = c_2$ on IM (13), then IMs (25), (26) have to also satisfy the latter condition.

5 On Stability of the Stationary Solutions

In this section, we investigate the stability of the above found stationary solutions by the 2nd Lyapunov method [11] and on the base of the Lyapunov stability theorem for linear approximation [12]. The 2nd Lyapunov method requires constructing a function (Lyapunov's function) possessing special properties. There is no method for obtaining such function, besides some approaches. We construct a Lyapunov's function from the first integrals of the problem. Here, e.g., the following problem arises. To find a combination of the integrals which provides "softest" sufficient stability conditions, i.e. these conditions are close to necessary ones. For this purpose, it is necessary to make a series of computational experiments. In this case, computer algebra tools provide essential help.

5.1 On Stability of the Permanent Rotations

Let us investigate the stability for the elements of the 1st family of permanent rotations (6)

$$\begin{aligned} &p = -x_0\,\alpha_1\,r_0^{1/2}\,(\lambda - r_0)^{-1/2},\; q = -b\,\alpha_1\,r_0^{1/2}\,(\lambda - r_0)^{-1/2},\; r = r_0,\\ &\gamma_1 = -x_0\,\alpha_1^2,\; \gamma_2 = -b\,\alpha_1^2,\; \gamma_3 = [r_0\,(\lambda - r_0)]^{1/2}\alpha_1,\; \delta_1 = x_0\,\alpha_1^2,\\ &\delta_2 = b\,\alpha_1^2,\; \delta_3 = -[r_0\,(\lambda - r_0)]^{1/2}\alpha_1 \end{aligned} \tag{27}$$

by the 2nd Lyapunov method. From now and further, the denotations of the Subsect. 3.1 are used.

For obtaining a Lyapunov function, we shall use integral K_1 (12) at the following constraints on the parameters: $\lambda_0 = (\lambda\rho_1^2\lambda_1)/(2(\lambda - r_0))$, $\lambda_4 = 0$. Under these conditions, the integral takes the form:

$$2F_1 = \frac{\lambda\,(\rho_1^2 H + 2(x_0^2 V_2 + b^2 V_3)}{\lambda - r_0} - V_1.$$

Introduce

$$\begin{aligned} &y_1 = \delta_1 - x_0\,\alpha_1^2,\; y_2 = \delta_2 - b\,\alpha_1^2,\; y_3 = \delta_3 + [r_0\,(\lambda - r_0)]^{1/2}\alpha_1,\\ &y_4 = \gamma_1 + x_0\,\alpha_1^2,\; y_5 = \gamma_2 + b\,\alpha_1^2,\; y_6 = \gamma_3 - [r_0\,(\lambda - r_0)]^{1/2}\alpha_1,\\ &y_7 = p + x_0\,\alpha_1\,r_0^{1/2}\,(\lambda - r_0)^{-1/2}, y_8 = q + b\,\alpha_1\,r_0^{1/2}\,(\lambda - r_0)^{-1/2}, \end{aligned}$$

the deviations from the elements of family (27).

In the above deviations, the 2nd variation of the integral F_1 in the neighbourhood of the elements of the family under study on the linear manifold

$$\begin{aligned} &\delta H = (\lambda - r_0)^{1/2}\,(x_0 y_4 - b y_2) - 2\,r_0^{1/2}\alpha_1\,(x_0 y_7 + b y_8) = 0,\\ &\delta V_3 = 2^{1/2}\rho_1\,(b^2 + x_0^2)\,[\alpha_1\,(x_0 y_1 + b y_2) - [r_0\,(\lambda - r_0)]^{1/2} y_3] = 0 \end{aligned}$$

can be written as:

$$\begin{aligned} \delta^2 F_1 \; &= a_{11}y_1^2 + a_{12}y_1y_2 + a_{22}y_2^2 + (1 - \lambda\,r_0^{-1})\,x_0^2 y_1 y_4 + a_{24}y_2y_4 + a_{44}y_4^2\\ &+b x_0 y_1 y_5 + x_0^2 y_2 y_5 - x_0^3\,b^{-1} y_4 y_5 + (\lambda + r_0)\,x_0^2\,[2(\lambda - r_0)]^{-1} y_5^2\\ &+\lambda x_0^2(\lambda - r_0)^{-1} y_6^2 + a_{17}y_1y_7 + a_{27}y_2y_7 + a_{47}y_4y_7 + a_{57}y_5y_7\\ &+2\lambda x_0 y_6 y_7 + a_{77}y_7^2. \end{aligned} \tag{28}$$

Here the coefficients a_{ij} depend on the parameters b, x_0, r_0, λ (see Appendix).

The conditions for the quadratic form $\delta^2 F_1$ to be positive definite are sufficient for the stability of the elements of family (27). In the form of the Sylvester inequalities, they are:

$$\begin{aligned} &\lambda D > 0,\; \lambda\,(\lambda + r_0) D^2 > 0,\; \frac{\lambda D^2}{2\,b^2}\Big[\lambda^2\,x_0^2\,(\lambda^2 - r_0^2) + 2\sigma_1\Big] > 0,\\ &\frac{\lambda^2 D^3}{b^2\,r_0}\Big[(\lambda - r_0)\,[b^2\lambda\,(\lambda - r_0)^2 - 2 r_0\,x_0^2\,(\lambda - 2r_0)(\lambda + r_0)] + 4\alpha_1 r_0 \sigma_2\Big] > 0, \end{aligned}$$

$$\frac{\lambda^3 D^5}{r_0\, x_0^2}\left[(\lambda - 2\,r_0)\,(\lambda - r_0)\,[\alpha_1^2\,(b^2\,(\lambda - 2\,r_0) - r_0\,x_0^2) - r_0^2\,(\lambda - r_0)] + \alpha_1^4\sigma_2\right] > 0,$$

$$-\frac{b^2\,\lambda^4 D^6}{r_0\,x_0^4}\,(\lambda - 2r_0)\,[2\,r_0\,(\lambda - r_0) + \rho_1^2]^2 > 0, \tag{29}$$

where $D = x_0^2\,(\lambda - r_0)^{-1}$, $\sigma_1 = \lambda\,(b^2 + x_0^2)^2 + 4b^2 r_0 x_0^2$,
$\sigma_2 = \lambda\,(b^2 - x_0^2)^2 - 2x_0^4\,(\lambda - r_0) + 4b^2 r_0 x_0^2$.

The solutions under study are real when $\lambda < 0$ and $\lambda < r_0 \leq 0$ or $\lambda > 0$ and $0 \leq r_0 < \lambda$. With these conditions we find that inequalities (29) are compatible at the following constraints on the parameters b, x_0, r_0, λ:

$$b \neq 0,\ x_0 \neq 0 \text{ and } ((\lambda < 0,\ 2\lambda < r_0 < \lambda) \text{ or } (\lambda > 0,\ \lambda < r_0 < 2\lambda)).$$

These isolate some subfamily from the family of solutions (27), the elements of which are stable.

The above conditions are also sufficient for the stability of the elements of the 2nd family of solutions (6) when the same integral (the integral F_1) is used to construct a Lyapunov function. In a similar manner, we have analyzed the stability for the elements of the families of permanent rotations (7)–(9).

Using the integrals $V_2 + V_3 \pm 2V_4$ for obtaining Lyapunov functions, it is possible to investigate the stability of IMs (3), (4) which the families of solutions (6)–(9) belong to.

For the equations of perturbed motion, the variation of the integral $F = V_2 + V_3 + 2V_4$ in the neighbourhood of IM (3) is:

$$2\Delta F = (y_1 + y_4)^2 + (y_2 + y_5)^2 + (y_3 + y_6)^2,$$

where $y_1 = \delta_1 - x_0\,\bar{\alpha}_1^2$, $y_2 = \delta_2 - b\,\bar{\alpha}_1^2$, $y_3 = \delta_3 + [r\,(\lambda - r)]^{1/2}\bar{\alpha}_1$, $y_4 = \gamma_1 + x_0\,\bar{\alpha}_1^2$, $y_5 = \gamma_2 + b\,\bar{\alpha}_1^2$, $y_6 = \gamma_3 - [r\,(\lambda - r)]^{1/2}\bar{\alpha}_1$, $y_7 = p + x_0\,\bar{\alpha}_1\,r^{1/2}\,(\lambda - r)^{-1/2}$, $y_8 = q + b\,\bar{\alpha}_1\,r^{1/2}\,(\lambda - r)^{-1/2}$ are the deviations from the IM under study, $\bar{\alpha}_1 = \bar{\rho}_1\,\beta$, $\bar{\rho}_1 = [r\,(r - \lambda) + (4(b^2 + x_0^2) + (\lambda - r)^2 r^2)^{1/2}]^{1/2}$.

Since the quadratic form ΔF is sign-definite for the variables appearing in it, then IM (3) is stable with respect to the variables $\delta_1 + \gamma_1$, $\delta_2 + \gamma_2$, $\delta_3 + \gamma_3$. The latter means that the IM keeps stability when the directions of the force fields are perturbed. Analogously, the stability of IM (4) with respect to part of the variables is proved.

5.2 On Stability of the Equilibria

Now, we analyze the stability for the elements of the 1st family of equilibria (15)

$$p = 0,\ q = 0,\ r = 0, \gamma_1 = -(x_0^2 - b^2 \delta_1^{0^2})^{1/2} x_0{}^{-1},\ \gamma_2 = -b\,\delta_1^0 x_0^{-1},$$
$$\gamma_3 = 0,\ \delta_1 = \delta_1^0,\ \delta_2 = (1 - \delta_1^{0^2})^{1/2},\ \delta_3 = 0 \tag{30}$$

by the 2nd Lyapunov method.

For constructing a Lyapunov function, we use integral K_2 (17) at the following constraints: $\lambda_2 = -2\lambda_1 x_0^2$, $\lambda_4 = 0$. Under these conditions, the integral is:

$$2F_2 = 2\,[zH + x_0^2 V_2 + b^2 V_3] - V_1.$$

From now and further, the denotations of the Subsect. 3.2 are used.

For the equations of perturbed motion, the 2nd variation of the integral F_2 in the neighbourhood of the elements of the family under study on the linear manifold

$$\delta H = x_0 y_4 - b y_2 = 0,\ \delta V_1 = 2\,[2b\,\delta_1^0\,(b y_1 - x_0 y_5) + (b y_2 + x_0 y_4)\,\bar{z}] = 0,$$
$$\delta V_2 = -2\,[(x_0^2 - b^2 \delta_1^{0^2})^{1/2}\,y_4 + b\,\delta_1^0 y_5] = 0,\ \delta V_3 = 2\,[\delta_1^0 y_1 + (1-\delta_1^{0^2})^{1/2} y_2] = 0$$

can be written as: $\delta^2 F_2 = Q_1 + Q_2$, where

$$Q_1 = b^2 y_3^2 + x_0^2 y_6^2 + 2\lambda(x_0 y_6 y_7 - b y_3 y_8) + [\lambda^2 + 2b(1-\delta_1^{0^2})^{1/2}]\,y_7^2$$
$$-4b\,\delta_1^0 y_7 y_8 + [\lambda^2 + 2(x_0^2 - b^2\delta_1^{0^2})^{1/2}]\,y_8^2,\ 2Q_2 = \frac{z^2 y_2^2}{\delta_1^{0^2}} + z y_9^2.$$

Here y_i $(i=1,\ldots,9)$ are the deviations from the elements of family (30).

The conditions for the quadratic form $\delta^2 F_2$ to be positive definite

$$\Delta_1 = b^2 > 0,\ \Delta_2 = b^3 x_0^2 > 0,\ \Delta_3 = b^3 x_0^2\,(1-\delta_1^{0^2})^{1/2} > 0,$$
$$\Delta_4 = b^3 x_0^2\,[(1-\delta_1^{0^2})^{1/2}(x_0^2 - b^2\delta_1^{0^2})^{1/2} - b\,\delta_1^{0^2}] > 0,\ z > 0 \text{ and } \delta_1^0 \neq 0 \quad (31)$$

are sufficient for the stability of the elements of the family under consideration.

Taking into account the conditions for solution (30) to be real, we find that inequalities (31) are compatible when $b > 0$, $x_0 \neq 0$ and $((-\sigma < \delta_1^0 < 0$ or $0 < \delta_1^0 < \sigma))$. Here $\sigma = |x_0|\,(b^2 + x_0^2)^{-1/2}$.

As mentioned before, the integral V_4 takes the form $-[b\,(1-\delta_1^2)^{1/2} + (x_0^2 - b^2\delta_1^2)^{1/2}]\,\delta_1\,x_0^{-1} = c_2$ on solutions (30). This relation together with the above constraints sets the boundaries of varying the angles between the directions of the force fields, within of which the elements of family (30) are stable. Similar stability conditions have been obtained for the elements of the 1st family of solutions (16).

For the rest of the families of the equilibria, we have derived the conditions of their instability on the base of the Lyapunov stability theorem for linear approximation.

Consider the 2nd family of solutions (15). We introduce the deviations

$$y_1 = \delta_1 - \delta_1^0,\ y_2 = \delta_2 - (1-\delta_1^{0^2})^{1/2},\ y_3 = \delta_3,\ y_4 = \gamma_1 - (x_0^2 - b^2\delta_1^{0^2})^{1/2} x_0^{-1},$$
$$y_5 = \gamma_2 + b\delta_1^0 x_0^{-1},\ y_6 = \gamma_3,\ y_7 = p,\ y_8 = q,\ y_9 = r$$

from the elements of the family under study, and write down the equations of 1st approximation in the neighbourhood of the elements of the family:

$$\begin{aligned}
&\dot{y}_1 = (1-\delta_1^{0\,2})^{1/2}\, y_9,\ \dot{y}_2 = -\delta_1^0 y_9,\ \dot{y}_3 = \delta_1^0 y_8 - (1-\delta_1^{0\,2})^{1/2}\, y_7,\\
&\dot{y}_4 = -b\,\delta_1^0\, x_0^{-1} y_9,\ \dot{y}_5 = -(x_0^2 - b^2\delta_1^{0\,2})^{1/2} x_0^{-1} y_9,\\
&\dot{y}_6 = [b\,\delta_1^0 y_7 + (x_0^2 - b^2\delta_1^{0\,2})^{1/2}\, y_9]\, x_0^{-1},\ 2\dot{y}_7 = b\, y_3 - \lambda y_8,\\
&2\dot{y}_8 = x_0 y_6 + \lambda y_7,\ \dot{y}_9 = -(b y_1 + x_0 y_5).
\end{aligned} \tag{32}$$

The characteristic equation of system (32) has the form:

$$\begin{aligned}
&x^3\,[4x^6 + x^4(6\bar{z} + \lambda^2) + \frac{1}{2}x^2\,[(5\bar{z} + 2\lambda^2)\,\bar{z} - (b^2 + x_0^2)] - (2\,b^2\,\delta_1^{0^2} - x_0^2)\,\bar{z}\\
&-(b^2 - x_0^2)\,(x_0^2 - b^2\delta_1^{0^2})^{1/2}] = 0.
\end{aligned} \tag{33}$$

It is split up into two equations:

$$\begin{aligned}
&x^3 = 0,\\
&4\zeta^6 + \zeta^4(6\bar{z} + \lambda^2) + \frac{1}{2}\zeta^2\,[(5\bar{z} + 2\lambda^2)\,\bar{z} - (b^2 + x_0^2)] - (2\,b^2\,\delta_1^{0^2} - x_0^2)\,\bar{z}\\
&-(b^2 - x_0^2)\,(x_0^2 - b^2\delta_1^{0^2})^{1/2} = 0,
\end{aligned} \tag{34}$$

where $\zeta = x^2$.

Consider the free term of the latter equation:

$$R = -(2\,b^2\,\delta_1^{0^2} - x_0^2)\,\bar{z} - (b^2 - x_0^2)\,(x_0^2 - b^2\delta_1^{0^2})^{1/2}.$$

Taking into consideration the conditions for the solutions under study to be real, we find that $R > 0$ when $0 < b < |x_0|$and $|\delta_1^0| \leq 1$. The latter means that Eq. (34) (and also (33)) has no less than one positive root. So, the elements of the family under consideration are unstable over the above range of varying the parameter δ_1^0. Similar conditions of instability have been obtained for the elements of the 2nd family of solutions (16).

The above algorithms for the stability analysis of stationary solutions by the Lyapunov methods have been encoded in "Mathematica" as the software package mentioned before. This package is intended for the qualitative analysis of phase spaces of dynamical systems having first integrals. The package contains programs for the stability analysis of stationary solutions on the base of Lyapunov's linear stability theorems and the 2nd Lyapunov method. In the considered stability problems, using a solution under study and a combination of the first integrals as input data, the package returns the conditions for the sign-definiteness of the quadratic form (Sylvester's inequalities). The resulting inequalities are analyzed with the aid of the corresponding computer algebra tools. In a similar manner, the package is employed for the stability analysis of stationary solutions on the base of Lyapunov's stability theorems for linear approximation.

6 Conclusion

With the aid of computer algebra methods and the software package developed on the base of "Mathematica" CAS, the analysis of the stationary motions of the Kowalewski gyrostat in two force fields has been performed. Such motions have been found directly from the equations of motion by Gröbner basis method. These represent the families of permanent rotations and equilibria. It has been revealed that the elements of these families correspond to both the parallel force fields and to the fields of special directions. It has been also shown that all the found solutions belong to an intersection of two IMs of codimension 2 of the equations of motion.

The stability of the stationary solutions has been analyzed on the base of Lyapunov's stability theorems. For the elements of the families of permanent rotations, the sufficient conditions of their stability have been obtained. For the elements of the families of equilibria, both the sufficient conditions of their stability, and the conditions of their instability have been derived.

The obtained results as well as the approaches and methods which were applied in this paper, can be used in the analysis of similar problems, in particular, at the stage of preliminary design of satellite systems.

Acknowledgements. This work was supported by the RFBR (Project 16-07-00201a) and the Program for the Leading Scientific Schools of the Russian Federation (NSh-8081.2016.9).

7 Appendix

The coefficients of Eq. (26):

$$\begin{aligned}
\sigma &= b^2\,[(\delta_1^2+\delta_2^2)\,r-2\delta_3\,(p\,\delta_1+q\delta_2)]+2b\,x_0\,[p\,(\delta_3\gamma_2-\delta_2\gamma_3)+q\,(\delta_1\gamma_3-\delta_3\gamma_1)\\
&\quad+r\,(\delta_2\gamma_1-\delta_1\gamma_2)]+x_0^2\,[(\gamma_1^2+\gamma_2^2)\,r-2\gamma_3\,(p\gamma_1+q\gamma_2)],\\
\varrho_0 &= r\,(b^2\,\delta_1\delta_3+b\,x_0\,(\delta_3\gamma_2-\delta_2\gamma_3)+x_0^2\gamma_1\gamma_3),\\
\varrho_1 &= b^2\,[p\,r\,(\delta_1^2-\delta_2^2-\delta_3^2)+\delta_1(\delta_3\,(p^2+q^2+r^2)+2q\,r\,\delta_2)-\delta_3^2(p\,r-x_0\gamma_3)]\\
&\quad+b\,x_0\,(p^2+q^2-r^2)(\delta_3\gamma_2-\delta_2\gamma_3)+x_0^2\,[p\,r\,(\gamma_1^2-\gamma_2^2-\gamma_3^2)+\gamma_1\,(\gamma_3\,(p^2+q^2\\
&\quad+r^2)+2q\,r\gamma_2)-\gamma_3^2(p\,r-x_0\gamma_3)],\\
\varrho_2 &= b^3\delta_3^2\,(\delta_2p-\delta_1q)+b^2\,[(p^2+q^2)\,[p\,(\delta_1^2-\delta_2^2)+2\,(q\delta_1\delta_2-p\,\delta_3^2)]\\
&\quad+\delta_1\delta_3r\,(q^2-p^2)+p\,r\,(r\,(\delta_1^2+\delta_2^2)-2\delta_2\delta_3q)]+b^2x_0\,[\,\delta_3\,(2\gamma_3(\delta_1p+\delta_2q)\\
&\quad+\delta_3\,(p\gamma_1+q\gamma_2))-r\gamma_3(\delta_1^2+\delta_2^2)]+b\,x_0\,r\,[(p^2-q^2)\,(\delta_3\gamma_2-\delta_2\,\gamma_3)\\
&\quad+2p\,[\gamma_1\,(r\,\delta_2-q\,\delta_3)+\delta_1\,(q\,\gamma_3-r\,\gamma_2)]\,]+b\,x_0^2\,\gamma_3\,[\,3\gamma_3\,(p\delta_2-q\delta_1)\\
&\quad-2[\delta_3\,(p\,\gamma_2-q\,\gamma_1)+r\,(\gamma_1\,\delta_2-\gamma_2\,\delta_1)]\,]\\
&\quad+x_0^2\,[(p^2+q^2)\,[p\,(\gamma_1^2-\gamma_2^2-\gamma_3^2)+2q\,\gamma_1\gamma_2]+r\,\gamma_3\,(\gamma_1\,(q^2-p^2)-2p\,q\gamma_2)\\
&\quad+p\,r^2\,((\gamma_1^2+\gamma_2^2)-\gamma_3^2(p^2+q^2))]+x_0^3\,r\,\gamma_3\,[\,3\gamma_3(p\gamma_1+q\gamma_2)-(\gamma_1^2+\gamma_2^2)],\\
\bar{\varrho}_0 &= -r(b^2\,\delta_2\delta_3+b\,x_0\,(\delta_1\gamma_3-\delta_3\gamma_1)+x_0^2\gamma_2\gamma_3),\\
\bar{\varrho}_1 &= b^2\,[q\,r\,(\delta_1^2-\delta_2^2+\delta_3^2)-\delta_2\,(\delta_3\,(p^2+q^2+r^2)+2p\,r\,\delta_1)+\delta_3^2\,(q\,r+b\,\delta_3)]\\
&\quad+b\,x_0\,(p^2+q^2-r^2)(\delta_3\gamma_1-\delta_1\gamma_3)+x_0^2\,[q\,r\,(\gamma_1^2-\gamma_2^2+\gamma_3^2)-\gamma_2\,(\gamma_3\,(p^2+q^2\\
&\quad+r^2)+2p\,r\,\gamma_1)+\gamma_3^2\,(q\,r+b\,\delta_3)],\\
\bar{\varrho}_2 &= b^3\,\delta_3\,[\,3\,\delta_3\,(p\,\delta_1+q\,\delta_2)-r\,(\delta_1^2+\delta_2^2)]+b^2\,[(p^2+q^2)\,[\,q\,(\delta_1^2-\delta_2^2+\delta_3^2)
\end{aligned}$$

$$
\begin{aligned}
&-2p\,\delta_1\delta_2] + r\,\delta_3\,[\delta_2\,(q^2-p^2) + 2p\,q\,\delta_1] - q\,[r^2\,(\delta_1^2+\delta_2^2) - \delta_3^2\,(p^2+q^2)]\,]\\
&+b^2 x_0\,\delta_3\,[2\gamma_3\,(p\,\delta_2 - q\,\delta_1) + 3\,\delta_3\,(2r\,(\delta_1\gamma_2 - \delta_2\gamma_1) - (p\,\gamma_2 - q\,\gamma_1))]\\
&+b\,x_0\,r\,[(p^2-q^2)\,(\delta_1\gamma_3 - \delta_3\gamma_1) + 2q\,(\gamma_2\,(r\,\delta_1 - p\,\delta_3) + \delta_2\,(p\,\gamma_3 - r\,\gamma_1))]\\
&+b\,x_0^2\,[\,\gamma_3\,[\gamma_3\,(p\,\delta_1 + q\,\delta_2) + 2\delta_3\,(p\,\gamma_1 + q\,\gamma_2)] - r\,\delta_3\,(\gamma_1^2+\gamma_2^2)]\\
&+x_0^2\,[\,(p^2+q^2)\,[q\,(\gamma_1^2-\gamma_2^2+\gamma_3^2) - 2p\,\gamma_1\gamma_2] + r\,\gamma_3\,[\gamma_2\,(q^2-p^2) + 2p\,q\,\gamma_1]\\
&-q\,[r^2\,(\gamma_1^2+\gamma_2^2) - \gamma_3^2\,(p^2+q^2)]\,] + x_0^3\gamma_3^2(q\gamma_1 - p\gamma_2).
\end{aligned}
$$

The coefficients of quadratic form (28):

$$
\begin{aligned}
a_{11} &= \frac{b^2}{2(\lambda - r_0)}\Big[\lambda + r_0 + \frac{2\lambda x_0^2\alpha_1^2}{r_0\,(\lambda - r_0)}\Big],\; a_{22} = \frac{b^2}{r_0\,(\lambda - r_0)}\Big[(\lambda - r_0)^2\\
&\quad + r_0^2 + \frac{b^2\alpha_1^2\,\lambda}{\lambda - r_0}\Big] + \frac{(\lambda - r_0)}{2r_0}\Big[x_0^2 + 2^{-1}\lambda\,(\lambda - r_0)\,\alpha_1^{-2}\Big],\\
a_{44} &= x_0^2\Big[\frac{\lambda}{\lambda - r_0} + \frac{\lambda - r_0}{2\,b^2 r_0}\Big(x_0^2 + 2^{-1}\lambda\,(\lambda - r_0)\,\alpha_1^{-2}\Big)\Big],\\
a_{77} &= \frac{b^2 + x_0^2}{b^2}\,[\lambda\,(\lambda - r_0) + \rho_1^2],\; a_{12} = \frac{\lambda\,b\,x_0}{r_0}\Big[\Big(\frac{2^{-1}b^2\,\alpha_1^2}{(\lambda - r_0)^2} + \frac{r_0}{\lambda}\Big) - 1\Big],\\
a_{17} &= 2^{1/2}\alpha_1\,[b^2 r_0 + x_0^2(\lambda - r_0)]\,[r_0\,(\lambda - r_0)]^{-1/2},\\
a_{24} &= -\frac{2x_0}{b\,r_0\,\alpha_1^2}\,[\lambda\,(\lambda - r_0)^2 + 2\alpha_1^2\,(b^2\lambda + x_0^2\,(\lambda - r_0))],\\
a_{27} &= x_0\,[\lambda\,(\lambda - r_0)^2 + 2\alpha_1^2\,(b^2\,(2\lambda - 3r_0) + x_0^2\,(\lambda - r_0))]\,(b\,\alpha_1)^{-1}\\
&\quad \times[r_0\,(\lambda - r_0)]^{-1/2},\\
a_{47} &= -x_0^2\,[\lambda\,(\lambda - r_0)^2 + 2\alpha_1^2\,(b^2\,(\lambda + r_0) + x_0^2\,(\lambda - r_0))]\,(b^2\alpha_1)^{-1}\\
&\quad \times[r_0\,(\lambda - r_0)]^{-1/2},\; a_{57} = 2\,r_0^{1/2}x_0\,\alpha_1\,(x_0^2 - b^2)\,b^{-1}(\lambda - r_0)^{-1/2}.
\end{aligned}
$$

References

1. Kowalewski, S.: Sur le probleme de la rotation d'un corps solide autour d'un point fixe. Acta Math. **12**, 177–232 (1889)
2. Sarychev, V.A., Gutnik, S.A.: Dynamics of a satellite subject to gravitational and aerodynamic torques. Investigation of equilibrium positions. Cosm. Res. **53**(6), 449–457 (2015)
3. Adler, V.E., Marikhin, V.G., Shabat, A.B.: Quantum tops as examples of commuting differential operators. Theoret. Math. Phys. **3**(172), 1187–1205 (2012)
4. Bobenko, A.I., Reyman, A.G., Semenov-Tian-Shansky, M.A.: The Kowalewski top 99 years later: a lax pair, generalizations and explicit solutions. Commun. Math. Phys. **122**, 321–354 (1989)
5. Kharlamov, M.P.: Critical subsystems of the Kowalevski gyrostat in two constant fields. J. Nonlin. Dyn. **3**(3), 331–348 (2007)
6. Kharlamov, M.P., Ryabov, P.E., Savushkin, A.Y., Smirnov, G.E.: Types of critical points of the Kowalevski gyrostat in double field. NAS of Ukraine. Mech. Solids **41**, 26–37 (2011)
7. Banshchikov, A.V., Burlakova, L.A., Irtegov, V.D., Titorenko, T.N.: Software package for finding and stability analysis of stationary sets. Certificate of State Registration of Software Programs. FGU-FIPS. No. 2011615235 (2011)

8. Komarov, I.V.: A generalization of the Kovalevskaya top. Phys. Lett. A **1**(123), 14–15 (1997)
9. Irtegov, V.D., Titorenko, T.N.: On one approach to investigation of mechanical systems. The institute of mathematics of NAS of Ukraine. Electron. J. Symmetry Integr. Geom.: Methods Appl. **2**, 049 (2006)
10. Irtegov, V., Titorenko, T.: Qualitative analysis of the Reyman – Semenov–Tian–Shansky integrable case of the generalized Kowalewski top. In: Gerdt, V.P., Koepf, W., Seiler, W.M., Vorozhtsov, E.V. (eds.) CASC 2016. LNCS, vol. 9890, pp. 289–304. Springer, Cham (2016). doi:10.1007/978-3-319-45641-6_19
11. Lyapunov, A.M.: On permanent helical motions of a rigid body in fluid. Collected works. USSR Acad. Sci. Moscow-Leningrad **1**, 276–319 (1954)
12. Lyapunov, A.M.: The general problem of motion stability. Collected works. USSR Acad. Sci. Moscow-Leningrad **2**, 7–263 (1956)

Computing the Integer Points of a Polyhedron, I: Algorithm

Rui-Juan Jing[1,2(✉)] and Marc Moreno Maza[2]

[1] KLMM, UCAS, Academy of Mathematics and Systems Science, Chinese Academy of Sciences, Beijing, China
[2] University of Western Ontario, London, Canada
rjing8@uwo.ca, moreno@csd.uwo.ca

Abstract. Let K be a polyhedron in $\mathbb{R}^d$, given by a system of m linear inequalities, with rational number coefficients bounded over in absolute value by L. In this series of two papers, we propose an algorithm for computing an irredundant representation of the integer points of K, in terms of "simpler" polyhedra, each of them having at least one integer point. Using the terminology of W. Pugh: for any such polyhedron P, no integer point of its grey shadow extends to an integer point of P. We show that, under mild assumptions, our algorithm runs in exponential time w.r.t. d and in polynomial w.r.t m and L. We report on a software experimentation. In this series of two papers, the first one presents our algorithm and the second one discusses our complexity estimates.

1 Introduction

The integer points of polyhedral sets are of interest in many areas of mathematical sciences, see for instance the landmark textbooks of Schrijver [19] and Barvinok [3], as well as the compilation of articles [4]. One of these areas is the analysis and transformation of computer programs. For instance, integer programming [7] is used by Feautrier in the scheduling of for-loop nests [8] and Barvinok's algorithm [2] for counting integer points in polyhedra is adapted by Köppe and Verdoolaege in [16] to answer questions like how many memory locations are touched by a for-loop nest. In [17], Pugh proposes an algorithm, called the *Omega Test*, for testing whether a polyhedron has integer points. In the same paper, Pugh shows how to use the Omega Test for performing dependence analysis [17] in for-loop nests. Then, in [18], he uses the Omega Test for deciding Presburger arithmetic formulas.

In [18], Pugh also suggests, without stating a formal algorithm, that the Omega Test could be used for quantifier elimination on Presburger formulas. This observation is a first motivation for the work presented in this series of two papers: we adapt the Omega Test so as to describe the integer points of a polyhedron via a *projection* scheme, thus performing elimination of existential quantifiers on Presburger formulas. Projections of polyhedra and parametric

V.P. Gerdt et al. (Eds.): CASC 2017, LNCS 10490, pp. 225–241, 2017.
DOI: 10.1007/978-3-319-66320-3_17

programming are tightly related problems, see [13]. Since the latter is essential to the parallelization of for-loop nests [7], which is of interest to the authors [5], we had here a second motivation for developing the proposed algorithm.

In [9], Fischer and Rabin show that any algorithm for deciding Presburger arithmetic formulas has a worst case running time which is doubly exponential in the length of the input formula. However, this worst case scenario is based on a formula alternating existential and universal quantifiers. Meanwhile, in practice, the original *Omega Test* (for testing whether a polyhedron has integer points) can solve "difficult problems" as shown by Pugh in [18] and others, e.g. Wonnacott in [22]. This observation brings our third motivation: determining realistic assumptions under which our algorithm, based on the Omega Test, could run in a single exponential time.

Our algorithm takes as input a system of linear inequalities $\mathbf{A}\mathbf{x} \leq \mathbf{b}$ where $\mathbf{A}$ is a matrix over $\mathbb{Z}$ with m rows and d columns, $\mathbf{x}$ is the unknown vector and $\mathbf{b}$ is a vector of m coefficients in $\mathbb{Z}$. The points $\mathbf{x} \in \mathbb{R}^d$ satisfying $\mathbf{A}\mathbf{x} \leq \mathbf{b}$ form a polyhedron K and our algorithm decomposes its integer points (that is, $K \cap \mathbb{Z}^d$) into a disjoint union $(K_1 \cap \mathbb{Z}^{d_1}) \uplus \cdots \uplus (K_e \cap \mathbb{Z}^{d_e})$, where $K_1, \ldots, K_e$ are "simpler" polyhedra such that $K_i \cap \mathbb{Z}^d \neq \varnothing$ holds and d_i is the dimentions of K_i, for $1 \leq i \leq e$. To use the terminology introduced by W. Pugh for the Omega test, no integer point of the grey shadow of any polyhedron K_i extends to an integer point of K_i. As a consequence, applying our algorithm to K_i would return K_i itself, for $1 \leq i \leq e$. Let us present the key principles and features of our algorithm through an example. Consider the polyhedron K of $\mathbb{R}^4$ given below:

$$\begin{cases} 2x + 3y - 4z + 3w \leq 1 \\ -2x - 3y + 4z - 3w \leq -1 \\ -13x - 18y + 24z - 20w \leq -1 \\ -26x - 40y + 54z - 39w \leq 0 \\ -24x - 38y + 49z - 31w \leq 5 \\ 54x + 81y - 109z + 81w \leq 2 \end{cases}.$$

A first procedure, called IntegerNormalize, detects implicit equations and solves them using techniques based on *Hermite normal form*, see Sects. 3 and 4.1. In our example $2x + 3y - 4z + 3w = 1$ is an implicit equation and IntegerNormalize($\mathbf{A}\mathbf{x} \leq \mathbf{b}$) returns a triple $(\mathbf{t}, \mathbf{x} = \mathbf{P}\mathbf{t} + \mathbf{q}, \mathbf{M}\mathbf{t} \leq \mathbf{v})$ where $\mathbf{t}$ is a new unknown vector, the linear system $\mathbf{x} = \mathbf{P}\mathbf{t} + \mathbf{q}$ gives the general form of an integer solution of the implicit equation(s) and $\mathbf{M}\mathbf{t} \leq \mathbf{v}$ is obtained by substituting $\mathbf{x} = \mathbf{P}\mathbf{t} + \mathbf{q}$ into $\mathbf{A}\mathbf{x} \leq \mathbf{b}$. In our example, the systems $\mathbf{x} = \mathbf{P}\mathbf{t} + \mathbf{q}$ and $\mathbf{M}\mathbf{t} \leq \mathbf{v}$ are given by:

$$\begin{cases} x = -3t_1 + 2t_2 - 3t_3 + 2 \\ y = 2t_1 + t_3 - 1 \\ z = t_2 \\ w = t_3 \end{cases} \quad \text{and} \quad \begin{cases} 3t_1 - 2t_2 + t_3 \leq 7 \\ -2t_1 + 2t_2 - t_3 \leq 12 \\ -4t_1 + t_2 + 3t_3 \leq 15 \\ -t_2 \leq -25 \end{cases}.$$

A second procedure, called DarkShadow, takes $\mathbf{Mt} \leq \mathbf{v}$ as input and returns a couple $(\mathbf{t}', \Theta)$ where $\mathbf{t}'$ stands for all $\mathbf{t}$-variables except t_1, and Θ is a linear system in the $\mathbf{t}'$-variables such that any integer point solving Θ extends to an integer point solving $\mathbf{Mt} \leq \mathbf{v}$. In our example, $\mathbf{t}' = \{t_2, t_3\}$ and Θ is given by:

$$\begin{cases} 2t_2 - t_3 \leq 48 \\ -5t_2 + 13t_3 \leq 67 \\ -t_2 \leq -25 \end{cases}.$$

The polyhedron D of $\mathbb{R}^2$ defined by Θ, and the inequalities of $\mathbf{Mt} \leq \mathbf{v}$ not involving t_1, is called the *dark shadow* of the polyhedron defined by $\mathbf{Mt} \leq \mathbf{v}$.

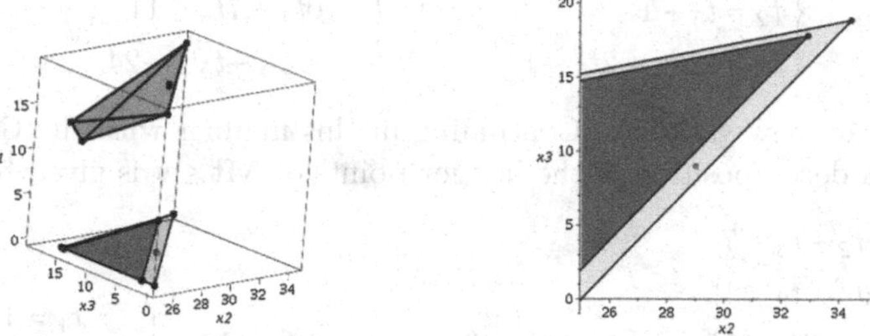

Fig. 1. The real, the dark and the grey shadows of a polyhedron.

On the left-hand side of Fig. 1, one can see the polyhedron defined in $\mathbb{R}^3$ by $\mathbf{Mt} \leq \mathbf{v}$ together with its dark shadow D (shown in dark grey) as well as its projection on the (t_2, t_3)-plane, denoted by R and called *real shadow* by W. Pugh. The right-hand side of Fig. 1 gives a planar view of D and R. As we will see in Sect. 4.4, if $\mathbf{M}'\mathbf{t}' \leq \mathbf{v}'$ is the linear system generated by applying *Fourier-Motzkin elimination* (without removing redundant inequalities) to $\mathbf{Mt} \leq \mathbf{v}$ (in order to eliminate t_1) then Θ is given by a linear system of the form $\mathbf{M}'\mathbf{t}' \leq \mathbf{w}'$. This explains why, on the right-hand side of Fig. 1, each facet of the dark shadow D is parallel to a facet of the real shadow R. While this property is observed on almost all practical problems, in particular in the area of analysis and transformation of computer programs, it is possible to build examples where this property does not hold. We have examples in Sect. 5 of the second paper.

On the right-hand side of Fig. 1, one observes that the region $R \setminus D$, called *grey shadow*, contains integer points. Some of them, like $(t_2, t_3) = (29, 9)$, do not extend to an integer solution of $\mathbf{Mt} \leq \mathbf{v}$. Indeed, plugging $(t_2, t_3) = (29, 9)$ into $\mathbf{Mt} \leq \mathbf{v}$ yields $\frac{37}{2} \leq t_1 \leq \frac{56}{3}$, which has no integer solutions. However, other integer points of $R \setminus D$ may extend to integer solutions of $\mathbf{Mt} \leq \mathbf{v}$. In order to determine them, a third procedure, called GreyShadow, considers in turn the negation of each inequality θ of Θ. However, for each θ of Θ, instead of simply making a recursive call to the entire algorithm applied to $\mathbf{Mt} \leq \mathbf{v} \cup \{\theta\}$, simplifications (involving θ and the inequalities from which θ is derived) permit to replace this recursive call by several ones in lower dimension, thus guaranteeing termination of the whole algorithm. Details are given in Sects. 4.5 and 4.6.

Returning to our example, the negation of the inequality $2t_2 - t_3 \le 48$ from Θ, combined with the system $\mathbf{Mt} \le \mathbf{v}$, yields the following

$$\begin{cases} -2t_1 + 2t_2 - t_3 = 12 \\ 3t_1 - 2t_2 + t_3 \le 7 \\ -4t_1 + t_2 + 3t_3 \le 15 \\ -t_2 \le -25 \end{cases},$$

which, by means of IntegerNormalize, rewrites to:

$$\begin{cases} t_1 = t_4 \\ t_2 = t_5 + 1 \\ t_3 = -2t_4 + 2t_5 + 1 \end{cases}, \quad \text{and} \quad \begin{cases} t_4 \le 8 \\ -10t_4 + 7t_5 \le 11 \\ -t_5 \le -24 \end{cases},$$

where t_4, t_5 are new variables. Continuing in this manner with the GreyShadow procedure, a decomposition of the integer points of $\mathbf{Mt} \le \mathbf{v}$ is given by:

$$\begin{cases} 3t_1 - 2t_2 + t_3 \le 7 \\ -2t_1 + 2t_2 - t_3 \le 12 \\ -4t_1 + t_2 + 3t_3 \le 15 \\ 2t_2 - t_3 \le 48 \\ -5t_2 + 13t_3 \le 67 \\ -t_2 \le -25 \\ 2 \le t_3 \le 17 \end{cases}, \begin{cases} t_1 = 15 \\ t_2 = 27 \\ t_3 = 16 \end{cases}, \begin{cases} t_1 = 18 \\ t_2 = 33 \\ t_3 = 18 \end{cases}, \begin{cases} t_1 = 14 \\ t_2 = 25 \\ t_3 = 15 \end{cases}, \begin{cases} t_1 = 19 \\ t_2 = 50 + t_6 \\ t_3 = 50 + 2t_6 \\ -25 \le t_6 \le -16. \end{cases}.$$

Denoting these 5 systems respectively by $S_1, \ldots, S_5$, the integer points of K are finally given by the union of the integer points of the systems $\mathbf{x} = \mathbf{Pt} + \mathbf{q} \cup S_i$, for $1 \le i \le 5$. The systems $S_2, \ldots, S_5$ look simple enough to be considered as solution sets. What about S_1? The system S_1, as well as $S_2, \ldots, S_5$, satisfies a "back-substitution" property which is similar to that of a *regular chain* in the theory of polynomial system solving [1]. This property (formally stated in Sect. 4.2), when applied to S_1, says that for all $2 \le i \le 3$, every integer point of $\mathbb{R}^{4-i}$ solving all the inequalities of S_1 involving $t_i, \ldots, t_3$ only, extends to an integer point of $\mathbb{R}^{5-i}$ solving all the inequalities of S_1 involving $t_{i-1}, \ldots, t_3$.

With respect to the original Omega Test [17], our contributions are as follows.

1. We turn the decision procedure of the Omega Test into an algorithm decomposing all the integer points of a polyhedron.
2. Our decomposition is disjoint whereas the recursive calls in the original Omega Test may search for integer points in intersecting polyhedral regions.
3. The original Omega Test uses an ad-hoc routine for computing the integer solutions of linear equation systems, while we rely on Hermite normal form for this task. Consequently, we deduce complexity estimates for that task.

4. We also provide complexity estimates for the procedures GreyShadow and DarkShadow under realistic assumptions. From there, we derive complexity estimates for the entire algorithm, whereas no complexity estimates were known for the original Omega Test.

We report our work in a series of two papers. The present one describes and proves our algorithm. The second one establishes our complexity estimates.

2 Polyhedral Sets

This section is a review of the theory of polyhedral sets. It is based on the books of Grünbaum [10] and Schrijver [19], where proofs of the statements below can be found.

Given a positive integer d, we consider the d-dimensional Euclidean space $\mathbb{R}^d$ equipped with the Euclidean topology. Let K be a subset of $\mathbb{R}^d$. The *dimension* $\dim(K)$ of K is $a-1$ where a is the maximum number of affinely independent points in K. Let $\mathbf{a} \in \mathbb{R}^d$, let $b \in \mathbb{R}$ and denote by H the hyperplane defined by $H = \{\mathbf{x} \in \mathbb{R}^d \mid \mathbf{a}^T\mathbf{x} = b\}$. We say that the hyperplane H *supports* K if either $\sup\{\mathbf{a}^T\mathbf{x} \mid \mathbf{x} \in K\} = b$ or $\inf\{\mathbf{a}^T\mathbf{x} \mid \mathbf{x} \in K\} = b$ holds, but not both.

From now on, let us assume that K is convex. A set $F \subseteq K$ is a *face* if either $F = \varnothing$ or $F = K$, or if there exists a hyperplane H supporting K such that we have $F = K \cap H$. The set of all faces of K is denoted by $\mathcal{F}(K)$. We say that $F \in \mathcal{F}(K)$ is proper if we have $F \neq \varnothing$ or $F \neq K$. We note that the intersection of any family of faces of K is itself a face of K.

We say that K is a *polyhedral set* or a *polyhedron* if it is the intersection of finitely many closed half-spaces of $\mathbb{R}^d$. We say that K is *full-dimensional*, if we have $\dim(K) = d$, that is, if the interior of K is not empty. The proper faces of K that are $\subseteq$-maximal are called *facets* and those of dimension zero are called *vertices*. We observe that every face of K is also a polyhedral set.

Let $H_1, \ldots, H_m$ be closed half-spaces such that the intersection $\cap_{i=1}^{i=m} H_i$ is *irredundant*, that is, $\cap_{i=1}^{i=m} H_i \neq \cap_{i=1, j\neq i}^{i=m} H_i$ for all $1 \leq j \leq m$. We observe that this intersection is closed and convex. For each $i = 1 \cdots m$, let $\mathbf{a}_i \in \mathbb{R}^d$ and $b_i \in \mathbb{R}$ such that H_i is defined by $\mathbf{a}_i^T\mathbf{x} \leq b_i$. We denote by $\mathbf{A}$ the $m \times d$ matrix $(\mathbf{a}_i^T, 1 \leq i \leq m)$ and by $\mathbf{b}$ the vector $(b_1, \ldots, b_d)^T$.

From now on, we assume that $K = \cap_{i=1}^{i=m} H_i$ holds. Such an irredundant decomposition of a polyhedral set can be computed from an arbitrary intersection of finitely many closed half-spaces, in time polynomial in both d and m, using linear programming, see Khachian in [15]. The following property is essential. For every face F of K, there exists a subset I of $\{1, \ldots, m\}$ such that F corresponds to the set of solutions to the system of equations and inequalities

$$\mathbf{a}_i^T\mathbf{x} = b_i \quad \text{for } i \in I, \quad \text{and} \quad \mathbf{a}_i^T\mathbf{x} \leq b_i \quad \text{for } i \notin I.$$

This latter property has several important consequences. For each $i = 1 \cdots m$, the set $F_i = K \cap \{\mathbf{a}_i^T\mathbf{x} = b_i\}$ is a facet of K and the border of K equals $\cup_{i=1}^{i=m} F_i$. In particular, each proper face of K is contained in a facet of K. Each facet of a facet of K is the intersection of two facets of K. Moreover, if the $(m \times d)$-matrix A has full column rank, then the $\subseteq$-minimal faces are the vertices. The set $\mathcal{F}(K)$ is finite and has at most 2^m elements.

For $\mathbf{a} \in \mathbb{R}^d$ and $b \in \mathbb{R}$, we say that $\mathbf{a}^T\mathbf{x} \leq b$ is an *implicit equation* in $\mathbf{Ax} \leq \mathbf{b}$ if for all $\mathbf{x} \in \mathbb{R}^d$ we have

$$\mathbf{Ax} \leq \mathbf{b} \implies \mathbf{ax} = b. \tag{1}$$

Following [19], we denote by $\mathbf{A}^=$ (resp. $\mathbf{A}^+$) and $\mathbf{b}^=$ (resp. $\mathbf{b}^+$) the rows of $\mathbf{A}$ and $\mathbf{b}$ corresponding to the implicit (resp. non-implicit) equations. The following properties are easy to prove. If K is not empty, then there exists $\mathbf{x} \in K$ satisfying both

$$\mathbf{A}^=\mathbf{x} = b^= \text{ and } \mathbf{A}^+\mathbf{x} < b^+.$$

The facets of K are in 1-to-1 correspondence with the inequalities of $\mathbf{A}^+\mathbf{x} \leq \mathbf{b}^+$. In addition, if K is full-dimensional, then $\mathbf{A}^+ = \mathbf{A}$ and $\mathbf{b}^+ = \mathbf{b}$ both hold; moreover the system of inequalities $\mathbf{Ax} \leq b$ is a unique representation of K, up to multiplication of inequalities by positive scalars.

From now on and in the sequel of this paper, we assume that variables are ordered as $x_1 > \cdots > x_d$. We call initial coefficient, or simply *initial*, of an inequality $\mathbf{a}_i^T\mathbf{x} \leq b_i$, for $1 \leq i \leq m$, the coefficient of $a_i^T\mathbf{x}$ in its largest variable. Following the terminology of Pugh in [17], if v is the largest variable of the inequality $\mathbf{a}_i^T\mathbf{x} \leq b_i$, we say that this inequality is an *upper* (resp. *lower*) *bound* of v whenever the initial c of $\mathbf{a}_i^T\mathbf{x} \leq b_i$ is positive (resp. negative); indeed, we have $v \leq \frac{\gamma}{c}$ (resp. $v \geq \frac{\gamma}{c}$) where $\gamma = b_i - \mathbf{a}_i^T\mathbf{x} - cv$.

Canonical Representation. Recall that we assume that none of the inequalities of $Ax \leq b$ is redundant. If K is full-dimensional and if the initial of each inequality in $Ax \leq b$ is 1 or -1, then we call $Ax \leq b$ the *canonical representation* of K w.r.t. the variable ordering $x_1 > \cdots > x_d$ and we denote it by $\mathsf{can}(K; x_1, \ldots, x_d)$.

We observe that the notion of *canonical representation* can also be expressed in a more geometrical and less algebraic way, that is, independently of any coordinate system. Assume again that K is full-dimensional and that the intersection $\cap_{i=1}^{i=n} H_i = K$ of closed half-spaces $H_1, \ldots, H_n$ is irredundant. Since K is full-dimensional, the supporting hyperplane of each facet of K must be the frontier of one half-space among $H_1, \ldots, H_n$. Clearly, two (or more) half-spaces among $H_1, \ldots, H_n$ may not have the same frontier without contradicting one of our hypotheses (K is full-dimensional, $\cap_{i=1}^{i=n} H_i$ is irredundant). Therefore, the half-spaces $H_1, \ldots, H_n$ are in one-to-one correspondence with the facets of K. This implies that there is a unique irredundant intersection of closed half-spaces equaling K and we denote it by $\mathsf{can}(K)$.

Projected Representation. Let again $\mathbf{Ax} \leq \mathbf{b}$ be the *canonical representation* of the polyhedral set K w.r.t. the variable ordering $x_1 > \cdots > x_d$. We denote by $\mathbf{A}^{x_1}$ (resp. $\mathbf{A}^{<x_1}$) and $\mathbf{b}^{x_1}$ (resp. $\mathbf{b}^{<x_1}$) the rows of $\mathbf{A}$ and $\mathbf{b}$ corresponding to the inequalities whose largest variable is x_1 (resp. less than x_1). For each upper bound $cx_1 \leq \gamma$ of x_1 and each lower bound $-ax_1 \leq -\alpha$ of x_1 (where $c > 0$, $a > 0$, $\gamma \in \mathbb{R}[x_2, \ldots, x_d]$ and $\alpha \in \mathbb{R}[x_2, \ldots, x_d]$ hold), we have a new inequality $c\alpha - a\gamma \leq 0$. Augmenting $\mathbf{A}^{<x_1}$ with all inequalities obtained in this way, we obtain

a new linear system which represents a polyhedral set which is the standard projection of K on the $d-1$ least coordinates of $\mathbb{R}^d$, namely $(x_2,\ldots,x_d)$; hence we denote this latter polyhedral set by $\Pi^{x_2,\ldots,x_d}K$ and we call it the *real shadow* of K, following the terminology of [17]. The procedure by which $\Pi^{x_2,\ldots,x_d}K$ is computed from K is the well-known Fourier-Motzkin elimination procedure, see [15]. We call *projected representation* of K w.r.t. the variable ordering $x_1 > \cdots > x_d$ and denote by $\mathsf{proj}(K; x_1,\ldots,x_d)$ the linear system given by $\mathbf{A}^{x_1}\mathbf{x} \le \mathbf{b}^{x_1}$ if $d=1$ and, by the conjunction of $\mathbf{A}^{x_1}\mathbf{x} \le \mathbf{b}^{x_1}$ and $\mathsf{proj}(\Pi^{x_2,\ldots,x_d}K; x_2,\ldots,x_d)$, otherwise.

3 Integer Solutions of Linear Equation Systems

We review how Hermite normal forms [6,19] can be used to represent the integer solutions of systems of linear equations. Let $\mathbf{A} = (a_{i,j})$ and $H = (h_{i,j})$ be two matrices over $\mathbb{Z}$ with m rows and d columns, and let $\mathbf{b}$ be a vector over $\mathbb{Z}$ with d coefficients. We denote by r the rank of $\mathbf{A}$ and by h the maximum bit size of coefficients in the matrix $[\mathbf{A}\ \mathbf{b}]$. Definition 1 is taken from [14], see also [12].

Definition 1. *The matrix H is called a* column Hermite normal form (abbr. column HNF) *if there exists a strictly increasing map f from $[d-r+1,d]\cap\mathbb{Z}$ to $[1,m]\cap\mathbb{Z}$ satisfying the following properties for all $j\in[d-r+1,d]\cap\mathbb{Z}$:*

1. *for all integer i such that $1\le i\le m$ and $i > f(j)$ both hold, we have $h_{i,j}=0$,*
2. *for all integer k such that $j<k\le d$ holds, we have $h_{f(j),j} > h_{f(j),k} \ge 0$,*
3. *the first $d-r$ columns of H are equal to zero.*

We say that H is the column Hermite normal form *of $\mathbf{A}$ if H is a column Hermite normal form and there exists a uni-modular $d\times d$-matrix U over $\mathbb{Z}$ such that we have $H=\mathbf{A}U$. When those properties hold, we call $\{f(d-r+1),\ldots,f(d)\}$ the* pivot row set *of $\mathbf{A}$.*

Remark 1. The matrix $\mathbf{A}$ admits a unique column Hermite normal form. Let H be this column Hermite normal form and let U be the uni-modular $(d\times d)$-matrix given in Definition 1. Let us decompose U as $U=[U_L,U_R]$ where U_L (resp. U_R) consist of the first $d-r$ (resp. last r) columns of U. Then we define $H_L := \mathbf{A}U_L$ and $H_R := \mathbf{A}U_R$. We have $H_L = \mathbf{0}^{m,d-r}$, where $\mathbf{0}^{m,d-r}$ is the zero-matrix with m rows and $d-r$ columns. We observe that U_R is a full column-rank matrix. Moreover, if $\mathbf{A}$ is full row-rank, that is, if $r=m$ holds, then H_R is non-singular.

Lemma 2 shows how to compute the integer solutions of the system of linear equations $\mathbf{A}\mathbf{x}=\mathbf{b}$ when $\mathbf{A}$ is full row-rank. In the general case, one can use Lemma 1 to reduce to the hypothesis of Lemma 2. While the construction of this latter lemma relies on the HNF, alternative approaches are available. For instance, one can use the *equation elimination procedure* of the Omega Test [17], However, no running-time estimates are known for that procedure.

Notation 1. *For $I \subseteq \{1, \ldots, m\}$, we denote by $\mathbf{A}_I$ (resp. $\mathbf{b}_I$) the sub-matrix (resp. vector) of $\mathbf{A}$ (resp. $\mathbf{b}$) consisting of the rows of $\mathbf{A}$ (coefficients of $\mathbf{b}$) with indices in I.*

Lemma 1. *Let I be the pivot row set of $\mathbf{A}$, as given in Definition 1. Assume that $\mathbf{A}\mathbf{x} = \mathbf{b}$ admits at least one solution in $\mathbb{R}^d$. Then, for any $\mathbf{x} \in \mathbb{R}^d$, we have*

$$\mathbf{A}\mathbf{x} = \mathbf{b} \iff \mathbf{A}_I\mathbf{x} = \mathbf{b}_I.$$

Proof. We clearly have $\{\mathbf{x} \mid \mathbf{A}\mathbf{x} = \mathbf{b}\} \subseteq \{\mathbf{x} \mid \mathbf{A}_I\mathbf{x} = \mathbf{b}_I\}$. We prove the reversed inclusion. Since I is the pivot row set of $\mathbf{A}$, one can check that $\text{rank}(\mathbf{A}) = \text{rank}(\mathbf{A}_I)$ holds. Since $\mathbf{A}\mathbf{x} = \mathbf{b}$ admits solutions, we have $\text{rank}(\mathbf{A}) = \text{rank}([\mathbf{A}\ \mathbf{b}])$. Similarly, we have $\text{rank}(\mathbf{A}_I) = \text{rank}([\mathbf{A}_I\ \mathbf{b}_I])$. Therefore, we have $\text{rank}([\mathbf{A}\ \mathbf{b}]) = \text{rank}([\mathbf{A}_I\ \mathbf{b}_I])$. Hence, any equation $\mathbf{a}^T\mathbf{x} = b$ in $\mathbf{A}\mathbf{x} = \mathbf{b}$ is a linear combination of the equations of $\mathbf{A}_I\mathbf{x} = \mathbf{b}_I$, thus $\{\mathbf{x} \mid \mathbf{A}_I\mathbf{x} = \mathbf{b}_I\} \subseteq \{\mathbf{x} \mid \mathbf{A}\mathbf{x} = \mathbf{b}\}$ holds.

Lemma 2. *We use the same notations as in Definition 1 and Remark 1. We assume that H_R is non-singular. Then, the system $\mathbf{A}\mathbf{x} = \mathbf{b}$ has an integer solution if and only if $H_R^{-1}\mathbf{b}$ is integral. In this case, all integral solutions to $\mathbf{A}\mathbf{x} = \mathbf{b}$ are given by $\mathbf{x} = \mathbf{P}\,\mathbf{t} + \mathbf{q}$ where*

1. *the columns of $\mathbf{P}$ consist of a $\mathbb{Z}$-basis of the linear space $\{\mathbf{x} : \mathbf{A}\mathbf{x} = \mathbf{0}\}$,*
2. *$\mathbf{q}$ is a particular solution of $\mathbf{A}\mathbf{x} = \mathbf{b}$, and*
3. *$\mathbf{t} = (t_1, \ldots, t_{d-r})$ is a vector of $d - r$ unknowns.*

The maximum absolute value of any coefficient in $\mathbf{P}$ (resp. $\mathbf{q}$) can be bounded over by $r^{r+1}L^{2r}$ (resp. $r^{r+1}L^{2r}$), where L is the maximum absolute value of any coefficient in $\mathbf{A}$ (resp. in either $\mathbf{A}$ or $\mathbf{b}$). Moreover, $\mathbf{P}$ and $\mathbf{q}$ can be computed within $O(mdr^2(\log r + \log L)^2 + r^4(\log r + \log L)^3)$ bit operations.

Proof. Except for the coefficient bound and running time estimates, we refer to [11] for a proof of this lemma. The running time estimate follows from Theorem 19 of [20] whereas the coefficient bound estimates are taken from [21]. □

Example 1. Let $\mathbf{A}$, H and U be as follows:

$$\mathbf{A} = \begin{pmatrix} 3 & 4 & -4 & -1 \\ 2 & -2 & 8 & 4 \\ 5 & 2 & 4 & 3 \\ 3 & 5 & -5 & -2 \\ 2 & -3 & 9 & 5 \end{pmatrix}, \quad H = \begin{pmatrix} 0 & -18 & -1 & -15 \\ 0 & \mathbf{18} & \mathbf{2} & \mathbf{16} \\ 0 & 0 & 1 & 1 \\ 0 & \mathbf{0} & \mathbf{1} & \mathbf{0} \\ 0 & \mathbf{0} & \mathbf{0} & \mathbf{1} \end{pmatrix}, \quad U = \begin{pmatrix} -1 & 30 & -3 & -25 \\ 1 & -37 & 4 & 31 \\ 0 & -19 & 2 & 16 \\ 1 & 0 & 0 & 0 \end{pmatrix}.$$

The matrix H is the column HNF of $\mathbf{A}$, with unimodular matrix U and pivot row set $[2, 4, 5]$. We denote by H_R the sub-matrix of H whose coefficients are in bold fonts. Applying Lemma 1, we deduce that for any vector $\mathbf{b}$ such that $\mathbf{A}\mathbf{x} = \mathbf{b}$ admits one rational solution, we have:

$$\begin{cases} 3x_1 + 4x_2 - 4x_3 - x_4 = b_1 \\ 2x_1 - 2x_2 + 8x_3 + 4x_4 = b_2 \\ 5x_1 + 2x_2 + 4x_3 + 3x_4 = b_3 \\ 3x_1 + 5x_2 - 5x_3 - 2x_4 = b_4 \\ 2x_1 - 3x_2 + 9x_3 + 5x_4 = b_5 \end{cases} \Leftrightarrow \begin{cases} 2x_1 - 2x_2 + 8x_3 + 4x_4 = b_2 \\ 3x_1 + 5x_2 - 5x_3 - 2x_4 = b_4 \\ 2x_1 - 3x_2 + 9x_3 + 5x_4 = b_5 \end{cases}. \tag{2}$$

We apply Lemma 2: if $\mathbf{A}\mathbf{x} = \mathbf{b}$ is consistent over $\mathbb{Q}$ and if $H_R^{-1}[b_2, b_4, b_5]^T$ is integral, then all the integer solutions of the second equation system in Relation (2) are given by $\mathbf{x} = \mathbf{P}\mathbf{t} + \mathbf{q}$, where $\mathbf{P} = [-1, 1, 0, 1]^T$, $\mathbf{q} = [\frac{5}{3}b_2 - \frac{19}{3}b_4 - \frac{155}{3}b_5, -\frac{37}{18}b_2 + \frac{73}{9}b_4 + \frac{575}{9}b_5, -\frac{19}{18}b_2 + \frac{37}{9}b_4 + \frac{296}{9}b_5]^T$, $\mathbf{t} = (t_1)$ and t_1 is a new variable.

4 Integer Solutions of Linear Inequality Systems

In this section, we present an algorithm for computing the integer points of a polyhedron $K \subseteq \mathbb{R}^d$, that is, the set $K \cap \mathbb{Z}^d$. To do so, we adapt the Omega Test invented by Pugh [17] for deciding whether or not a polyhedral set has an integer point. Our algorithm decomposes the set $K \cap \mathbb{Z}^d$ into a disjoint union $(K_1 \cap \mathbb{Z}^d) \cup \cdots \cup (K_s \cap \mathbb{Z}^d)$, where $K_1, \ldots, K_s$ are polyhedral sets in $\mathbb{R}^d$, for which the integer points can be represented in a sense specified in Sect. 4.2. Section 4.3 states the specifications of the main procedure while Sects. 3, 4.1, 4.4, 4.5 and 4.6. describe its main subroutines and its proof. We use the same notations as in Sect. 2. However, from now on, we assume that all matrix and vector coefficients are integer numbers, that is, elements of $\mathbb{Z}$. To be precise, we have the following.

Notation 2. *We consider a polyhedral set $K \subseteq \mathbb{R}^d$ given by an irredundant intersection $K = \cap_{i=1}^{i=m} H_i$ of closed half-spaces $H_1, \ldots, H_m$ such that, for each $i = 1, \ldots, m$, the half-space H_i is defined by $\mathbf{a}_i^T\mathbf{x} \le b_i$, with $\mathbf{a}_i \in \mathbb{Z}^d$ and $b_i \in \mathbb{Z}$. The conjunction of those inequalities forms a system of linear inequalities that we denote by $\mathbf{A}\mathbf{x} \le \mathbf{b}$, as well as Σ. We do not assume that K is full-dimensional.*

4.1 Normalization of Linear Inequality Systems

The purpose of the procedure IntegerNormalize, presented below, is to solve the system consisting of the equations of $\mathbf{A}\mathbf{x} \le \mathbf{b}$ and substitute its solutions into the system consisting of the inequalities of $\mathbf{A}\mathbf{x} \le \mathbf{b}$. This process is performed by Steps (S2) to (S6) and relies on Lemmas 1 and 2; this yields Proposition 1, which provides the output specification of IntegerNormalize. Step (S1) is an optimization: performing it is not needed, but improves performance in practice.

When applied to $\mathbf{A}\mathbf{x} \le \mathbf{b}$, the IntegerNormalize procedure proceeds as follows.

(S1) It computes $\mathsf{proj}(K; x_1, \ldots, x_d)$, obtaining a new system of linear inequalities that we denote again by Σ; if this proves that K has no rational points, then the procedure stops and returns $(\varnothing, \varnothing, \varnothing)$ implying that $K \cap \mathbb{Z}^d$ is empty,

(S2) for every inequality $\mathbf{a}\mathbf{x} \le b$, let g be the absolute value of the GCD of coefficients in $\mathbf{a}$: if $g > 1$, replace $\mathbf{a}\mathbf{x} \le b$ by $\frac{\mathbf{a}}{g}\mathbf{x} \le \lfloor \frac{b}{g} \rfloor$.

(S3) Every pair of inequalities of the form $(\mathbf{a}_i^T\mathbf{x} \le b_i, -\mathbf{a}_i^T\mathbf{x} \le -b_i)$ is replaced by the equivalent equation, that is, $\mathbf{a}_i^T\mathbf{x} = b_i$; Every pair of inequalities of the form $(\mathbf{a}_i^T\mathbf{x} \le b_i, \mathbf{a}_i^T\mathbf{x} \le b_j)$ is replaced by $\mathbf{a}_i^T\mathbf{x} \le \min(b_i, b_j)$.

(S4) Equations and inequalities form, respectively, a system of linear equations $\mathbf{A}^{=}\mathbf{x} = \mathbf{b}^{=}$ and a system of linear inequalities $\mathbf{A}^{\le}\mathbf{x} \le \mathbf{b}^{\le}$, as specified in Notation 3, so that the conjunction of these two systems is equivalent to Σ.

(S5) If $\mathbf{A}^{=}\mathbf{x} = \mathbf{b}^{=}$ is empty, that is, if Σ has no equations, then the procedure stops returning $(\mathbf{x}, \varnothing, \mathbf{A}^{\leq}\mathbf{x} \leq \mathbf{b}^{\leq})$.

(S6) Proposition 1 is applied to $\mathbf{A}^{=}\mathbf{x} = \mathbf{b}^{=}$; if this proves that this latter system has no integer solutions, then the procedure stops returning $(\varnothing, \varnothing, \varnothing)$, otherwise the change of variables given by (3) is applied to $\mathbf{A}^{\leq}\mathbf{x} \leq \mathbf{b}^{\leq}$; as a result, the output of the IntegerNormalize procedure is the triple $(\mathbf{t}, \mathbf{x} = \mathbf{P}\,\mathbf{t} + \mathbf{q}, \mathbf{M}\mathbf{t} \leq \mathbf{v})$, where $\mathbf{t}, \mathbf{P}, \mathbf{q}, \mathbf{M}, \mathbf{v}$ are defined in Proposition 1.

Notation 3. *From now we consider an equation system $\mathbf{A}^{=}\mathbf{x} = \mathbf{b}^{=}$ and an inequality system $\mathbf{A}^{\leq}\mathbf{x} \leq \mathbf{b}^{\leq}$. The matrices $\mathbf{A}^{=}, \mathbf{A}^{\leq}$ as well as the vectors $\mathbf{b}^{=}, \mathbf{b}^{\leq}$ have integer coefficients. The total number of rows in both $\mathbf{A}^{=}$ and $\mathbf{A}^{\leq}$ is m, each of $\mathbf{A}^{=}, \mathbf{A}^{\leq}$ has d columns, and $\mathbf{A}^{\leq}$ has e rows. We denote by L and h the maximum absolute value and maximal bit size of any coefficient in the matrix in either $[\mathbf{A}^{=}\ \mathbf{b}^{=}]$ or $[\mathbf{A}^{\leq}\ \mathbf{b}^{\leq}]$ respectively. We define $r := \mathrm{rank}(\mathbf{A}^{=})$.*

Proposition 1. *One can decide whether or not $\mathbf{A}^{=}\mathbf{x} = \mathbf{b}^{=}$ has integer solutions. If this system has integer solutions, then, for any $\varepsilon > 0$, one can compute*

1. *a matrix $\mathbf{P} \in \mathbb{Z}^{d\times(d-r)}$ within $O(m\,d\,r^{2+\varepsilon}\,h^3)$ bit operations,*
2. *a vector $\mathbf{q} \in \mathbb{Z}^{d}$ within $O(mdr^{2+\varepsilon}\,h^3)$ bit operations,*
3. *a matrix $\mathbf{M} \in \mathbb{Z}^{e\times(d-r)}$, whose coefficients can be bounded over by $dr^{r+1}L^{2r+1}$, within $O(m\,d^2\,r^{1+\varepsilon}h^3)$ bit operations,*
4. *a vector $\mathbf{v} \in \mathbb{Z}^{e}$, whose coefficients can be bounded over by $2dr^{r+1}L^{2r+1}$, within $O(m\,d^2\,r^{1+\varepsilon}h^3)$ bit operations,*

such that an integer point $(x_1, \ldots, x_d) \in \mathbb{Z}^d$ solves $\mathbf{A}^{=}\mathbf{x} = \mathbf{b}^{=}$ and $\mathbf{A}^{\leq}\mathbf{x} \leq \mathbf{b}^{\leq}$ if and only if there exists an integer point $(t_1, \ldots, t_{d-r}) \in \mathbb{Z}^{d-r}$ such that we have

$$\begin{cases} (x_1, \ldots, x_d)^T = \mathbf{P}(t_1, \ldots, t_{d-r})^T + (q_1, \ldots, q_d)^T \\ \mathbf{M}(t_1, \ldots, t_{d-r})^T \leq (v_1, \ldots, v_e)^T \end{cases}. \tag{3}$$

That is, one can perform the IntegerNormalize *procedure within $O(m\,d^2\,r^{1+\varepsilon}\,h^3)$ bit operations.*

Proof. We first observe that one can decide whether or not $\mathbf{A}^{=}\mathbf{x} = \mathbf{b}^{=}$ has solutions in $\mathbb{R}^d$, using standard techniques, say Gaussian elimination. If $\mathbf{A}^{=}$ is not full row-rank, this observation allows us to apply Lemma 1 and thus to reduce to the case where $\mathbf{A}^{=}$ is full row-rank, via the computation of the column HNF of $\mathbf{A}^{=}$. Hence, from now on, we assume that $\mathbf{A}^{=}$ is full row-rank. We apply Lemma 2 which yields the matrix $\mathbf{P}$ and the vector $\mathbf{q}$. Next, we compute $\mathbf{M}$ and $\mathbf{v}$ as follows: $\mathbf{M} := \mathbf{A}^{\leq}\mathbf{P}$ and $\mathbf{v} := -\mathbf{A}^{\leq}\mathbf{q} + \mathbf{b}$. The coefficient bounds and cost estimates for $\mathbf{M}$ and $\mathbf{v}$ follow easily from Lemma 2 and the inequality $r \leq d$. □

4.2 Representing the Integer Points

Applying IntegerNormalize to $\mathbf{A}\mathbf{x} \leq \mathbf{b}$ produces a triple $(\mathbf{t}, \mathbf{x} = \mathbf{P}\,\mathbf{t} + \mathbf{q}, \mathbf{M}\mathbf{t} \leq \mathbf{v})$, with $\mathbf{P}, \mathbf{q}, \mathbf{M}, \mathbf{v}$ as in Proposition 1. Assume $\mathbf{t} \neq \varnothing$. Since the system $\mathbf{x} = \mathbf{P}\,\mathbf{t} + \mathbf{q}$ solves the $\mathbf{x}$-variables as functions of the $\mathbf{t}$-variables, we turn our attention to $\mathbf{M}\mathbf{t} \leq \mathbf{v}$. Definition 2 states conditions on $\mathbf{M}$ under which we view $(\mathbf{x} = \mathbf{P}\,\mathbf{t} + \mathbf{q}, \mathbf{M}\mathbf{t} \leq \mathbf{v})$ as a "solved system", that is, a system describing its integer solutions.

Definition 2. *Let $\widehat{K}$ be the polyhedron of $\mathbb{Z}^{2d-r}$ defined by the system of linear equations and inequalities given by $\mathbf{x} = \mathbf{P}\,\mathbf{t} + \mathbf{q}$ and $\mathbf{M}\mathbf{t} \le \mathbf{v}$, in Relation (3). We say that this system is a* representation of the integer points *of the polyhedron $\widehat{K}$ whenever $\mathbf{M}$ has the following form:*

$$\begin{pmatrix} M_{11} & M_{12} & \cdots & M_{1,\ell-1} & M_{1,\ell} \\ & M_{22} & \cdots & M_{2,\ell-1} & M_{2,\ell} \\ & & \ddots & \vdots & \vdots \\ & & & M_{\ell-1,\ell-1} & M_{\ell-1,\ell} \\ & & & & M_{\ell,\ell} \end{pmatrix}, \tag{4}$$

where for each i, j with $1 \le i, j \le \ell$, the block $M_{i,j}$ has m_i rows and k_j columns such that the following six assertions hold:

(i) $k_1, \ldots, k_{\ell-1} \ge 1$, $k_\ell \ge 0$ *and* $k_1 + \cdots + k_\ell = d - r$*;*
(ii) $m_1, \ldots, m_{\ell-1} \ge 2$ *and* $m_\ell \ge 0$*;*
(iii) for $1 \le i < \ell$*, each column in* $M_{i,i}$ *has both positive coefficients and negative coefficients, but no null coefficients;*
(iv) if $m_\ell > 0$ *holds, then in each column of* $M_{\ell,\ell}$*, all coefficients are non-zero and have the same sign;*
(v) (Consistency) *the system* $\mathbf{M}\mathbf{t} \le \mathbf{v}$ *admits at least one integer point in* $\mathbb{Z}^{d-r}$*;*
(vi) (Extensibility) *for all* $1 < i < d - r$*, every integer point of* $\mathbb{R}^{d-r-i}$ *solving all the inequalities of* $\mathbf{M}\mathbf{t} \le \mathbf{v}$ *involving* $t_{i+1}, \ldots, t_{d-r}$ *only extends to an integer point of* $\mathbb{R}^{d-r-i+1}$ *solving all the inequalities of* $\mathbf{M}\mathbf{t} \le \mathbf{v}$ *involving* $t_i, \ldots, t_{d-r}$*.*

More generally, we say that $\mathbf{x} = \mathbf{P}\,\mathbf{t} + \mathbf{q}$ and $\mathbf{M}\mathbf{t} \le \mathbf{v}$ form a representation of the integer points of $\widehat{K}$ if $\mathbf{M}$ satisfies (i) to (vi) up to a permutation of its columns.

Remark 2. Assume that the above matrix $\mathbf{M}$ satisfies the properties (i) to (vi) of Definition 2. Then, the values of the first $k_1+\cdots+k_{\ell-1}$ (resp. last k_ℓ) variables of $\mathbf{t}$ are bounded (resp. unbounded) in the polyhedron given by $\mathbf{M}\mathbf{t} \le \mathbf{v}$. For these reasons, we call those variables *bounded* and *unbounded* in $\mathbf{M}\mathbf{t} \le \mathbf{v}$, respectively. Clearly, the original polyhedron $\mathbf{A}\mathbf{x} \le \mathbf{b}$ is bounded if and only if $m_\ell = k_\ell = 0$.

4.3 The IntegerSolve Procedure: Specifications

We are ready to specify the main algorithm presented in this paper. This procedure, called IntegerSolve will be formally stated in Sect. 4.6. When applied to $\mathbf{A}\mathbf{x} \le \mathbf{b}$, with the assumptions of Notation 2, IntegerSolve produces a decomposition of the integer points of the polyhedron K in the sense of the following.

Definition 3. *Let $\mathbf{A}, \mathbf{x}, \mathbf{b}, K$ be as in Notation 2. A sequence of pairs $(\mathbf{y}_1, \Sigma_1)$, ..., $(\mathbf{y}_s, \Sigma_s)$ is called a* decomposition of the integer points of the polyhedron K *whenever the following conditions hold:*

(i) $\mathbf{y}_i$ *is a sequence of* $d_i \ge d$ *independent variables* $x_1, \ldots, x_d, x_{d+1}, \ldots, x_{d_i}$ *thus starting with* $\mathbf{x}$*,*

(ii) Σ_i *is a system of linear inequalities with* $\mathbf{y}_i$ *as unknown,*
(iii) Σ_i *is a representation of the integer points of a polyhedral set* K_i*,*

and we have $V_{\mathbb{Z}}(\Sigma) = V_{\mathbb{Z}}(\Sigma_1, \mathbf{x}) \cup \cdots \cup V_{\mathbb{Z}}(\Sigma_s, \mathbf{x})$*, where* $V_{\mathbb{Z}}(\Sigma)$ *denotes the set of the integer points of* Σ *and where* $V_{\mathbb{Z}}(\Sigma_i, \mathbf{x})$ *is defined as the set of the points* $(x_1, \ldots, x_d) \in \mathbb{Z}^d$ *such that there exists a point* $(x_{d+1}, \ldots, x_{d_i}) \in \mathbb{Z}^{d_i - d}$ *such that* $(x_1, \ldots, x_d, x_{d+1}, \ldots, x_{d_i})$ *solves* Σ_i*.*

In the sequel of Sect. 4, we shall propose and prove an algorithm satisfying the above specifications. The construction is by induction on $d \geq 1$. We observe that the case $d = 1$ is trivial. Indeed, in this case, K is necessarily an interval of the real line. Then, either $K \cap \mathbb{Z}$ is empty and IntegerSolve(Σ) returns the empty set, or $K \cap \mathbb{Z}$ is not empty and the system Σ is clearly a representation of the integer points of K in the sense of Definition 2. The case $d > 1$ will be treated in Sect. 4.6, after presenting the main subroutines of the IntegerSolve procedure.

4.4 The DarkShadow Procedure

Let $\mathbf{M}, \mathbf{v}$ be as in Proposition 1. Recall that we write $\mathbf{t} = (t_1, \ldots, t_{d-r})$ and assume $0 \leq r < d$. The system $\mathbf{Mt} \leq \mathbf{v}$ represents a polyhedral set that we denote by $K_{\mathbf{t}}$. We order the variables as $t_1 > \cdots > t_{d-r}$. We call DarkShadow the procedure stated by Algorithm 1, for which Proposition 2 serves as output specification. In Algorithm 1, the polyhedral set represented by $\mathbf{M}^{<t_1}\mathbf{t} \leq \mathbf{v}^{<t_1}$ (resp. Θ) is called the *dark shadow* of $K_{\mathbf{t}}$, denoted as $\mathbf{D}_{t_1}$ when **case** 1 (resp. **case** 2) holds.

Algorithm 1. DarkShadow($\mathbf{Mt} \leq \mathbf{v}$)

case 1: for all $1 \leq i \leq d - r$, the inequalities in t_i are either all lower bounds of t_i or all upper bounds of t_i
    return $((t_2, \ldots, t_{d-r}), \mathbf{M}^{<t_1}\mathbf{t} \leq \mathbf{v}^{<t_1})$.
case 2: otherwise
    re-order the variables, such that t_1 has both lower bounds and upper bounds.
    initialize Δ to the empty set.
    for each upper bound $ct_1 \leq \gamma$ of t_1, where $c > 0$, $\gamma \in \mathbb{Z}[t_2, \ldots, t_{d-r}]$ **do**
        for each lower bound $-at_1 \leq -\alpha$ of t_1, where $a > 0$, $\alpha \in \mathbb{Z}[t_2, \ldots, t_{d-r}]$ **do**
            let $\Delta := \Delta \cup \{c\alpha - a\gamma \leq -(c-1)(a-1)\}$.
        end for
    end for
    Let $\Theta_0 := \Delta \cup \mathbf{M}^{<t_1}\mathbf{t} \leq \mathbf{v}^{<t_1}$
    Let Θ be the system obtained by removing from Θ_0 all redundant inequalities.
    return $((t_2, \ldots, t_{d-r}), \Theta)$.

For the inequalities in the set Δ in Algorithm 1, we have the following.

Lemma 3 Pugh [17]**.** *Let* $ct_1 \leq \gamma$ *be an upper bound of* t_1 *and* $-at_1 \leq -\alpha$ *be a lower bound of* t_1*, where* $c > 0$*,* $a > 0$*,* $\gamma \in \mathbb{Z}[t_2, \ldots, t_{d-r}]$ *and* $\alpha \in \mathbb{Z}[t_2, \ldots, t_{d-r}]$ *hold. Then, every integer point* $(t_2, \ldots, t_{d-r})$ *satisfying* $c\alpha - a\gamma \leq -(c-1)(a-1)$ *extends to an integer point* $(t_1, t_2, \ldots, t_{d-r})$ *satisfying both* $ct_1 \leq \gamma$ *and* $-at_1 \leq \alpha$*.*

Proposition 2. *Let* $((t_2, \ldots, t_{d-r}), \Theta)$ *be the output of the* DarkShadow *procedure. Then, every integer point of* $V_{\mathbb{Z}}(\Theta, (t_2, \ldots, t_{d-r}))$ *extends to an integer point solving* $\mathbf{Ax} \leq \mathbf{b}$.

Proof. If the DarkShadow procedure returns at Line 2 of Algorithm 1, the claim holds easily. Lemma 3 shows that any integer point $(t_2, \ldots, t_{d-r})$ solving Δ can be extended to an integer point solving $\mathbf{Mt} \leq \mathbf{v}$, thus with Proposition 1, to an integer point solving $\mathbf{Ax} \leq \mathbf{b}$. Therefore, if the DarkShadow procedure returns at Line 13, the claim also holds.

4.5 The GreyShadow Procedure

Let $\mathbf{M}$, $\mathbf{t}$, $\mathbf{v}$, $K_{\mathbf{t}}, \mathbf{D}_{t_1}$ be as in Sect. 4.4. We call *grey shadow* of $K_{\mathbf{t}}$, denoted by $\mathbf{G}_{t_1}$, the set-theoretic difference $\left(\Pi^{t_2,\ldots,t_{d-r}} K_{\mathbf{t}}\right) \setminus \mathbf{D}_{t_1}$. Algorithm 2 states the GreyShadow procedure, for which Lemma 4 serves as output specification.

Lemma 4. *Let* $\mathcal{G} = \{(\mathbf{u}_1, \mathbf{t} = \mathbf{P}_1\mathbf{u}_1 + \mathbf{q}_1, \mathbf{M}_1\mathbf{u}_1 \leq \mathbf{v}_1), \ldots, (\mathbf{u}_s, \mathbf{t} = \mathbf{P}_s\mathbf{u}_s + \mathbf{q}_s, \mathbf{M}_s\mathbf{u}_s \leq \mathbf{v}_s)\}$ *be the output of Algorithm 2. Then, the disjoint union* $\biguplus_{1 \leq i \leq s} V_{\mathbb{Z}}(\mathbf{t} = \mathbf{P}_i\mathbf{u}_i + \mathbf{q}_i \cup \mathbf{M}_i\mathbf{u}_i \leq \mathbf{v}_i, \mathbf{t})$ *forms the set of the integer points of the grey shadow* $\mathbf{G}_{t_1}$.

Proof. The correctness of **case** 1 follows from the fact that $\mathbf{G}_{t_1}$ is empty when all $\mathbf{t}$-variables are unbounded. From now on, we consider **case** 2. At Line 12, all the $\mathbf{t}$-variables are solved by IntegerNormalize as functions of new variables $\mathbf{u}_i$. The fact that $\bigcup_{1 \leq i \leq s} V_{\mathbb{Z}}(\mathbf{t} = \mathbf{P}_i\mathbf{u}_i + \mathbf{q}_i \cup \mathbf{M}_i\mathbf{u}_i \leq \mathbf{v}_i, \mathbf{t})$ equals $\mathbf{G}_{t_1}$ follows

---

Algorithm 2. GreyShadow($\mathbf{Mt} \leq \mathbf{v}$)

```
1: case 1: for all 1 ≤ i ≤ d − r, the inequalities in t_i are either all lower bounds of t_i
   or all upper bounds of t_i
2:   return (∅, ∅, ∅)
3: case 2: otherwise
4:   Re-order the variables, such that t_1 has both lower bounds and upper bounds.
5:   Initialize both Υ and G to the empty set; the former set will be a set of linear
   inequalities while the latter will form the result of the procedure.
6:   for each upper bound c t_1 ≤ γ of t_1, where c > 0, γ ∈ Z[t_2, ..., t_h] do
7:     for each lower bound −a t_1 ≤ −α of t_1, where a > 0, α ∈ Z[t_2, ..., t_h] do
8:       let Θ_2 := Υ ∪ Mt ≤ v ∪ {cα − aγ > −(c − 1)(a − 1)},
9:       for each non-negative integer i ≤ (ca−c−a)/c do
10:        check whether a t_1 = α + i is consistent over Z using Lemma 2,
11:        case no: move to the next iteration,
12:        case yes: let G := G ∪ IntegerNormalize({a t_1 = α + i} ∪ Θ_2),
13:      end for
14:      let Υ := Υ ∪ {cα − aγ ≤ −(c − 1)(a − 1)}.
15:    end for
16:    return G.
17:  end for
```

from Sect. 2.3.1. of [17]. Now, at Line 8 of Algorithm 2, we add the constraint $c\alpha - a\gamma > -(c-1)(a-1)$ to Θ_2, while at Line 14, we use $c\alpha - a\gamma \le -(c-1)(a-1)$ to construct Υ in the next loop iteration. From that construction of Θ_2 and Υ, we easily deduce that the above union is disjoint.

4.6 The IntegerSolve Procedure: Algorithm

We are ready to state an algorithm satisfying the specifications of IntegerSolve introduced in Sect. 4.3. The recursive nature of this algorithm leads us to define an "inner procedure", called $\mathsf{IntegerSolve}_0$, of which IntegerSolve is a wrapper function. The procedure $\mathsf{IntegerSolve}_0$ takes as input the system to be solved, namely $\mathbf{Ax} \le \mathbf{b}$, together with another system of linear equations and inequalities, denoted by E, see Notation 4. This second system E keeps track of the relations between those variables that have already been solved and those that remain to be solved. To be more precise, the procedure $\mathsf{IntegerSolve}_0$, see Algorithm 3, relies on IntegerNormalize and thus introduces new variables when solving systems of linear equations over $\mathbb{Z}$. For this reason, variables appearing in E may not be present in $\mathbf{x}$ and we need another vector of variables, namely $\mathbf{y} = (y_1, \ldots, y_{d'})$, to denote the unknowns of E that are regarded as "solved".

Notation 4. *We denote by E a second system of linear equations and inequalities, with coefficients in $\mathbb{Z}$ and with $\mathbf{y} \oplus \mathbf{x}$ as "unknown" vector, where $\mathbf{y} \oplus \mathbf{x}$ denotes the concatenated vector $(y_1, \ldots, y_{d'}, x_1, \ldots, x_d)$. In fact, the variables of $\mathbf{y}$ are regarded as solved by the equations and inequalities of E, meanwhile those of $\mathbf{x}$ remain to be solved. Hence, we can view the conjunction of the systems $\mathbf{Ax} \le \mathbf{b}$ and E as a system of linear equations and inequalities with $\mathbf{y} \oplus \mathbf{x}$ as unknown vector, defining a polyhedron K^E in $\mathbb{R}^{d'+d}$.*

Theorem 1 states, that Algorithm 3 returns a decomposition (in the sense of Definition 3) of the integer points of the polyhedron K^E, defined in Notation 4. From Algorithm 3, we easily implement the IntegerSolve procedure (as specified in Sect. 4.3) with the call $\mathsf{IntegerSolve}_0(\{\ \}, \{\ \}, \mathbf{x}, \mathbf{Ax} \le \mathbf{b})$.

Theorem 1. *Algorithm 3 terminates and returns a decomposition of the integer points of the polyhedron K^E.*

Proof. We first prove termination. Lines 1 to 21 in Algorithm 3 handle the case where $\mathbf{Ax} \le \mathbf{b}$ has a single unknown. This is simply done by case inspection. Consider now the case where $\mathbf{Ax} \le \mathbf{b}$ has more than one variable. The calls to the procedures DarkShadow and GreyShadow at Lines 29 and 32 generate the input to the recursive calls. From Lines 2 and 13 of Algorithm 1, and Lines 2 and 12 of Algorithm 2, we deduce that the number of unknowns decreases at least by one after each recursive call. Therefore, Algorithm 3 terminates.

Next we prove that Algorithm 3 is correct. Let $(\mathbf{y}_1, \Sigma_1)$, ..., $(\mathbf{y}_s, \Sigma_s)$ be the output of Algorithm 3 where each Σ_i is a system of linear inequalities with $\mathbf{y}_i$ as unknown. The fact that each Σ_i is a representation of the integer points of

Algorithm 3. $\mathsf{IntegerSolve}_0(\mathbf{y}, E, \mathbf{x}, \mathbf{Ax} \le \mathbf{b})$

1: Let d be the cardinality of $\mathbf{x}$;
2: **case** $d = 1$
3: let $\mathbf{x} = \{x\}$, solve $\mathbf{Ax} \le \mathbf{b}$ over $\mathbb{R}$,
4: **case** only lower bounds of x exist in $\mathbf{Ax} \le \mathbf{b}$
5: the solution to $\mathbf{Ax} \le \mathbf{b}$ over $\mathbb{R}$ is $\{x : -x \le q_1\}$ for some $q_1 \in \mathbb{R}$,
6: $\mathbf{y} := \mathbf{y} \oplus \mathbf{x}$ and $E := E \cup \{-x \le \lfloor q_1 \rfloor\}$;
7: **return** $\{(\mathbf{y}, E)\}$
8: **case** only upper bounds of x exist in $\mathbf{Ax} \le \mathbf{b}$
9: the solution to $\mathbf{Ax} \le \mathbf{b}$ over $\mathbb{R}$ is $\{x : x \le q_2\}$ for some $q_2 \in \mathbb{R}$,
10: $\mathbf{y} := \mathbf{y} \oplus \mathbf{x}$ and $E := E \cup \{x \le \lfloor q_2 \rfloor\}$;
11: **return** $\{(\mathbf{y}, E)\}$
12: **case** both lower bounds and upper bounds of x exist in $\mathbf{Ax} \le \mathbf{b}$
13: the solution to $\mathbf{Ax} \le \mathbf{b}$ over $\mathbb{R}$ is $\{x : x \le q_3$ and $-x \le q_4\}$ for some $q_3, q_4 \in \mathbb{R}$,
14: **case** $\lfloor q_3 \rfloor > -\lfloor q_4 \rfloor$
15: $\mathbf{y} := \mathbf{y} \oplus \mathbf{x}$ and $E := E \cup \{x \le \lfloor q_3 \rfloor, -x \le \lfloor q_4 \rfloor\}$;
16: **return** $\{(\mathbf{y}, E)\}$
17: **case** $\lfloor q_3 \rfloor = -\lfloor q_4 \rfloor$
18: $\mathbf{y} := \mathbf{y} \oplus \mathbf{x}$, $E := \mathsf{eval}(E, x = \lfloor q_3 \rfloor) \cup \{x = \lfloor q_3 \rfloor\}$,
19: **return** $\{(\mathbf{y}, E)\}$
20: **case** $\lfloor q_3 \rfloor < -\lfloor q_4 \rfloor$
21: **return** $\{(\varnothing, \varnothing)\}$
22: **case** $d > 1$
23: $(\mathbf{t}, \mathbf{x} = \mathbf{Pt} + \mathbf{q}, \mathbf{Mt} \le \mathbf{v}) := \mathsf{IntegerNormalize}(\mathbf{Ax} \le \mathbf{b})$,
24: **case** $(\mathbf{t}, \mathbf{x} = \mathbf{Pt} + \mathbf{q}, \mathbf{Mt} \le \mathbf{v}) = (\varnothing, \varnothing, \varnothing)$
25: **return** $\{(\varnothing, \varnothing)\}$
26: **case** $(\mathbf{t}, \mathbf{x} = \mathbf{Pt} + \mathbf{q}, \mathbf{Mt} \le \mathbf{v}) \ne (\varnothing, \varnothing, \varnothing)$
27: $\mathbf{y} := \mathbf{y} \oplus \mathbf{x}$, $E := \mathsf{eval}(E, \mathbf{x} = \mathbf{Pt} + \mathbf{q}) \cup \mathbf{x} = \mathbf{Pt} + \mathbf{q} \cup \mathbf{M}^{t_1}\mathbf{t} \le \mathbf{v}^{t_1}$,
28: $\mathcal{G} := \varnothing$,
29: $(\mathbf{t}', \Theta) := \mathsf{DarkShadow}(\mathbf{Mt} \le \mathbf{v})$,
30: $\mathbf{y} := \mathbf{y} \oplus \{t_1\}$,
31: $\mathcal{G} := \mathcal{G} \cup \mathsf{IntegerSolve}_0(\mathbf{y}, E, \mathbf{t}', \Theta)$;
32: **for** $(\mathbf{u}, E_u, \mathbf{M}_u\mathbf{u} \le \mathbf{v}_\mathbf{u}) \in \mathsf{GreyShadow}(\mathbf{Mt} \le \mathbf{v})$ **do**
33: $\mathcal{G} := \mathcal{G} \cup \mathsf{IntegerSolve}_0(\mathbf{y} \cup \mathbf{t}, E \cup E_\mathbf{u}, \mathbf{u}, \mathbf{M}_\mathbf{u}\mathbf{u} \le \mathbf{v}_\mathbf{u})$
34: **end for**
35: **return** $\mathcal{G}$

the polyhedron it defines, can be established by induction on the length of $\mathbf{y}_i$. To give more details, the properties required by Definition 2 are easy to check in the case $d = 1$. For the cases $d > 1$, these properties, in particular the consistency and the extensibility, follow from the way the set E is incremented at Lines 27 and 33, as well as from Proposition 2. Finally, the fact that the integer points of the input system of the initial call to Algorithm 3 are given by the integer points of $\Sigma_1, \ldots, \Sigma_s$ can be established by induction on the length of $\mathbf{y}_i$, thanks to Lemma 4.

Software

We have implemented the algorithm presented in the first paper within the `Polyhedra` library in MAPLE. This library is publicly available in source on the download page of the `RegularChains` library at www.regularchains.org.

Acknowledgements. The authors would like to thank IBM Canada Ltd (CAS project 880) and NSERC of Canada (CRD grant CRDPJ500717-16), as well as the University of Chinese Academy of Sciences, UCAS Joint PhD Training Program, for supporting their work.

References

1. Aubry, P., Lazard, D., Moreno Maza, M.: On the theories of triangular sets. J. Symb. Comput. **28**, 105–124 (1999)
2. Barvinok, A.I.: A polynomial time algorithm for counting integral points in polyhedra when the dimension is fixed. Math. Oper. Res. **19**(4), 769–779 (1994)
3. Barvinok, A.I.: Integer Points in Polyhedra. Contemporary Mathematics. European Mathematical Society (2008)
4. Beck, M.: Integer Points in Polyhedra-Geometry, Number Theory, Representation Theory, Algebra, Optimization, Statistics: AMS-IMS-SIAM Joint Summer Research Conference, 11–15 June 2006, Snowbird. Utah. Contemporary mathematics - Amer. Math, Soc. (2008)
5. Chen, C., Chen, X., Keita, A., Moreno Maza, M., Xie, N.: MetaFork: a compilation framework for concurrency models targeting hardware accelerators and its application to the generation of parametric CUDA kernels. In: Proceedings of CASCON 2015, pp. 70–79 (2015)
6. Cohen, H.: A Course in Computational Algebraic Number Theory, vol. 138. Springer Science & Business Media, Heidelberg (2013)
7. Feautrier, P.: Parametric integer programming. RAIRO Recherche Opérationnelle **22** (1988). http://citeseerx.ist.psu.edu/viewdoc/download?doi=10.1.1.30.9957&rep=rep.1&type=pdf
8. Feautrier, P.: Automatic parallelization in the polytope model. In: Perrin, G.-R., Darte, A. (eds.) The Data Parallel Programming Model. LNCS, vol. 1132, pp. 79–103. Springer, Heidelberg (1996). doi:10.1007/3-540-61736-1_44. http://dl.acm.org/citation.cfm?id=647429.723579
9. Fischer, M.J., Fischer, M.J., Rabin, M.O.: Super-exponential complexity of presburger arithmetic. Technical report, Cambridge, MA, USA (1974)
10. Grünbaum, B.: Convex Polytops. Springer, New York (2003)
11. Hung, M.S., Rom, W.O.: An application of the hermite normal form in integer programming. Linear Algebra Appl. **140**, 163–179 (1990)
12. Jing, R.-J., Yuan, C.-M., Gao, X.-S.: A polynomial-time algorithm to compute generalized hermite normal form of matrices over $\mathbb{Z}[x]$. CoRR, abs/1601.01067 (2016)
13. Jones, C.N., Kerrigan, E.C., Maciejowski, J.M.: On polyhedral projection and parametric programming. J. Optim. Theory Appl. **138**(2), 207–220 (2008)
14. Kannan, R., Bachem, A.: Polynomial algorithms for computing the smith and hermite normal forms of an integer matrix. SIAM J. Comput. **8**(4), 499–507 (1979)

15. Khachiyan, L.: Fourier-motzkin elimination method. In: Floudas, C.A., Pardalos, P.M. (eds.) Encyclopedia of Optimization, 2nd edn, pp. 1074–1077. Springer, Heidelberg (2009). doi:10.1007/978-0-387-74759-0_187
16. Köppe, M., Verdoolaege, S.: Computing parametric rational generating functions with a primal Barvinok algorithm. Electr. J. Comb. **15**(1), R16 (2008)
17. Pugh, W.: The omega test: a fast and practical integer programming algorithm for dependence analysis. In: Martin, J.L. (ed.) Proceedings Supercomputing 1991, Albuquerque, NM, USA, 18–22 November 1991, pp. 4–13. ACM (1991)
18. Pugh, W.: Counting solutions to presburger formulas: how and why. In: Sarkar, V., Ryder, B.G., Soffa, M.L. (eds.) Proceedings of the ACM SIGPLAN 1994 Conference on Programming Language Design and Implementation (PLDI), Orlando, Florida, USA, 20–24 June 1994, pp. 121–134. ACM (1994)
19. Schrijver, A.: Theory of Linear and Integer Programming. Wiley, New York (1986)
20. Storjohann, A.: A fast practical deterministic algorithm for triangularizing integer matrices. Citeseer (1996)
21. Storjohann, A.: Algorithms for matrix canonical forms. Ph.D. thesis, Swiss Federal Institute of Technology Zürich (2000)
22. Wonnacott, D.: Omega test. In: Encyclopedia of Parallel Computing, pp. 1355–1365 (2011)

Computing the Integer Points of a Polyhedron, II: Complexity Estimates

Rui-Juan Jing[1,2(✉)] and Marc Moreno Maza[2]

[1] KLMM, UCAS, Academy of Mathematics and Systems Science, Chinese Academy of Sciences, Beijing, China
[2] University of Western Ontario, London, Canada
rjing8@uwo.ca, moreno@csd.uwo.ca

Abstract. Let K be a polyhedron in $\mathbb{R}^d$, given by a system of m linear inequalities, with rational number coefficients bounded over in absolute value by L. In this series of two papers, we propose an algorithm for computing an irredundant representation of the integer points of K, in terms of "simpler" polyhedra, each of them having at least one integer point. Using the terminology of W. Pugh: for any such polyhedron P, no integer point of its grey shadow extends to an integer point of P. We show that, under mild assumptions, our algorithm runs in exponential time w.r.t. d and in polynomial w.r.t m and L. We report on a software experimentation. In this series of two papers, the first one presents our algorithm and the second one discusses our complexity estimates.

1 Introduction

In the first paper of that series of two, we have presented an algorithm, called IntegerSolve, for decomposing the set of integer points of a polyhedron. See Sect. 4 of the first paper. This second paper is dedicated to complexity estimates considering both running time and output size. Our main result is Theorem 1, which states an exponential time complexity[1] for IntegerSolve, under Hypothesis 1, that we call *Pugh's assumption*. Before discussing this hypothesis and stating the theorem, we set up some notations.

Notation 1. *Recall that we consider a polyhedral set $K \subseteq \mathbb{R}^d$ given by an irredundant intersection $K = \cap_{i=1}^{i=m} H_i$ of closed half-spaces $H_1, \ldots, H_m$ such that, for each $i = 1, \ldots, m$, the half-space H_i is defined by $\mathbf{a}_i^T \mathbf{x} \leq b_i$, with $\mathbf{a}_i \in \mathbb{Z}^d$ and $b_i \in \mathbb{Z}$. The conjunction of those inequalities forms a system of linear inequalities that we denote by $\mathbf{A}\mathbf{x} \leq \mathbf{b}$. Let L (resp. h) be the maximum absolute value (resp. maximum bit size) of a coefficient in either $\mathbf{A}$ or $\mathbf{b}$. Thus $h = \lfloor \log_2(L) \rfloor + 2$.*

Hypothesis 1. We assume that during the execution of the function call IntegerSolve(K), for any polyhedral set K', input of a recursive call, each facet of the dark shadow[2] of K' is parallel to a facet of the real shadow of K'.

[1] To be precise, in the EXP complexity class.

[2] The notions of real shadow, dark shadow and grey shadow are presented in Sect. 3 of the first paper.

V.P. Gerdt et al. (Eds.): CASC 2017, LNCS 10490, pp. 242–256, 2017.
DOI: 10.1007/978-3-319-66320-3_18

The figure in the introduction of the first paper shows a polyhedron for which each facet of its dark shadow is parallel to a facet of its real shadow. This property is commonly observed in practice, see Sect. 5. In [9], W. Pugh observes that it is possible to build polyhedra K that challenge the Omega Test by generating many recursive calls when searching for integer points of K that extend integer points of its grey shadow. But he notices that, in practice, this combinatorial explosion is rare, due to the fact that the grey shadow of K is often empty (or at least for most of the recursive calls of the Omega test, when searching for integer points in K). This experimental observation leads us to Hypothesis 1 which is less strong than the property observed by W. Pugh, while being sufficient to guarantee that our algorithm runs in exponential time.

We believe that this running estimate could still hold with the following even weaker hypothesis: during the execution of the function call IntegerSolve(K), for any polyhedral set K', input of a recursive call, the number of facets of the dark shadow of K' is in "big-O" of the number of facets of its real shadow. Investigating this question is left for future work.

To state our main result, we need a notation for the running time of solving a linear program. Indeed, linear programming is an essential tool for removing redundant inequalities generated by Fourier-Motzkin elimination, see [7].

Notation 2. *For an input linear program with total bit size H and with d variables, we denote by $\mathsf{LP}(d, H)$ an upper bound for the number of bit operations required for solving this linear program. For instance, in the case of Karmarkar's algorithm [6], we have $\mathsf{LP}(d, H) \in O(d^{3.5}H^2 \cdot \log H \cdot \log\log H)$.*

Theorem 1. *Under Hypothesis 1, the call function IntegerSolve(K) runs within $O(m^{2d^2} d^{4d^3} L^{4d^3} \mathsf{LP}(d, m^d d^4(\log d + \log L)))$ bit operations.*

The running time estimate in Theorem 1 is exponential w.r.t. d but polynomial w.r.t m and L. Since our algorithm transforms the Omega Test from a decision procedure into a system solving algorithm, our result also holds for the original Omega Test. To our knowledge, this is the first complexity estimate for the whole Omega Test procedure.

The proof follows from a series of results established in Sects. 2, 3 and 4. We believe that some of them are interesting on their own.

Section 2 deals with the following problem. Let F be a k-dimensional face of K, for $0 \leq k < d$. What is the computational cost of projecting F onto a k-dimensional linear subspace of $\mathbb{R}^d$?

Section 3 gives complexity estimates for Fourier-Motzkin elimination (FME). While it is known that FME can run in single exponential time [5,7], we are not aware of running time estimates for FME in the literature. Thanks to Hypothesis 1, our FME estimates applies to the DarkShadow sub-routine of IntegerSolve.

Section 4 gathers results for completing the proof of Theorem 1. The recursive nature of this algorithm leads us to give upper bounds for three quantities: the number of nodes in the tree of the recursive calls, the number of facets of each polyhedron input of a recursive call, the maximum absolute value of a coefficient in a linear system defining such a polyhedron.

2 Properties of the Projection of Faces of a Polyhedron

This section gathers preliminary results towards the complexity analysis of the IntegerSolve algorithm. Some of these results are probably not new, but we could not find a reference for them in the literature.

Definition 1. *Let I be a subset of $\{1, \ldots, m\}$ and denote by $\mathbf{B}_I$ the affine space $\{\mathbf{x} \in \mathbb{R}^d \mid \mathbf{a}_i^T \mathbf{x} = b_i \, for \;\; i \in I\}$. If $\mathbf{B}_I \cap K$ is not empty, then $\mathbf{B}_I \cap K$ is a face of K. We call such an index set I a* defining index set *of the face $\mathbf{B}_I \cap K$.*

Let F be a k-dimensional face of K for an integer $0 \leq k < d$ and let I be a defining index set of F with maximum cardinality. Consider the set O_I given by:

$$O_I \;\; = \;\; \mathbf{B}_I \cap \{\mathbf{x} \mid \mathbf{a}_i^T \mathbf{x} < b_i, i \notin I\}. \tag{1}$$

Proposition 1. *The set O_I is not empty.*

Proof. The assumption on I implies that for all $i \notin I$ the equality $\mathbf{a}_i^T \mathbf{x} = b_i$ is not an implicit equation of F. Indeed, if $\mathbf{a}_i^T \mathbf{x} = b_i$ were an implicit equation of F, then the set $\{i\} \cup I$ would be a defining index set of F as well, a contradiction. From Sect. 8.1 of [10] and since no equation $\mathbf{a}_i^T \mathbf{x} = b_i$ for $i \notin I$ is an implicit equation of F defined by I, we deduce that the set O_I is not empty. □

Using Gaussian elimination, we can compute a parametric representation of O_I where $\dim(\mathbf{B}_I)$ variables are treated as parameters; we denote by $\mathbf{x}'$ those parameters. The other $d - \dim(\mathbf{B}_I)$ variables are referred as *main variables* or *leading variables*, following the terminology of the theory of regular chains [2]. Once we substitute the main variables by their linear forms in the parameters (solved from $\mathbf{B}_I$) into the system $\{\mathbf{x} \mid \mathbf{a}_i^T \mathbf{x} < b_i, i \notin I\}$, we obtain a consistent strict inequality system in the parameters, whose solution set, that we call O_o, is of dimension $\dim(\mathbf{B}_I)$ in the parameter space.

Proposition 2. *We have* $\dim(\mathbf{B}_I) = \dim(O_I) = \dim(O_o) = \dim(F) = k$.

Proof. Note that the set O_o is the image of O_I in the standard projection onto the parameter space and that O_o is open in that space (equipped with the Euclidean topology). Hence, we have $\dim(\mathbf{B}_I) = \dim(O_I) = \dim(O_o)$. In fact, this elimination-and-substitution process shows that O_I is the solution set of a so-called *regular semi-algebraic system* [4] where the regular chain part is given by a regular chain of height $d - \dim(\mathbf{B}_I)$. Meanwhile, we have $\dim(\mathbf{B}_I) \geq \dim(F) \geq \dim(O_I)$, since $\mathbf{B}_I \supseteq F \supseteq O_I$ holds by definition. Moreover, we have $\dim(O_I) \geq \dim(O_o) = \dim(\mathbf{B}_I)$ since O_o is the image of O_I. Finally, since $\dim(F) = k$ holds by assumption, we deduce $\dim(\mathbf{B}_I) = \dim(F) = k$. □

The following lemma was found by the authors independently of the work of Imbert [5] but it is likely that our result could be derived from that paper.

Lemma 1. *Let F be a k-dimensional face of K for some integer $0 \leq k < d$. Then, the face F admits a defining index set with cardinality $d - k$.*

Proof. First, we shall prove that there exists a defining index set with cardinality at least $d-k$. Assume that I is a defining index set of F with maximum cardinality. From Proposition 2, we have $\dim(\mathbf{B}_I) = k$, hence I has at least $d-k$ elements. Assume $I = \{i_1, i_2, \ldots, i_t\}$, with $t \geq d-k$. Since $\dim(\mathbf{B}_I) = k$ holds, one can easily deduce that the rank of the matrix

$$(\mathbf{a}_{i_1}^T, \mathbf{a}_{i_2}^T, \ldots, \mathbf{a}_{i_t}^T)$$

and the rank of the matrix

$$(\mathbf{a}_{i_1}^T, \mathbf{a}_{i_2}^T, \ldots, \mathbf{a}_{i_t}^T, (b_{i_1}, b_{i_2}, \ldots, b_{i_t})^T)$$

are both $d-k$. Thus, we can further assume w.l.o.g. that

$$(\mathbf{a}_{i_1}^T, \mathbf{a}_{i_2}^T, \ldots, \mathbf{a}_{i_{d-k}}^T)$$

has rank $d-k$. Then clearly, the set $I^* = \{i_1, i_2, \ldots, i_{d-k}\}$ is also a defining index set of F. That is, the k-dimensional face F admits a defining index set with cardinality $d-k$. □

Corollary 1 follows immediately from Lemma 1.

Corollary 1. *Let $0 \leq k < d$ be an integer. Let $f_{d,m,k}$ be the number of k-dimensional faces of K. Then, we have*

$$f_{d,m,k} \leq \binom{m}{d-k}.$$

Therefore, we have

$$f_{d,m,0} + f_{d,m,1} + \cdots + f_{d,m,d-1} \leq m^d.$$

Note that, from now on, when we say *a defining index set of a k-dimensional face of K*, we shall always refer to one with cardinality $d-k$. Let F_P be the closure of O_o in the Euclidean topology. Then, F_P is the projection of F on the coordinates $\mathbf{x}'$, where $\mathbf{x}'$ stand for the parameters introduced above. Thus, F_P is a polyhedron and Corollary 2 gives upper-bound estimates on a representation of F_P with a system of linear inequalities.

Corollary 2. *One can compute a matrix $\mathbf{C}$ over $\mathbb{Z}$ and a vector $\mathbf{d}$ over $\mathbb{Z}$ such that the integer points of F_P are given by $\mathbf{C}\mathbf{x}' \leq \mathbf{d}$ and the maximum absolute value of a coefficient in either $\mathbf{C}$ or $\mathbf{d}$ is no more than $(d-k+1)^{\frac{d-k+1}{2}} L^{d-k+1}$, where L is the maximum absolute value of a coefficient in either $\mathbf{A}$ or $\mathbf{b}$.*

Proof. Without loss of generality, assume $I = \{1, \ldots, d-k\}$. From Proposition 2, we have $\dim(F_P) = k$. From the proof of Lemma 1, the rank of the matrix $(\mathbf{a}_1^T, \mathbf{a}_2^T, \ldots, \mathbf{a}_{d-k}^T)$ and the rank of matrix $(\mathbf{a}_1^T, \mathbf{a}_2^T, \ldots, \mathbf{a}_{d-k}^T, (b_1, b_2, \ldots, b_{d-k})^T)$ are both equal to $d-k$. Without loss of generality, assume that the first $d-k$ rows of each of the above two matrices are linearly independent. Therefore, we have $\mathbf{x}' = [x_{d-k+1}, \ldots, x_d]^T$. It follows that F_P can be defined by the inequality system $\mathbf{C}\mathbf{x}' \leq \mathbf{d}$ obtained by Fraction-Free Gaussian Elimination, where $\mathbf{C}$ and $\mathbf{d}$ are given by:

$$\mathbf{C} = (c_{i,j})_{d-k<i,j\leq d}, \text{ where } c_{i,j} \text{ is the determinant of } \begin{pmatrix} a_{11} & \cdots & a_{1,d-k} & a_{1,j} \\ \vdots & \vdots & \vdots & \vdots \\ a_{d-k,1} & \cdots & a_{d-k,d-k} & a_{d-k,j} \\ a_{i,1} & \cdots & a_{i,d-k} & a_{i,j} \end{pmatrix},$$

$$\mathbf{d} = [d_{d-k+1}, \ldots, d_d]^T, \text{ where } d_i \text{ is the determinant of } \begin{pmatrix} a_{11} & \cdots & a_{1,d-k} & b_1 \\ \vdots & \vdots & \vdots & \vdots \\ a_{d-k,1} & \cdots & a_{d-k,d-k} & b_{d-k} \\ a_{i,1} & \cdots & a_{i,d-k} & b_i \end{pmatrix}.$$

Using Hadamard's inequality, the absolute value of any $c_{i,j}$ and d_j can be bounded by $(d-k+1)^{\frac{d-k+1}{2}} L^{d-k+1}$. □

3 Complexity Estimates for Fourier-Motzkin Elimination

Proposition 4 states a running time estimate for computing the linear inequality system $\mathsf{proj}(K; x_1, \ldots, x_d)$ defined in Sect. 2 of the first paper. Note that the article [7] states that Fourier-Motzkin elimination can be run in single exponential time but without giving a running time estimate. Let $k < d$ be a positive integer. Following the notations of Sect. 2 of the first paper, we denote by $\Pi^{x_{k+1},\ldots,x_d}$ the standard projection from $\mathbb{R}^d$ to $\mathbb{R}^{d-k}$ mapping $(x_1, \ldots, x_d)$ to $(x_{k+1}, \ldots, x_d)$.

Proposition 3. *Assume that K is full-dimensional. Then, we have:*

(i) The projected polyhedron $\Pi^{x_{k+1},\ldots,x_d} K$ admits at most $\binom{m}{d-k-1}$ facets.
(ii) Any facet of $\Pi^{x_{k+1},\ldots,x_d} K$ can be given by a system consisting of one linear equation and $m-k-1$ linear inequalities, all in $\mathbb{R}^{d-k}$, such that the absolute value of any coefficient in those constraints is at most $(k+1)^{\frac{k+1}{2}} L^{k+1}$.

Proof. Let G be a facet of $\Pi^{x_{k+1},\ldots,x_d} K$. There exists a face F of K such that G is the projection of F. Since K is full-dimensional, it is clear that $\Pi^{x_{k+1},\ldots,x_d} K$ is full-dimensional as well. Hence, we have $\dim(G) = d - k - 1 \leq \dim(F)$. Clearly, choosing F with minimum dimension implies $d-k-1 = \dim(F)$. With Corollary 1, we deduce (i). Now we prove (ii). It follows from Lemma 1 that one can choose a defining index set I of F with cardinality $d - (d-k-1) = k+1$. Thus, we have $\mathbf{B}_I \cap K = F$, with $\mathbf{B}_I$ given in Definition 1. Consider, then, the set O_I given by (1). We know from Proposition 1 that O_I is not empty and from Proposition 2 that $\dim(\mathbf{B}_I) = d-k-1$. Consider now the system of linear equations given by:

$$G_I \quad = \quad \mathbf{B}_I \cap \{\mathbf{x} \mid \mathbf{a}_i^T \mathbf{x} = b_i, i \notin I\}. \tag{2}$$

Using Fraction-Free Gaussian Elimination on G_I and since $\dim(\mathbf{B}_I) = d-k-1$ holds, one can use the $k+1$ equations defining $\mathbf{B}_I$ to eliminate $x_1, \ldots, x_k$ from the inequalities $\{\mathbf{x} \mid \mathbf{a}_i^T\mathbf{x} < b_i, i \notin I\}$ and, in addition, obtain one equation involving the variables $x_{k+1}, \ldots, x_d$ only. Clearly, the resulting inequalities and equation exactly define G. Using Hadamard's inequality as in the proof of Corollary 2, we deduce (*ii*). □

Definition 2. *Let θ be an inequality in the irredundant representation of the projected polyhedron $\Pi^{x_{k+1},\ldots,x_d}K$. Let G be the facet of $\Pi^{x_{k+1},\ldots,x_d}K$ associated with θ. There exists a $(k+1)$-dimensional face G' of K such that $\Pi^{x_{k+1},\ldots,x_d}G' = G$ holds. We call* defining index set *of θ any defining index set of G'.*

Lemma 2. *Let $\mathbf{v} \in \mathbb{R}^d$ and $s \in \mathbb{R}$ such that h is also the maximum bit size of any coefficient in $\mathbf{v}$ and s. Hence, the total bit size of the linear program $\sup\{-(\mathbf{v}\mathbf{x} - s) \mid \mathbf{A}\mathbf{x} \le \mathbf{b}\}$ is $H \in O(h\,m\,d)$. Moreover, deciding whether the inequality $\mathbf{v}\mathbf{x} \le s$ is implied by $\mathbf{A}\mathbf{x} \le \mathbf{b}$ or not can be done within $O(\mathsf{LP}(d, H))$ bit operations.*

Proof. The estimate $H \in O(h\,m\,d)$ clearly holds. On the other hand, the inequality $\mathbf{v}\mathbf{x} \le s$ is implied by $\mathbf{A}\mathbf{x} \le \mathbf{b}$ if only if $\sup\{-(\mathbf{v}\mathbf{x} - s) \mid \mathbf{A}\mathbf{x} \le \mathbf{b}\}$ is zero. □

Proposition 4. *Within $O(d^2\,m^{2d}\,\mathsf{LP}(d, 2^d h d^2 m^d))$ bit operations, the projected representation $\mathsf{proj}(K; x_1, \ldots, x_d)$ of K can be computed.*

Proof. Following the notations of Sect. 2 of the first paper, the process of eliminating x_1 in $\mathbf{A}\mathbf{x} \le \mathbf{b}$ generates at most $\frac{m^2}{4}$ new inequalities. Augmenting $\mathbf{A}^{<x_1}$ with all these new inequalities and, making this augmented system irredundant, we obtain a total number of inequalities that we denote by c_2. We define $c_1 := m$, $m_1 := c_1$ and $m_2 := c_1 + c_2$. We observe that:

1. generating all the new inequalities (irredundant or not) amounts to at most $O(\frac{m_1^2}{4}\,d\,h_1^2)$ bit operations, and
2. removing the redundant ones amounts to at most to $O(\frac{m_1^2}{4}\,\mathsf{LP}(d, h_1\,d\,m_1))$ bit operations, thanks to Lemma 2.

Similarly, during the process of eliminating x_2, we observe that:

1. generating all the new inequalities (irredundant or not) amounts to at most $O(\frac{c_2^2}{4}\,d\,h_2^2)$ bit operations, and
2. removing the redundant ones amounts to at most to $O(\frac{c_2^2}{4}\,\mathsf{LP}(d, h_2\,d\,m_2))$ bit operations and yields a total number of c_3 inequalities in x_3.

Continuing in this manner, we deduce that for successively eliminating $x_1, \ldots, x_{d-1}$,

1. generating all the new inequalities (irredundant or not) amounts to at most

$$O(\frac{c_1^2}{4}\,d\,h_1^2 + \cdots + \frac{c_{d-1}^2}{4}\,d\,h_{d-1}^2), \tag{3}$$

2. removing the redundant ones amounts to at most to

$$O(\frac{c_1^2}{4}\mathsf{LP}(d, h_1\, d\, m_1) + \cdots + \frac{c_{d-1}^2}{4}\mathsf{LP}(d, h_{d-1}\, d\, m_{d-1})), \quad (4)$$

where $m_i := c_1 + \cdots + c_i$, for $1 \leq i < d$, as well as $h_0 := h$ and $h_{i+1} \leq 2\, h_i + 1$, for $0 \leq i < d$. We observe that c_i is bounded over by the number of facets of $\Pi^{x_i,\ldots,x_d} K$, for $1 \leq i < d$. Observe also that, for $1 < i < d$, each facet of $\Pi^{x_i,\ldots,x_d} K$ is the projection of a face of $\Pi^{x_{i-1},\ldots,x_d} K$. Using Lemma 1, we deduce that, for all $1 \leq i < d$, we have: $c_i \leq m^d$. Therefore, the running time estimates of (3) and (4) can be bounded over by $O(d^2\, m^{2d}\, d\, (2^d h)^2)$ and $O(d^2\, m^{2d}\, \mathsf{LP}(d, (2^d h) d (d m^d)))$. The latter dominates the former; the conclusion follows. □

4 Proof of Theorem 1

We use Fig. 1 and Notation 3 to provide further explanation on Algorithm IntegerSolve$_0$, presented in the first paper.

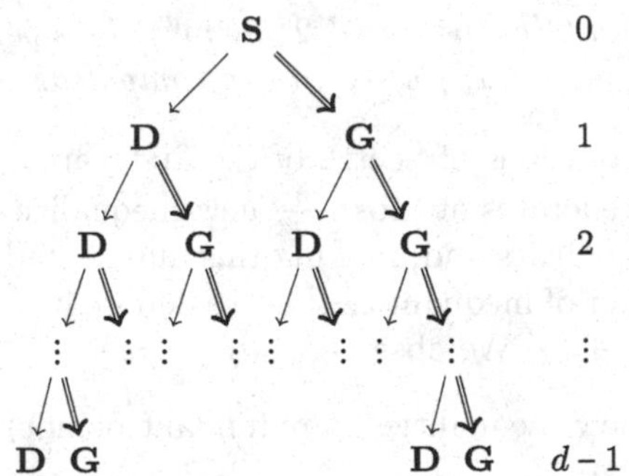

Fig. 1. Diagram

Notation 3. *Fig. 1 illustrates the tree of recursive calls for the* IntegerSolve$_0$ *procedure. The root of the tree is labelled with* **S**, *which stands for the input system. The left (resp. right) child of a node, other than a leaf, is labelled by* **D** *(resp.* **G***) which stands for the output of the* DarkShadow *procedure (resp. the* GreyShadow *procedure). Since the* DarkShadow *procedure generates one input system for* IntegerSolve, *we use a simple* $\rightarrow$ *arrow as an edge to a* **D***-node. However, the* GreyShadow *procedure may generate several linear inequality systems, leading to several recursive calls to* IntegerSolve$_0$. *Thus, we use a* $\Rightarrow$ *arrow as an edge to a* **G***-node. The numbers on the right-hand side of Fig. 1 stand for the levels in the tree.*

Let $\mathbf{Ax} \leq \mathbf{b}, m, d$ be as in Notation 1. Let L and h denote the maximum absolute value and height of any coefficient in either $\mathbf{A}$ or $\mathbf{b}$.

Notation 4. *Recall that Fig. 1 depicts the tree of recursive calls in Algorithm* IntegerSolve$_0$. *Let* $\mathbf{N}$ *denote any node in that tree, whether it is labelled* $\mathbf{S}$, $\mathbf{D}$ *or* $\mathbf{G}$. *If* $\mathbf{N}$ *is labelled with* $\mathbf{S}$ *or* $\mathbf{D}$, *it is associated with a single linear system denoted by* $\mathbf{M_N t_N} \leq \mathbf{v_N}$. *If* $\mathbf{N}$ *is labelled with* $\mathbf{G}$, *it is associated with a sequence of linear systems produced by the* GreyShadow *procedure and we denote by* $\mathbf{M_N t_N} \leq \mathbf{v_N}$ *any of those systems. For any linear system* $\mathbf{M_N t_N} \leq \mathbf{v_N}$ *(whether* $\mathbf{N}$ *is labelled* $\mathbf{S}$, $\mathbf{D}$ *or* $\mathbf{G}$*), we denote by* $m_\mathbf{N}$ *and* $d_\mathbf{N}$ *the number of rows and columns of* $\mathbf{M_N}$. *We denote by* $L_\mathbf{N}$ *(resp.* $\ell_\mathbf{N}$*) be the maximum absolute value of any coefficient in* $\mathbf{M_N}$ *(resp. in either* $\mathbf{M_N}$ *or* $\mathbf{v_N}$*). We denote by* $h_\mathbf{N} = \lfloor \log_2 \ell_\mathbf{N} \rfloor + 1$ *the maximum bit size of a coefficient in either* $\mathbf{M_N}$ *or* $\mathbf{v_N}$. *The system* $\mathbf{M_N t_N} \leq \mathbf{v_N}$ *encodes a polyhedron* $K_\mathbf{N}$ *in* $\mathbb{R}^{d_\mathbf{N}}$ *and we denote by* $F_\mathbf{N}$ *an arbitrary facet of* $K_\mathbf{N}$. *Every path from the root to a leaf* $\mathbf{N}_r$ *in the tree depicted in Fig. 1 can be labelled* $\mathbf{S} \to \mathbf{N}_1 \to \cdots \to \mathbf{N}_r$ *for some* $r \leq d - 1$. *Note that a leaf (that is, a node with no children) may have level less than* $d - 1$. *For simplicity, for the node* $\mathbf{N}_r$, *we write* d_r, L_r, ℓ_r, h_r, $\mathbf{t}_r$, $\mathbf{M}_r$, $\mathbf{v}_r$, K_r, F_r *instead of* $d_{\mathbf{N}_r}$, $L_{\mathbf{N}_r}$, $\ell_{\mathbf{N}_r}$, $h_{\mathbf{N}_r}$, $\mathbf{t}_{\mathbf{N}_r}$, $\mathbf{M}_{\mathbf{N}_r}$, $\mathbf{v}_{\mathbf{N}_r}$, $K_{\mathbf{N}_r}$, $F_{\mathbf{N}_r}$ *respectively, when there is no ambiguity.*

In particular, let d_0, L_0, ℓ_0, h_0, $\mathbf{t}_0$, $\mathbf{M}_0$, $\mathbf{v}_0$, K_0, F_0 *denote the corresponding values of node* $\mathbf{S}$.

Without loss of generality, we assume the polyhedron K is full-dimensional, that is, $\dim(K) = d$ and, thus, that the input system $\mathbf{S}$ has no implicit equations. Then, each call to the DarkShadow or GreyShadow procedures at level 1 reduces the dimension of the ambient space by one. Similarly, at every level, we assume that the input system of inequalities of IntegerSolve$_0$ (that is, the fourth argument of this procedure) is full-dimensional. Hence at Line 23 of Algorithm IntegerSolve$_0$, the output of IntegerNormalize($\mathbf{Ax} \leq \mathbf{b}$) is ($\varnothing, \varnothing$, $\mathbf{Ax} \leq \mathbf{b}$).

This full-dimensionality assumption has two consequences. First, along any path $\mathbf{S} \to \mathbf{N}_1 \to \cdots \to \mathbf{N}_r$ we have $d_{k+1} = d_k - 1$, for $1 \leq k < r$, and thus, we have $d_k = d - k$. Second, at node $\mathbf{N}_k$, the input system is $\mathbf{M}_{k-1}\mathbf{t}_{k-1} \leq \mathbf{v}_{k-1}$ (while the output is $\mathbf{M}_k\mathbf{t}_k \leq \mathbf{v}_k$).

It is easy to see that this full-dimensionality assumption is a worst case scenario as far as running time is concerned. Indeed, when this assumption does not hold, for at least one path $\mathbf{S} \to \mathbf{N}_1 \to \cdots \to \mathbf{N}_r$, implicit equations will be discovered at Line 23 of Algorithm IntegerSolve$_0$ in the first paper, and dimension will drop by more than one at one node of that path.

To prove Theorem 1 we shall establish a series of intermediate results. Lemmas 5, 7, 8 provide upper bounds for the absolute values of any coefficient in the systems $\mathbf{M_N t_N} \leq \mathbf{v_N}$ while Lemmas 3, 4, 9, 10 deal with running time estimates. We start with Lemmas 3 and 4, which give running time estimates for the DarkShadow and GreyShadow procedures at level k. The proof of Lemma 3 follows that of Proposition 4.

Lemma 3. *For any non-negative integer* $k < d - 1$, *the* DarkShadow *procedure at level* $k+1$ *runs within* $O(\frac{m_k^2}{4}\mathsf{LP}(d_k, d_k h_k m_k))$ *bit operations.*

Proof. The input system of $\mathbf{D}_{k+1}$ is $\mathbf{M}_k\mathbf{t}_k \leq \mathbf{v}_k$, which has m_k inequalities and h_k as maximum coefficient size in either $\mathbf{M}_k$ or $\mathbf{v}_k$. The process of

(E) eliminating the first variable of $\mathbf{t}_k$ in $\mathbf{M}_k\mathbf{t}_k \leq \mathbf{v}_k$,
(A) adding the at most $\frac{m_k^2}{4}$ resulting inequalities to those of $\mathbf{M}_k\mathbf{t}_k \leq \mathbf{v}_k$ where the first variable of $\mathbf{t}_k$ does not appear, and
(R) removing all redundant inequalities,

yields $\mathbf{M}_k\mathbf{t}_k \leq \mathbf{v}_k$, see Algorithm DarkShadow. Observe that Steps (E) and (R) amount to at most $O(\frac{m_k^2}{4}\, d_k\, h_k^2)$ and $O(\frac{m_k^2}{4}\; \mathsf{LP}(d_k, h_k\, d_k\, m_k))$ bit operations, respectively. The latter dominates the former. The conclusion follows. □

Lemma 4. *For any non-negative integer $k < d-1$, the* GreyShadow *procedure at level $k+1$ runs within $O(m_k^2 d_k^{3+\varepsilon} h_k^3)$ bit operations.*

Proof. For the GreyShadow procedure at level $k+1$, we need to call at most m_k times the IntegerNormalize procedure. Then, the lemma follows from Proposition 1 in the first paper. □

Lemma 5. *Consider a path of the form*

$$\mathbf{S} \to \mathbf{D}_1 \to \cdots \to \mathbf{D}_r, \tag{5}$$

where all nodes, except the first one, are labelled by $\mathbf{D}$. *Then, for all $1 \leq k \leq r$, we have $m_k \leq m^{k+1}$ and the maximum absolute value L_k of any coefficient in $\mathbf{M}_k$ is no more than $(k+1)^{\frac{k+1}{2}} L^{k+1}$.*

Proof. Let $1 \leq k \leq r$. Under Hypothesis 1, each facet of K_k is parallel to a facet of the real shadow of K_{k-1}. Inductively, each facet of K_k is parallel to a facet of the projection $\Pi^{x_{k+1},\ldots,x_d} K$. By Proposition 3, we have $m_k \leq m^{k+1}$ and $L_k \leq (k+1)^{\frac{k+1}{2}} L^{k+1}$. □

Next, we will consider an arbitrary path:

$$\mathbf{S} \to \mathbf{D}_1 \to \cdots \to \mathbf{D}_{j_1-1} \to \mathbf{G}_{j_1} \to \cdots \to \mathbf{G}_{j_s} \to \mathbf{D}_{j_s+1} \to \cdots \to \mathbf{D}_r. \tag{6}$$

In the path (6), only the subscripts $j_1, j_2, \ldots, j_s$ correspond to the GreyShadow procedures.

To make things simpler, instead of setting $\Theta_2 := \Upsilon \cup \mathbf{M}\mathbf{t} \leq \mathbf{v} \cup \{c\alpha - a\gamma > -(c-1)(a-1)\}$ in Line 8 of Algorithm GreyShadow in the first paper, we let $\Theta_2 := \mathbf{M}\mathbf{t} \leq \mathbf{v}$. This simplification cannot guarantee that $V_{\mathbb{Z}}(\mathbf{t} = \mathbf{P}_k\mathbf{u}_k + \mathbf{q}_k \cup \mathbf{M}_k\mathbf{u}_k \leq \mathbf{v}_k,\ \mathbf{t})$ for $k = 1, \ldots, s$ form a disjoint union. However, it will endow $\mathbf{M}_k$ with good structural properties, as we will see later. Actually, since all the inequalities in Υ and the negation of $c\alpha - a\gamma > -(c-1)(a-1)$ can be obtained by the DarkShadow procedure and since we are doing the worst case complexity analysis, all the coming conclusions apply to our algorithm as it was originally stated in the first paper.

First, we consider the sub-path of (6): $\mathbf{S} \to \mathbf{D}_1 \to \cdots \to \mathbf{D}_{j_1-1} \to \mathbf{G}_{j_1}$. We assume the variable order is $x_1 > x_2 > \cdots > x_d$. Thus, we can denote the variable set $\mathbf{t}_{j_1-1}$ for the input system of node $\mathbf{D}_{j_1-1}$ as: $\mathbf{t}_{j_1-1} = [x_{j_1}, x_{j_1+1}, \ldots, x_d]^T$ since $\mathbf{t}_{j_1-1} \subset \mathbf{x}$. For the node $\mathbf{G}_{j_1}$, we need to add one equation based on the output system $\mathbf{M}_{j_1-1}\mathbf{t}_{j_1-1} \le \mathbf{v}_{j_1-1}$ of node $\mathbf{D}_{j_1-1}$. Without loss of generality, we assume the new equation is $\mathbf{m}\mathbf{t}_{j_1-1} = v + i$ for some non-negative integer $i \le L_{j_1-1}$, where $\mathbf{m}\mathbf{t}_{j_1-1} \le v$ is the first inequality in the system $\mathbf{M}_{j_1-1}\mathbf{t}_{j_1-1} \le \mathbf{v}_{j_1-1}$. Let I be the defining index set of $\mathbf{m}\mathbf{t}_{j_1-1} \le v$, which has cardinality j_1.

Recall that $\mathbf{M}_0\mathbf{t}_0 \le \mathbf{v}_0$ is the input system of node $\mathbf{D}_1$. Let $\mathbf{M}_0^{(1)}$ and $\mathbf{M}_0^{(2)}$ be the sub-matrices of $\mathbf{M}_0$ consisting of the first $j_1 - 1$ columns and the last $d - j_1 + 1$ columns, respectively. Denote by $(\mathbf{v}_0)_I$ (resp. $(\mathbf{M}_0)_I$) the sub-vector (resp. sub-matrix) of $\mathbf{v}_0$ (resp. $\mathbf{M}_0$) with index (resp. row index) I. Let $\mathbf{Q}_{j_1-1}$ be a matrix whose columns consist of a $\mathbb{Z}$-basis of the space $\{\mathbf{x} : (\mathbf{M}_0)_I\,\mathbf{x} = \mathbf{0}\}$. We have assumed that the input polyhedron K is full-dimensional, which implies that the rank of $\mathbf{M}_0$ is d. By the definition of defining index set I, we can easily deduce that the rank of $(\mathbf{M}_0)_I$ is j_1, that is, $\mathbf{Q}_{j_1-1}$ is an integer matrix with d rows and $d - j_1$ columns. Let $\mathbf{Q}'_{j_1-1}$ be the sub-matrix consisting of the last $d-j_1+1$ rows of $\mathbf{Q}_{j_1-1}$. Let $\mathbf{V}_1 := [\mathbf{e}_1, \ldots, \mathbf{e}_{j_1-1}, \binom{\mathbf{0}}{\mathbf{Q}'_{j_1-1}}] \in \mathbb{Z}^{d\times(d-1)}$. Let $\mathbf{S}_1$ be a node associated with the system $\mathbf{M}_0^{(1)}[x_1, \ldots, x_{j_1-1}] + \mathbf{M}_0^{(2)}\mathbf{Q}'_{j_1-1}\mathbf{t}_{j_1+1} \le \mathbf{v}'_0$, *i.e.* $\mathbf{M}_0\mathbf{V}_1[x_1, \ldots, x_{j_1-1}, \mathbf{t}_{j_1+1}]^T \le \mathbf{v}'_0$. For $j_1 \le k < j_2$, let $\mathbf{M}'_k\mathbf{t}'_k \le \mathbf{v}'_k$ be the output system of the node $\mathbf{D}_{k-1}$ in the path: $\mathbf{S}_1 \to \mathbf{D}_1 \to \cdots \to \mathbf{D}_{k-1}$.

Lemma 6. *With the above notations, we have* $\mathbf{M}_{j_1} = \mathbf{M}'_{j_1}$. *Consequently,* $\mathbf{M}_k = \mathbf{M}'_k$ *for* $j_1 \le k < j_2$.

Proof. The second statement will follow once the first lemma is valid.

Following the algorithm DarkShadow in the first paper, there exists a matrix $U \in \mathbb{Z}^{m_{j_1-1}\times m_0}$, such that $U\mathbf{M}_0^{(1)} = \mathbf{0}$ and $\mathbf{M}_{j_1-1} = U_d U\mathbf{M}_0^{(2)}$, where $U_d = \mathrm{DiagonalMatrix}(\frac{1}{gcd_1}, \ldots, \frac{1}{gcd_{m_{j_1-1}}})$ and gcd_i is the gcd of all the coefficients in the i-th row of $U\mathbf{M}_0^{(2)}$ for $1 \le i \le m_{j_1-1}$. Let $\mathbf{u} \in \mathbb{Z}^{m_0}$ be the first row of U. Then, $\mathbf{m} = \frac{1}{gcd_1}\mathbf{u}\mathbf{M}_0^{(2)}$ since $\mathbf{m}\mathbf{t}_{j_1-1} \le v$ is the first inequality of $\mathbf{M}_{j_1-1}\mathbf{t}_{j_1-1} \le \mathbf{v}_{j_1}$. Then, $\mathbf{m} = \frac{1}{gcd_1}\mathbf{u}_I(\mathbf{M}_0^{(2)})_I$. Solving the equation $\mathbf{m}\mathbf{t}_{j_1-1} = v+i$ by Lemma 2 in the first paper, we have $\mathbf{t}_{j_1-1} = \mathbf{P}_{j_1-1}\,\mathbf{t}_{j_1} + \mathbf{q}_{j_1-1}$, where $\mathbf{P}_{j_1-1} \in \mathbb{Z}^{(d-j_1+1)\times(d-j_1)}$ whose columns consist of a $\mathbb{Z}$-basis for $\{\mathbf{y} : \mathbf{m}\mathbf{y} = 0\} = \{\mathbf{y} : \mathbf{u}_I(\mathbf{M}_0^{(2)})_I\mathbf{y} = 0\}$. Therefore, $\mathbf{M}_{j_1}\mathbf{t}_{j_1} \le \mathbf{v}_{j_1}$ comes from $\mathbf{M}_{j_1-1}\mathbf{P}_{j_1-1}\mathbf{t}_{j_1} \le \mathbf{v}_{j_1-1} - \mathbf{M}_{j_1-1}\mathbf{q}_{j_1-1}$, *i.e.* $U_d U\mathbf{M}_0^{(2)}\mathbf{P}_{j_1-1}\mathbf{t}_{j_1} \le \mathbf{v}_{j_1-1} - \mathbf{M}_{j_1-1}\mathbf{q}_{j_1-1}$.

Next, we will show that $\mathbf{P}_{j_1-1}$ can be replaced by $\mathbf{Q}'_{j_1-1}$ introduced above. Since $\mathbf{u}_I(\mathbf{M}_0^{(1)})_I = 0$, we have any $\mathbf{y} \in \mathbb{Z}^d$ satisfying $\mathbf{u}_I(\mathbf{M}_0)_I\mathbf{y} = \mathbf{0}$ is equivalent to $\mathbf{u}_I(\mathbf{M}_0^{(2)})_I\mathbf{y}^{(2)} = 0$, where $\mathbf{y}^{(2)}$ is the last $d - j_1 + 1$ elements of $\mathbf{y}$. Thus, $[\mathbf{e}_1, \ldots, \mathbf{e}_{j_1-1}, \binom{\mathbf{0}}{\mathbf{P}_{j_1-1}}]$ is a $\mathbb{Z}$-basis for the space $\{\mathbf{y} : \mathbf{u}(\mathbf{M}_0)_I\mathbf{y} = 0\}$. For any row vector $\mathbf{y} \in \mathbb{Z}^d$ such that $\mathbf{u}_I(\mathbf{M}_0)_I\mathbf{y} = 0$, either $\mathbf{0} \ne (\mathbf{M}_0)_I\mathbf{y} \in \{\mathbf{z} : \mathbf{u}_I\mathbf{z} = 0\}$ or $(\mathbf{M}_0)_I\mathbf{y} = \mathbf{0}$. For the first case, $\mathbf{e}_1, \ldots, \mathbf{e}_{j_1-1}$ is a $\mathbb{Z}$-basis for the solutions of $\mathbf{y}$, where $\mathbf{e}_k \in \mathbb{Z}^d$ is the k-th standard basis for $1 \le k \le j_1 - 1$. For the second case, columns of $\mathbf{Q}_{j_1-1}$ consisting of a $\mathbb{Z}$-basis for the solutions of $\mathbf{y}$. Thus, $[\mathbf{e}_1, \ldots, \mathbf{e}_{j_1-1}, \mathbf{Q}_{j_1-1}]$ is a $\mathbb{Z}$-basis for the space $\{\mathbf{y} : \mathbf{u}_I(\mathbf{M}_0)_I\mathbf{y} = 0\}$. Consequently, $\mathbf{P}_{j_1-1}$ is equivalent to $\mathbf{Q}'_{j_1-1}$, which is the last $d - j_1 + 1$ rows of

$\mathbf{Q}_{j_1-1}$. That is, the integer solutions to $\mathbf{m}\,\mathbf{t}_{j_1-1} = v + i$ can be represented by $\mathbf{t}_{j_1-1} = \mathbf{Q}'_{j_1-1}\,\mathbf{t}_{j_1} + \mathbf{q}_{j_1-1}$, where $|\mathbf{Q}'_{j_1-1}| \leq j_1^{j_1+1} L^{2j_1}$. Therefore, we can make $\mathbf{M}_{j_1} = U_d U \mathbf{M}_0^{(2)} \mathbf{Q}'_{j_1-1}$.

Remember that $\mathbf{S}_1$ is associated with the system $\mathbf{M}_0 \mathbf{V}_1 [x_1, \ldots, x_{j_1-1}, \mathbf{t}_{j_1+1}]^T \leq \mathbf{v}_0$, where $\mathbf{V}_1 := [\mathbf{e}_1, \ldots, \mathbf{e}_{j_1-1}, \binom{\mathbf{0}}{\mathbf{Q}'_{j_1-1}}]$, and $\mathbf{M}'_k \mathbf{t}'_k \leq \mathbf{v}'_k$ is the output system of the node $\mathbf{D}_{k-1}$ in the path: $\mathbf{S}_1 \rightarrow \mathbf{D}_1 \rightarrow \cdots \rightarrow \mathbf{D}_{k-1}$. We have $\mathbf{M}'_{j_1} = U_d U \mathbf{M}_0^{(2)} \mathbf{Q}'_{j_1}$. Consequently, $\mathbf{M}'_k = \mathbf{M}_k$ for any integer $k : j_1 \leq k < j_2$. □

Then, we have the following lemma:

Lemma 7. *For $j_1 \leq k < j_2$, the maximum absolute value of any coefficient in $\mathbf{M}_k$ can be bounded over by $d^k k^{2k^2} L^{3k^2}$. Moreover, we have $m_k \leq m^{k+1}$.*

Proof. By Lemma 5, the maximum absolute value of any coefficient in $\mathbf{M}'_k = \mathbf{M}_k$ can be bounded over by $|\mathbf{M}_k| \leq k^{\frac{k}{2}} (d - j_1 + 1)^k j_1^{kj_1+k} L^{2kj_1+k} \leq d^k k^{2k^2} L^{3k^2}$.

Moreover, $m_k \leq m^{k+1}$ follows from the equivalent path $\mathbf{S}_1 \rightarrow \mathbf{D}_1 \rightarrow \cdots \rightarrow \mathbf{D}_{k-1}$ for integer $k : j_1 \leq k < j_2$. □

For any $1 \leq t \leq s$, we assume that the new equation is $\mathbf{m}_t \mathbf{t}_{j_t-1} = v_t + i_t$ for some non-negative integer $i_t \leq L_{j_t-1}$, where $\mathbf{m}_t \mathbf{t}_{j_t-1} \leq v_t$ comes from the input system $\mathbf{M}_{j_t-1} \mathbf{t}_{j_t-1} \leq \mathbf{v}_{j_t-1}$ of the node $\mathbf{G}_{j_t}$. Let I_t be the defining index set of the inequality $\mathbf{m}_t \mathbf{t}_{j_t-1} \leq v_t$, with cardinality j_t. Let $\mathbf{Q}_t \in \mathbb{Z}^{d \times (d-j_t)}$ consist of the columns of a $\mathbb{Z}$-basis of space $\{\mathbf{y} : (\mathbf{M}_0)_{I_t} \mathbf{y} = 0\}$. For any $1 \leq t \leq s$, we define $\mathbf{V}_t = [\mathbf{e}_1, \ldots, \mathbf{e}_{j_1-1}, \mathbf{Q}_1^{(1)}, \ldots, \mathbf{Q}_{t-1}^{(t-1)}, \mathbf{Q}_t^{(t)}] \in \mathbb{Z}^{d \times (d-t)}$ and $\mathbf{t}_t := [x_1, \ldots, x_{j_1}, \mathbf{t}_{j_1-1}^{(1)}, \ldots, \mathbf{t}_{j_t-1}^{(t)}]^T$ as follows:

1. When $k < t$, we let $\mathbf{Q}'_k$ be the sub-matrix consisting of the last $d - j_k + 1$ rows and $(j_{k+1} - j_k - 1)$ columns of $\mathbf{Q}_k$. Let $\mathbf{Q}_k^{(k)} \in \mathbb{Z}^{d \times (j_{k+1} - j_k - 1)}$ be the matrix $\binom{\mathbf{0}}{\mathbf{Q}'_k}$, where $\mathbf{0}$ is a zero matrix which has $j_k - 1$ rows and $j_{k+1} - j_k - 1$ columns.
2. When $k = t$, we let $\mathbf{Q}'_t$ be the sub-matrix consisting of the last $d - j_t + 1$ rows of $\mathbf{Q}_t$. Let $\mathbf{Q}_t^{(t)} \in \mathbb{Z}^{d \times (d-j_t)}$ be the matrix $\binom{\mathbf{0}}{\mathbf{Q}'_t}$, where $\mathbf{0}$ is a zero matrix which has $j_t - 1$ rows and $d - j_t$ columns.
3. Denote by $\mathbf{t}_{j_k-1}^{(k)}$ (resp. $\mathbf{t}_{j_t-1}$) the set of $j_{k+1} - j_k - 1$ (resp. $d - j_t$) variables. and the variables in $\mathbf{t}_t$ are independent variables.

Let $\mathbf{S}_t$ be the system represented by $\mathbf{M}_0 \mathbf{V}_t \mathbf{t}_t \leq \mathbf{v}_0$ for $1 \leq t \leq s$.

Lemma 8. *The maximum absolute value of any coefficient in $|\mathbf{M}_k|$ (resp. $|\mathbf{v}_k|$) can be bounded by $d^k k^{2k^2} L^{3k^2}$ (resp. $d^{3k^2} k^{4k^3} L^{6k^3}$) for $1 \leq k \leq r$. Moreover, we have $m_k \leq m^{k+1}$.*

Proof. Similar to the notations defined before Lemma 6, for $1 \leq t \leq s$ and $j_t \leq k < j_{t+1}$, let $\mathbf{M}'_k \mathbf{t}'_k \leq \mathbf{v}'_k$ be the output system of the path: $\mathbf{S}_t \rightarrow \mathbf{D}_1 \rightarrow \cdots \rightarrow \mathbf{D}_{k-t}$, where j_{s+1} is defined as $r + 1$.

We claim $\mathbf{M}_{j_t} = \mathbf{M}'_{j_t}$ for any $1 \le t \le s$, where $\mathbf{M}_{j_t}\mathbf{t}_{j_t} \le \mathbf{v}_{j_t}$ is the output system of the path: $\mathbf{S} \to \mathbf{D}_1 \to \cdots \to \mathbf{D}_{j_1-1} \to \mathbf{G}_{j_1} \to \cdots \to \mathbf{G}_{j_t}$. This claim is valid if $t = 1$ by Lemma 6. We suppose it is valid for $t = 1, \ldots, s-1$. Then, we have $\mathbf{M}_{j_s-1} = \mathbf{M}'_{j_s-1}$, where $\mathbf{M}'_{j_s-1}\mathbf{t}'_{j_s-1} \le \mathbf{v}'_{j_s-1}$ is the output system of the node $\mathbf{D}_{j_s-s}$ in the path: $\mathbf{S}_{s-1} \to \mathbf{D}_1 \to \cdots \to \mathbf{D}_{j_s-s}$. Let $\mathbf{M}''_{j_s}\mathbf{t}''_{j_s} \le \mathbf{v}''_{j_s}$ be the output system of node $\mathbf{G}_{j_s-s+1}$ in the path: $\mathbf{S}_{s-1} \to \mathbf{D}_1 \to \cdots \to \mathbf{D}_{j_s-s} \to \mathbf{G}_{j_s-s+1}$. Note that $\mathbf{S}_{s-1}$ is associated with $\mathbf{M}_0\mathbf{V}_{s-1}\mathbf{t}_{s-1} \le \mathbf{v}_0$ and the input system of node $\mathbf{G}_{j_s-s+1}$ is $\mathbf{M}'_{j_s-1}\mathbf{t}'_{j_s-1} \le \mathbf{v}'_{j_s-1}$, *i.e.* $\mathbf{M}_{j_s-1}\mathbf{t}'_{j_s-1} \le \mathbf{v}'_{j_s-1}$. We have $\mathbf{M}''_{j_s} = \mathbf{M}_{j_s}$ immediately, since both of them come from the output system of node $\mathbf{G}_{j_s-s+1}$ of the path: $\mathbf{S}_{s-1} \to \mathbf{D}_1 \to \cdots \to \mathbf{D}_{j_s-s} \to \mathbf{G}_{j_s-s+1}$. By the proof of Lemma 6, $\mathbf{M}''_{j_s}$ can be obtained from the output system of the path: $\mathbf{S}'_s \to \mathbf{D}_1 \to \cdots \to \mathbf{D}_{j_s-s}$, where $\mathbf{S}'_s$ is associated with $\mathbf{M}_0[\mathbf{V}_{s-1}[\mathbf{e}_1, \ldots, \mathbf{e}_{j_s-s}], \mathbf{Q}_s]\mathbf{t}_s \le \mathbf{v}_0$, *i.e.* $\mathbf{M}_0\mathbf{V}_s\mathbf{t}_s \le \mathbf{v}_0$, which associates to the label $\mathbf{S}_s$. Then, we have $\mathbf{M}''_{j_s} = \mathbf{M}'_{j_s}$. The claim is valid. That is, $\mathbf{M}_{j_s}$ can be obtained from the output system of the path: $\mathbf{S}_s \to \mathbf{D}_1 \to \cdots \to \mathbf{D}_{j_s-s}$.

By Proposition 1 of the first paper, we know that the maximum absolute valuc of any coefficient in $\mathbf{M}_0\mathbf{V}_t$ can be bounded by $dj_t^{j_t+1}L^{2j_t+1}$. Thus, by Lemma 5, for any $1 \le t \le s$ and $j_t \le k < j_{t+1}$, we have the maximum absolute value of any coefficient in $\mathbf{M}_k$ can be bounded by $(k-t)^{\frac{k-t}{2}}(dj_t^{j_t+1}L^{2j_t+1})^{k-t} \le d^k k^{2k^2} L^{3k^2}$. The first statement is valid.

Let $1 \le k \le r$. For the node $\mathbf{D}_k$, we have $|\mathbf{v}_k| \le L_{k-1}^2 + 2L_{k-1}|\mathbf{v}_{k-1}|$. For the node $\mathbf{G}_k$, we have $|\mathbf{v}_k| \le 2d_k L_{k-1}^2|\mathbf{v}_{k-1}|$ since we only need to solve one equation. That is, for any node $\mathbf{N}_k$, we will have $|\mathbf{v}_k| \le 2d_k L_{k-1}^2|\mathbf{v}_{k-1}|$. Thus, $|\mathbf{v}_k| \le 2^k d^k L_{k-1}^2 \cdots L_1^2 |\mathbf{v}_0|^2 \le d^{3k^2}k^{4k^3}L^{6k^3}$ for any $1 \le k \le r$. □

Until now, we can safely say that any coefficient in $\mathbf{M}_r$ (resp.$\mathbf{v}_r$) produced by each path in Fig. 1 can be bounded over by $L_r \le d^r r^{2r^2} L^{3r^2}$ (resp. $\ell_r \le d^{3r^2} r^{4r^3} L^{6r^3}$). That is, the coefficient size associated with the node $\mathbf{N}_r$ can be bounded over by $h_r \le 6r^3(\log d + \log L)$. Moreover, we can have at most m_r inequalities in $\mathbf{M}_r\mathbf{t}_r \le \mathbf{v}_r$. The following lemma shows the complexity estimates for implementing each path of the tree in Fig. 1:

Lemma 9. *The path (6) can be implemented within* $O(m^{2r+2}d^{3+\varepsilon}r^{10}(\log d + \log L)^3) + O(rm^{2r+2}\mathsf{LP}(d, dm^r r^3(\log d + \log L)))$ *bit operations.*

Proof. By Lemma 3 (resp. Lemma 4), each node $\mathbf{D}_k$ (resp. $\mathbf{G}_k$) can be implemented with $O(\frac{m_k^2}{4}\mathsf{LP}(d_k, d_k h_k m_k))$ (resp. $O(m_k^2 d_k^{3+\varepsilon} h_k^3)$) bit operations. Thus, the path (6) can be implemented within

$$r \cdot O(m_r^2 d_r^{3+\varepsilon} h_r^3) + r \cdot O(\frac{m_r^2}{4}\mathsf{LP}(d, dh_r m_r))$$
$$\le O(m^{2r+2}d^{3+\varepsilon}r^{10}(\log d + \log L)^3) + O(rm^{2r+2}\mathsf{LP}(d, dm^{2r+2}r^3(\log d + \log L)))$$

bit operations. □

Let T_r be the total number of nodes in the r-th level. In particular, we have $T_0 = 1$, $T_1 \le mL$. We have the following lemma:

Lemma 10. *We have:* $T_{r+1} \leq m^{r+1} d^r r^{2r^2} L^{3r^2} T_r$ *for* $r = 0, \ldots, d-2$. *Thus, we have* $T_{d-1} \leq m^{d^2} d^{3d^3} L^{3d^3}$.

Proof. By Lemma 8, each node can have at most m^{r+1} inequalities as the input and each inequality has coefficient bound L_r. Following the Algorithm $\mathsf{IntegerSolve_0}$ and Fig. 1, each node can give out at most $m^{r+1} L_r$ branches. Considering we have T_r nodes in the r-th level, we can easily deduce that $T_{r+1} \leq m^{r+1} L_r T_r \leq m^{r+1} d^r r^{2r^2} L^{3r^2} T_r$. The second statement follows easily. □

Now we give the proof for Theorem 1:

Proof. Under Hypothesis 1, by Lemmas 9 and 10, the complexity estimates for $\mathsf{IntegerSolve}(K)$ can be bounded over

$$\begin{aligned}
&T_{d-1} O(m^{2r+2} d^{3+\varepsilon} r^{10} (\log d + \log L)^3) + \\
&T_{d-1} O(m^{2r+2} r \mathsf{LP}(d, dm^{r+1} r^3 (\log d + \log L))) \\
\leq &O(m^{2d^2} d^{4d^3} L^{4d^3} \mathsf{LP}(d, m^d d^4 (\log d + \log L))) \text{ bit operations, since } r < d.
\end{aligned}$$

The theorem is valid. □

5 Experimentation

We have implemented the algorithm presented in the first paper within the `Polyhedra` library in MAPLE. This library is publicly available in source on the download page of the `RegularChains` library at www.regularchains.org.

We have used test-cases coming from various application areas: regular polytopes (first 5 examples in Table 1), examples from Presburger arithmetic (next 5 examples in Table 1), random polytopes (next 5 examples in Table 1), random unbounded polyhedra (next 5 examples in Table 1), examples from text-books (next 3 examples in Table 1) and examples from research articles on automatic parallelization of for-loop nests (last 4 examples in Table 1).

For each example, Table 1 gives the number of defining inequalities (Column m), the number of variables (Column d), the maximum absolute value of an input coefficient (Column L), the number of polyhedra returned by IntegerSolve (Column m_o), the maximum absolute value of an output coefficient (Column L_o) and whether Hypothesis 1 holds or not (Column ?Hyp).

Recall from Sect. 4.1 of the first paper that Step ($S4$) of the IntegerNormalize procedure can use either the HNF method introduced in Lemma 2 of the first paper, or the method introduced by Pugh in [9]. We implemented both of them. It is important to observe that Pugh's method does not solve systems of linear equations according to our prescribed variable order, in contrast to the HNF method. In fact, Pugh's method determines a variable order dynamically, based on coefficient size considerations. In Table 1, the columns t_{H} and t_{P} correspond to the timings for the HNF and Pugh's method, respectively.

Table 1. Implementation

Example	m	d	L	m_o	L_o	?Hyp	t_H	t_P
Tetrahedron	4	3	1	1	1	Yes	0.695	0.697
Cuboctahedron	14	3	2	1	2	Yes	1.855	1.846
Octahedron	8	3	1	1	1	Yes	1.357	1.357
TruncatedOctahedr.	14	3	3	1	1	Yes	1.995	1.977
TruncatedTetrahedr.	8	3	1	1	1	Yes	1.461	1.468
Presburger 1	3	2	2	1	1	Yes	0.083	0.082
Presburger 2	3	2	20	1	20	Yes	0.184	0.182
Presburger 3	3	2	18	3	4	Yes	0.287	0.260
Presburger 4	3	4	5	2	12	Yes	0.706	0.871
Presburger 6	4	5	89	6	35	Yes	0.893	0.746
Bounded 5	6	3	19	4	224	Yes	16.433	15.091
Bounded 7	8	3	19	3	190	No	138.448	239.637
Bounded 8	4	3	25	5	67	Yes	6.462	3.821
Bounded 9	6	3	18	6	74	No	23.574	16.763
Bounded 10	4	3	15	1	176	Yes	0.559	0.558
Unbounded 2	3	4	10	61	2255	No	0.547	0.600
Unbounded 3	4	4	20	1	20	No	0.981	0.987
Unbounded 4	6	5	2	1	2	No	0.722	0.510
Unbounded 5	5	4	8	1	8	No	1.321	1.319
Unbounded 6	10	4	8	1	8	No	1.494	1.479
P91	12	3	96	5	96	No	19.318	15.458
Sys_1	6	3	15	2	67	Yes	2.413	1.915
Sys_3	8	3	1	1	1	Yes	1.481	1.479
Automatic	8	2	999	1	999	Yes	0.552	0.549
Automatic2	6	4	1	1	2	Yes	1.115	1.113
Automatic3	3	4	1	1	1	Yes	0.130	0.135
Automatic4	3	5	1	1	1	Yes	0.227	0.232

From Table 1, we make a few observations:

1. Hypothesis 1 holds for most examples while it usually does not hold for random ones.
2. For 16 out of 27 examples, IntegerSolve produces a single component, which means that each such input polyhedron has no integer points in its grey shadow; this is, in particular, the case for regular polytopes and for examples from automatic parallelization.
3. When a decomposition consists of more than one component, most of those components are points; for example, the decomposition of Unbounded 2 has 61 components and 46 of them are points.

4. Coefficients of the output polyhedra are usually not much larger than the coefficients of the corresponding input polyhedron.
5. Among the challenging problems, some of them are solved faster when IntegerNormalize is based on HNF (e.g. Bounded 7) while others are solved faster when IntegerNormalize is based on Pugh's method (e.g. Bounded 9) which suggests that having both approaches at hand is useful.

To the best of our knowledge, there are two other published software libraries which are capable of describing the integer points of a polyhedron: one is `4ti2` [1] and the other is `Normaliz` [3]. Both softwares rely on Motzkin's theorem [8] which expresses any rational polyhedron as the Minkowski sum of a rational polytope and a rational cone. Hence, they do not decompose a polyhedron in the sense of our algorithm `IntegerSolve`.

Acknowledgements. The authors would like to thank IBM Canada Ltd (CAS project 880) and NSERC of Canada (CRD grant CRDPJ500717-16), as well as the University of Chinese Academy of Sciences, UCAS Joint PhD Training Program, for supporting their work.

References

1. 4ti2 team. 4ti2–a software package for algebraic, geometric and combinatorial problems on linear spaces. www.4ti2.de
2. Aubry, P., Lazard, D., Moreno Maza, M.: On the theories of triangular sets. J. Symb. Comput. **28**, 105–124 (1999)
3. Bruns, W., Ichim, B., Römer, T., Sieg, R., Söger, C.: Normaliz. Algorithms for rational cones and affine monoids. https://www.normaliz.uni-osnabrueck.de
4. Chen, C., Davenport, J.H., May, J.P., Moreno Maza, M., Xia, B., Xiao, R.: Triangular decomposition of semi-algebraic systems. J. Symb. Comput. **49**, 3–26 (2013)
5. Imbert, J.-L.: Fourier's elimination: which to choose? pp. 117–129 (1993)
6. Karmarkar, N.: A new polynomial-time algorithm for linear programming. In: Proceedings of the Sixteenth Annual ACM Aymposium on Theory of Computing. STOC 1984, pp. 302–311. ACM, New York, NY, USA (1984)
7. Khachiyan, L.: Fourier-motzkin elimination method. In: Floudas, C.A., Pardalos, P.M. (eds.) Encyclopedia of Optimization, pp. 1074–1077. Springer, Heidelberg (2009). doi:10.1007/978-0-387-74759-0_187
8. Motzkin, T.S.: Beiträge zur Theorie der linearen Ungleichungen. Azriel Press, Jerusalem (1936)
9. Pugh, W.: The omega test: a fast and practical integer programming algorithm for dependence analysis. In: Martin, J.L. (ed.), Proceedings Supercomputing 1991, Albuquerque, NM, USA, 18–22 November 1991, pp. 4–13. ACM (1991)
10. Schrijver, A.: Theory of Linear and Integer Programming. Wiley, New York (1986)

Non-linearity and Non-convexity in Optimal Knots Selection for Sparse Reduced Data

Ryszard Kozera[1,2(✉)] and Lyle Noakes[3]

[1] Faculty of Applied Informatics and Mathematics, Warsaw University of Life Sciences-SGGW, Nowoursynowska Str. 159, 02-776 Warsaw, Poland
ryszard.kozera@gmail.com

[2] School of Computer Science and Software Engineering, The University of Western Australia, 35 Stirling Highway, Crawley, Perth, WA 6009, Australia

[3] School of Mathematics and Statistics, The University of Western Australia, 35 Stirling Highway, Crawley, Perth, WA 6009, Australia
lyle.noakes@uwa.edu.au

Abstract. The problem of fitting sparse reduced data in arbitrary Euclidean space is discussed in this work. In our setting, the unknown interpolation knots are determined upon solving the corresponding optimization task. This paper outlines the non-linearity and non-convexity of the resulting optimization problem and illustrates the latter in examples. Symbolic computation within *Mathematica* software is used to generate the relevant optimization scheme for estimating the missing interpolation knots. Experiments confirm the theoretical input of this work and enable numerical comparisons (again with the aid of *Mathematica*) between various schemes used in the optimization step. Modelling and/or fitting reduced sparse data constitutes a common problem in natural sciences (e.g. biology) and engineering (e.g. computer graphics).

Keywords: Reduced sparse data · Optimization · Interpolation · Knots selection · Symbolic computation

1 Problem Formulation

A sequence of interpolation points $\mathcal{M} = \{x_0, x_1, x_2, \ldots, x_n\}$ (here $n \geq 2$) in Euclidean space E^m is called *reduced data* if the corresponding interpolation knots $\{t_i\}_{i=0}^n$ are not given (see e.g. [6,10,12,13,20,23,25,29,30]). Let the class of *admissible curves* γ (denoted by $\mathcal{I}_T$) form the set of piece-wise C^2 curves $\gamma : [0,T] \to E^m$ interpolating $\mathcal{M}$ with the ordered *free unknown admissible knots* $\{t_i\}_{i=0}^n$ satisfying $\gamma(t_i) = x_i$. Here $t_i < t_{i+1}$ are free with, upon re-scaling $t_0 = 0$ and $t_n = T$ set to an arbitrary constant $T > 0$. More precisely, for each choice of ordered knots, the curve γ is assumed to be C^2 except of being only at least C^1 over $\{t_i\}_{i=0}^n$. The analysis to follow is not restricted to a thinner class

V.P. Gerdt et al. (Eds.): CASC 2017, LNCS 10490, pp. 257–271, 2017.
DOI: 10.1007/978-3-319-66320-3_19

of $\gamma \in C^2([t_0, t_n])$ due to the ultimate choice of computational scheme (called herein *Leap-Frog* - see [18,24,27,28]) which effectively deals with the optimization problem (1). However, the computed optimum by Leap-Frog belongs to the tighter class of functions coinciding with $C^2([t_0, t_n])$ as addressed in [17,18].

Assume now, we search for *an optimal* $\gamma_{opt} \in \mathscr{I}_T$ to minimize:

$$\mathscr{J}_T(\gamma) = \int_{t_0}^{T} \|\ddot{\gamma}(t)\|^2 dt = \sum_{i=0}^{n-1} \int_{t_i}^{t_{i+1}} \|\ddot{\gamma}(t)\|^2 dt. \tag{1}$$

The latter defines an infinite dimensional optimization task over $\mathscr{I}_T$. The unknown interpolation knots $\{t_i\}_{i=0}^n$ ($t_0 = 0$ and $0 < t_n = T$ can be fixed) belong to:

$$\Omega_{t_0}^T = \{(t_1, t_2, \ldots, t_{n-1}) \in \mathbb{R}^{n-1} : t_0 = 0 < t_1 < t_2 < \ldots < t_{n-1} < t_n = T < \infty\}. \tag{2}$$

For any affine reparameterization $\phi : [0, T] \to [0, \tilde{T}]$ defined as $\phi(t) = t\tilde{T}/T$ (with $t = \phi^{-1}(s) = sT/\tilde{T}$) $\phi^{-1'} \equiv T/\tilde{T}$ and $\phi^{-1''} \equiv 0$, formula (1), for $\tilde{\gamma}(s) = (\gamma \circ \phi^{-1})(s)$ reads:

$$\begin{aligned}\mathscr{J}_{\tilde{T}}(\tilde{\gamma}) &= \sum_{i=0}^{n-1} \int_{s_i}^{s_{i+1}} \|\ddot{\tilde{\gamma}}(s)\|^2 ds = \frac{T^3}{\tilde{T}^3} \sum_{i=0}^{n-1} \int_{\tilde{t}_i}^{\tilde{t}_{i+1}} \phi^{-1'}(s) \|(\ddot{\gamma} \circ \phi^{-1})(s)\|^2 ds \\ &= \frac{T^3}{\tilde{T}^3} \mathscr{J}_T(\gamma).\end{aligned} \tag{3}$$

Thus, a curve $\gamma_{opt} \in \mathscr{I}_T$ is optimal to $\mathscr{J}_T$ if and only if a corresponding $\tilde{\gamma}_{opt} \in \mathscr{I}_{\tilde{T}}$ is optimal for $\mathscr{J}_{\tilde{T}}$. Hence $t_n = T$ can be taken as arbitrary, and with the additional affine mapping $\phi(t) = t - t_0$, one can also set $t_0 = 0$.

Recall now *a cubic spline interpolant* $\gamma_{\mathscr{T}}^{C_i} = \gamma_{\mathscr{T}}^{C}|_{[t_i, t_{i+1}]}$ (see e.g. [3]), which for given temporarily fixed admissible interpolation knots $\mathscr{T} = (t_0, t_1, \ldots, t_{n-1}, t_n)$ reads as:

$$\gamma_{\mathscr{T}}^{C_i}(t) = c_{1,i} + c_{2,i}(t - t_i) + c_{3,i}(t - t_i)^2 + c_{4,i}(t - t_i)^3, \tag{4}$$

and fulfills (for $i = 0, 1, 2, \ldots, n-1$; $c_{j,i} \in \mathbb{R}^m$, where $j = 1, 2, 3, 4$)

$$\gamma_{\mathscr{T}}^{C_i}(t_{i+k}) = x_{i+k}, \quad \dot{\gamma}_{\mathscr{T}}^{C_i}(t_{i+k}) = v_{i+k}, \quad k = 0, 1$$

with the assumed unknown velocities $v_0, v_1, v_2, \ldots, v_{n-1}, v_n \in \mathbb{R}^m$. The coefficients $c_{j,i}$ (with $\Delta t_i = t_{i+1} - t_i$) are defined as follows:

$$\begin{aligned} &c_{1,i} = x_i, \qquad c_{2,i} = v_i, \\ &c_{4,i} = \frac{v_i + v_{i+1} - 2\frac{x_{i+1} - x_i}{\Delta t_i}}{(\Delta t_i)^2}, \qquad c_{3,i} = \frac{\frac{(x_{i+1} - x_i)}{\Delta t_i} - v_i}{\Delta t_i} - c_{4,i} \Delta t_i. \end{aligned} \tag{5}$$

Adding $n-1$ conditions $\ddot{\gamma}_{\mathscr{T}}^{C_{i-1}}(t_i) = \ddot{\gamma}_{\mathscr{T}}^{C_i}(t_i)$ over $x_1, x_2, \ldots, x_{n-1}$ yields m tridiagonal linear systems (see [3]) of $n-1$ equations in $n+1$ vector unknowns

$v_0, v_1, \ldots, v_n \in \mathbb{R}^m$:

$$v_{i-1}\Delta t_i + 2v_i(\Delta t_{i-1} + \Delta t_i) + v_{i+1}\Delta t_{i-1} = b_i,$$
$$b_i = 3\left(\Delta t_i \frac{x_i - x_{i-1}}{\Delta t_{i-1}} + \Delta t_{i-1}\frac{x_{i+1} - x_i}{\Delta t_i}\right). \tag{6}$$

In case of the so-called *natural cubic spline interpolant* (denoted as $\gamma^C_{\mathscr{T}} = \gamma^{NS}_{\mathscr{T}}$), two extra constraints involving v_0 and v_n stipulate that $\ddot{\gamma}^C_{\mathscr{T}}(0) = \ddot{\gamma}^C_{\mathscr{T}}(T) = 0$ which leads to:

$$2v_0 + v_1 = 3\frac{x_1 - x_0}{\Delta t_0}, \quad v_{n-1} + 2v_n = 3\frac{x_n - x_{n-1}}{\Delta t_{n-1}}. \tag{7}$$

The resulting m linear systems (i.e. (6) and (7)), each of size $(n+1)\times(n+1)$, determine unique vectors $v_0, v_1, v_2, \ldots, v_n$ (see [3, Chap. 4]), which when fed into (5) and then passed to (4) determine explicitly *a natural cubic spline* $\gamma^{NS}_{\mathscr{T}}$ (with fixed $\mathscr{T}$). Visibly all computed velocities $\{v_i\}_{i=0}^m$ (and, thus, $\gamma^{NS}_{\mathscr{T}}$) with the aid of the above procedure depend in fact on the interpolation knots $\{t_i\}_{i=0}^m$ and fixed data $\mathscr{M}$. It is well known (see e.g. [3]) that if the respective knots $\{t_i\}_{i=0}^m$ are frozen the optimization task (1) is minimized by a unique natural spline $\gamma^{NS}_{\mathscr{T}}$ defined by $\{t_i\}_{i=0}^n$ and $\mathscr{M}$. Therefore, upon *relaxing all internal knots* $\{t_i\}_{i=1}^{n-1}$ in (1) (for arbitrarily fixed terminal knots to e.g. $t_0 = 0$ and $t_n = T$) one arrives at the following (see [3,17–19]):

Theorem 1. *For a given $\mathscr{M}$ with points in Euclidean space E^m, the subclass of natural splines $\mathscr{I}^{NS} \subset \mathscr{I}_T$ satisfies*

$$\min_{\gamma\in\mathscr{I}_T} \mathscr{J}_T(\gamma) = \min_{\gamma^{NS}\in\mathscr{I}^{NS}} \mathscr{J}_T(\gamma^{NS}), \tag{8}$$

which reduces to the finite dimensional optimization in $\hat{\mathscr{T}} = (t_1, t_2, \ldots, t_{n-1})$ over non-compact $\Omega_{t_0}^T$ introduced in (2):

$$\begin{aligned}\mathscr{J}_T(\gamma^{NS}_{opt}) &= \min_{\hat{\mathscr{T}}\in\Omega_{t_0}^T} \mathscr{J}^F_T(t_1, t_2, \ldots, t_{n-1})\\ &= \min_{\hat{\mathscr{T}}\in\Omega_{t_0}^{T_c}} 4\sum_{i=0}^{n-1}\Big(\frac{-1}{(\Delta t_i)^3}(-3\|x_{i+1} - x_i\|^2 + 3\langle v_i + v_{i+1}|x_{i+1} - x_i\rangle\Delta t_i\\ &\qquad -(\|v_i\|^2 + \|v_{i+1}\|^2 + \langle v_i|v_{i+1}\rangle)(\Delta t_i)^2\Big),\end{aligned} \tag{9}$$

for which at least one global minimum $\hat{\mathscr{T}}_{opt} = (t_1^{opt}, t_2^{opt}, \ldots, t_{n-1}^{opt}) \in \Omega_{t_0}^T$ exists.

We take here the computed optimal values of $\hat{\mathscr{T}}_{opt}$, as estimates $\{\hat{t}_i\}_{i=0}^m \approx \{t_i\}_{i=0}^m$. In this paper, we demonstrate strong non-linearity and non-convexity effects built-in the optimization scheme (9). The relevant examples and theoretical insight is supplemented to justify the latter. Sufficient conditions for convexity (or unimodality) of (9) are proved at least for $n = 2$. The complexity of the optimization scheme (9) not only impedes its theoretical analysis but also impacts on the choice of feasible numerical scheme handling computationally (9). Finally, this work is supplemented with illustrative examples and numerical tests used to fit input sparse reduced data $\mathcal{M}$ for various n and $m = 2, 3$.

Related work on fitting reduced data $\mathcal{M}$ (sparse or dense) can also be found in [8,9,15,16,21,22,26,33,34]. Some applications in computer vision and graphics, image processing, engineering, physics, and astronomy are discussed e.g. in [1,2,5,7,11,21,31,32].

2 Non-Linearity of $\mathcal{J}_T^F$ and Numerical Difficulties

First we demonstrate a high non-linearity featuring the optimization task (9). This is accomplished by generating an explicit formula for (9) whose complexity is substantial even for n small and gets complicated for n incremented. The latter is illustrated by the next two examples followed by pertinent computational tests.

Example 1. Consider four data points (i.e. here $n = 3$) $\mathcal{M} = \{x_0, x_1, x_2, x_3\}$ in E^m. Formula for $\mathcal{J}_T^F$ (see (9)) reads here as $\mathcal{J}_{T_c}^{F,3}(\hat{\mathcal{T}}) = \mathcal{J}_0^3 + \mathcal{J}_1^3 + \mathcal{J}_2^3$ (with $\hat{\mathcal{T}} = (t_0, t_1, t_2, t_3)$ and $t_0 = 0$ and e.g. $t_3 = T = T_c$ - see (12)), where

$$\begin{aligned}
\mathcal{J}_0^3 &= \frac{1}{(t_0 - t_1)^3}(-3\|x_0\|^2 - 3\|x_1\|^2 + (t_0 - t_1)(3\langle v_0|x_0\rangle - 3\langle v_0|x_1\rangle + 3\langle v_1|x_0\rangle \\
&\quad -3\langle v_1|x_1\rangle + (\|v_0\|^2 + \|v_1\|^2 + \langle v_0|v_1\rangle)(t_1 - t_0)) + 6\langle x_0|x_1\rangle), \\
\mathcal{J}_1^3 &= \frac{1}{(t_1 - t_2)^3}(-3\|x_1\|^2 - 3\|x_2\|^2 + (t_1 - t_2)(3\langle v_1|x_1\rangle - 3\langle v_1|x_2\rangle + 3\langle v_2|x_1\rangle \\
&\quad -3\langle v_2|x_2\rangle + (\|v_1\|^2 + \|v_2\|^2 + \langle v_1|v_2\rangle)(t_2 - t_1)) + 6\langle x_1|x_2\rangle), \\
\mathcal{J}_2^3 &= \frac{1}{(t_2 - t_3)^3}(-3\|x_2\|^2 - 3\|x_3\|^2 + (t_2 - t_3)(3\langle v_2|x_2\rangle - 3\langle v_2|x_3\rangle + 3\langle v_3|x_2\rangle \\
&\quad -3\langle v_3|x_3\rangle + (\|v_2\|^2 + \|v_3\|^2 + \langle v_2|v_3\rangle)(t_3 - t_2)) + 6\langle x_2|x_3\rangle). \qquad (10)
\end{aligned}$$

The missing velocities $\{v_0, v_1, v_2, v_3\}$ for natural spline $\gamma_{\mathcal{T}}^{NS}$ (see (4)) are determined here by the following four matrix equations, with $i = 1, \ldots, m$ (see (6) and (7))

$$\begin{pmatrix} 2 & 1 & 0 & 0 \\ t_2 - t_1 & 2(t_2 - t_0) & t_1 - t_0 & 0 \\ 0 & t_3 - t_2 & 2(t_3 - t_1) & t_2 - t_1 \\ 0 & 0 & 1 & 2 \end{pmatrix} \begin{pmatrix} v_0^i \\ v_1^i \\ v_2^i \\ v_3^i \end{pmatrix} = \begin{pmatrix} 3\frac{x_1^i - x_0^i}{t_1 - t_0} \\ 3(\frac{(t_2-t_1)(x_1^i - x_0^i)}{t_1 - t_0} + \frac{(t_1-t_0)(x_2^i - x_1^i)}{t_2 - t_1}) \\ 3(\frac{(t_3-t_2)(x_2^i - x_1^i)}{t_2 - t_1} + \frac{(t_2-t_1)(x_3^i - x_2^i)}{t_3 - t_2}) \\ 3\frac{x_3^i - x_2^i}{t_3 - t_2} \end{pmatrix}$$

yielding (with the aid of symbolic computation in *Mathematica* - see [35]) a unique solution. For the sake of this example, we consider exclusively the case of $m = 1$. This can be easily extended to $m > 1$, since both square of norms and dot products (appearing in non-reduced form of $\mathcal{J}_{T_c}^F(\hat{\mathcal{T}})$) are additive by each vector component. Upon substituting computed velocities from the last matrix equations into $\mathscr{J}^{F,3}_{T_c}$ (as previously we set $t_0 = 0$ and $t_3 = T_c$ - see (12)) and taking into account that $m = 1$, *Mathematica FullSimplify* (see [35]) function yields an explicit formula for

$$\mathscr{J}^{F,3}_{T_c}(t_1,t_2) = N_3(t_1,t_2)/(t_1^2(t_1-t_2)^2(t_2-T_c)^2((t_1+t_2)^2-4T_ct_2)),$$

where

$$\begin{aligned}N_3(t_1,t_2) = {}&(3(-T_c^3t_2^2(x_0-x_1)^2+2T_c^2t_2^3(x_0-x_1)^2+T_ct_2^4(x_0-x_1)(x_1-x_0)\\&+t_1^3(-T_c(x_0+x_1-2x_2)+t_2(x_0+x_1-2x_3))(T_c(x_2-x_0)+t_2(x_0-x_3))\\&-t_1^2(T_c^3(x_0-x_2)^2-3T_ct_2^2(x_0-x_2)(x_0-x_3)\\&+t_2^3(x_0-x_3)(2x_0-x_2-x_3))+t_1(2T_c^3t_2(x_0-x_1)(x_0-x_2)\\&-3T_c^2t_2^2(x_0-x_1)(x_0-x_2)+t_2^4(x_0-x_1)(x_0-x_3)\\&-T_ct_2^3(x_0-x_1)(x_2-x_3))-t_1^4(T_c(x_2-x_0)+t_2(x_0-x_3))(x_2-x_3))).\end{aligned}$$

Note that $N_3(t_1,t_2)$ is a 5th order polynomial in t_1 and t_2. □

Example 2. Let five data points (i.e. here $n = 4$) $\mathscr{M} = \{x_0, x_1, x_2, x_3, x_4\}$ be given in E^m. Formula (9) reads here $\mathscr{J}^{F,4}_{T_c}(\hat{\mathscr{T}}) = \mathscr{J}^4_0 + \mathscr{J}^4_1 + \mathscr{J}^4_2 + \mathscr{J}^4_3$ (for $\hat{\mathscr{T}} = (t_0, t_1, t_2, t_3, t_4)$ with $t_0 = 0$ and $t_4 = T_c$ - see (12)), where $\mathscr{J}^4_k = \mathscr{J}^3_k$, for $k = 0, 1, 2$ (see (10)) and

$$\begin{aligned}\mathscr{J}^4_3 = {}&\frac{1}{(t_3-t_4)^3}(-3\|x_3\|^2-3\|x_4\|^2+(t_3-t_4)(3\langle v_4|x_3\rangle-3\langle v_3|x_4\rangle+3\langle v_4|x_3\rangle\\&-3\langle v_4|x_4\rangle+(\|v_3\|^2+\|v_4\|^2+\langle v_3|v_4\rangle)(t_4-t_3))+6\langle x_3|x_4\rangle).\end{aligned}$$

Again the missing velocities $\{v_0, v_1, v_2, v_3, v_4\}$ for the natural spline $\gamma^{NS}_{\mathscr{T}}$ defined by (4) are determined here by five matrix equations, with $i = 1, \ldots, m$ (see (6) and (7)):

$$\begin{pmatrix} 2 & 1 & 0 & 0 & 0\\ t_2-t_1 & 2(t_2-t_0) & t_1-t_0 & 0 & 0\\ 0 & t_3-t_2 & 2(t_3-t_1) & t_2-t_1 & 0\\ 0 & 0 & t_4-t_3 & 2(t_4-t_2) & t_3-t_2\\ 0 & 0 & 0 & 1 & 2\end{pmatrix}\begin{pmatrix} v_0^i\\ v_1^i\\ v_2^i\\ v_3^i\\ v_4^i\end{pmatrix} = B_i, \tag{11}$$

where

$$B_i = \begin{pmatrix} 3\frac{x_1^i-x_0^i}{t_1-t_0}\\ 3(\frac{(t_2-t_1)(x_1^i-x_0^i)}{t_1-t_0}+\frac{(t_1-t_0)(x_2^i-x_1^i)}{t_2-t_1})\\ 3(\frac{(t_3-t_2)(x_2^i-x_1^i)}{t_2-t_1}+\frac{(t_2-t_1)(x_3^i-x_2^i)}{t_3-t_2})\\ 3(\frac{(t_4-t_3)(x_3^i-x_2^i)}{t_3-t_2}+\frac{(t_3-t_2)(x_4^i-x_3^i)}{t_4-t_3})\\ 3\frac{x_4^i-x_3^i}{t_4-t_3}\end{pmatrix}.$$

Again the system (11) renders a unique solution $\{v_0, v_1, v_2, v_3, v_4\}$ (found e.g. upon using *Mathematica* software - see [35]). As previously only the case of $m = 1$

is here considered. Upon substituting computed velocities from (11) into $\mathcal{J}_{T_c}^{F,4}$ and setting $t_0 = 0$ and $t_4 = T_c$ (see (12)) *Mathematica FullSimplify* function yields an explicit formula for $\mathcal{J}_{T_c}^{F,4}(t_1, t_2, t_3) =$

$$\frac{N_4(t_1,t_2,t_3)}{4t_1^2(t_1-t_2)^2(t_2-t_3)^2(t_3-T_c)^2(t_2(t_3-t_1)(t_1+2t_2+t_3)+T_c((t_1+t_2)^2-4t_2t_3))}.$$

It can be checked that $N_4(t_1, t_2, t_3)$ is an 8th order polynomial in t_1, t_2 and t_3. We omit here to present a full explicit formula for $N_4(t_1, t_2, t_3)$ since it takes more than one A4 format page size. □

Examples 1 and 2 indicate the growing complexity of the *non-linearity* in (9) while n increases. Thus, in a search for global minimum of (9), any numerical optimization scheme relying on derivative computation (irrespectively of an initial guess) faces the computational difficulties for n getting bigger. The latter is demonstrated in the next Example 3 for $n = 7$, where *Mathematica FindMinimum* applied with *Newton Method* fails (see [35]). Similar effects appear when *Mathematica Minimize[f,constraints,variables]* is invoked (see [35]) which works efficiently for both minimized function and imposed constraints (such as inequality or equations) expressed as polynomials. The latter happens with the numerator of the derivative of (9). To alleviate this problem and to efficiently optimize (9) we invoke first a multidimensional version of the *Secant Method* (not relying on derivative computation) given in *Mathematica* software e.g. for two free variables as $FindMinimum[f, \{\{var1, 2num1\}, \{var, 2num2\}\}]$ - see [35]. Its super-linear convergence order (e.g. for $m = 1$ equal to $((1+\sqrt{5})/2) \approx 1.618$) though slower than Newton quadratic rate, makes it still both faster and computationally feasible as opposed to most standard optimization techniques based on derivative calculation. In the last section of this paper, we compare the *Secant Method* with a *Leap-Frog Algorithm* (see [18,27,28])). One of the advantages of *Leap-Frog* over the *Secant Method* is a faster execution time (see also [17,18]).

In order to set up a computationally feasible numerical optimization scheme a good initial guess is needed. In particular, for the *Secant Method* for each free variable, two numbers are needed to be selected. A possible choice is the so-called *cumulative chord* $\mathcal{T}_c = \{t_i^c\}_{i=0}^n$ (see e.g. [13,14,20,25]):

$$t_0^c = 0, \qquad t_{i+1}^c = t_i^c + \|x_{i+1} - x_i\|, \quad i = 0, \ldots, n-1, \tag{12}$$

with $T^c = \sum_{i=0}^{n-1} \|x_{i+1} - x_i\|$. Cumulative chord parameterization in a normalized form $\mathcal{T}_{cc}$ reads $t_i^{cc} = t_i^c / T^c$ (for $i = 0, 1, \ldots, n$). Here an additional assumption about reduced data $\mathcal{M}$ i.e. $x_i \neq x_{i+1}$ is also drawn. For the *Secant Method* and each free knot t_i appearing in (9) (here $i = 1, 2, \ldots, n-1$) we choose two starting numbers as $t_i^c - \varepsilon$ and $t_i^c + \varepsilon$, with some prescribed small value for ε.

The next example illustrates expected computational difficulties in optimizing (9).

Example 3. (a) Consider four 2D reduced data points (i.e. here $n = 3$):

$$\mathcal{M}_3 = \{(-4, 0), (-0.5, -4), (0.5, -3), (-0.5, 4)\}.$$

The cumulative chord knots (see (12)) based on $\mathcal{M}_3$ coincide with $\mathcal{T}_c = \{0, 5.31507, 6.72929, 13.8004\}$. In fact, here we have only two free variables $\{t_1, t_2\}$ corresponding to the unknown knots at points x_i ($i = 1, 2$). *FindMinimum* (*Newton Method*) applied to (9) with the initial guess as cumulative chords $\mathcal{T}_c$ yields the following optimal knots (with optimal energy value $\mathcal{J}^F_{T_c}(\hat{\mathcal{T}}_{opt}) = 0.741614$):

$$\mathcal{T}_{opt} = \{0, 5.3834, 8.2118, 13.8004\}, \tag{13}$$

where $\hat{\mathcal{T}}_{opt} = \{5.3834, 8.2118\}$. The execution time $T^N_{\mathcal{M}_3} = 3.012858\,\text{s}$. *FindMinimum* for the *Secant Method* (with two numbers associated to each free variables taken as $\varepsilon = \pm 0.5$ variation of (12)) gives exactly the same optimal knots (13) (with the same $\mathcal{J}^F_{T_c}(\hat{\mathcal{T}}_{opt}) = 0.741614$) but shorter execution time $T^S_{\mathcal{M}_3} = 0.647756\,\text{s}$. Finally, *Minimize* with constraints $0 < t_1 < t_2 < 13.8004$ gives optimal knots (13) with $\mathcal{J}^F_{T_c}(\hat{\mathcal{T}}_{opt}) = 0.741614$. The execution time amounts here $T^M_{\mathcal{M}_3} = 9.229526\,\text{s} > T^N_{\mathcal{M}_3} > T^S_{\mathcal{M}_3}$.

(b) Consider now six 2D reduced data points (i.e. here $n = 5$):

$$\mathcal{M}_5 = \{(0, 0), (-0.5, -4), (0.5, -4), (-0.5, 4), (0.5, 4), (-1, 3.8)\}.$$

The resulting cumulative chord knots (based on (12) and $\mathcal{M}_5$) are equal here to $\mathcal{T}_c = \{0, 4.03113, 5.03113, 13.0934, 14.0934, 15.6067\}$ with internal knots $\hat{\mathcal{T}}_c = \{4.03113, 5.03113, 13.0934, 14.0934\}$. In this case, there are four free variables $\{t_1, t_2, t_3, t_4\}$ corresponding to the unknown knots at points x_i ($i = 1, \ldots, 4$). *FindMinimum* (*Newton Method*) applied to (9) with initial guess as $\mathcal{T}_c$ yields the following optimal knots (with optimal energy value $\mathcal{J}^F_{T_c}(\hat{\mathcal{T}}_{opt}) = 4.65476$):

$$\mathcal{T}_{opt} = \{0, 2.9185, 5.12397, 11.1964, 13.507\}, \tag{14}$$

where $\hat{\mathcal{T}}_{opt} = \{2.9185, 5.12397, 11.1964\}$. The execution time $T^N_{\mathcal{M}_5} = 29.946006\,\text{s}$. *FindMinimum* (*Secant Method*) (here again $\varepsilon = \pm 0.5$ is added to cumulative chord initial guess along each free knot t_i) yields exactly the same optimal knots (14) (with $\mathcal{J}^F_{T_c}(\hat{\mathcal{T}}_{opt}) = 4.65476$) and again shorter execution time $T^S_{\mathcal{M}_5} = 6.385922\,\text{s}$. As previously, *Minimize* with constraints $0 < t_1 < t_2 < t_3 < t_4 < 13.507$ yields optimal knots (14) and $\mathcal{J}^F_{T_c}(\hat{\mathcal{T}}_{opt}) = 4.65476$. The execution time here reads $T^M_{\mathcal{M}_5} = 358.390915\,\text{s} >> T^N_{\mathcal{M}_5} > T^S_{\mathcal{M}_5}$.

(c) Finally, consider now eight 2D reduced data points (i.e. here $n = 7$):

$$\mathcal{M}_6 = \{(0, 0), (-0.5, -4), (0.5, -4), (-0.5, 4), \\ (0.5, 4), (-1, 3.8), (0.3, 0.3), (0.5, 0.5)\}.$$

By (12) $\mathcal{T}_c = \{0, 4.03113, 5.03113, 13.0934, 14.0934, 15.6067, 19.3403, 19.6231\}$. As previously $\hat{\mathcal{T}}_c = \{4.03113, 5.03113, 13.0934, 14.0934, 15.6067, 19.3403\}$. For both optimization schemes *FindMinimum* (*Newton Method*) and *Minimize* no result was reported within 60 min. *FindMinimum* (*Secant Method*) (with as previously $\varepsilon = \pm 0.5$ variations of cumulative chords for each free variable) yields optimal knots (with energy $\mathcal{J}^F_{T_c}(\hat{\mathcal{T}}_{opt}) = 8.27118$):

$$\mathcal{T}_{opt} = \{0, 2.67713, 4.69731, 10.3221, 12.3943, 14.8132, 19.0316, 19.6231\},$$

where $\mathcal{T}_{opt} = \{2.67713, 4.69731, 10.3221, 12.3943, 14.8132\}$. The execution time is $T^S_{\mathcal{M}_7} = 35.708519$ s. □

The above experiments illustrate that for $n \geq 7$, *FindMinimum* (*Secant Method*) offers a feasible computational scheme to optimize (9). In Sect. 4 of this paper, we compare the performance of already discussed *Secant Method* with *Leap-Frog Algorithm* (see [17,18,27,28]).

3 Non-Convexity of $\mathcal{J}^F_T$

We demonstrate in this section that $\mathcal{J}^F_{T_c}$ introduced in (9) *may not be convex*. In doing so, a simple degenerate case of (9) with $n = 2$ is examined. Three points $\mathcal{M} = \{x_0, x_1, x_2\}$ are admitted with one relaxed internal knot $t_1 \in [t_0 = 0, t_2 = T_c]$ (see (12)).

Example 4. For $n = 2$ and arbitrary natural $m \geq 1$ (using *Mathematica* package), by (9) the energy $\mathcal{J}^F_{T_c}(t_1) = (T_c - t_0)^{-3}(\tilde{\mathcal{E}}_{deg} \circ \phi^{-1})(t_1)$, where $\tilde{\mathcal{E}}_{deg}(s_1) = 3\|\frac{x_0 - x_1}{s_1} + \frac{x_2 - x_1}{1 - s_1}\|^2$ - here $t_1 \in (t_0, t_2 = T_c)$, and $s_1 = \phi(t_1) = (t - t_0)(T_c - t_0)^{-1} \in (0, 1)$. Obviously since $\phi^{-1'} \equiv T_c - t_0 > 0$ and $\phi^{-1''} \equiv 0$ the convexity (non-convexity) of $\tilde{\mathcal{E}}_{deg}$ is inherited by $\mathcal{J}^F_{T_c}$. Take now for $m = 1$ the following points $x_0 = -1$, $x_1 = 0$ and $x_2 = 20$ (here $(x_0 - x_1)(x_2 - x_1) = -20 < 0$). The graph of the energy $\tilde{\mathcal{E}}_{deg}(s_1) = 3(\frac{-1}{s_1} + \frac{20}{1 - s_1})^2$ over the interval $(0, 1)$ is plotted in Fig. 1(a). The non-convexity is better visible in Fig. 1(b) with the graph of the energy $\tilde{\mathcal{E}}_{deg}$ localized over the sub-interval $(0.05, 0.35)$. Finally the change of sign in the second derivative $\tilde{\mathcal{E}}''_{deg}$ is also illustrated in Fig. 1(c). In fact, $\tilde{\mathcal{E}}''_{deg}(0.21) = 1338.7$ and $\tilde{\mathcal{E}}''_{deg}(0.20) = -1640.63$. Thus, $\tilde{\mathcal{E}}_{deg}$ is not convex which also implies non-convexity of $\mathcal{J}^F_{T_c}$ in the general case. □

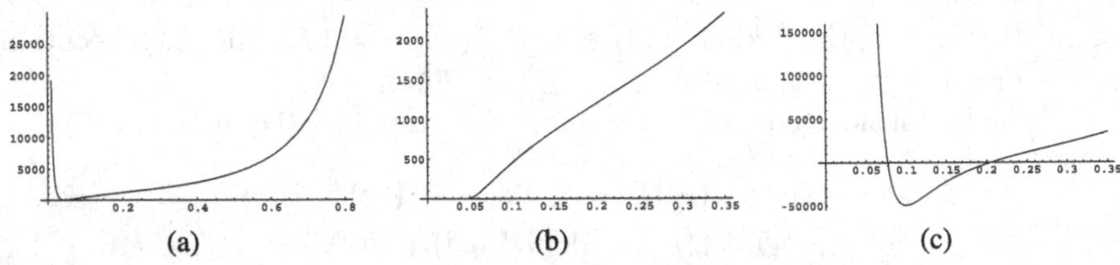

Fig. 1. The graph of the non-convex energy $\tilde{\mathcal{E}}_{deg}$ for $x_0 = -1$, $x_1 = 0$, and $x_2 = 20$ (a) over the interval $(0, 1)$, (b) in the proximity of non-convexity sub-interval $(0.05, 0.35)$ (c) and the graph of the corresponding changing signs second derivative $\tilde{\mathcal{E}}''_{deg}$ in the proximity of $(0.05, 0.35)$

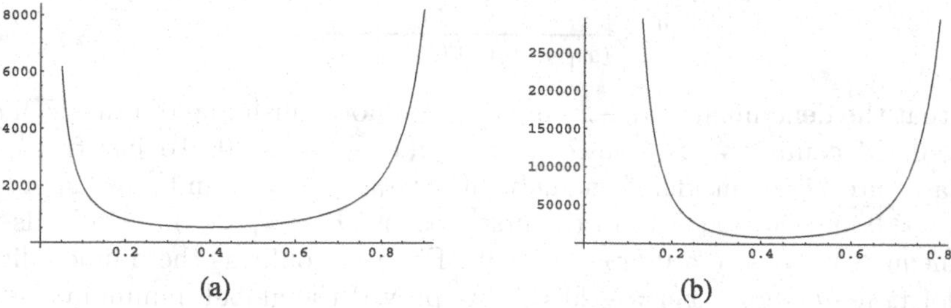

Fig. 2. The graph of the convex energy $\tilde{\mathscr{E}}_{deg}$ for $x_0 = 2$, $x_1 = 0$, and $x_2 = 5$ (a) over the interval $(0,1)$, (b) and the graph of the corresponding second derivative $\tilde{\mathscr{E}}''_{deg} \geq 0$ over $(0,1)$

The next example formulates *sufficient conditions to enforce convexity* of $\mathscr{J}^F_{T_c}$, but only for $m = 1$ and $n = 2$. The latter can be extended to the general case of $m \geq 1$ and $n = 2$. Such general case is here omitted due to the paper length limitation.

Example 5. (i) For $m = 1$ it is easy to show that $\tilde{\mathscr{E}}_{deg}$ is convex (and so thus $\mathscr{J}^F_{T_c}$) if $(x_0 - x_1)(x_2 - x_1) \geq 0$ - under this constraint, we have exactly one critical point for $\tilde{\mathscr{E}}_{deg}$. Indeed, recalling that $\tilde{\mathscr{E}}_{deg}(s_1) = 3f^2(s_1)$ with $f(s_1) = \frac{x_0 - x_1}{s_1} + \frac{x_2 - x_1}{1 - s_1}$ it suffices to show that f is either convex and $f \geq 0$ or it is concave and $f \leq 0$, for $s_1 \in (0,1)$. Indeed the latter combined with $\tilde{\mathscr{E}}''_{deg} = 3(f^2)'' = 6(f')^2 + 6ff''$ yields the convexity of $\mathscr{J}^F_{T_c}$ given non-negativity of ff'' over $s_1 \in (0,1)$ which follows from $(x_0 - x_1)(x_2 - x_1) \geq 0$ applied both to $f''(s_1) = \frac{x_0 - x_1}{s_1^3} + \frac{x_2 - x_1}{(1 - s_1)^3}$ and f. Figure 2(a) shows that convexity of $\tilde{\mathscr{E}}_{deg}(s_1) = 3(\frac{2}{s_1} + \frac{5}{1 - s_1})^2$ indeed follows for $x_0 = 2$, $x_1 = 0$ and $x_2 = 5$ with $(x_0 - x_1)(x_2 - x_1) \geq 0$ clearly fulfilled. As expected $\tilde{\mathscr{E}}''_{deg} \geq 0$ over $(0,1)$ - see Fig. 2(b). The corresponding sufficient conditions guaranteeing the convexity of $\tilde{\mathscr{E}}_{deg}$ for $m \geq 1$ and $n = 2$ can also be formulated (though omitted here) - see [19]. As it turns out, the vector generalization $\langle x_0 - x_1 | x_2 - x_1 \rangle \geq 0$ of scalar inequality $(x_0 - x_1)(x_2 - x_1) \geq 0$ assures the convexity of $\tilde{\mathscr{E}}_{deg}$ and thus of $\mathscr{J}^F_{T_c}$.

(ii) In case of scalar data (i.e. when $m = 1$) if $(x_0 - x_1)(x_2 - x_1) < 0$ holds then the existence of exactly one critical point (and thus one global minimum - see Theorem 1) of $\tilde{\mathscr{E}}_{deg}$ (and so of $\mathscr{J}^F_{T_c}$) follows, which yields the *unimodality* of $\tilde{\mathscr{E}}_{deg} = 3f^2$ (see [4]). Indeed, assume that $(x_0 - x_1)(x_2 - x_1) < 0$. Since now $x_0 \neq x_1$ and $x_1 \neq x_2$ we have $x_0 - x_1 \neq 0$ and $x_2 - x_1 \neq 0$. To show unimodality of f^2 we need to prove the existence of exactly one critical point $s_1 \in (0,1)$ satisfying $(f^2)'(s_1) = 2f(s_1)f'(s_1) = 0$. Taking into account $(x_0 - x_1)(x_2 - x_1) < 0$ we have that $f'(s_1) = -\frac{x_0 - x_1}{s_1^2} + \frac{x_2 - x_1}{(1 - s_1)^2}$ is either always positive or negative over $(0,1)$. Hence for unimodality of f^2 it suffices to show that $f(s_1) = 0$ has one root $s_1^0 \in (0,1)$ defining a unique global minimum of $f^2 = 0$ (and of $\tilde{\mathscr{E}}_{deg}$). In this case as $\tilde{\mathscr{E}}_{deg}(s_1) = 3f^2(s_1)$ the energy $\tilde{\mathscr{E}}_{deg}$ also vanishes at s_1^0. The latter may not be the case for the convexity case, since another factor $f'(s_1) = 0$ may contribute to $(f^2)'(s_1) = 0$. A simple inspection shows that

$$s_1^0 = \frac{(x_0 - x_1)}{(x_0 - x_1) - (x_2 - x_1)}. \tag{15}$$

Note that the denominator $x_0 - x_2$ in (15) does not vanish due to $(x_0 - x_1)(x_2 - x_1) < 0$. Of course, $s_1^0 > 0$ since $(x_0 - x_1)(x_2 - x_1) < 0$. To justify $s_1^0 < 1$ two cases are here considered, namely either $x_0 - x_1 > 0$ and $x_2 - x_1 < 0$ or $x_0 - x_1 < 0$ and $x_2 - x_1 > 0$. In the first (second) case $s_1^0 < 1$ in (15) leads to a true inequality $x_2 - x_1 < 0$ ($x_2 - x_1 > 0$). Figure 1 confirms the unimodality of $\mathscr{E}_{deg}$ for $(x_0 - x_1)(x_2 - x_1) = -20 < 0$. As proved the global minimum of (9) is attained at $s_1^0 = 1/21 \approx 0.047619$ nullifying $\mathscr{J}^F_{T_c}$.

Note that in case of convexity (enforced by $(x_0 - x_1)(x_2 - x_1) \geq 0$), the unique global minimum s_1^0 can also be found in analytic form. Indeed as $x_0 \neq x_1$ and $x_1 \neq x_2$ it suffices to assume a stronger inequality $(x_0 - x_1)(x_2 - x_1) > 0$. The latter results in either $f > 0$ or $f < 0$. Consequently for $6f'f$ to vanish we need to solve $f'(s_1) = 0$ over $(0, 1)$ which leads to $s_1^0 = \frac{\sqrt{|x_0 - x_1|}}{\sqrt{|x_2 - x_1|} + \sqrt{|x_0 - x_1|}} \in (0, 1)$. Thus, f^2 is unimodal (and so $\tilde{\mathscr{E}}_{deg}$) since $(f^2)' = 2ff'$ vanishes at exactly one point $s_1^0 \in (0, 1)$. The unimodality of $\tilde{\mathscr{E}}_{deg}$ can also be proved in case of $\langle x_0 - x_1 | x_2 - x_1 \rangle < 0$ for arbitrary data with $m \geq 1$ and $n = 2$. □

4 Numerical Experiments for Fitting Sparse Reduced Data

All experiments are conducted in *Mathematica* - see [35]. The numerical tests compare the *Leap-Frog* algorithm (see [17,18]) with the *Secant Method* both used to optimize (9). Only sparse reduced data points $\mathscr{M}$ in $E^{2,3}$ are admitted here, though the entire setting is applicable for arbitrary m, i.e. for reduced data $\mathscr{M}$ in arbitrary Euclidean space.

The first example admits reduced data $\mathscr{M}$ in E^2 (i.e. for $m = 2$).

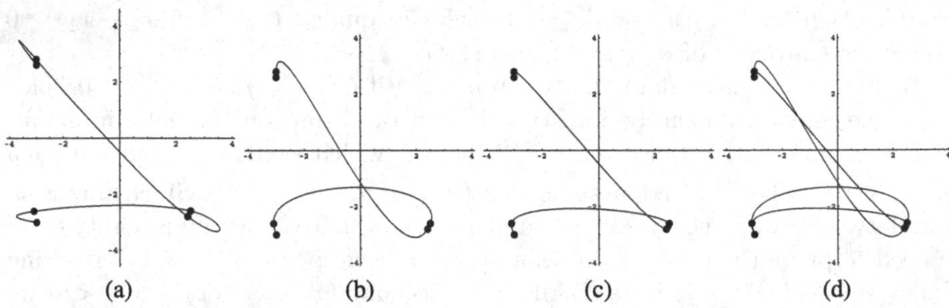

Fig. 3. Natural splines interpolating data points $\mathscr{M}_{2D3}$ (a) $\gamma^{NS}_{\mathscr{T}_{uni}}$ with uniform knots $\mathscr{T}_{uni}$, (b) $\gamma^{NS}_{\mathscr{T}_c}$ with cumulative chords $\mathscr{T}_c$, (c) $\gamma^{NS}_{\mathscr{T}^{LF}_{opt}}$ with optimal knots $\mathscr{T}^{LF}_{opt} = \mathscr{T}^{SM}_{opt}$ (thus $\gamma^{NS}_{\mathscr{T}^{LF}_{opt}} = \gamma^{NS}_{\mathscr{T}^{SM}_{opt}}$) (d) $\gamma^{NS}_{\mathscr{T}^{LF}_{opt}}$ and $\gamma^{NS}_{\mathscr{T}_c}$ plotted together

Example 6. Assume for $n = 6$, the following 2D points (see dotted points in Fig. 3):

$$\mathscr{M}_{2D3} = \{(-3,-3),(-3.1,-2.6),(2.5,-2.6),(2.4,-2.8),(-3,2.8),(-3,2.6)\}.$$

The *uniform interpolation knots*, $\{\hat{t}_i = \frac{i}{6}T_c\}_{i=0}^6$ (rescaled to T_c - see (12)) taken as a blind guess of $\{t_i\}_{i=0}^6$, read as:

$$\mathscr{T}_{uni} = \{0, 2.84308, 5.68615, 8.52923, 11.3723, 14.2154\}$$

and the initial guess based on *cumulative chord* $\mathscr{T}_c$ (see (12)) coincides with:

$$\mathscr{T}_c = \{0, 0.412311, 6.01231, 6.23592, 14.0154, 14.2154\}.$$

Here $\hat{\mathscr{T}}_{uni}$ (and $\hat{\mathscr{T}}_c$) is defined as $\mathscr{T}_{uni}$ (and $\mathscr{T}_c$) stripped from its terminal values. The natural splines $\gamma^{NS}_{\mathscr{T}_{uni}}$ (based on $\mathscr{T}_{uni}$) and $\gamma^{NS}_{\mathscr{T}_c}$ (based on $\mathscr{T}_c$) yield the following energies $\mathscr{J}^F_{\mathscr{T}_c}(\hat{\mathscr{T}}_{uni}) = 15.4253 > \mathscr{J}^F_{\mathscr{T}_c}(\hat{\mathscr{T}}_c) = 8.51108$. Both interpolants $\gamma^{NS}_{\mathscr{T}_{uni}}$ and $\gamma^{NS}_{\mathscr{T}_c}$ are shown in Fig. 3(a) and (b), respectively.

One expects that the *Secant Method* with two initial numbers $t_i^c \pm 0.5$ may produce a bad solution as $|t_0^c - t_1^c| < 0.5$, $|t_2 - t_3| < 0.5$ and $|t_4 - t_5| < 0.5$. Indeed the *Secant Method* returns $t_1^{opt} = -8.2211 < t_0 = 0$, which is disallowed. Upon adjusting $t_i^c \pm 0.05$ the *Secant Method* yields (for (9)) the optimal knots $\hat{\mathscr{T}}^{SM}_{opt}$ augmented by terminal times $t_0 = 0$ and $t_5 = T_c$ as:

$$\mathscr{T}^{SM}_{opt} = \{0, 0.737027, 6.07314, 7.14642, 13.5208, 14.2154\}$$

with the optimal energy $\mathscr{J}^F_{\mathscr{T}_c}(\hat{\mathscr{T}}^{SM}_{opt}) = 5.04331$. The execution time amounts to $T^{SM} = 9.204045$ s. The resulting curve $\gamma^S_{\mathscr{T}^{SM}_{opt}}$ is plotted in Fig. 3(c). In fact, for general data it is safer for each free variable optimized by the *Secant Method* to choose a pair of numbers $t_i^c \pm 0.5 \min_{0 \leq i \leq n-1}\{|t_{i+1}^c - t_i^c|\}$.

Leap-Frog decreases the initial energy to $\mathscr{J}^F_{\mathscr{T}_c}(\hat{\mathscr{T}}^{LF}_{opt}) = \mathscr{J}^F_{\mathscr{T}_c}(\hat{\mathscr{T}}^{SM}_{opt})$ (as for the *Secant Method*) with the iteration stopping conditions $\hat{\mathscr{T}}^{LF}_{opt} = \hat{\mathscr{T}}^{SM}_{opt}$ (up to 6th decimal point) upon 38 iterations. The respective *execution time* amounts to $T^{LF} = 3.247230 < T^{SM}$. The 0th (i.e. $\mathscr{J}^F_{\mathscr{T}_c}(\hat{\mathscr{T}}_c)$), 1st, 2nd, 10th, 18th, and 38th iterations of *Leap-Frog* decrease the energy to:

$$\{8.51108, 5.91959, 5.23031, 5.04455, 5.04331, 5.04331\}$$

with again only the first three iterations contributing to the major correction of the initial guess knots $\mathscr{T}_c$. The resulting natural spline $\gamma^{NS}_{\mathscr{T}^{LF}_{opt}}$ (clearly the same as $\gamma^{NS}_{\mathscr{T}^{SM}_{opt}}$ yielded by the *Secant Method*) based on $\mathscr{T}^{LF}_{opt}$ is shown in Fig. 3(c) and also visually compared with $\gamma^{NS}_{\mathscr{T}_c}$ in Fig. 3(d).

Again if *Leap-Frog* iteration bound condition is changed e.g. to make current *Leap-Frog* energy equal to $\mathscr{J}^F_{\mathscr{T}_c}(\hat{\mathscr{T}}_c^{SM})$ (say up to 5th decimal place) then only 18 iterations are needed here with shorter execution time $T_E^{LF} = 1.785042 < T^{SM}$ and with optimal times

$$\mathscr{T}^{LF_E}_{opt} = \{0, 0.736394, 6.0697, 7.14349, 13.5205, 14.2154\}.$$

We miss out here a bit on a precise estimation of the optimal knots but we accelerate the *Leap-Frog* execution time by obtaining almost the same interpolating curve as the optimal one (as $\hat{\mathcal{T}}_{opt}^{LF_E} \approx \hat{\mathcal{T}}_{opt}^{SM}$). For other iteration stopping criteria accelerating the execution of *Leap-Frog* at almost no cost in difference between computed and optimal curve see [19]. □

We pass now to an example of reduced data in E^3 (i.e. with $m = 3$).

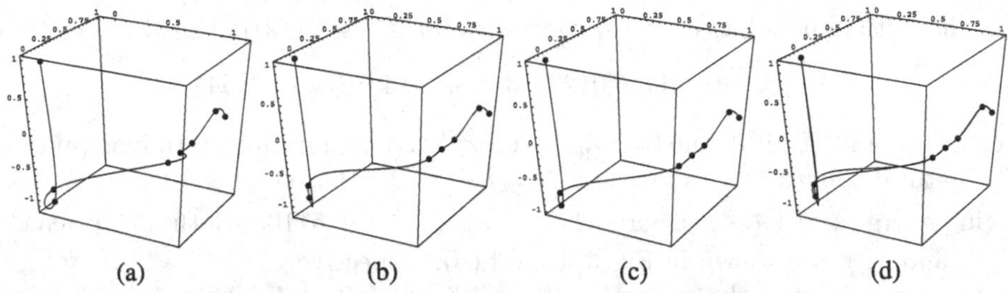

Fig. 4. Natural splines interpolating data points $\mathcal{M}_{3D3}$ (a) $\gamma_{\mathcal{T}_{uni}}^{NS}$ with uniform knots $\mathcal{T}_{uni}$, (b) $\gamma_{\mathcal{T}_c}^{NS}$ with cumulative chords $\mathcal{T}_c$, (c) $\gamma_{\mathcal{T}_{opt}^{LF}}^{NS}$ with optimal knots $\mathcal{T}_{opt}^{LF} = \mathcal{T}_{opt}^{SM}$ (thus $\gamma_{\mathcal{T}_{opt}^{LF}}^{NS} = \gamma_{\mathcal{T}_{opt}^{SM}}^{NS}$) (d) $\gamma_{\mathcal{T}_{opt}^{LF}}^{NS}$ and $\gamma_{\mathcal{T}_c^{SM}}^{NS}$ plotted together

Example 7. Consider for $n = 7$ the following 3D points (see dotted points in Fig. 4):

$$\mathcal{M}_{3D3} = \{(0,0,1),(0,0,-1),(0,0,-0.8),(1,0,0),(1,0.2,0),(1,0.4,0),(1,0.8,0.2),(1,1,0)\}.$$

The *uniform interpolation knots* $\{\hat{t}_i = \frac{i}{7}T_c\}_{i=0}^7 \approx \{t_i\}_{i=0}^7$ (rescaled to $t_0 = 0$ and to T_c – see (12)) read as:

$$\mathcal{T}_{uni} = \{0, 0.658669, 1.31734, 1.97601, 2.63467, 3.29334, 3.95201, 4.61068\}$$

and the initial guess based on *cumulative chords* $\mathcal{T}_c$ is equal to:

$$\mathcal{T}_c = \{0, 2, 2.2, 3.48062, 3.68062, 3.88062, 4.32784, 4.61068\}.$$

Here $\hat{\mathcal{T}}_{uni} = \{0.658669, 1.31734, 1.97601, 2.63467, 3.29334, 3.95201\}$, while the other one $\hat{\mathcal{T}}_c = \{0, 2, 2.2, 3.48062, 3.68062, 3.88062, 4.32784, 4.61068\}$. The natural splines $\gamma_{\mathcal{T}_{uni}}^{NS}$ (based on $\mathcal{T}_{uni}$) and $\gamma_{\mathcal{T}_c}^{NS}$ (based on $\mathcal{T}_c$) yields the following energies $\mathcal{J}_{\mathcal{T}_c}^F(\hat{\mathcal{T}}_{uni}) = 46.7919 > \mathcal{J}_{\mathcal{T}_c}^F(\hat{\mathcal{T}}_c) = 22.3564$. Both interpolants $\gamma_{\mathcal{T}_{uni}}^{NS}$ and $\gamma_{\mathcal{T}_c}^{NS}$ are shown in Fig. 4(a) and(b), respectively. Noticeably the energy based on blind guess of knots (i.e. for uniform knots) is far from the optimal one.

The *Secant Method* yields for (9) the optimal knots $\hat{\mathcal{T}}_{opt}^{SM}$ (augmented by terminal knots $t_0 = 0$ and $t_7 = T_c$ - see (12))

$$\mathcal{T}_{opt}^{SM} = \{0, 1.34728, 1.82093, 3.12718, 3.39487, 3.62307, 4.19613, 4.61068\}$$

with the optimal energy $\mathcal{J}^F_{\mathcal{T}_c}(\hat{\mathcal{T}}^{SM}_{opt}) = 15.407$. The *execution time* amounts to $T^{SM} = 128.804084$ s. The resulting curve $\gamma^{NS}_{\mathcal{T}^{SM}_{opt}}$ is plotted in Fig. 4(c). Note that for each free variable, the *Secant Method* uses here two initial numbers $t^c_i \pm 0.1$.

Leap-Frog decreases the initial energy to $\mathcal{J}^F_{\mathcal{T}_c}(\hat{\mathcal{T}}^{LF}_{opt}) = \mathcal{J}^F_{\mathcal{T}_c}(\hat{\mathcal{T}}^{SM}_{opt})$ (as for the *Secant Method*) with the iteration stopping conditions $\hat{\mathcal{T}}^{LF}_{opt} = \hat{\mathcal{T}}^{SM}_{opt}$ (up to 5th decimal point) upon 620 iterations. The *execution time* amounts to $T^{LF} = 73.749111\,\mathrm{s} < T^{SM}$. The 0th (i.e. $\mathcal{J}^F_{\mathcal{T}_c}(\hat{\mathcal{T}}_c)$), 1st, 2nd, 10th, 50th, 40th and 100th, 200th, 281th and 620th iterations of *Leap-Frog* decrease the energy to:

$$\{22.3564, 18.5598, 18.274, 17.4628, 15.8596.15.5049, 15.409115.407, 15.407\}$$

with again only the first three iterations contributing to major correction of the initial guess knots $\mathcal{T}_c$. The resulting natural spline $\gamma^{NS}_{\mathcal{T}^{LF}_{opt}}$ (clearly the same as $\gamma^{NS}_{\mathcal{T}^{SM}_{opt}}$ yielded by the *Secant Method*) based on $\mathcal{T}^{LF}_{opt_2}$ is shown in Fig. 4(c) and also visually compared with $\gamma^{NS}_{\mathcal{T}_c}$ in Fig. 4(d). The optimal curve does not vary too much from the initial guess curve based on cumulative chord knots.

Again if *Leap-Frog* iteration bound condition is changed e.g. to make current *Leap-Frog* energy equal to $\mathcal{J}^F_{\mathcal{T}_c}(\hat{\mathcal{T}}^{SM}_c)$ (say up to 4th decimal place) then only 281 iterations are needed here with shorter execution time $T^{LF}_E = 33.931990\,\mathrm{s} < T^{SM}$ and with optimal knots:

$$\mathcal{T}^{LF_E}_{opt} = \{0, 1.348043, 1.82195, 3.12892, 3.39651, 3.62453, 4.19679, 4.61068\}.$$

As previously, we lose here slightly on a precise estimation of the optimal knots but we accelerate the *Leap-Frog* execution time by obtaining almost the same interpolating curve as the optimal one (as $\mathcal{T}^{LF_E}_{opt} \approx \mathcal{T}^{SM}_{opt}$). For other iteration stopping criteria accelerating the execution of *Leap-Frog* at almost no cost in difference between computed curve and optimal curve see [19]. □

5 Conclusions

In this paper, we discuss the method of estimating the unknown interpolation knots $\{t_i\}_{i=0}^n$ by $\{\hat{t}_i\}_{i=0}^n$ to fit reduced sparse data $\mathcal{M} = \{q_i\}_{i=0}^n$ with the natural cubic spline in arbitrary Euclidean space E^m. As indicated here, the above task is transformed into the corresponding finite-dimensional constrained optimization task (9) in $(t_1, t_2, \ldots, t_{n-1})$-variables, subject to the satisfaction of the inequalities $t_0 < t_1 < t_2, < \ldots < t_{n-1} < t_n$. We first demonstrate a high non-linearity and possible non-convexity of (9) - Sects. 1, 2, and 3. Consequently, the latter hinders the use of standard optimization techniques like *Newton Method* to deal with such optimization task. Finally, two computationally feasible schemes are implemented, i.e. *Leap-Frog* and the *Secant Method* to examine the quality of the reconstructed interpolants. The derivation of the explicit formula in (9) including its particular forms examined in Examples 1 and 2 relies on *Mathematica* symbolic computation - see [35]. All numerical computations performed in

Examples 3, 6 and 7 resort to the numerical functions supplied by *Mathematica* software (see [35]). In addition, sufficient conditions to enforce the convexity (or unimodality) of (9) are generated for the special case of $n = 2$ - see Example 5. Future work involves the analysis of a more general case i.e. when n is arbitrary. Alternatively one may also consider to derive a similar to (9) optimization task set for a complete spline interpolant (see [3]), with the initial velocities v_0 and v_n either a priori given or approximated according to [15].

References

1. Bézier, P.E.: Numerical Control: Mathematics and Applications. Wiley, New York (1972)
2. Boehm, E., Farin, G., Kahmann, J.: A survey of curve and surface methods in CAGD. Comput. Aided Geom. Des. **1**(1), 1–60 (1988)
3. de Boor, C.: A Practical Guide to Spline. Springer, New York (1985)
4. Boyd, S., Vandenberghe, L.: Convex Optimization. Cambridge University Press, Cambridge (2004)
5. Budzko, D.A., Prokopenya, A.N.: On the stability of equilibrium positions in the circular restricted four-body problem. In: Gerdt, V.P., Koepf, W., Mayr, E.W., Vorozhtsov, E.V. (eds.) CASC 2011. LNCS, vol. 6885, pp. 88–100. Springer, Heidelberg (2011). doi:10.1007/978-3-642-23568-9_8
6. Epstein, M.P.: On the influence of parameterization in parametric interpolation. SIAM J. Numer. Anal. **13**, 261–268 (1976)
7. Farin, G.: Curves and Surfaces for Computer Aided Geometric Design. Academic Press, San Diego (1993)
8. Farouki, R.T.: Optimal parameterizations. Comput. Aided Geom. Des. **14**(2), 153–168 (1997)
9. Floater, M.S.: Chordal cubic spline interpolation is fourth order accurate. IMA J. Numer. Anal. **26**, 25–33 (2006)
10. Hoschek, J.: Intrinsic parametrization for approximation. Comput. Aided Geom. Des. **5**(1), 27–31 (1988)
11. Janik, M., Kozera, R., Kozioł, P.: Reduced data for curve modeling - applications in graphics, computer vision and physics. Adv. Sci. Tech. **7**(18), 28–35 (2013)
12. Kocić, L.M., Simoncinelli, A.C., Della, V.B.: Blending parameterization of polynomial and spline interpolants. Facta Universitatis (NIŠ), Ser. Math. Inform. **5**, 95–107 (1990)
13. Kozera, R.: Curve modelling via interpolation based on multidimensional reduced data. Stud. Informatica **25**, 1–140 (2004). (4B(61))
14. Kozera, R., Noakes, L.: Piecewise-quadratics and exponential parameterizations for reduced data. Appl. Maths Comput. **221**, 620–638 (2013)
15. Kozera, R., Noakes, L.: C^1 Interpolation with cumulative chord cubics. Fundamenta Informaticae **61**(3–4), 285–301 (2004)
16. Noakes, L., Kozera, R.: Interpolating sporadic data. In: Heyden, A., Sparr, G., Nielsen, M., Johansen, P. (eds.) ECCV 2002. LNCS, vol. 2351, pp. 613–625. Springer, Heidelberg (2002). doi:10.1007/3-540-47967-8_41
17. Kozera, R., Noakes, L.: Optimal knots selection for sparse reduced data. In: Huang, F., Sugimoto, A. (eds.) PSIVT 2015. LNCS, vol. 9555, pp. 3–14. Springer, Cham (2016). doi:10.1007/978-3-319-30285-0_1

18. Kozera, R., Noakes, L.: Modeling reduced sparse data. In: Romaniuk, R.S. (ed.) Photonics, Applications in Astronomy, Communications, Industry, and High-Energy Physics Experiments 2016. SPIE 2016, vol. 10031. Society of Photo-Optical Instrumentation Engineers, Bellingham (2016)
19. Kozera, R., Noakes, L.: Fitting Data via Optimal Interpolation Knots. (Submitted)
20. Kvasov, B.I.: Methods of Shape-Preserving Spline Approximation. World Scientific, Singapore (2000)
21. Kuznetsov, E.B., Yakimovich, A.Y.: The best parameterization for parametric interpolation. J. Comp. Appl. Maths **191**, 239–245 (2006)
22. Marin, S.P.: An approach to data parameterization in parametric cubic spline interpolation problems. J. Approx. Theory **41**, 64–86 (1984)
23. Mørken, K., Scherer, K.: A general framework for high-accuracy parametric interpolation. Math. Comput. **66**(217), 237–260 (1997)
24. Noakes, L.: A global algorithm for geodesics. J. Math. Austral. Soc. Ser. A **64**, 37–50 (1999)
25. Noakes, L., Kozera, R.: Cumulative chords piecewise-quadratics and piecewise-cubics. In: Klette, R., Kozera, R., Noakes, L., Weickert, J. (eds.) Geometric Properties from Incomplete Data. Computational Imaging and Vision, vol. 31, pp. 59–75. Springer, The Netherlands (2006)
26. Noakes, L., Kozera, R.: More-or-less uniform sampling and lengths of curves. Quar. Appl. Maths **61**(3), 475–484 (2003)
27. Noakes, L., Kozera, R.: Nonlinearities and noise reduction in 3-source photometric stereo. J. Math. Imag. Vis. **18**(3), 119–127 (2003)
28. Noakes, L., Kozera, R.: 2D leap-frog algorithm for optimal surface reconstruction. In: Latecki, M.J. (ed.) SPIE 1999. Vision Geometry VIII, vol. 3811, pp. 317–328. Society of Industrial and Applied Mathematics, Bellingham (1999)
29. Lee, E.T.Y.: Corners, cusps, and parameterization: variations on a theorem of Epstein. SIAM J. Numer. Anal. **29**, 553–565 (1992)
30. Lee, E.T.Y.: Choosing nodes in parametric curve interpolation. Comput. Aided Geom. Des. **21**, 363–370 (1989)
31. Piegl, L., Tiller, W.: The NURBS Book. Springer, Berlin (1997)
32. Prokopenya, A.N.: Hamiltonian normalization in the restricted many-body problem by computer algebra methods. Program. Comput. Softw. **38**(3), 156–166 (2012)
33. Rababah, A.: High order approximation methods for curves. Comput. Aided Geom. Des. **12**, 89–102 (1995)
34. Schaback, R.: Optimal geometric Hermite interpolation of curves. In: Dæhlen, M., Lyche, T., Schumaker, L. (eds.) Mathematical Methods for Curves and Surfaces II, pp. 1–12. Vanderbilt University Press, Nashville (1998)
35. Wolfram, S.: The Mathematica Book, 5th edn. Wolfram Media, Champaign (2003)

The Convergence Conditions of Interval Newton's Method Based on Point Estimates

Zhe Li[1,2](✉), Baocheng Wan[2,3], and Shugong Zhang[2]

[1] School of Science, Changchun University of Science and Technology, Changchun 130022, China
lizhe@amss.ac.cn

[2] School of Mathematics, Key Laboratory of Symbolic Computation and Knowledge Engineering, Jilin University, Changchun 130012, China
wanbaocheng@163.com, sgzh@jlu.edu.cn

[3] Jilin Agricultural University, Changchun 130118, China

Abstract. Both Smale's alpha theory and Rump's interval theorem provide the conditions which guarantee the existence of a simple solution of a square nonlinear system. In this paper, we generalize the conclusion provided by Rall to reveal the relationship between Smale's alpha theory and Rump's interval theorem. By point estimates, we propose the conditions under which the condition of Rump's interval theorem holds. Furthermore, using only the information of the given system at the initial approximate point, we give the convergence conditions of interval Newton's algorithm proposed by Rump.

Keywords: Newton iteration · Alpha theory · Interval algorithm · Point estimate · Verification

1 Introduction

Solving a nonlinear system in the form $\boldsymbol{f}(\boldsymbol{x}) = \boldsymbol{0}$ with $\boldsymbol{f} = (f_1, f_2, \ldots, f_n)^T$ and $\boldsymbol{x} = (x_1, \ldots, x_n)^T$ is one of the most fundamental problems in scientific computing. In this paper, we assume that $\boldsymbol{f} : \mathbb{R}^n \to \mathbb{R}^n$ and $f_1, f_2, \ldots, f_n$ have all order continuous partial derivatives. Denote the Jacobian matrix of $\boldsymbol{f}$ at $\boldsymbol{x}$ by $\boldsymbol{f}'(\boldsymbol{x})$.

Newton's method and its modifications have long played an important role in solving nonlinear systems. Under certain conditions, Newton's method constructs a sequence of iteration points that will converge to a solution of the given nonlinear system. In 1948, the author of [2] established the Kantorovich theorem based on the assumption that the Jacobian matrix of the nonlinear

Z. Li—This research was supported by Chinese National Natural Science Foundation under Grant Nos. 11601039, 11671169, 11501051, by the open fund Key Lab. of Symbolic Computation and Knowledge Engineering under Grant No. 93K172015K06, and by the Education Department of Jilin Province, "13th Five-Year" science and technology project under Grant No. JJKH20170618KJ.

V.P. Gerdt et al. (Eds.): CASC 2017, LNCS 10490, pp. 272–284, 2017.
DOI: 10.1007/978-3-319-66320-3_20

system is Lipschitz continuous on some domain. The Kantorovich theorem first gives the condition to ensure that a simple solution exists close to the initial point and the Newton iteration sequence quadratically converges to this simple solution. Using the technique of point estimates, Smale et al. [15–17] developed the alpha theory to locate and approximate simple solutions. The alpha theory requires only information concerning the nonlinear system at the initial point of the Newton iteration. By introducing the dominating sequence technique, Wang and Han [18] improved both the condition and conclusion of the alpha theory. With the aid of Schröder operator, Giusti et al. [3] provided a criterion for locating clusters of solutions of univariate nonlinear functions. Later on, Giusti et al. [4] generalized their results to locate breadth-one singular solutions of multivariate nonlinear systems. For the performance of the alpha theory, Hauenstein and Sottile [6] described the program alphaCertified to certify solutions of polynomial systems. Recently, Hauenstein and Levandovskyy [5] extended the program alphaCertified to verify solutions to polynomial-exponential systems.

Interval arithmetic is another important tool of verification methods. In 1960s, Krawczyk [9] first introduced an interval version of Newton's method for verifying the existence of simple solutions. Moore [10] proposed computationally verifiable sufficient condition for interval Newton's method given by Krawczyk. Rump [12] made interval Newton's method perform better in practice, which is called Rump's interval theorem and included in **verifynlss** function in INTLAB toolbox [13] in Matlab. Based on the deflation technique using smoothing parameters, Rump and Graillat [14] described a verification algorithm to verify multiple solutions of univariate nonlinear functions and double solutions of multivariate nonlinear systems. Further, Li and Zhi [8] provided an algorithm to verify breadth-one singular solutions of polynomial systems, which had been generalized to deal with the verification of isolated singular solutions in [7]. By combining interval algorithms with some other methods, Yang et al. [19] investigated the verification for real solutions of positive-dimensional polynomial systems.

In 1980, Rall [11] exhibited the relationship between the Kantorovich theorem and Moore's interval theorem. By the quantities of the Kantorovich theorem, Rall provided the conditions under which Moore's verifiable sufficient condition holds. For an initial approximate $\boldsymbol{x}^{(0)} \in \mathbb{R}^n$ and a radius $\rho > 0$, let $\boldsymbol{X}_\rho$ denote the set $\{\boldsymbol{x} : \|\boldsymbol{x} - \boldsymbol{x}^{(0)}\|_\infty < \rho\}$, and let η, B, κ be the constants such that

$$\begin{aligned}
&\|\boldsymbol{f}'(\boldsymbol{x}^{(0)})^{-1}\boldsymbol{f}(\boldsymbol{x}^{(0)})\|_\infty \le \eta, \\
&\|\boldsymbol{f}'(\boldsymbol{x}^{(0)})^{-1}\|_\infty \le B, \\
&\|\boldsymbol{f}'(\boldsymbol{u}) - \boldsymbol{f}'(\boldsymbol{v})\|_\infty \le \kappa\|\boldsymbol{u} - \boldsymbol{v}\|, \qquad \boldsymbol{u}, \boldsymbol{v} \in \Omega,
\end{aligned}$$

where Ω is a sufficiently large region containing $\boldsymbol{x}^{(0)}$. Rall's conclusion is that if

$$h = B\kappa\eta < \frac{1}{4},$$

then

$$\boldsymbol{x}^{(0)} - \boldsymbol{f}'(\boldsymbol{x}^{(0)})^{-1}\boldsymbol{f}(\boldsymbol{x}^{(0)}) + (I - \boldsymbol{f}'(\boldsymbol{x}^{(0)})^{-1}\boldsymbol{f}'(\boldsymbol{X}_\rho))(\boldsymbol{X}_\rho - \boldsymbol{x}^{(0)}) \subset \boldsymbol{X}_\rho$$

holds for any ρ satisfying the inequality

$$\frac{1-\sqrt{1-4h}}{2h}\eta \le \rho \le \frac{1+\sqrt{1-4h}}{2h}\eta.$$

Since the alpha theory and Rump's interval theorem are respectively the generalization of the Kantorovich theorem and Moore's interval theorem, we generalize Rall's conclusion to discuss the relationship between the alpha theory and Rump's interval theorem in this paper. By only the information of the given system at the initial approximate point, we provide the conditions to guarantee that we can obtain an approximate point after finite Newton's iterations, where this approximate iteration point corresponds to an interval solution satisfying the condition of Rump's interval theorem. The next section will give some notation and background results.

2 Notation and Preliminaries

First of all, we emphasize that the norm $\|\cdot\|$ of the vector and the matrix in this paper are both the infinite norm $\|\cdot\|_\infty$ since the metric for the interval vector is closely related to the infinite norm.

Henceforward, we use boldface letters to express tuples and denote their entries by the same letter with subscripts, for example $\boldsymbol{\alpha} = (\alpha_1, \ldots, \alpha_n)^T$. Denote the usual product order on $\mathbb{R}^n$ by $\le$, that is, for arbitrary $\boldsymbol{\alpha}, \boldsymbol{\beta} \in \mathbb{R}^n$, $\boldsymbol{\alpha} \le \boldsymbol{\beta}$ if and only if $\alpha_i \le \beta_i$ for $1 \le i \le n$.

For $\boldsymbol{x} \in \mathbb{R}^n$, if $\boldsymbol{f}'(\boldsymbol{x})$ is nonsingular, then define

$$\alpha(\boldsymbol{f}, \boldsymbol{x}) = \beta(\boldsymbol{f}, \boldsymbol{x})\gamma(\boldsymbol{f}, \boldsymbol{x}),$$
$$\beta(\boldsymbol{f}, \boldsymbol{x}) = \|\boldsymbol{f}'(\boldsymbol{x})^{-1}\boldsymbol{f}(\boldsymbol{x})\|,$$
$$\gamma(\boldsymbol{f}, \boldsymbol{x}) = \sup_{k\ge 2}\left\|\boldsymbol{f}'(\boldsymbol{x})^{-1}\frac{\boldsymbol{f}^{(k)}(\boldsymbol{x})}{k!}\right\|^{1/(k-1)}.$$

Given the initial approximate $\boldsymbol{x}^{(0)}$ with the associated simple root $\boldsymbol{x}^*$ of $\boldsymbol{f}$, we let α, β and γ to stand for $\alpha(\boldsymbol{f}, \boldsymbol{x}^{(0)})$, $\beta(\boldsymbol{f}, \boldsymbol{x}^{(0)})$ and $\gamma(\boldsymbol{f}, \boldsymbol{x}^{(0)})$, respectively. Applying Newton's method for $\boldsymbol{f}$ can get the Newton iteration sequence $\{\boldsymbol{x}^{(k)}\}$, that is,

$$\boldsymbol{x}^{(k+1)} = \boldsymbol{x}^{(k)} - \boldsymbol{f}'(\boldsymbol{x}^{(k)})^{-1}\boldsymbol{f}(\boldsymbol{x}^{(k)}), \qquad k \in \mathbb{N}.$$

The alpha theory provides the convergence conditions which ensure the sequence $\{\boldsymbol{x}^{(k)}\}$ converges to $\boldsymbol{x}^*$ only with the values α, β and γ. The dominating sequence technique is a powerful tool for improving the alpha theory. The dominating sequence $\{t^{(k)}\}$ is produced by the Newton iteration with the initial approximate $t^{(0)} = 0$ for the univariate function

$$h(t) = \beta - t + \frac{\gamma t^2}{1-\gamma t},$$

where the equation $h(t) = 0$ has the following two solutions

$$t^* = \frac{2\beta}{1+\alpha+\sqrt{1-6\alpha+\alpha^2}}, \quad t^{**} = \frac{1+\alpha+\sqrt{1-6\alpha+\alpha^2}}{4\gamma}. \tag{1}$$

The following theorem is a version of the alpha theory given by Wang and Han, where the condition on the quantity α is best possible.

Theorem 1. [4,18] *If $0 < \alpha < 3 - 2\sqrt{2}$, then for any $t^* \le \rho < t^{**}$, the system $\boldsymbol{f}(\boldsymbol{x}) = \boldsymbol{0}$ has exactly one simple root $\boldsymbol{x}^*$ in $\overline{B}(\boldsymbol{x}^{(0)}, \rho)$. In addition, the Newton iteration sequence $\{\boldsymbol{x}^{(k)}\}$ converges quadratically to $\boldsymbol{x}^*$, and the dominating sequence $\{t^{(k)}\}$ increases and converges quadratically to t^*. Furthermore, for all $k \in \mathbb{N}$,*

$$\|\boldsymbol{x}^{(k+1)} - \boldsymbol{x}^{(k)}\| \le t^{(k+1)} - t^{(k)},$$
$$\|\boldsymbol{x}^{(k+1)} - \boldsymbol{x}^{(k)}\| \le q(\alpha)^{2^k-1}\beta$$

with

$$q(\alpha) = \frac{4\alpha}{\left(1-\alpha+\sqrt{1-6\alpha+\alpha^2}\right)^2}. \tag{2}$$

Denote the set of intervals by $\mathbb{IR}$. An interval vector $\mathbf{X} = [\underline{\boldsymbol{x}}, \overline{\boldsymbol{x}}] \in \mathbb{IR}^n$ with $\underline{\boldsymbol{x}}, \overline{\boldsymbol{x}} \in \mathbb{R}^n$ and $\underline{\boldsymbol{x}} \le \overline{\boldsymbol{x}}$ is defined by

$$\mathbf{X} = [\underline{\boldsymbol{x}}, \overline{\boldsymbol{x}}] = \{\boldsymbol{x} \in \mathbb{R}^n : \underline{\boldsymbol{x}} \le \boldsymbol{x} \le \overline{\boldsymbol{x}}\}.$$

For $\boldsymbol{x} \in \mathbb{R}^n$, $\mathbf{X} = [\underline{\boldsymbol{x}}, \overline{\boldsymbol{x}}] \in \mathbb{IR}^n$, $\boldsymbol{x} + \mathbf{X} = [\boldsymbol{x} + \underline{\boldsymbol{x}}, \boldsymbol{x} + \overline{\boldsymbol{x}}]$. Let $\mathbf{Y}_\rho = \{\boldsymbol{y} \in \mathbb{R}^n : \|\boldsymbol{y}\| \le \rho\}$, then $\overline{B}(\boldsymbol{x}, \rho) = \boldsymbol{x} + \mathbf{Y}_\rho$.

The norm of the interval vector $\mathbf{X} = [\underline{\boldsymbol{x}}, \overline{\boldsymbol{x}}]$ is defined by

$$\|\mathbf{X}\| = \|[\underline{\boldsymbol{x}}, \overline{\boldsymbol{x}}]\| = \max\{\|\boldsymbol{x}\| : \boldsymbol{x} \in \mathbf{X}\}.$$

Besides int$(\mathbf{X})$ designates the interior of the interval vector $\mathbf{X}$. Given a set $\mathbf{Z} \subset \mathbb{R}^n$, the interval hull of $\mathbf{Z}$ is the narrowest interval vector containing $\mathbf{Z}$, namely,

$$\text{hull}(\mathbf{Z}) = \bigcap\{\mathbf{X} \in \mathbb{IR}^n : \mathbf{X} \supseteq \mathbf{Z}\}.$$

Given a continuous mapping $\boldsymbol{g} : \mathbb{R}^n \to \mathbb{R}^m$ and an interval vector $\mathbf{X}$, the interval vector $\boldsymbol{g}(\mathbf{X}) \in \mathbb{IR}^m$ is defined as

$$\boldsymbol{g}(\mathbf{X}) = \text{hull}\{\boldsymbol{g}(\boldsymbol{x}) : \boldsymbol{x} \in \mathbf{X}\}.$$

Given an interval matrix $\mathbf{A} \in \mathbb{IR}^{m\times n}$ and an interval vector $\mathbf{X} \in \mathbb{IR}^n$, the interval vector $\mathbf{AX}$ is defined by

$$\mathbf{AX} = \text{hull}\{A\boldsymbol{x} : A \in \mathbf{A}, \boldsymbol{x} \in \mathbf{X}\}.$$

Specially, the norm of the interval matrix $\mathbf{A}$ is defined as

$$\|\mathbf{A}\| = \max\{\|A\| : A \in \mathbf{A}\}.$$

The following theorem is a version of the interval Newton's method given by Rump.

Theorem 2. [12] *Given* $\boldsymbol{x}^{(0)} \in \mathbb{R}^n$, $\mathbf{Y} \in \mathbb{IR}^n$ *with* $\mathbf{0} \in \mathbf{Y}$, $R \in \mathbb{R}^{n\times n}$, *if*

$$S(\mathbf{Y}, \boldsymbol{x}^{(0)}) := -R\boldsymbol{f}(\boldsymbol{x}^{(0)}) + (I - R\boldsymbol{f}'(\boldsymbol{x}^{(0)} + \mathbf{Y}))\mathbf{Y} \subseteq \mathrm{int}(\mathbf{Y}),$$

then there exists a unique $\boldsymbol{x}^* \in \boldsymbol{x}^{(0)} + \mathbf{Y}$ *such that* $\boldsymbol{f}(\boldsymbol{x}^*) = \mathbf{0}$.

3 Main Results

To give the main results of this paper, we need the following functions,

$$\begin{aligned}
\psi(u) &= 2u^2 - 4u + 1,\\
g_1(\alpha) &= \alpha^2 - 4\alpha + 10,\\
g_2(\alpha) &= \alpha^3 - 6\alpha^2 + 21\alpha + 28,\\
g_3(\alpha) &= 4\alpha^3 - 25\alpha^2 + 88\alpha - 8,\\
\theta(\alpha) &= \frac{1}{3}\arccos(\frac{g_2(\alpha)}{\sqrt{g_1(\alpha)^3}}),\\
\omega_1(\alpha) &= \sqrt{g_1(\alpha)}\cos(\theta(\alpha) + \frac{2\pi}{3}) + 2 + \frac{\alpha}{2},\\
\omega_2(\alpha) &= \sqrt{g_1(\alpha)}\cos(\theta(\alpha) + \frac{4\pi}{3}) + 2 + \frac{\alpha}{2},\\
\omega_3(\alpha) &= \sqrt{g_1(\alpha)}\cos(\theta(\alpha)) + 2 + \frac{\alpha}{2}.
\end{aligned}$$

Here we give some lemmas and one proposition from which the main theorem will easily follow.

Lemma 1. [15] *Given* $\widetilde{\boldsymbol{x}} \in \mathbb{R}^n$, *if* $\gamma\|\widetilde{\boldsymbol{x}} - \boldsymbol{x}^{(0)}\| < 1 - \sqrt{2}/2$, *then*

$$\gamma(\boldsymbol{f}, \widetilde{\boldsymbol{x}}) \leq \frac{\gamma}{\psi(\gamma\|\widetilde{\boldsymbol{x}} - \boldsymbol{x}^{(0)}\|)(1 - \gamma\|\widetilde{\boldsymbol{x}} - \boldsymbol{x}^{(0)}\|)}.$$

Lemma 2. *If* $0 < \alpha < 3 - 2\sqrt{2}$, *then for all* $k \geq 1$,

$$\gamma\|\boldsymbol{x}^{(k)} - \boldsymbol{x}^{(0)}\| < 1 - \frac{\sqrt{2}}{2}, \tag{3}$$

$$\gamma(f, \boldsymbol{x}^{(k)}) \leq \frac{\gamma}{\psi(\gamma\|\boldsymbol{x}^{(k)} - \boldsymbol{x}^{(0)}\|)(1 - \gamma\|\boldsymbol{x}^{(k)} - \boldsymbol{x}^{(0)}\|)}. \tag{4}$$

Proof. If $0 < \alpha < 3 - 2\sqrt{2}$, then by Theorem 1,

$$\begin{aligned}
\|\boldsymbol{x}^{(k)} - \boldsymbol{x}^{(0)}\| &\leq \sum_{j=1}^{k}\|\boldsymbol{x}^{(j)} - \boldsymbol{x}^{(j-1)}\| \leq \sum_{j=1}^{k}(t^{(j)} - t^{(j-1)})\\
&\leq t^{(k)} \leq t^*.
\end{aligned}$$

Therefore

$$\gamma\|\boldsymbol{x}^{(k)}-\boldsymbol{x}^{(0)}\| \le \frac{2\alpha}{1+\alpha+\sqrt{1-6\alpha+\alpha^2}}.$$

Since the right hand side of the above inequality monotonously increases from 0 to $1-\sqrt{2}/2$ as α goes from 0 to $3-2\sqrt{2}$, it follows that (3) holds. By means of Lemma 1, (4) follows. □

Lemma 3. *If*

$$0<\rho<\frac{1-\sqrt{2}/2}{\gamma}, \tag{5}$$

$$\beta+\rho(\frac{1}{(1-\gamma\rho)^2}-1)<\rho, \tag{6}$$

then

$$S(\boldsymbol{Y}_\rho, \boldsymbol{x}^{(0)}) \subseteq \mathrm{int}(\boldsymbol{Y}_\rho). \tag{7}$$

Proof. Given an arbitrary real vector $\boldsymbol{y} \in \mathbf{Y}_\rho$, we expand the Jacobian matrix $\boldsymbol{f}'(\boldsymbol{x}^{(0)}+\boldsymbol{y})$ into power series and get

$$\begin{aligned}\boldsymbol{f}'(\boldsymbol{x}^{(0)})^{-1}\boldsymbol{f}'(\boldsymbol{x}^{(0)}+\boldsymbol{y}) &= \boldsymbol{f}'(\boldsymbol{x}^{(0)})^{-1}(\boldsymbol{f}'(\boldsymbol{x}^{(0)})+\sum_{k=2}^{\infty}\boldsymbol{f}^{(k)}(\boldsymbol{x}^{(0)})\frac{\boldsymbol{y}^{k-1}}{(k-1)!})\\ &= I+\sum_{k=2}^{\infty}k\boldsymbol{f}'(\boldsymbol{x}^{(0)})^{-1}\boldsymbol{f}^{(k)}(\boldsymbol{x}^{(0)})\frac{\boldsymbol{y}^{k-1}}{k!}.\end{aligned}$$

If (5) holds, then

$$\begin{aligned}\|I-\boldsymbol{f}'(\boldsymbol{x}^{(0)})^{-1}\boldsymbol{f}'(\boldsymbol{x}^{(0)}+\boldsymbol{y})\| &\le \sum_{k=2}^{\infty}k\|\boldsymbol{f}'(\boldsymbol{x}^{(0)})^{-1}\frac{\boldsymbol{f}^{(k)}(\boldsymbol{x}^{(0)})}{k!}\|\|\boldsymbol{y}\|^{k-1}\\ &\le \sum_{k=2}^{\infty}k(\gamma\rho)^{k-1}\\ &= \frac{1}{(1-\gamma\rho)^2}-1.\end{aligned}$$

Suppose that (5) (6) hold, then for an arbitrary real vector $\boldsymbol{y} \in \mathbf{X}_\rho$, we can infer that

$$\|-\boldsymbol{f}'(\boldsymbol{x}^{(0)})^{-1}\boldsymbol{f}(\boldsymbol{x}^{(0)})+(I-\boldsymbol{f}'(\boldsymbol{x}^{(0)})^{-1}\boldsymbol{f}'(\boldsymbol{x}^{(0)}+\boldsymbol{y}))\boldsymbol{y}\|<\rho,$$

which implies that (7) holds. □

Lemma 4. *Let*

$$\begin{aligned}\alpha^* &= -\frac{1}{12}(22247+1320\sqrt{330})^{1/3}+\frac{431}{12(22247+1320\sqrt{330})^{1/3}}+\frac{25}{12} \qquad (8)\\ &\approx 0.093347623,\end{aligned}$$

then for an arbitrary $0 < \alpha < \alpha^*$, *we have*

$$\begin{aligned}
&g_1(\alpha^*) < g_1(\alpha) < g_1(0),\\
&g_2(0) < g_2(\alpha) < g_2(\alpha^*),\\
&g_3(0) < g_3(\alpha) < g_3(\alpha^*) = 0,\\
&\theta(\alpha^*) < \theta(\alpha) < \theta(0),\\
&\omega_1(0) < \omega_1(\alpha) < \omega_1(\alpha^*),\\
&\omega_2(\alpha^*) < \omega_2(\alpha) < \omega_2(0) < 3 - \frac{3\sqrt{2}}{2},\\
&\omega_3(\alpha) > 3 - \frac{3\sqrt{2}}{2}.
\end{aligned}$$

Proof. According to Cartan's root-finding formula, the equation $g_3(\alpha) = 0$ has only a positive real root α^* as in (8). Obviously, $g_1'(\alpha) < 0$ and $g_2'(\alpha) > 0$ for all $0 < \alpha < \alpha^*$. Thus both $\theta(\alpha)$ and $\omega_2(\alpha)$ monotonously decrease on the interval $(0, 3 - 2\sqrt{2})$. A routine computation gives rise to

$$\omega_1'(\alpha) = \frac{(4-2\alpha)\sqrt{-3g_3(\alpha)}\sin(\theta(\alpha)+\frac{\pi}{6}) + (84-36\alpha)\cos(\theta(\alpha)+\frac{\pi}{6})}{2\sqrt{-3g_1(\alpha)g_3(\alpha)}} + \frac{1}{2},$$

then for all $0 < \alpha < \alpha^*$, $\omega_1'(\alpha) > 0$. This lemma follows immediately from what we have proved. □

Proposition 1. *Under the condition (5), the inequality (6) holds if and only if*

$$0 < \alpha < \alpha^*, \tag{9}$$

$$\frac{\omega_1(\alpha)}{3\gamma} < \rho < \frac{\omega_2(\alpha)}{3\gamma}. \tag{10}$$

Proof. Define $\Delta = (q/2)^2 + (p/3)^3$ with

$$\begin{aligned}
p &= -\frac{1}{3}\left(\frac{4+\beta\gamma}{-2\gamma}\right)^2 + \frac{1+2\beta\gamma}{2\gamma^2},\\
q &= 2\left(\frac{4+\beta\gamma}{-6\gamma}\right)^3 + \frac{\beta}{2\gamma^2} - \frac{(4+\beta\gamma)(1+2\beta\gamma)}{-12\gamma^3},
\end{aligned}$$

then

$$p = -\frac{g_1(\alpha)}{12\gamma^2}, \quad q = -\frac{g_2(\alpha)}{108\gamma^3}, \quad \Delta = \frac{g_3(\alpha)}{1728\gamma^6}.$$

It follows by Lemma 4 that for all $\alpha > 0$, $p < 0$ and $q < 0$, then we have three cases to consider.

Case I: $\alpha > \alpha^*$. In this case, $\Delta > 0$ and the equation

$$\rho^3 - \frac{4+\beta\gamma}{2\gamma}\rho^2 + \frac{1+2\beta\gamma}{2\gamma^2}\rho - \frac{\beta}{2\gamma^2} = 0 \tag{11}$$

has only one real solution

$$\rho^* = \sqrt[3]{-\frac{q}{2} + \sqrt{\Delta}} + \sqrt[3]{-\frac{q}{2} - \sqrt{\Delta}}.$$

An easy computation yields that for all $\alpha > \alpha^*$, $\rho^* > (1 - \sqrt{2}/2)/\gamma$. Therefore, in this case, (6) holds if and only if $\rho > \rho^*$, which contradicts the condition (5).

Case II: $\alpha = \alpha^*$. In this case, $\Delta = 0$ and (11) has only two unequal real solutions

$$-\sqrt[3]{\frac{-q}{2}}, \quad 2\sqrt[3]{\frac{-q}{2}}.$$

Clearly,

$$-\sqrt[3]{\frac{-q}{2}} < 0, \quad 2\sqrt[3]{\frac{-q}{2}} > \frac{1 - \sqrt{2}/2}{\gamma}.$$

Hence in this case, (6) can not hold under the condition (5).

Case III: $0 < \alpha < \alpha^*$. In this case, $\Delta < 0$ and (11) has three unequal real solutions

$$\frac{\omega_1(\alpha)}{3\gamma}, \quad \frac{\omega_2(\alpha)}{3\gamma}, \quad \frac{\omega_3(\alpha)}{3\gamma}.$$

Recalling Lemma 4, we know that for all $0 < \alpha < \alpha^*$,

$$\frac{\omega_3(\alpha)}{3\gamma} > \frac{1 - \sqrt{2}/2}{\gamma},$$
$$\frac{\omega_2(\alpha)}{3\gamma} < \frac{1 - \sqrt{2}/2}{\gamma},$$
$$0 < \frac{\omega_1(\alpha)}{3\gamma} < \frac{\omega_2(\alpha)}{3\gamma}.$$

Thus in this case, (6) holds if and only if (10) holds.

As a whole, under the condition (5), the inequality (6) holds if and only if (9) and (10) hold. □

On the basis of α, β and γ, the following theorem provides the conditions such that (7) holds, which is an immediate conclusion of Proposition 1.

Theorem 3. *If $0 < \alpha < \alpha^*$, then for any ρ satisfying the inequality*

$$\frac{\omega_1(\alpha)}{3\gamma} < \rho < \frac{\omega_2(\alpha)}{3\gamma},$$

the condition (7) holds.

The following corollary indicates that the alpha theory is of greater precision than Rump's interval theorem, which can be directly deduced by an easy computation.

Corollary 1. *If* $0 < \alpha < \alpha^*$, *then*

$$\frac{\omega_1(\alpha)}{3\gamma} > t^*.$$

For the Newton iteration sequence $\{\boldsymbol{x}^{(k)}\}$, define

$$\rho^*(\boldsymbol{f}, \boldsymbol{x}^{(k)}) = \frac{\omega_1(\alpha(\boldsymbol{f}, \boldsymbol{x}^{(k)}))}{3\gamma(\alpha(\boldsymbol{f}, \boldsymbol{x}^{(k)}))}, \quad \rho^{**}(\boldsymbol{f}, \boldsymbol{x}^{(k)}) = \frac{\omega_2(\alpha(\boldsymbol{f}, \boldsymbol{x}^{(k)}))}{3\gamma(\alpha(\boldsymbol{f}, \boldsymbol{x}^{(k)}))}. \tag{12}$$

In view of the quantity α of the alpha theory proposed by Wang and Han [18], we give the following convergence condition of interval Newton's algorithm proposed by Rump.

Proposition 2. *Let* $\lceil\cdot\rceil$ *be the integer ceiling function,* $p(\alpha)$ *be defined by*

$$p(\alpha) = \frac{2\alpha}{1 + \alpha + \sqrt{1 - 6\alpha + \alpha^2}},$$

and $q(\alpha)$ *be defined in (2). If* $0 < \alpha < 3 - 2\sqrt{2}$, *then for any*

$$k \geq \lceil \log_2(\frac{\ln \alpha^* + \ln(1 - p(\alpha)) + \ln \psi(p(\alpha)) - \ln \alpha}{\ln q(\alpha)} + 1)\rceil, \tag{13}$$

the condition

$$S(\boldsymbol{Y}_{\rho^{(k)}}, \boldsymbol{x}^{(k)}) \subseteq \mathrm{int}(\boldsymbol{Y}_{\rho^{(k)}}) \tag{14}$$

holds for any $\rho^{(k)}$ *satisfying the inequality*

$$\rho^*(\boldsymbol{f}, \boldsymbol{x}^{(k)}) < \rho^{(k)} < \rho^{**}(\boldsymbol{f}, \boldsymbol{x}^{(k)}). \tag{15}$$

Proof. By Theorem 1 and Lemma 2, we know that if $0 < \alpha < 3 - 2\sqrt{2}$, then for any $k \in \mathbb{N}$,

$$\alpha(\boldsymbol{f}, \boldsymbol{x}^{(k)}) \leq \frac{\alpha q(\alpha)^{2^k - 1}}{\psi(\gamma\|\boldsymbol{x}^{(k)} - \boldsymbol{x}^{(0)}\|)(1 - \gamma\|\boldsymbol{x}^{(k)} - \boldsymbol{x}^{(0)}\|)}.$$

If $0 < \alpha < 3 - 2\sqrt{2}$, then for all $k \in \mathbb{N}$,

$$\gamma\|\boldsymbol{x}^{(k)} - \boldsymbol{x}^{(0)}\| \leq p(\alpha) < 1 - \frac{\sqrt{2}}{2},$$

which implies that for all $k \in \mathbb{N}$,

$$\psi(\gamma\|\boldsymbol{x}^{(k)} - \boldsymbol{x}^{(0)}\|)(1 - \gamma\|\boldsymbol{x}^{(k)} - \boldsymbol{x}^{(0)}\|) \geq \psi(p(\alpha))(1 - p(\alpha)).$$

Hence if $0 < \alpha < 3 - 2\sqrt{2}$, then for all $k \in \mathbb{N}$,

$$\alpha(\boldsymbol{f}, \boldsymbol{x}^{(k)}) \leq \frac{\alpha q(\alpha)^{2^k - 1}}{\psi(p(\alpha))(1 - p(\alpha))},$$

which implies that $0 < \alpha(\boldsymbol{f}, \boldsymbol{x}^{(k)}) < \alpha^*$ holds if the iteration number k satisfies the inequality (13). Our conclusion will follow from Theorem 3. □

With the aid of the quantity α of the alpha theory given in [1], the conclusion of the above proposition can be improved as follows.

Proposition 3. *If* $0 < \alpha \leq (13 - 3\sqrt{17})/4$, *then for any* $k \geq 3$, *the condition (14) holds for any* $\rho^{(k)}$ *satisfying the inequality (15).*

Proof. Since the function $\alpha/\psi(\alpha)^2$ monotonously increases on the interval $[0, 1 - \sqrt{2}/2)$, it follows that $\alpha/\psi(\alpha)^2 < 1$ for any $0 < \alpha \leq (13 - 3\sqrt{17})/4$. Recalling Proposition 1 in [15], we can deduce that if $0 < \alpha \leq (13 - 3\sqrt{17})/4$, then for any $k \in \mathbb{N}$,

$$\alpha(\boldsymbol{f}, \boldsymbol{x}^{(k)}) \leq \left(\frac{\alpha}{\psi(\alpha)^2}\right)^{2^k - 1} \alpha.$$

As a result, if $0 < \alpha \leq (13 - 3\sqrt{17})/4$, then $0 < \alpha(\boldsymbol{f}, \boldsymbol{x}^{(3)}) < \alpha^*$. □

4 Example

In this section, we propose some examples to illustrate our conclusion, which are done in Matlab R2012a with INTLAB V6 under Windows 7. In these examples, the true interval of the display as $-0.0059\underline{}$ is obtained by subtracting and adding 1 to the last displayed digit, namely,

$$-0.0059\underline{} = [-0.00600000000000, -0.00580000000000].$$

Example 1. Let

$$f_1 = x_1^2 + x_2 - 2 = 0,$$
$$f_2 = x_1 - x_2 = 0,$$

and $\boldsymbol{x}^{(0)} = (1.8, 1.8)^T$. The program of the alpha theory computes

$$\alpha^* < \alpha = 0.1436672967 < \frac{13 - 3\sqrt{17}}{4},$$

then by Proposition 3, we can immediately deduce that $0 < \alpha(\boldsymbol{f}, \boldsymbol{x}^{(3)}) < \alpha^*$. The values of $\alpha(\boldsymbol{f}, \boldsymbol{x}^{(k)})$, $\rho^*(\boldsymbol{f}, \boldsymbol{x}^{(k)})$, $\rho^{**}(\boldsymbol{f}, \boldsymbol{x}^{(k)})$, $k = 1, 2, 3$, are shown in Table 1, and the values of $\boldsymbol{x}^{(k)}$, $\rho^{(k)}$, $S(\mathbf{Y}_{\rho^{(k)}}, \boldsymbol{x}^{(k)})$, $k = 1, 2, 3$, are shown in Table 2.

Example 2. [11] Let

$$f_i(\boldsymbol{x}) = x_i - 0.7 x_i \sum_{j=1}^{9} a_{i,j} x_j - 1 = 0, \qquad i = 1, 2 \ldots, 9,$$

with

$$a_{i,j} = \frac{3}{4} \frac{t_i (1 - t_j^2)^2 w_j}{t_i + t_j},$$

Table 1. The values of $\alpha(\boldsymbol{f}, \boldsymbol{x}^{(k)})$, $\rho^*(\boldsymbol{f}, \boldsymbol{x}^{(k)})$ and $\rho^{**}(\boldsymbol{f}, \boldsymbol{x}^{(k)})$ about Example 1

k	$\alpha(\boldsymbol{f}, \boldsymbol{x}^{(k)})$	$\rho^*(\boldsymbol{f}, \boldsymbol{x}^{(k)})$	$\rho^{**}(\boldsymbol{f}, \boldsymbol{x}^{(k)})$
1	0.04063913776	0.1474393618	0.8654682610
2	0.001956685388	0.005916476543	0.8785582303
3	0.000003858607108	0.00001157539288	0.8785582303

Table 2. The values of $\boldsymbol{x}^{(k)}$, $\rho^{(k)}$ and $S(\mathbf{Y}_{\rho^{(k)}}, \boldsymbol{x}^{(k)})$ about Example 1

k	$\boldsymbol{x}^{(k)}$	$\rho^{(k)}$	$S(\mathbf{Y}_{\rho^{(k)}}, \boldsymbol{x}^{(k)})$
1	$\begin{pmatrix} 1.13913043490000 \\ 1.13913043490000 \end{pmatrix}$	0.147440361800000	$\begin{pmatrix} [-0.14648800759416, -0.11996338231131] \\ [-0.14648800759416, -0.11996338231131] \end{pmatrix}$
2	$\begin{pmatrix} 1.00590473980000 \\ 1.00590473980000 \end{pmatrix}$	0.005917476543000	$\begin{pmatrix} -0.0059\_\_\_\_\_\_\_\_ \\ -0.0059\_\_\_\_\_\_\_\_ \end{pmatrix}$
3	$\begin{pmatrix} 1.00001157619900 \\ 1.00001157619900 \end{pmatrix}$	1.257539288000000e-05	$1.0\text{e-}005 * \begin{pmatrix} -0.11576\_\_\_\_\_\_\_\_ \\ -0.11576\_\_\_\_\_\_\_\_ \end{pmatrix}$

where $t_i, w_i, i = 1, 2, \ldots, 9$, are respectively the nodes and weights of Gaussian integration rule of order 17 on the interval $[0, 1]$ (See Table 3).

For the initial approximate

$$\boldsymbol{x}^{(0)} = [1.36, 1.36, 1.36, 1.36, 1.36, 1.36, 1.36, 1.36, 1.36]^T,$$

we have

$$\alpha^* < \alpha = 0.144327213206543 < \frac{13 - 3\sqrt{17}}{4}.$$

It follows by Proposition 3 that $0 < \alpha(\boldsymbol{f}, \boldsymbol{x}^{(3)}) < \alpha^*$. To be precise,

$$\alpha(\boldsymbol{f}, \boldsymbol{x}^{(3)}) = 9.771724278220212\text{e-}10,$$
$$\rho^*(\boldsymbol{f}, \boldsymbol{x}^{(3)}) = 2.106476714805001\text{e-}09,$$
$$\rho^{**}(\boldsymbol{f}, \boldsymbol{x}^{(3)}) = 0.721018979786819.$$

Choose $\rho^{(3)} = 2.206476714805001\text{e-}09$, then $S(\mathbf{Y}_{\rho^{(3)}}, \boldsymbol{x}^{(3)}) \subseteq \text{int}(\mathbf{Y}_{\rho^{(3)}})$. The values of $\boldsymbol{x}^{(3)}$ and $S(\mathbf{Y}_{\rho^{(3)}}, \boldsymbol{x}^{(3)})$ are shown in Table 4, where i stands for the coordinate index.

Example 3. Let

$$f_i = x_i^2 + x_{i+1} - 2 = 0, \qquad i = 1, 2, \ldots, 99,$$
$$f_{100} = x_{99} - x_{100} = 0.$$

Given $\boldsymbol{x}^{(0)}$ with $x_i^{(0)} = 1.28$, $i = 1, 2, \ldots, 100$, it follows that

$$\frac{13 - 3\sqrt{17}}{4} < \alpha = 0.165370210314031 < 3 - 2\sqrt{2}.$$

Table 3. The values of t_i, w_i of Example 2

i	t_i	w_i
1	0.015919880000000	0.040637193262940
2	0.081984445000000	0.090324080584283
3	0.193314285000000	0.130305351576427
4	0.337873290000000	0.156173536255424
5	0.500000000000000	0.165119676661738
6	0.662126710000000	0.156173536254610
7	0.806685715000000	0.130305351576465
8	0.918015555000000	0.090324080583571
9	0.984080120000000	0.040637193262741

Table 4. The values of $\boldsymbol{x}^{(3)}$ and $S(\mathbf{Y}_{\rho^{(3)}}, \boldsymbol{x}^{(3)})$ about Example 2

i	$\boldsymbol{x}^{(3)}$	$S(\mathbf{Y}_{\rho^{(3)}}, \boldsymbol{x}^{(3)})$
1	1.03266743259998	1.0e-008 $*$ -0.2175469960____
2	1.10583043469589	1.0e-008 $*$ 0.058000438_____
3	1.17693975483393	1.0e-008 $*$ 0.074880771_____
4	1.23474234890867	1.0e-008 $*$ -0.046156608_____
5	1.27801354526920	1.0e-008 $*$ 0.033335094_____
6	1.30888887603008	1.0e-008 $*$ -0.052842004_____
7	1.32995480221212	1.0e-008 $*$ 0.015004650_____
8	1.34328756776955	1.0e-008 $*$ -0.029283159_____
9	1.35027189327776	1.0e-008 $*$ 0.057999652_____

Recalling Proposition 2, we know that there exists $k \in \mathbb{N}$ such that $\alpha(\boldsymbol{f}, \boldsymbol{x}^{(k)}) < \alpha^*$. Indeed, for the first-step iteration point $\boldsymbol{x}^{(1)}$, we have

$$\begin{aligned}\alpha(\boldsymbol{f}, \boldsymbol{x}^{(1)}) &= 0.0209408111113277 < \alpha^*,\\ \rho^*(\boldsymbol{f}, \boldsymbol{x}^{(1)}) &= 0.0229019461130693,\\ \rho^{**}(\boldsymbol{f}, \boldsymbol{x}^{(1)}) &= 0.291551749088838.\end{aligned}$$

Choose $\rho^{(1)} = 0.0229019461130694$, then $S(\mathbf{Y}_{\rho^{(1)}}, \boldsymbol{x}^{(1)}) \subseteq \mathrm{int}(\mathbf{Y}_{\rho^{(1)}})$.

References

1. Blum, L., Cucker, F., Shub, M., Smale, S.: Complexity and Real Computation. Springer, New York (1998)
2. Kantorovich, L.V.: Functional analysis and applied mathematics. Uspehi. Mat. Nauk. **3**(6), 89–185 (1948)

3. Giusti, M., Lecerf, G., Salvy, B., Yakoubsohn, J.C.: On location and approximation of clusters of zeros of analytic functions. Found. Comput. Math. **5**(3), 257–311 (2005)
4. Giusti, M., Lecerf, G., Salvy, B., Yakoubsohn, J.C.: On location and approximation of clusters of zeros: case of embedding dimension one. Found. Comput. Math. **7**(1), 1–58 (2007)
5. Hauenstein, J.D., Levandovskyy, V.: Certifying solutions to square systems of polynomial-exponential equations. J. Symb. Comput. **79**(3), 575–593 (2015)
6. Hauenstein, J.D., Sottile, F.: Algorithm 921: alphacertified: certifying solutions to polynomial systems. ACM Trans. Math. Softw. **38**(4), 1–20 (2011)
7. Li, N., Zhi, L.: Verified error bounds for isolated singular solutions of polynomial systems: case of breadth one. Theor. Comput. Sci. **479**, 163–173 (2013)
8. Li, N., Zhi, L.: Verified error bounds for isolated singular solutions of polynomial systems. SIAM J. Numer. Anal. **52**(4), 1623–1640 (2014)
9. Krawczyk, R.: Newton-algorithmen zur bestimmung von nullstellen mit fehlerschranken. Computing **4**, 187–201 (1969)
10. Moore, R.E.: A test for existence of solutions to nonlinear system. SIAM J. Numer. Anal. **14**(4), 611–615 (1977)
11. Rall, L.B.: A comparison of the existence theorems of Kantorvich and Moore. SIAM J. Numer. Anal. **17**(1), 148–161 (1980)
12. Rump, S.M.: Solving algebraic problems with high accuracy. In: Kulisch, W.L., Miranker, W.L. (eds.) A New Approach to Scientific Computation, pp. 51–120. Academic Press, San Diego (1983)
13. Rump, S.M.: INTLAB-Interval Laboratory. Springer, Netherlands, Berlin (1999)
14. Rump, S.M., Graillat, S.: Verified error bounds for multiple roots of systems of nonlinear equations. Numer. Algorithm **54**, 359–377 (2009)
15. Smale, S.: Newton method estimates from data at one point. The Merging of Disciplines: New Directions in Pure, Applied and Computational Mathematics, pp. 185C–196C. Springer, Berlin (1986)
16. Shub, M., Smale, S.: Computational complexity: on the geometry of polynomials and a theory of cost. I. Ann. Sci. Ècole Norm. Sup. **18**(1), 107–142 (1985)
17. Shub, M., Smale, S.: Computational complexity: on the geometry of polynomials and a theory of cost. II. SIAM J. Comput. **15**(1), 145–161 (1986)
18. Wang, X.H., Han, D.F.: On dominating sequence method in the point estimate and Smale theorem. Sci. China Ser. A **33**(2), 135–144 (1990)
19. Yang, Z., Zhi, L., Zhu, Y.: Verfied error bounds for real solutions of positive-dimensional polynomial systems. In: The 2013 International Symposium on Symbolic and Algebraic Computation, pp. 371–378. ACM press, San Jose (2013)

Normalization of Indexed Differentials Based on Function Distance Invariants

Jiang Liu(✉)

Department of Systems Science, University of Shanghai for Science and Technology, Shanghai 200093, China
jliu113@126.com

Abstract. This paper puts forward the method of function distance invariant, and develops an efficient normalization algorithm for indexed differentials. The algorithm allows us to determine the equivalence of indexed differentials in $\mathcal{R}_2[\not\partial]$, and is mainly based on two algorithms. One is an index replacement algorithm. The other is a normalization algorithm with respect to monoterm symmetries, whose complexity is lower than known algorithms.

1 Introduction

Differential geometry often involves massive calculation of indexed differentials, such as tensor verification problem or the problem of finding transformation rules of indexed functions under the transformation of local coordinates. The following examples are typical in differential geometry.

Example 1. Let h^i_j be a (1, 1)-typed tensor. Prove that

$$H^i_{jk} = -\frac{1}{8}(h^p_j\partial_p h^i_k - h^p_k\partial_p h^i_j) + \frac{1}{8}(h^i_p\partial_j h^p_k - h^i_p\partial_k h^p_j)$$

is a (1, 2)-typed tensor, i.e., it satisfies the following transformation rule of (1, 2)-typed tensor:

$$H^{i'}_{j'k'} - \not\partial^{i'}_{i}\not\partial^{j}_{j'}\not\partial^{k}_{k'}H^i_{jk} = 0. \tag{1}$$

Example 2. A Riemannian manifold M^n admits an almost complex structure $\mathbf{J}^i_j$. Let H^i_{jk} be 1/8 times the Nijenhuis tensor of $\mathbf{J}^i_j$. The "torsional derivative" of a tensor field on M^n is defined as follows:

$$T^{i_1\cdots i_a}_{j_1\cdots j_b||rs} = H^p_{rs}\frac{\partial T^{i_1\cdots i_a}_{j_1\cdots j_b}}{\partial x^p} + \sum_{u=1}^{a} T^{i_1\cdots i_{u-1}pi_{u+1}\cdots i_a}_{j_1\cdots j_b}h^{i_u}_{prs} - \sum_{v=1}^{b} T^{i_1\cdots i_a}_{j_1\cdots j_{v-1}pj_{v+1}\cdots j_b}h^p_{j_v rs},$$

where h^i_{jrs} has the following "rather strange" definition:

$$2h^i_{jrs} = \mathbf{J}^i_p(\mathbf{J}^q_j\frac{\partial H^p_{rs}}{\partial x^q} - H^q_{rs}\frac{\partial \mathbf{J}^p_j}{\partial x^q} + H^p_{qs}\frac{\partial \mathbf{J}^q_j}{\partial x^r} - H^p_{qr}\frac{\partial \mathbf{J}^q_j}{\partial x^s}) - \frac{\partial H^i_{rs}}{\partial x^j}.$$

What is the transformation rule of h^i_{jrs} under coordinate transformation?

V.P. Gerdt et al. (Eds.): CASC 2017, LNCS 10490, pp. 285–300, 2017.
DOI: 10.1007/978-3-319-66320-3_21

However, the simplification of indexed differential expressions is tricky and cumbersome to perform by manual calculations. For example, it is not easy to prove that the following two monomials are equivalent:

$$\frac{\partial^2 x^f}{\partial x^f \partial x^{c'}} \frac{\partial^2 x^{c'}}{\partial x^{a'} \partial x^b} \frac{\partial^2 x^r}{\partial x^q \partial x^{p'}} \frac{\partial^2 x^{p'}}{\partial x^{s'} \partial x^r} \frac{\partial x^q}{\partial x^{q'}} \frac{\partial^2 x^{s'}}{\partial x^{l'} \partial x^s} \frac{\partial x^s}{\partial x^{k'}} \frac{\partial^2 x^b}{\partial x^j \partial x^{d'}} \frac{\partial^2 x^{d'}}{\partial x^{g'} \partial x^h} \frac{\partial^2 x^j}{\partial x^m \partial x^{n'}};$$

$$\frac{\partial^2 x^{f'}}{\partial x^{f'} \partial x^c} \frac{\partial^2 x^c}{\partial x^{b'} \partial x^{a'}} \frac{\partial^2 x^r}{\partial x^{s'} \partial x^{q'}} \frac{\partial^2 x^{s'}}{\partial x^r \partial x^s} \frac{\partial^2 x^s}{\partial x^{l'} \partial x^{k'}} \frac{\partial^2 x^{b'}}{\partial x^{j'} \partial x^d} \frac{\partial^2 x^d}{\partial x^h \partial x^{g'}} \frac{\partial^2 x^{j'}}{\partial x^{n'} \partial x^m}.$$

Then a natural problem is that in computer algebra, how can we judge the equivalence of indexed differential expressions?

Symbolic manipulation of indexed expressions, e.g. tensor expressions, is one of the oldest research topics in computer algebra [1–13]. It remains to be a challenging problem, for the reason as follows: In order to compute the canonical form of an indexed polynomial, we need to find a finite Gröbner basis for the ideal generated by the basic syzygies. But unfortunately, the ideal cannot be finitely generated, mainly due to the property of dummy index renaming. Efforts have been made to describe algorithms for simplifying tensor expressions. For example, Refs. [5,6] presented algorithms to put tensor expressions into canonical forms with respect to monoterm symmetries and cyclic symmetry. However, those algorithms are not applicable for indexed differential expressions, because index elimination is indispensable for simplifying them.

Liu [14] presented a normalization algorithm for indexed differential expressions in $\mathcal{R}_2[\partial\!\!\!/]$, which consists of two parts. In the first part, a polynomial is rewritten modulo monoterm symmetries. In the second part, we compute the canonical form of a polynomial with respect to monoterm symmetries. Then two polynomials are equal if and only if they have the same canonical forms.

However, the algorithm with respect to monoterm symmetries in [14] has a factorial complexity. More precisely, suppose that f is a monomial indexed with i, j, k, where i is the number of functions, j is the number of pairs of dummy indices, and k is the number of pairs of commutable lower indices. Since we need to compare all the monomials equivalent to f, the complexity is at least $\mathrm{O}(i! \times j! \times 2^k)$. An other method for computing canonical forms with respect to monoterm symmetries is due to [6]: First replacing the dummy indices by numbers according to the index positions, the tensor names and index classes, then finding the smallest element in the equivalence class, i.e., finding the smallest from $i!n_1! \dots n_k!$ objects. So the complexity is at least $\mathrm{O}(i!n_1! \dots n_k!)$, where k is the number of groups of pairwise interchangeable lower indices, and n_i is the number of indices in each group.

The reason for such factorial complexity in some of these algorithms is that dummy indices within an indexed polynomial f are described either by original letters or by integers according to the order of their appearance. Neither of the descriptions is invariant. In other words, they vary when we rewrite f. Consequently, to find the canonical form, we need to list and compare all the elements in the equivalence class of f.

So a question arises: How do we describe an invariant (with respect to monoterm symmetries), and provide an efficient normalization algorithm based on the invariants?

Besides, the normalization algorithm in [14] depends on a skillful and tricky classification of 2nd-order partial differential functions according to their connections (e.g. circle, chain, maximum lower tree and so on). But the classification is no longer valid for higher order. It appears infeasible to classify partial differential functions for each order.

Then another question arises: How can we provide a normalization algorithm that is independent of function classifications?

To answer the questions, we first define the distance from one indexed function to another, and define the type list. Then we prove that they are both invariants. Next, we present an index replacement algorithm, and use it to develop a normalization algorithm with respect to monoterm symmetries for polynomials in $\mathcal{R}[\emptyset]$, whose complexity is less than the factorial complexity of existing algorithms, and reduces to at most $O(i^2)$ or $O(\prod_{j=1}^{k} C_{n_j}^2)$. Finally, by the method of index replacement, a normalization algorithm is provided for polynomials in $\mathcal{R}_2[\emptyset]$, which is independent of function classifications.

2 Indexed Differential Polynomial Ring

In this section we briefly review some notions in [14].

An *indexed function* is composed of four parts: a function name, a sequence of upper indices, a sequence of lower indices, and variables. For example, the Christoffel symbol $\Gamma^i_{kh} = \Gamma^i_{kh}(\mathbf{x})$ has the function name Γ, upper index i, lower indices k, h, and variable $\mathbf{x}$. An *indexed monomial* is the product of indexed functions and obeys Einstein summation convention. A *sub-monomial* refers to the product of some indexed functions within an indexed monomial. In an indexed monomial, a *free index* occurs only once, and a *dummy index* occurs twice, as an upper index and a lower one respectively. If a free index occurs as a lower (or upper) index, then it is called a *covariant* (or *contravariant*) free index.

Einstein summation convention has a basic property of renaming dummy indices (Ren): If $f(i)$ is an indexed monomial taking i as a dummy index, then for any j which does not occur in $f(i)$,

$$f(i) = f(j). \tag{2}$$

Let $\mathcal{A}$ be a non-associative ring generated by indexed functions, and I be the two-sided ideal generated by (2). The quotient ring $\mathcal{A}/I$ is called *Einstein summation ring*. The multiplication induced in $\mathcal{A}/I$ is called *Einstein multiplication*.

Let two overlapping local coordinate neighborhoods on an n-dimensional differentiable manifold be $(x^1, x^2, \dots, x^n)$ and $(x^{1'}, x^{2'}, \dots, x^{n'})$. The former coordinate component indices are denoted by small letters while the latter by small letters with apostrophes. If there is no need to make a distinction between different coordinate systems, the indices are denoted by capital letters.

A partial differential function $\eth^B_{A_1\ldots A_r}$ is defined by

$$\eth^B_{A_1\ldots A_r} := \partial_{A_1\ldots A_r} x^B.$$

The product of partial differential functions of order at most r is called an *rth order partial differential polynomial.*

All partial differential polynomials form a commutative ring under Einstein multiplication, which is denoted by $\mathcal{R}[\eth]$. The subring composed of partial differential polynomials of order at most r is denoted by $\mathcal{R}_r[\eth]$. In particular, $\mathcal{M}[\eth]$ and $\mathcal{M}_r[\eth]$ denote the monoids of monomials in $\mathcal{R}[\eth]$ and $\mathcal{R}_r[\eth]$, respectively.

As an algebraic ring, $\mathcal{R}[\eth]$ can be formally defined by the following syzygies.

(i) **Evaluations** (Eval): If A_r and B are indices of the same coordinate system, then $\eth^B_{A_1\ldots A_r}$ equals 0 if $r > 1$, and equals $\delta^B_{A_1}$ if $r = 1$.

(ii) **Unifying symmetry** (US): If $A_{i_1}, A_{i_1+1}, \ldots, A_{i_1+s}$ are indices in the same coordinate system, $1 \le i_1 \le r - s$, and $A_{j_0}, A_{j_1}, \ldots, A_{j_s}$ is a permutation of $A_{i_1}, A_{i_1+1}, \ldots, A_{i_1+s}$, then

$$\eth^B_{A_1\ldots A_r} = \eth^B_{A_1\ldots A_{j_0} A_{j_1}\ldots A_{j_s}\ldots A_r}. \tag{3}$$

(iii) **Kronecker rule** (Kron): For any indexed monomial $M^{B_1\ldots B_r}_{A_1\ldots A_s}$, if $1 \le i \le r$ and $1 \le j \le s$, then

$$\delta^{A_i}_{B_j} M^{B_1\ldots B_r}_{A_1\ldots A_s} = M^{B_1\ldots B_r}_{A_1\ldots A_{i-1} B_j A_{i+1}\ldots A_s} = M^{B_1\ldots B_{j-1} A_i B_{j+1}\ldots B_r}_{A_1\ldots A_s}.$$

(iv) **Jacobi rule**:

$$\eth^{j'}_i \, \eth^B_{j' A_1\ldots A_s} = \eth^B_{i A_1\ldots A_s}.$$

(v) **Leibniz rule**: For any sequence of indices $I = C_1 C_2 \ldots C_t$,

$$\sum_{i=0}^{t} \sum_{(i,t-i)\vdash I} \eth^{j'}_{I_{(1)} i} \; \eth^B_{I_{(2)} j' A_1\ldots A_s} = \eth^B_{I i A_1\ldots A_s}.$$

On the other hand, $\mathcal{R}[\eth]$ obviously is also a differential ring, with the partial differential operator formally defined by the following properties.

(i') **Evaluations** (Eval): If f is a constant, *e.g.* $f = \delta^j_i$, then $\partial_A f = 0$.

(ii') **Unifying symmetry** (US): If for some $1 \le i_1 < i_2 \le r$, $A_{i_1}, A_{i_1+1}, \ldots, A_{i_2}$ are indices of the same coordinate system, then $\partial_{A_1\ldots A_r}$ is symmetric in $A_{i_1}, A_{i_1+1}, \ldots, A_{i_2}$.

(iii') **Jacobi rule**:

$$\eth^{j'}_i \partial_{j'} = \partial_i.$$

(iv') **Leibniz rule**: For any differentiable functions f, g, and any sequence of indices $I = C_1C_2 \ldots C_t$,

$$\partial_I(fg) = \sum_{i=0}^{t} \sum_{(i,t-i)\vdash I} (\partial_{I_{(1)}} f)(\partial_{I_{(2)}} g).$$

The following is a generated syzygy.

(v') **Bottom antisymmetry** (BS):

$$\partial\!\!\!/^{j'}_{k'i}\partial_{j'} = -\partial\!\!\!/^{j}_{ik'}\partial_j. \tag{4}$$

By commutativity of multiplication (Com), we mean

$$\partial\!\!\!/^{B}_{A_1A_2\ldots A_m}\partial\!\!\!/^{D}_{C_1C_2\ldots C_n} = \partial\!\!\!/^{D}_{C_1C_2\ldots C_n}\partial\!\!\!/^{B}_{A_1A_2\ldots A_m}, \tag{5}$$

where $\partial\!\!\!/^{B}_{A_1A_2\ldots A_m}$, $\partial\!\!\!/^{D}_{C_1C_2\ldots C_n}$ are two partial differential functions in an indexed monomial.

In what follows, we refer to Ren, US and Com as *monoterm symmetries.*

3 Distances Between Indexed Functions

In this section, we prove that the type lists and the distances are both invariants with respect to monoterm symmetries.

Definition 1. *Let $\overline{f} = \partial\!\!\!/^{B}_{A_1A_2\ldots A_m}$ be a partial differential indexed function. Suppose $\overline{f}$ has i lower dummy indices and j upper dummy indices. Define the type list of $\overline{f}$ as*

$$(m, i, j, L^{F_{dn}}, L^{F'_{dn}}, L^{F_{up}}, L^{F'_{up}}),$$

and denote it by $TL(\overline{f})$, where $L^{F_{dn}}$ (or $L^{F'_{dn}}$) denotes the sequence of all the covariant free indices without (or with) apostrophes in alphabetical order, and $L^{F_{up}}$ (or $L^{F'_{up}}$) denotes the sequence with respect to contravariant free indices.

Definition 2. *Suppose $f \in \mathcal{M}[\partial\!\!\!/]$, and the indexed functions from left to right of f are $\overline{f}_1$, $\overline{f}_2$, $\ldots$, $\overline{f}_n$. If $\overline{f}_i$ has m lower dummy indices occurring as upper ones in $\overline{f}_j$, then define the distance from $\overline{f}_i$ to $\overline{f}_j$ as m, and denote it by $d\langle\overline{f}_i, \overline{f}_j\rangle = m$.*

Suppose $\{\overline{g}_1, \overline{g}_2, \cdots, \overline{g}_t\}$ is an indexed function sequence. We define the adjacency matrix of the sequence by letting the (i, j)th element of the matrix be $d\langle\overline{g}_i, \overline{g}_j\rangle$ $(i, j = 1, 2, \ldots, t)$.

For any two $n \times n$ adjacency matrices A and B, if there exist $i_0, j_0 \leqslant n$, s.t. $a_{i_0j_0} < b_{i_0j_0}$; for $i = i_0, j < j_0$ and for $i < i_0$, $a_{ij} = b_{ij}$, then define $A < B$.

Definition 3. *Let $\overline{f}_i$ and $\overline{f}_j$ be two indexed functions. $(d\langle\overline{f}_i, \overline{f}_j\rangle, d\langle\overline{f}_j, \overline{f}_i\rangle)$ is called the distance pair between $\overline{f}_i$ and $\overline{f}_j$, and denoted by $D(\overline{f}_i, \overline{f}_j)$.*

Example 3. Consider the monomial $\not{\partial}^{b'}_{b'c}\not{\partial}^{c}_{eas'}\not{\partial}^{a}_{d'}\not{\partial}^{x}_{tpqr'}\not{\partial}^{w'}_{xl}\not{\partial}^{r'}_{w'zg}$. The indexed functions from left to right are denoted by $\overline{f}_1, \overline{f}_2, \ldots, \overline{f}_6$ respectively. According to Definition 2, we get the following nonzero distances

$$d\langle\overline{f}_1, \overline{f}_1\rangle = 1, d\langle\overline{f}_1, \overline{f}_2\rangle = 1, d\langle\overline{f}_2, \overline{f}_3\rangle = 1,$$
$$d\langle\overline{f}_4, \overline{f}_6\rangle = 1, d\langle\overline{f}_5, \overline{f}_4\rangle = 1, d\langle\overline{f}_6, \overline{f}_5\rangle = 1.$$

Notation. If $f_1, f_2 \in \mathcal{M}[\not{\partial}]$ are equivalent with respect to monoterm symmetries, i.e., $f_1 - f_2 \in \bar{0} = 0 + \mathcal{S}_{\text{mon}}$, where $\mathcal{S}_{\text{mon}}$ is the ideal generated by (2), (3) and (5), then denote it by $f_1 \overset{\text{mon}}{\sim} f_2$. Similarly, we write $f_1 \overset{\text{mon,BS}}{\sim} f_2$ if they are equivalent with respect to BS and monoterm symmetries.

Lemma 1. *If $f_1, f_2 \in \mathcal{M}[\not{\partial}]$ and $f_1 \overset{mon,BS}{\sim} f_2$, then f_1 can be rewritten as f_2 in a finite number of steps by Eqs. (2), (3), (4) and (5).*

Proof. We adapt the proof for $\mathcal{M}_2[\not{\partial}]$ in [15] to this case. In what follows, two indexed monomials are called like terms if they are identical except for constant coefficients.

Since f_1, f_2 are equivalent,

$$f_1 - f_2 = r_1(\bar{A}_1 - \bar{B}_1) + r_2(\bar{A}_2 - \bar{B}_2) + \ldots + r_n(\bar{A}_n - \bar{B}_n),$$

where $r_1, r_2, \ldots, r_n$ are nonzero monomials in $\mathcal{M}[\not{\partial}]$, $\bar{A}_i$ and $\bar{B}_i$ are the two sides of one of the Eqs. (2), (3), (4) and (5). Obviously, $r_i\bar{A}_i$ and $r_i\bar{B}_i$ are not like terms. Denote the set $\{r_i\bar{A}_i, -r_i\bar{B}_i | i = 1, \ldots, n\}$ by $\mathbf{E}$, and call $r_i\bar{A}_i$ the matching monomial of $-r_i\bar{B}_i$. For any subset $\mathbf{X}$ of $\mathbf{E}$, the matching monomials of all elements in $\mathbf{X}$ form a subset of $\mathbf{E}\backslash\mathbf{X}$, and we denote it by $\widetilde{\mathbf{X}}$.

If both f_1 and f_2 can be rewritten as 0, the conclusion holds obviously.

Otherwise, without loss of generality, assume that f_1 cannot be rewritten as 0. In $\mathbf{E}$, let $\mathbf{E}_1$ be $\{kf_1 \mid k \in \mathbb{C}\} \bigcap \mathbf{E}$, and let $\mathbf{E}_2$ be $\widetilde{\mathbf{E}_1}$. If none of the elements in $\mathbf{E}_2$ is a product of a constant and f_2, then we can construct $\mathbf{E}_3$ and $\mathbf{E}_4$ as follows. First in $\mathbf{E}\backslash \bigcup_{i=1}^{2} \mathbf{E}_i$, we can find the subset $\mathbf{E}_3$, such that the sum of all elements in $\mathbf{E}_3$ is the opposite of the sum of those in $\mathbf{E}_2$. Then, let $\mathbf{E}_4$ be $\widetilde{\mathbf{E}_3} \bigcap (\mathbf{E}\backslash\mathbf{E}_3)$.

Obviously $\mathbf{E}_4 \subseteq \mathbf{E}\backslash \bigcup_{i=1}^{3} \mathbf{E}_i$, since $\widetilde{\mathbf{E}_3} \subseteq \mathbf{E}\backslash \bigcup_{i=1}^{2} \mathbf{E}_i$.

We claim that $\mathbf{E}_4$ is not empty, for the reason as follows.

We find all the groups of like terms in $\mathbf{E}_2$, and denote them by $\sum_1, \sum_2, \ldots, \sum_p$. Let x_i be the term obtained by omitting the coefficient of any element in $\sum_i$ $(i = 1, \ldots, p)$. Since f_1 cannot be rewritten as 0, for any element $kx_i \in \mathbf{E}_2$ $(k \neq 0)$, there is a unique nonzero k_i such that the matching monomial of kx_i is $\frac{k}{k_i}f_1$. Let the sum of all elements in $\mathbf{E}_2$ be $a_1k_1x_1 + a_2k_2x_2 + \ldots + a_pk_px_p$, then by the definitions of $\mathbf{E}_1$ and $\mathbf{E}_2$, we get $a_1 + a_2 + \ldots + a_p = -1$. On the other hand, assume that $\mathbf{E}_4$ is empty. Then any element in $\mathbf{E}_3$ has its matching monomial lying in $\mathbf{E}_3$. Besides, the matching monomial of k_ix_i can only be k_jx_j $(i \neq j)$, therefore $a_1 + a_2 + \ldots + a_p = 0$, contradicting $a_1 + a_2 + \ldots + a_p = -1$. Hence $\mathbf{E}_4$ cannot be empty.

From the proof of the claim, we also get the following: Let $a'_1 k_1 x_1 + \ldots + a'_p k_p x_p$ be the result of collecting the like terms of $\widetilde{\mathbf{E}_4}$, then $a'_1 + \ldots + a'_p = 1$, since $(-a_1 - a'_1) + \ldots + (-a_p - a'_p) = 0$.

Similar to the process of constructing $\mathbf{E}_3$ and $\mathbf{E}_4$ from $\mathbf{E}_2$, if none of the elements in $\mathbf{E}_4$ is a product of a constant and f_2, we construct $\mathbf{E}_5$ and $\mathbf{E}_6$ from $\mathbf{E}_4$ ($\mathbf{E}_6 \subseteq \mathbf{E} \setminus \bigcup_{i=1}^{5} \mathbf{E}_i$). $\mathbf{E}_6$ must be non-empty, as otherwise any element in $\mathbf{E}_5$ has its matching monomial lying in $\mathbf{E}_5$, and it follows that $a'_1 + \ldots + a'_p = 0$, contradicting $a'_1 + \ldots + a'_p = 1$.

Generally, if none of the elements in $\mathbf{E}_{2n}$ ($n \in \mathbb{N}$) is a product of a constant and f_2, we can obtain a non-empty set $\mathbf{E}_{2n+2}$ ($\mathbf{E}_{2n+2} \subseteq \mathbf{E} \setminus \bigcup_{i=1}^{2n+1} \mathbf{E}_i$). Since $\mathbf{E}$ is finite, such a process cannot be endless, i.e., there exists m such that at least one element in $\mathbf{E}_{2m}$ is a product of a constant and f_2. Therefore by the definitions of $\mathbf{E}_i$, f_1 can be rewritten as cf_2 ($c \neq 0$). c must equal to 1, which can be proved by contradiction. Let us assume that c is unequal to 1. Then f_1 is equivalent to 0 since f_1 and f_2 are equivalent. Consequently, f_1 can be rewritten as a product of a constant and 0, contradicting that f_1 cannot be rewritten as 0.

Proposition 1. *Suppose* $f_1 \overset{mon}{\sim} f_2$.

(i) There is a bijection Ψ from the set of indexed functions of f_1 to that of f_2. For any indexed function $\overline{f}_i$ of f_1, $TL(\overline{f}_i) = TL(\Psi(\overline{f}_i))$.
(ii) For any two indexed functions $\overline{f}_i$ and $\overline{f}_j$, we have $d\langle \overline{f}_i, \overline{f}_j \rangle = d\langle \Psi(\overline{f}_i), \Psi(\overline{f}_j) \rangle$ (i may equal j).

Proof. According to Lemma 1, f_1 can be rewritten as f_2 in several steps. If the ith step is generated from US (see Eq. (3)), then define $\Psi_i(\not{\partial}^B_{A_1 \ldots A_r}) = \not{\partial}^B_{A_1 \ldots A_{j_0} A_{j_1} \ldots A_{j_s} \ldots A_r}$, and for any other indexed function $\overline{f}$, define $\Psi_i(\overline{f}) = \overline{f}$. Such Ψ_i is a bijection, and keeps invariant of type lists.

Since US does not involve interchange of indices among indexed functions, by the definition of distance, we have $d\langle \overline{f}_i, \overline{f}_j \rangle = d\langle \Psi(\overline{f}_i), \Psi(\overline{f}_j) \rangle$.

A similar argument is applied to Ren and Com.

Then the composition of all the mappings $\Psi_n \circ \ldots \circ \Psi_2 \circ \Psi_1$ forms a bijection denoted by Ψ, and $d\langle \overline{f}_i, \overline{f}_j \rangle = d\langle \Psi_1(\overline{f}_i), \Psi_1(\overline{f}_j) \rangle = d\langle \Psi_2 \circ \Psi_1(\overline{f}_i), \Psi_2 \circ \Psi_1(\overline{f}_j) \rangle = \ldots = d\langle \Psi(\overline{f}_i), \Psi(\overline{f}_j) \rangle$.

4 Normalization with Respect to Monoterm Symmetries

As mentioned in [14], to find the canonical form of a polynomial in $\mathcal{R}_2[\not{\partial}]$, it is sufficient to consider the problem in $\mathcal{M}_2[\not{\partial}]$. Hence in this section, we only consider $\mathcal{M}[\not{\partial}]$ and $\mathcal{M}_2[\not{\partial}]$.

According to Proposition 1, the type list and the distance are both invariants with respect to monoterm symmetries. In what follows, firstly we will sort indexed functions of a monomial based on the invariants, and then replace the indices with numbers according to the order on indexed functions. Finally an algorithm with respect to monoterm symmetries is presented, whose complexity is showed to be smaller than existing algorithms.

Notation. The result of applying Step 1 of the following Algorithm I to f is denoted by $f^{(\text{tri})}$.

Algorithm I. Indexed function rearrangement algorithm.
Input: All the indexed functions of f.
Output: A sequence of indexed functions.

Step 1. Let $\Omega = \emptyset$. Rewrite all the sub-monomials $\not\partial_A^A$ and $\not\partial_A^{A_1}\not\partial_{A_1}^{A_2}\ldots\not\partial_{A_r}^{A}$ (except for names of indices) $(r \geq 1)$ as 0 and n respectively, and denote the new monomial by $f^{(\text{tri})}$.

Step 2. Suppose $\overline{f}_i, \overline{f}_j$ are two indexed functions. Then $\overline{f}_i < \overline{f}_j$ if and only if one of the following conditions holds:
(a) $TL(\overline{f}_i) < TL(\overline{f}_j)$.
(b) $TL(\overline{f}_i) = TL(\overline{f}_j)$, and $d\langle\overline{f}_i, \overline{f}_i\rangle < d\langle\overline{f}_j, \overline{f}_j\rangle$.

Step 3. All the indexed functions of $f^{(\text{tri})}$ are classified as $\Gamma_1, \Gamma_2, \ldots, \Gamma_m$, such that the order on each class has not been defined, and if $i < j$, then all elements in Γ_i are smaller than those in Γ_j.

Step 4. For each element $\overline{f}_j^{(i1)}$ of Γ_{i1}, sort the elements of the set $\{D(\overline{f}_j^{(i1)}, \overline{f}_k^{(i2)}) | \forall \overline{f}_k^{(i2)} \in \Gamma_{i2}\}$ in ascending order, and get a sequence denoted by $L^{(D_j^{(i1,i2)})}$. Define $\overline{f}_j^{(i1)} < \overline{f}_l^{(i1)}$, if and only if

$$L^{(D_j^{(i1,1)})} < L^{(D_l^{(i1,1)})},$$

or

$$L^{(D_j^{(i1,1)})} = L^{(D_l^{(i1,1)})}, \ldots, L^{(D_j^{(i1,h-1)})} = L^{(D_l^{(i1,h-1)})}, L^{(D_j^{(i1,h)})} < L^{(D_l^{(i1,h)})}.$$

Step 5. If for each i, there is only one element in Γ_i, then put the sequence into Ω; If for each i, the order on Γ_i has not been defined, then go to Step 6. Otherwise, return to Step 3.

Step 6. Suppose the classes whose elements contain covariant free indices are $\Gamma_1^{(F_{dn})}, \Gamma_2^{(F_{dn})}, \ldots, \Gamma_s^{(F_{dn})}$. Consider all possible orders on each class, i.e., $m_1! \times m_2! \times \ldots \times m_s!$ possible orders, where m_i is the number of elements of $\Gamma_i^{(F_{dn})}$. Under each possibility, return to Step 3.

Step 7. Find the smallest among the adjacency matrices of the elements of Ω, and output the corresponding sequences, denoted by $[f]_1^{(\min)}, [f]_2^{(\min)}, \ldots, [f]_p^{(\min)}$.

Example 4. Sort the indexed functions of

$$f = \not\partial^{x'}_{ca}\not\partial^{y'}_{ab}\not\partial^{c}_{y'ghae'}\not\partial^{b}_{m'pqhx'}. \tag{6}$$

Step 1. Denote the indexed functions of f from left to right by $\overline{f}_1, \overline{f}_2, \overline{f}_3, \overline{f}_4$. By comparing type lists, we get $\overline{f}_1, \overline{f}_2 < \overline{f}_4 < \overline{f}_3$.
Step 2. The indexed functions of f are classified as $\Gamma_1=\{\overline{f}_1, \overline{f}_2\}$, $\Gamma_2=\{\overline{f}_4\}$, $\Gamma_3 = \{\overline{f}_3\}$.
Step 3. $L^{(D_1^{(1,2)})} = \{(0,1)\}$, $L^{(D_2^{(1,2)})} = \{(1,0)\}$. Hence, $L^{(D_1^{(1,2)})} < L^{(D_1^{(1,2)})}$, $\overline{f}_1 < \overline{f}_2$.

The following proposition prepares for the complexity analysis of the algorithms.

Proposition 2. *The set Ω in Step 7 of Algorithm I must be non-empty. In another word, if the order on $\Gamma_j^{(F_{dn})}$ $(j = 1, 2, \ldots, s)$ of Step 6 is defined, then by Algorithm I, the order on all indexed functions is defined.*

Proof. (a) Firstly, if $\overline{f}$ has a lower dummy index occurring in $\overline{g}$, and the order on the class containing $\overline{g}$ is defined, then by Algorithm I, the order on the class containing $\overline{f}$ can be defined. The reason is as follows. Suppose $\overline{f}'$ and $\overline{f}$ are in the same class, then $d\langle \overline{f}, \overline{g}\rangle = 1 \neq 0 = d\langle \overline{f}', \overline{g}\rangle$, consequently $D(\overline{f}, \overline{g}) \neq D(\overline{f}', \overline{g})$. Hence by Algorithm I, $\overline{f}$ and $\overline{f}'$ are ordered.

(b) If each element of a class Γ_i of Step 3 does not contain covariant free indices, then for any element $\overline{f}_0 \in \Gamma_i$, there must exist an sequence $\{\overline{f}_1, \overline{f}_2, \ldots, \overline{f}_r\}$, satisfying: $\overline{f}_i$ has a lower dummy index occurring in $\overline{f}_{i+1}$, $0 \leq i \leq r-1$, and $\overline{f}_r$ contains covariant free indices. Assume that such sequence does not exist, then it can be derived that $f^{(\text{tri})}$ must contain the submonomial $\not{\partial}_A^A$ or $\not{\partial}_A^{A_1} \not{\partial}_{A_1}^{A_2} \cdots \not{\partial}_{A_r}^{A}$ (except for index names), contradicting Step 1.

The class containing $\overline{f}_r$ belongs to $\{\Gamma_j^{(F_{dn})} | j = 1, 2, \ldots, s\}$, and the order on it is defined. Therefore, according to (a), the order on the class containing $\overline{f}_{r-1}$ can be defined by Algorithm I, and then it holds for $\overline{f}_{r-2}$, and so on, finally the order on the class Γ_i containing $\overline{f}_0$ is defined. This completes the proof.

Corollary 1. *If the covariant free indices in a monomial f have different names, then Step 6 and Step 7 can be skipped.*

Proof. Any two indexed functions have no covariant free indices in common, so their type lists must be different. Hence, the order on $\Gamma_i^{(F_{dn})}$ of Step 6 is defined by Step 2, and there is only one possibility in Step 6, which implies that Steps 6 and 7 can be skipped.

Algorithm II. Index replacement algorithm.
Input: $f \in \mathcal{M}[\not{\partial}]$, with an order on its indexed functions.
output: f.
Step 1. Replacement of the free indices of f.

1. Remove apostrophes from the free indices.
2. Replace the contravariant free indices alphabetically by $1, \ldots, N^{(F_{up})}$ (among which a positive integer may occur more than once, since some contravariant free indices may have identical names), $N^{(F_{up})} \geq 0$.

3. Replace the covariant free indices alphabetically by $1+N^{(F_{up})}$, $2+N^{(F_{up})}$, $\ldots$, $N^{(F_{dn})}+N^{(F_{up})}$, where $N^{(F_{dn})}$ is the number of covariant free indices.
4. Put apostrophes back.

Step 2. Replacement of the dummy indices of f.

1. Sort the indexed functions of f in ascending order.
2. Determine the order of replacement as follows. For any dummy index D, occurring as an upper index of the $i^{(D)}$th indexed function, as a lower one of the $j^{(D)}$th function, it corresponds to a pair $(i^{(D)}, j^{(D)})$. Then for any two dummy indices D_1, D_2, the replacement of D_2 is behind D_1 if and only if $i^{(D_1)} < i^{(D_2)}$, or $i^{(D_1)} = i^{(D_2)}$ and $j^{(D_1)} < j^{(D_2)}$.
3. Remove apostrophes from the dummy indices.
4. Replace the upper dummy indices in the order of replacement (determined in 2 of Step 2) by $N^{(F_{dn})}+N^{(F_{up})}+1, N^{(F_{dn})}+N^{(F_{up})}+2, \ldots, N^{(F_{dn})}+N^{(F_{up})}+N^{(D_{up})}$, where $N^{(D_{up})}$ is the number of upper dummy indices.
5. Replace the lower dummy indices in the order of replacement by $N^{(F_{dn})}+N^{(F_{up})}+N^{(D_{up})}+1, \ldots, N^{(F_{dn})}+N^{(F_{up})}+2N^{(D_{up})}$.
6. Put apostrophes back, and to avoid ambiguity, commas are added between any two neighbor lower indices within an indexed function.

Notation. In Algorithm II, each index is replaced by a positive integer n with or without an apostrophe. Denote the set $\{i, i' | i \in \mathbb{N}\}$ by $\mathfrak{N}'$.

Definition 4. *A total order $\prec$ on the set $\mathfrak{N}'$ is defined as follows. For any two positive integers a and b, define $a \prec b'$ if and only if $b - a > 0$.*

Definition 5. *Suppose $f \in \mathcal{M}[\mathcal{D}]$, and its indices are elements in $\mathfrak{N}'$. The numerical list of f, denoted by L_f, is identical to C, if Rep(f) is identical to a constant C. Otherwise list all the indices of f according to the order of appearance from left to right, and bottom to top.*

Definition 6. *A total order $\prec$ on the set of numerical lists is defined as follows. Suppose $f_1, f_2 \in \mathcal{M}[\mathcal{D}]$, and n_i is the number of elements of L_{f_i} $(i = 1, 2)$. $L_{f_1} \prec L_{f_2}$ if and only if*

$$n_1 < n_2$$

or

$$n_1 = n_2, L_{f_1}[1] < L_{f_2}[1]$$

or

$$n_1 = n_2, L_{f_1}[1] = L_{f_2}[1], \ldots, L_{f_1}[j] = L_{f_2}[j]; L_{f_1}[j+1] < L_{f_2}[j+1],$$

where $j \geq 1$, $L_{f_i}[k]$ denotes the kth element in the list L_{f_i}.

Algorithm III. Normalization with respect to monoterm symmetries.
Input: $f \in \mathcal{M}[\partial\!\!\!/]$.
Output: f.
Step 1. Apply Algorithm I to f to get indexed function sequences $[f]_1^{(\min)}$, $[f]_2^{(\min)}, \ldots, [f]_p^{(\min)}$.
Step 2. Apply Algorithm II to $[f]_1^{(\min)}, [f]_2^{(\min)}, \ldots, [f]_p^{(\min)}$, and get $f_1^{(\mathrm{Rep})}$, $f_2^{(\mathrm{Rep})}, \ldots, f_p^{(\mathrm{Rep})}$ respectively.
Step 3. For each $f_i^{(\mathrm{Rep})}$ $(1 \leq i \leq p)$, carry out the following steps to get $f_i^{(\mathrm{US})}$.

1. In each partial differential function $\partial\!\!\!/^{N_0}_{N_1 \ldots N_s}$, where $N_i \in \mathfrak{N}'$, find all the indices $N_{i1}, N_{i1+1}, \ldots, N_{i1+r}$ s.t. N_{i1+k} $(0 \leq k \leq r,\ i1 \geq 1)$ are in the same coordinate system, while N_{i1-1} and N_{i1+r+1} are in another one.
2. Sort $N_{i1}, N_{i1+1}, \ldots, N_{i1+r}$ in ascending order.

Step 4. Let $L_{f_q^{(\mathrm{US})}}$ be the smallest among $L_{f_i^{(\mathrm{US})}}$ $(1 \leq i \leq p)$. Output $f_q^{(\mathrm{US})}$.

Example 5. Normalize the monomial in (6) with respect to monoterm symmetries.

Step 1. By Example 4, $\overline{f}_1 < \overline{f}_2 < \overline{f}_4 < \overline{f}_3$.
Step 2. Replace the covariant free indices a, e', g, h, m', p, q alphabetically by $1, 2', 3, 4, 5', 6, 7$.
Step 3. The dummy indices c, x', b, y' correspond to $(4,1), (3,1), (3,2), (4,2)$ respectively. Hence $x' < b < c < y'$. Replace the upper indices x', b, c, y' by $8', 9, 10, 11'$ respectively, the lower ones by $12', 13, 14, 15'$.
Step 4. Rewrite $\partial\!\!\!/^{9}_{5',6,7,4,12'}, \partial\!\!\!/^{10}_{15',3,4,1,2'}$ as $\partial\!\!\!/^{9}_{5',4,6,7,12'}, \partial\!\!\!/^{10}_{15',1,3,4,2'}$ respectively.
Step 5. Output

$$\partial\!\!\!/^{8'}_{1,14} \partial\!\!\!/^{11'}_{1,13} \partial\!\!\!/^{9}_{5',4,6,7,12'} \partial\!\!\!/^{10}_{15',1,3,4,2'}.$$

Now we analyze the complexity of Algorithm III.

Let i be the number of indexed functions in a monomial, m the number of indexed functions that contain covariant free indices, k the number of groups of pairwise interchangeable lower indices, and n_i the number of indices in each group.

As mentioned in Sect. 1, the complexity of the algorithm in [6] is $\mathrm{O}(i! \prod_{j=1}^{k} n_j!)$ (which does not include the complexity of index replacement step in [6]). Hence, to compare with [6], we only need to consider Algorithm I and Step 3 of Algorithm III.

The indexed differential monomials that we have met in differential geometry all satisfy that the covariant free indices have different names, therefore we have the following consideration.

(i) When the covariant free indices have different names, by Corollary 1, only Steps 1–5 of Algorithm I are needed. Step 2 compares indexed functions in type list and distances, so the complexity is at most C_i^2. The order on the indexed functions that contain covariant free indices is defined in Step 2.

Step 4 compares indexed functions of the same class in distance pair sequence. It is carried out for at most $\frac{(i-m)}{2}$ times, since according to the proof of Proposition 2, once it is carried out, the order on at least one class (whose elements do not contain covariant free indices) is defined. Hence the complexity of Algorithm I is at most $C_i^2+\frac{(i-m)}{2}C_{i-m}^2$, or $O(i^3)$, much smaller than $O(i!\prod_{j=1}^{k} n_j!)$.

Step 3 of Algorithm III compares interchangeable indices inside partial differential functions, so the complexity is $\prod_{j=1}^{k} C_{n_j}^2$, also much smaller than $O(i!\prod_{j=1}^{k} n_j!)$.

(ii) When some covariant free indices have identical names, Step 6 of Algorithm I considers $\prod_{j=1}^{s} m_j!$ possible orders. Step 7 chooses the smallest from $\prod_{j=1}^{s} m_j!$ objects, whose complexity is $\prod_{j=1}^{s} m_j! - 1$. Therefore the complexity of Algorithm I is at most $C_i^2 + \prod_{j=1}^{s} m_j!\frac{(i-m)}{2}C_{i-m}^2 + \prod_{j=1}^{s} m_j! - 1)$. It must be less than $O(i!)$, since $\frac{(i-m)}{2}C_{i-m}^2 < (i-m)! < (i-m_1-m_2-\ldots-m_s)!$.

The complexity of Step 3 in Algorithm III is $p\prod_{j=1}^{k} C_{n_j}^2$, also smaller than $O(i!\prod_{j=1}^{k} n_j!)$, since $p < \prod_{j=1}^{s} m_j! < i!$.

Notation. The result of applying Algorithm III to f is denoted by $f^{(\mathrm{mon})}$.

Proposition 3. *Suppose $f_1, f_2 \in \mathcal{M}[\not\partial]$. If $f_1 \overset{mon}{\sim} f_2$, then $f_1^{(tri)} \overset{mon}{\sim} f_2^{(tri)}$.*

Proof. Since the sub-monomials $\not\partial_A^A$ or $\not\partial_A^{A_1}\not\partial_{A_1}^{A_2}\cdots\not\partial_{A_r}^A$ have no common indices with other sub-monomials, f_i $(i=1,2)$ can be divided into two independent parts. If $f_1 \overset{\mathrm{mon}}{\sim} f_2$, then the part composed of $\not\partial_A^A$ or $\not\partial_A^{A_1}\not\partial_{A_1}^{A_2}\ldots\not\partial_{A_r}^A$ in f_1 is equivalent to that in f_2 with respect to monoterm symmetries. So is the other part. This implies $f_1^{(\mathrm{tri})} \overset{\mathrm{mon}}{\sim} f_2^{(\mathrm{tri})}$.

Theorem 1. *Suppose $f_1, f_2 \in \mathcal{R}[\not\partial]$. If $f_1 \overset{mon}{\sim} f_2$, then f_1, f_2 are rewritten as identical forms by Algorithm III. Conversely, if two polynomials are rewritten as identical forms, they must be equivalent.*

Proof. By Proposition 3, $f_1^{(\mathrm{tri})} \overset{\mathrm{mon}}{\sim} f_2^{(\mathrm{tri})}$. Therefore by Proposition 1, there is a bijection Ψ from the set of indexed functions of $f_1^{(\mathrm{tri})}$ to that of $f_2^{(\mathrm{tri})}$, and Ψ keeps invariant of type lists and distances.

In Algorithm I, the order on indexed functions is determined only by type lists and distances. Therefore, if the sequence $\{\overline{f_{11}}, \ldots, \overline{f_{1k}}\}$ is among the output of applying Algorithm I to $f_1^{(\mathrm{tri})}$, i.e., it belongs to $\{[f_1]_1^{(\min)}, [f_1]_2^{(\min)}, \ldots, [f_1]_p^{(\min)}\}$, then

$$\{\Psi(\overline{f_{11}}), \ldots, \Psi(\overline{f_{1k}})\} \in \{[f_2]_1^{(\min)}, [f_2]_2^{(\min)}, \ldots, [f_2]_p^{(\min)}\}.$$

Let f_1' be the product of $\overline{f_{11}}$, …, $\overline{f_{1k}}$, and f_2' be the product of $\Psi(\overline{f_{11}})$, …, $\Psi(\overline{f_{1k}})$. Because the replacement of dummy indices in Algorithm II is uniquely determined by the order on functions, f_1', f_2' are rewritten as equivalent monomials with respect to US, and finally as identical forms by Step 3 of Algorithm III.

5 Normalization

In this section, we present a normalization algorithm for monomials in $\mathcal{M}_2[\not\partial]$ by using the method of index replacement.

According to Proposition 4.1 in [14], suppose f_1 and f_2 are equivalent, if we rewrite f_i as f_i' $(i = 1, 2)$ successively by elimination, mixed rewriting of unmixed interior circle, and coordinate system unification of self-restrained dummy indices, then $f_1' \overset{\mathrm{mon,BS}}{\sim} f_2'$. Hence in order to find the canonical form of any monomial, we only need to develop a normalization algorithm with respect to BS and monoterm symmetries.

Suppose $g \overset{\mathrm{mon,BS}}{\sim} f$. Since $g^{(\mathrm{mon})}$ and $f^{(\mathrm{mon})}$ have identical set of indices, the set $\left\{g^{(mon)} | g \overset{\mathrm{mon,BS}}{\sim} f\right\}$ is finite. Therefore we can choose the element associated with the smallest numerical list from the set as the canonical form of f, and have the following normalization algorithm.

Algorithm IV. Normalization with respect to BS and monoterm symmetries.
Input: $f \in \mathcal{M}_2[\not\partial]$, assuming all the covariant free indices have different names.
Output: The canonical form of f with respect to BS and monoterm symmetries.
Step 1. Apply Algorithm III to f to get $f^{(\mathrm{Rep})}$ and $f^{(\mathrm{mon})}$.
Step 2. Let $\mathcal{S}_1 = \{L_{f^{(\mathrm{mon})}}\}$, $\mathcal{S}_2 = \emptyset$, $\mathcal{R}_1 = \{(f^{(\mathrm{Rep})}, 0)\}$, $\mathcal{R}_2 = \emptyset$, $\Theta = 1$, and $N^{(D_{up})}$ be the number of upper dummy indices of $f^{(\mathrm{Rep})}$.
Step 3. Let $\mathcal{R}_2 = \mathcal{R}_1$, $\mathcal{R}_1 = \emptyset$, $\Theta = (-1)\Theta$.
Step 4. For the first component g of each element $(g, n^{(g)})$ in $\mathcal{R}_2$, carry out the following steps.

1. In g, find the pairs of functions in one of the four forms
   (a) $\not\partial^{n_3'}_{n_1' n_2}$ and $\not\partial^{N_6}_{n_4' N_5}$,
   (b) $\not\partial^{n_3'}_{n_1' n_2}$ and $\not\partial^{N_6}_{n_5' n_4'}$,
   (c) $\not\partial^{n_3}_{n_1 n_2'}$ and $\not\partial^{N_6}_{n_4 N_5}$,
   (d) $\not\partial^{n_3}_{n_1 n_2'}$ and $\not\partial^{N_6}_{n_5 n_4}$,
   such that n_i $(i = 1, \ldots, 5)$ is a positive integer, $n_3 \neq n^{(g)}$, $n_4 = n_3 + N^{(D_{up})}$, and $N_5, N_6 \in \mathfrak{N}'$.

2. Rewrite the couple functions by

$$\begin{aligned}
\not\partial^{n_3'}_{n_1' n_2} \not\partial^{N_6}_{n_4' N_5} &\longrightarrow -\not\partial^{n_3}_{n_2 n_1'} \not\partial^{N_6}_{n_4 N_5},\\
\not\partial^{n_3}_{n_1' n_2} \not\partial^{N_6}_{n_5' n_4'} &\longrightarrow -\not\partial^{n_3}_{n_2 n_1'} \not\partial^{N_6}_{n_4 n_5'},\\
\not\partial^{n_3}_{n_1 n_2'} \not\partial^{N_6}_{n_4 N_5} &\longrightarrow -\not\partial^{n_3}_{n_2' n_1} \not\partial^{N_6}_{n_4' N_5},\\
\not\partial^{n_3}_{n_1 n_2'} \not\partial^{N_6}_{n_5 n_4} &\longrightarrow -\not\partial^{n_3}_{n_2' n_1} \not\partial^{N_6}_{n_4' n_5}
\end{aligned}$$

respectively, and denote the new monomial by g'.

3. Apply Algorithm III to g', and get $g'^{(\mathrm{mon})}$.
4. If $\Theta = -1$ and $L_{g'^{(\mathrm{mon})}} \notin \mathcal{S}_2$, then add $L_{g'^{(\mathrm{mon})}}$ and the ordered pair (g', n_3) to $\mathcal{S}_2$ and $\mathcal{R}_1$ respectively.
5. If $\Theta = 1$ and $L_{g'^{(\mathrm{mon})}} \notin \mathcal{S}_1$, then add $L_{g'^{(\mathrm{mon})}}$ and the ordered pair (g', n_3) to $\mathcal{S}_1$ and $\mathcal{R}_1$ respectively.

Step 5. If $\mathcal{R}_1 = \emptyset$, let L_h be the smallest among $\mathcal{S}_1 \bigcup \mathcal{S}_2$, and output h. Otherwise, return to Step 3.

Remark 1

(i) In the above algorithm, Θ is the sign symbol of a monomial. Two monomials with different sign symbols are impossibly identical, hence are put into different sets $\mathcal{S}_1$ and $\mathcal{S}_2$.

(ii) Suppose f can be rewritten as g. When we rewrite g, to prevent the result from being f, we use the positive integer $n^{(g)}$ as a criterion.

Due to Algorithm IV, we directly have the following normalization algorithm.

Algorithm V. Normalization algorithm.
Input: $f \in \mathcal{M}_2[\not\partial]$.
Output: The canonical form of f.
Step 1. Apply the simplification algorithm in [14] to f.
Step 2. Carry out the mixed rewriting of unmixed interior circle and coordinate system unification of self-restrained dummy indices [14]:

$$\not\partial^{a_2'}_{b_1 a_1} \not\partial^{a_3}_{b_2' a_2'} \not\partial^{a_4'}_{b_3 a_3} \cdots \not\partial^{a_{2k}'}_{b_{2k-1} a_{2k-1}} \not\partial^{a_1}_{b_{2k}' a_{2k}'} \longrightarrow$$

$$\not\partial^{a_2}_{b_1 a_1'} \not\partial^{a_3'}_{b_2' a_2} \not\partial^{a_4}_{b_3 a_3'} \cdots \not\partial^{a_{2k}}_{b_{2k-1} a_{2k-1}'} \not\partial^{a_1'}_{b_{2k}' a_{2k}},$$

and

$$\not\partial^{d}_{dc'} \not\partial^{c'}_{b'a} \longrightarrow \not\partial^{d'}_{d'c} \not\partial^{c}_{ab'}.$$

Step 3. Divide the covariant free indices into groups by name. The order of replacement of indices in different groups are alphabetical, and the order within each group is arbitrary.
Step 4. Under all the possible orderings, apply Algorithm IV, and denote all the outputs by $f_1, \ldots, f_p$.

Step 5. Let f_s $(1 \leq s \leq p)$ be the monomial which has the smallest numerical list among L_{f_i} $(i = 1, \ldots, p)$. Output f_s.

Example 6. Put $f = \partial^{r'}_{a'_1 a_2} \partial^{t'}_{a'_3 a_4} \partial^{d}_{dt'} \partial^{c}_{r's'} \partial^{p}_{ce'} \partial^{l}_{k'}$ into canonical form.

Step 1. By the first two steps of Algorithm V, $\partial^{t'}_{a'_3 a_4} \partial^{d}_{dt'}$ is rewritten as $\partial^{t}_{a_4 a'_3} \partial^{d'}_{d't}$.
Step 2. According to Algorithm IV, rewrite f by BS. And among all the results, we find that $h^{(mon)} = \partial^{1}_{8'} \partial^{10}_{4,3'} \partial^{11}_{6,5'} \partial^{2}_{7',16'} \partial^{12'}_{9',14} \partial^{13'}_{17',15}$ has the smallest numerical list. Since the numbers from 1 to 9 denote the free indices $l, p, k', a'_1, a_2, a'_3, a_4, e', s'$, output $\partial^{l}_{k'} \partial^{10}_{a_2 a'_1} \partial^{11}_{a_4 a'_3} \partial^{p}_{e',16'} \partial^{12'}_{s',14} \partial^{13'}_{17',15}$.

Example 7. Prove that H^i_{jk} in Example 1 is a (1, 2)-typed tensor.

It suffices to verify Eq.(1).

Since h is a (1, 1)-typed tensor, $h^{i'}_{j'} = \partial^{j}_{j'} \partial^{i'}_{i} h^{i}_{j}$. Substituting the expressions of H^i_{jk}, $H^{i'}_{j'k'}$ (given in Example 1) and $h^{i'}_{j'}$ into the left side of (1), we get a polynomial with 16 terms, denoted by f. Each term of f is in $\mathcal{M}_2[\partial, h]$, which is a monoid composed of the partial derivatives of h and $\mathcal{M}_2[\partial]$ (see also [14]). Note that Algorithm III is independent of the differential function name ∂ (or we can take h as ∂). Besides, the simplification algorithm has been extended to $\mathcal{M}_2[\partial, h]$, as presented by [14]. Hence, Algorithm V can put f into canonical form. For each term of f, we find that there is another term such that the sum of their canonical forms is 0. For instance, the two terms $\partial^{i'}_{i} \partial^{s}_{p'} h^{i}_{s} \partial^{k}_{k'} \partial^{p'}_{j'p} h^{p}_{k}$ and $\partial^{m'}_{t} \partial^{k}_{k'} h^{t}_{k} \partial^{j}_{m'j'} \partial^{i'}_{i} h^{i}_{j}$ have the canonical forms $\mp \partial^{4}_{k'} \partial^{i'}_{9} h^{5}_{11} h^{6}_{8} \partial^{7}_{10,j'}$. Therefore, the canonical form of f is 0.

Acknowledgements. The author is grateful to the reviewers for helpful comments. This work was supported by Natural Science Foundation of Shanghai (15ZR1401600).

References

1. Fulling, S.A., King, R.C., Wybourne, B.G., Cummins, C.J.: Normal forms for tensor polynomials: I. The Riemann tensor. Class. Quantum Grav. **9**, 1151–1197 (1992)
2. Christensen, S., Parker, L.: MathTensor, A System for Performing Tensor Analysis by Computer. Addison-Wesley, Boston (1994)
3. Ilyin, V.A., Kryukov, A.P.: ATENSOR-REDUCE program for tensor simplification. Comput. Phys. Commun. **96**, 36–52 (1996)
4. Jaén, X., Balfagón, A.: TTC: symbolic tensor calculus with indices. Comput. Phys. **12**, 286–289 (1998)
5. Portugal, R.: An algorithm to simplify tensor expressions. Comput. Phys. Commun. **115**, 215–230 (1998)
6. Portugal, R.: Algorithmic simplification of tensor expressions. J. Phys. A: Math. Gen. **32**, 7779–7789 (1999)
7. Portugal, R.: The Riegeom package: abstract tensor calculation. Comput. Phys. Commun. **126**, 261–268 (2000)

8. Balfagón, A., Jaén, X.: Review of some classical gravitational superenergy tensors using computational techniques. Class. Quantum Grav. **17**, 2491–2497 (2000)
9. Manssur, L.R.U., Portugal, R., Svaiter, B.F.: Group-theoretic approach for symbolic tensor manipulation. Int. J. Mod. Phys. C. **13**, 859–880 (2002)
10. Manssur, L.R.U., Portugal, R.: The Canon package: a fast kernel for tensor manipulators. Comput. Phys. Commun. **157**, 173–180 (2004)
11. Martín-García, J.M., Portugal, R., Manssur, L.R.U.: The Invar tensor package. Comput. Phys. Commun. **177**, 640–648 (2007)
12. Martín-García, J.M., Yllanes, D., Portugal, R.: The Invar tensor package: differential invariants of Riemann. Comput. Phys. Commun. **179**, 586–590 (2008)
13. Liu, J., Li, H.B., Zhang, L.X.: A complete classification of canonical forms of a class of Riemann tensor indexed expressions and its applications in differential geometry (in Chinese). Sci. Sin. Math. **43**, 399–408 (2013)
14. Liu, J., Li, H.B., Cao, Y.H.: Simplification and normalization of indexed differentials involving coordinate transformation. Sci. China Ser. A. **52**, 2266–2286 (2009)
15. Liu, J.: Simplification and normalization of indexed polynomials. Ph.D. Thesis, Chinese Academy of Sciences, Beijing (2009)

Symbolic-Numeric Integration of the Dynamical Cosserat Equations

Dmitry A. Lyakhov[1(✉)], Vladimir P. Gerdt[3,4], Andreas G. Weber[5], and Dominik L. Michels[1,2]

[1] Visual Computing Center, King Abdullah University of Science and Technology, Al Khawarizmi Building, Thuwal 23955-6900, Kingdom of Saudi Arabia
{dmitry.lyakhov,dominik.michels}@kaust.edu.sa

[2] Department of Computer Science, Stanford University, 353 Serra Mall, Stanford, CA 94305, USA
michels@cs.stanford.edu

[3] Laboratory of Information Technologies, Joint Institute for Nuclear Research, 6 Joliot–Curie St., Dubna 141980, Russian Federation

[4] Peoples' Friendship University of Russia, 6 Miklukho–Maklaya St., Moscow 117198, Russian Federation
gerdt@jinr.ru

[5] Institute of Computer Science II, University of Bonn, Friedrich-Ebert-Allee 144, 53113 Bonn, Germany
weber@cs.uni-bonn.de

Abstract. We devise a symbolic-numeric approach to the integration of the dynamical part of the Cosserat equations, a system of nonlinear partial differential equations describing the mechanical behavior of slender structures, like fibers and rods. This is based on our previous results on the construction of a closed form general solution to the kinematic part of the Cosserat system. Our approach combines methods of numerical exponential integration and symbolic integration of the intermediate system of nonlinear ordinary differential equations describing the dynamics of one of the arbitrary vector-functions in the general solution of the kinematic part in terms of the module of the twist vector-function. We present an experimental comparison with the well-established generalized α-method illustrating the computational efficiency of our approach for problems in structural mechanics.

Keywords: Analytical solution · Cosserat rods · Dynamic equations · Exponential integration · Generalized α-method · Kinematic equations · Symbolic computation

1 Introduction

Deformable-body dynamics can be considered as a subarea of continuous mechanics that studies motion of deformable solids subject to the action of internal and external forces (cf. [11]). The equations describing the dynamics of

V.P. Gerdt et al. (Eds.): CASC 2017, LNCS 10490, pp. 301–312, 2017.
DOI: 10.1007/978-3-319-66320-3_22

such solids are nonlinear partial differential equations (PDEs) whose independent variables are three spatial coordinates and time. Given a particular deformable mechanical structure, to describe its dynamics, it is necessary to satisfy these equations at each point of the structure together with appropriate boundary conditions. For a mechanical structure having special geometric properties, it is worthwhile to exploit these properties to develop a simplified but geometrically exact mechanical model of the structure. Classical examples of such models are Cosserat theories of shells and rods; see e.g. [18] and references therein.

Rods are nearly one-dimensional structures whose dynamics can be described by the Cosserat theory of (elastic) rods (cf. [1], Chap. 8; [18], Chap. 5; and the original work [5]). This is a general and geometrically exact dynamical model that takes bending, extension, shear, and torsion into account, as well as rod deformations under external forces and torques. In this context, the dynamics of a rod is described by a governing system of twelve first-order nonlinear partial differential equations (PDEs) with a pair of independent variables (s, t), where s is the arc-length and t the time parameter. In this PDE system, the two kinematic vector equations ((9a)–(9b) in [1], Chap. 8) are parameter free and represent the compatibility conditions for four vector functions $\boldsymbol{\kappa}, \boldsymbol{\omega}, \boldsymbol{\nu}, \boldsymbol{v}$ in (s, t). Whereas the first vector equation only contains two vector functions $\boldsymbol{\kappa}$ and $\boldsymbol{\omega}$, the second one contains all four vector functions $\boldsymbol{\kappa}, \boldsymbol{\omega}, \boldsymbol{\nu}, \boldsymbol{v}$. The remaining two vector equations in the governing system are dynamical equations of motion and include two more dependent vector variables $\hat{\boldsymbol{m}}(s, t)$ and $\hat{\boldsymbol{n}}(s, t)$. Moreover, these dynamical equations contain parameters (or parametric functions of s) to characterize the rod and to include the external forces and torques. Studying the dynamics of Cosserat rods has various scientific and industrial applications, for example, in civil and mechanical engineering (cf. [2]), microelectronics and robotics (cf. [3]), biophysics (cf. [7] and references therein), and visual computing (cf. [15]).

Because of its inherent stiffness caused by different deformation modes, the treatment of the underlying equations usually requires the application of specific solvers; see e.g. [16]. In order to reduce the computational overhead caused by the stiffness, we employed Lie symmetry based integration methods (cf. [9,17]) and the theory of completion to involution (cf. [19]) to the two kinematic vector equations (cf. [1], Chap. 8, Eq. (9a)–(9b)) and constructed their general and analytic solution in [13,14], which depends on two arbitrary vector functions in (s, t).

In this contribution, we exploit the general analytic solution to the kinematic part of the governing Cosserat system constructed in [13,14] and develop a symbolic-numeric approach to the integration of the dynamical part of the system. Our approach combines the ideas of numerical exponential integration (see e.g. [8,12] and references therein) and symbolic integration of the intermediate system of nonlinear ordinary differential equations describing the dynamics of the arbitrary vector function in the general solution to the kinematic part in terms of the module of the twist vector function. The symbolic part of the integration is performed by means of Maple. We present an experimental comparison of the computational efficiency of our approach with that of the generalized α-method for the numerical integration of problems in structural mechanics (see e.g. [20]). This paper is organized as follows. In Sect. 2, we present the governing PDE system in the special Cosserat theory of rods and analytic solution to its kinematic part constructed in [13,14]. In Sect. 3, we show first that the (naive) straightforward numerical integration of the dynamical part of Cosserat system has a severe obstacle caused by a singularity in the system. Then we describe a symbolic-numeric method to integrate the dynamical equations based on the ideas of exponential integration and the construction of a closed form analytic solution to the underlying nonlinear dynamical system. In doing so, we show that this symbolic-numeric method is free of the singularity problem. In Sect. 4, we present an experimental comparison of our method with the generalized α-method. Some concluding remarks are given in Sect. 5.

2 Governing Cosserat Equations and the General Solution of Their Kinematic Part

The governing PDE system in the special Cosserat theory of rods (cf. [1,3,5,13,14]) can be written in the following form:

$$\boldsymbol{\kappa}_t = \boldsymbol{\omega}_s - \boldsymbol{\omega} \times \boldsymbol{\kappa}, \tag{1a}$$

$$\boldsymbol{\nu}_t = \boldsymbol{v}_s + \boldsymbol{\kappa} \times \boldsymbol{v} - \boldsymbol{\omega} \times \boldsymbol{\nu}, \tag{1b}$$

$$\rho \boldsymbol{J} \cdot \boldsymbol{\omega}_t = \hat{\boldsymbol{m}}_s + \boldsymbol{\kappa} \times \hat{\boldsymbol{m}} + \boldsymbol{\nu} \times \hat{\boldsymbol{n}} - \boldsymbol{\omega} \times (\rho \boldsymbol{J} \cdot \boldsymbol{\omega}) + \boldsymbol{L}, \tag{1c}$$

$$\rho A \boldsymbol{v}_t = \hat{\boldsymbol{n}}_s + \boldsymbol{\kappa} \times \hat{\boldsymbol{n}} - \boldsymbol{\omega} \times (\rho A \boldsymbol{v}) + \boldsymbol{F}. \tag{1d}$$

Here, the independent variable t denotes the time and another independent variable s the arc-length parameter identifying a *material cross section* of the rod, which consists of all material points whose reference positions are on the plane perpendicular to the rod at s. The *Darboux* vector-function $\boldsymbol{\kappa} = \sum_{k=1}^{3} \kappa_k \boldsymbol{d}_k$ and the *twist* vector-function $\boldsymbol{\omega} = \sum_{k=1}^{3} \omega_k \boldsymbol{d}_k$ are determined by the kinematic relations

$$\partial_s \boldsymbol{d}_k = \boldsymbol{\kappa} \times \boldsymbol{d}_k, \quad \partial_t \boldsymbol{d}_k = \boldsymbol{\omega} \times \boldsymbol{d}_k,$$

where the vectors $\boldsymbol{d}_1$, $\boldsymbol{d}_2$, and $\boldsymbol{d}_3 := \boldsymbol{d}_1 \times \boldsymbol{d}_2$ form a right-handed orthonormal moving frame. These vectors are called *directors*. The use of the triple $(\boldsymbol{d}_1, \boldsymbol{d}_2, \boldsymbol{d}_3)$ is natural for the intrinsic description of the rod deformation. Moreover, $\boldsymbol{r}$ describes the motion of the rod relative to the fixed frame $(\boldsymbol{e}_1, \boldsymbol{e}_2, \boldsymbol{e}_3)$. This is illustrated in Fig. 1.

In doing so, the motion of a rod is defined by the mapping

$$[a,b] \times \mathbb{R} \ni (s,t) \mapsto (\boldsymbol{r}(s,t), \boldsymbol{d}_1(s,t), \boldsymbol{d}_2(s,t), \boldsymbol{d}_3(s,t)) \in \mathbb{E}^3.$$

Furthermore, the governing system (1a)–(1d) includes additional vector-valued dependent variables: *linear strain* $\boldsymbol{\nu}$ of the rod and the *velocity* $\boldsymbol{v}$ of the material cross-section:

$$\boldsymbol{\nu} := \partial_s \boldsymbol{r} = \sum_{k=1}^{3} \nu_k \boldsymbol{d}_k, \quad \boldsymbol{v} := \partial_t \boldsymbol{r} = \sum_{k=1}^{3} v_k \boldsymbol{d}_k.$$

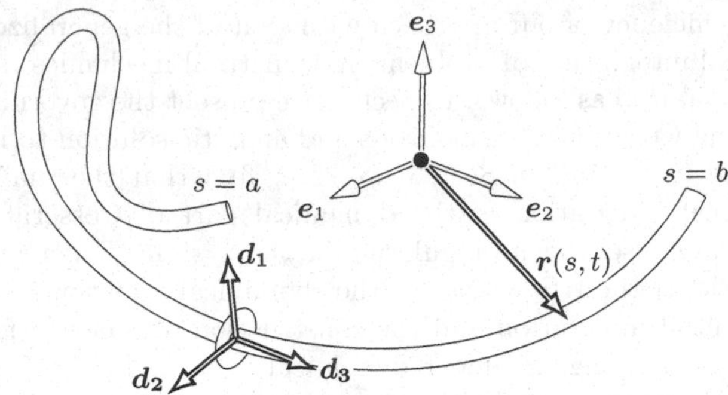

Fig. 1. The vector set $\{\boldsymbol{d}_1, \boldsymbol{d}_2, \boldsymbol{d}_3\}$ forms a right-handed orthonormal basis. The directors $\boldsymbol{d}_1$ and $\boldsymbol{d}_2$ span the local material cross-section, whereas $\boldsymbol{d}_3$ is perpendicular to the cross-section. Note that in the presence of shear deformations $\boldsymbol{d}_3$ is unequal to the tangent $\partial_s \boldsymbol{r}$ of the centerline of the rod.

The components of the *strain variables* $\boldsymbol{\kappa}$ and $\boldsymbol{\nu}$ describe the deformation of the rod: the flexure with respect to the two major axes of the cross section (κ_1, κ_2), torsion (κ_3), shear (ν_1, ν_2), and extension (ν_3).

The *kinematic part* of the governing Cosserat system consists of equations (1a)–(1b) ((9a)–(9b) in [1], Chap. 8). The remaining equations (1c)–(1d) ((9c)–(9d) in [1], Chap. 8) make up the *dynamical part* of the governing equations. For a rod density $\rho(s)$ and cross section $A(s)$, these equations follow from Newton's laws of motion:

$$\begin{aligned} \rho(s)A(s)\partial_t \boldsymbol{v} &= \partial_s \boldsymbol{n}(s,t) + \boldsymbol{F}(s,t), \\ \partial_t \boldsymbol{h}(s,t) &= \partial_s \boldsymbol{m}(s,t) + \boldsymbol{\nu}(s,t) \times \boldsymbol{n}(s,t) + \boldsymbol{L}(s,t), \end{aligned} \tag{2}$$

where $\boldsymbol{m}(s,t) = \sum_{k=1}^{3} m_k(s,t)\, \boldsymbol{d}_k(s,t)$ are the *contact torques*, $\boldsymbol{n}(s,t) = \sum_{k=1}^{3} n_k(s,t)\, \boldsymbol{d}_k(s,t)$ are the *contact forces*, $\boldsymbol{h}(s,t) = \sum_{k=1}^{3} h_k(s,t)\, \boldsymbol{d}_k(s,t)$ are the *angular momenta*, and $\boldsymbol{F}(s,t)$ and $\boldsymbol{L}(s,t)$ are the *external forces* and *torque densities*.

The contact torques $\boldsymbol{m}(s,t)$ and contact forces $\boldsymbol{n}(s,t)$ corresponding to the *internal stresses*, are related to the extension and shear strains $\boldsymbol{\nu}(s,t)$ as well as to the flexure and torsion strains $\boldsymbol{\kappa}(s,t)$ by the *constitutive relations*

$$\boldsymbol{m}(s,t) = \hat{\boldsymbol{m}}\left(\boldsymbol{\kappa}(s,t), \boldsymbol{\nu}(s,t), s\right), \quad \boldsymbol{n}(s,t) = \hat{\boldsymbol{n}}\left(\boldsymbol{\kappa}(s,t), \boldsymbol{\nu}(s,t), s\right).$$

Under certain reasonable assumptions (cf. [1,3,13]) on the structure of the right-hand sides in (2), they take the form (1c)–(1d) in which $\boldsymbol{J}$ is the inertia tensor of the cross section per unit length. Unlike the kinematic part, the dynamical part contains parameters characterizing the rod under consideration: ρ, A and $\boldsymbol{J}$ together with the external force $\boldsymbol{F}$ and torque $\boldsymbol{L}$, whereas the kinematic part is parameter free.

In our previous papers [13,14], by treating the kinematic Cosserat equations (1a)–(1b) with computer algebra aided methods of the modern Lie symmetry analysis (cf. [9,17]) and the theory of completion of partial differential systems to involution (cf. [19]), we constructed the following closed form of an analytical solution to the kinematic part and proved its generality:

$$\boldsymbol{\omega} = \boldsymbol{p}_t + \frac{p - \sin(p)}{p^3}\left(\boldsymbol{p}\,(\boldsymbol{p}\cdot\boldsymbol{p}_t) - p^2\,\boldsymbol{p}_t\right) - \frac{1-\cos(p)}{p^2}\,\boldsymbol{p}\times\boldsymbol{p}_t, \tag{3a}$$

$$\boldsymbol{\kappa} = \boldsymbol{p}_s + \frac{p - \sin(p)}{p^3}\left(\boldsymbol{p}\,(\boldsymbol{p}\cdot\boldsymbol{p}_s) - p^2\,\boldsymbol{p}_s\right) - \frac{1-\cos(p)}{p^2}\,\boldsymbol{p}\times\boldsymbol{p}_s, \tag{3b}$$

$$\boldsymbol{\nu} = \boldsymbol{q}\times\boldsymbol{\kappa} - \boldsymbol{q}_s, \tag{3c}$$

$$\boldsymbol{v} = \boldsymbol{q}\times\boldsymbol{\omega} - \boldsymbol{q}_t,$$

where $\boldsymbol{p}(s,t)$ and $\boldsymbol{q}(s,t)$ are arbitrary analytic vector functions, and

$$p = \sqrt{p_1^2 + p_2^2 + p_3^2}.$$

For the efficient numerical solving of the dynamical Cosserat equations (1c)–(1d), we use the following fact: the vector equation (3a) uniquely defines $\boldsymbol{p}_t$ in terms of $\boldsymbol{p}$ and $\boldsymbol{\omega}$. We formulate this fact as the following statement.

Proposition 1. *The temporal derivative $\boldsymbol{p}_t$ of the vector function $\boldsymbol{p}$, as a solution of* (3a), *reads*

$$\boldsymbol{p}_t = \frac{\boldsymbol{p}\cdot\boldsymbol{\omega}}{p^2}\,\boldsymbol{p} + \frac{1}{2}\,\boldsymbol{p}\times\boldsymbol{\omega} - \frac{p}{2}\cot\left(\frac{p}{2}\right)\cdot\frac{\boldsymbol{p}\times(\boldsymbol{p}\times\boldsymbol{\omega})}{p^2}. \tag{4}$$

Proof. The vector function $\boldsymbol{p}_t$ occurs linearly in (3a). In the component form, it is a linear system of three equations in three unknowns $(p_1)_t, (p_2)_t, (p_3)_t$ whose matrix has a non-singular determinant (cf. formula (11) in [14])

$$2\,\frac{\cos(p)-1}{p^2}. \tag{5}$$

The vector form of $\boldsymbol{p}_t$ as a solution to equality (3a) is given by (4). This can be verified either by hand computation or by using the routines of the Maple package *VectorCalculus* after the substitution of (4) into the right-hand side of (3a) and simplification of the obtained expression to $\boldsymbol{\omega}$. □

Instead of system (1a)–(1d) for unknowns $(\boldsymbol{\omega}, \boldsymbol{\kappa}, \boldsymbol{\nu}, \boldsymbol{v})$, we are going to solve the equivalent system

$$\boldsymbol{p}_t = \frac{\boldsymbol{p}\cdot\boldsymbol{\omega}}{p^2}\,\boldsymbol{p} + \frac{1}{2}\,\boldsymbol{p}\times\boldsymbol{\omega} - \frac{p}{2}\cot\left(\frac{p}{2}\right)\cdot\frac{\boldsymbol{p}\times(\boldsymbol{p}\times\boldsymbol{\omega})}{p^2}, \tag{6a}$$

$$\boldsymbol{q}_t = \boldsymbol{q}\times\boldsymbol{\omega} - \boldsymbol{v}, \tag{6b}$$

$$\rho\boldsymbol{J}\cdot\boldsymbol{\omega}_t = \hat{\boldsymbol{m}}_s + \boldsymbol{\kappa}\times\hat{\boldsymbol{m}} + \boldsymbol{\nu}\times\hat{\boldsymbol{n}} - \boldsymbol{\omega}\times(\rho\boldsymbol{J}\cdot\boldsymbol{\omega}) + \boldsymbol{L}, \tag{6c}$$

$$\rho A\boldsymbol{v}_t = \hat{\boldsymbol{n}}_s + \boldsymbol{\kappa}\times\hat{\boldsymbol{n}} - \boldsymbol{\omega}\times(\rho A\boldsymbol{v}) + \boldsymbol{F} \tag{6d}$$

for unknown vector functions $(\boldsymbol{p}, \boldsymbol{q}, \boldsymbol{\omega}, \boldsymbol{v})$, where $\boldsymbol{\kappa}$ and $\boldsymbol{\nu}$ are given by (3b)–(3c).

3 Symbolic-Numeric Integration Method

3.1 Naive Approach: Explicit Numerical Solving

Suppose we know the values of the vector-functions $\boldsymbol{\omega}$ and $\boldsymbol{p}$ on a time layer t. Then $\boldsymbol{p}_t$ in (6a) can be approximated by the forward Euler difference

$$\boldsymbol{p}_t \rightarrow \frac{\boldsymbol{p}(s,t+\triangle t)-\boldsymbol{p}(s,t)}{\triangle t}.$$

However, since Eq. (6a) has a singularity at $p = 2\pi$ related with vanishing (5), there is a restriction to the time step Δt caused by the condition

$$p(s,t+\triangle t) \in (0, 2\pi) \tag{7}$$

to be held for all values of s. This is a severe problem for the numerical solving of the governing Cosserat system, because in the course of solving, one must control the time step at every value of s to keep the values of $p(s,t)$ within the interval indicated in (7). This problem is resolved in the symbolic-numeric integration method described in the next subsection.

3.2 Advanced Approach Based on Exponential Integration

To avoid the problem of controlling the condition (7) we use the differential equation (6a) for $\boldsymbol{p}(t)$ and rewrite it in terms of p and the unit vector $\boldsymbol{e}$ where $\boldsymbol{p} = p\boldsymbol{e}$. It leads to the following differential system:

$$\begin{aligned} p_t &= \boldsymbol{e}\cdot\boldsymbol{\omega}, \\ 2\,\boldsymbol{e}_t &= \boldsymbol{e}\times\boldsymbol{\omega} - \cot\left(\frac{p}{2}\right)\boldsymbol{e}\times(\boldsymbol{e}\times\boldsymbol{\omega}). \end{aligned}$$

Now assume that the vector $\boldsymbol{\omega}$ is independent of t on the time interval Δt and choose the Cartesian coordinate system $\boldsymbol{e}_1, \boldsymbol{e}_2, \boldsymbol{e}_3$ such that $\boldsymbol{e}_3 || \boldsymbol{\omega}$:

$$\boldsymbol{e} = A_1\,\boldsymbol{e}_1 + A_2\,\boldsymbol{e}_2 + A_3\,\boldsymbol{e}_3, \quad \boldsymbol{\omega} = \omega\,\boldsymbol{e}_3, \quad \omega := \sqrt{\omega_1^2+\omega_2^2+\omega_3^2}.$$

Then, we obtain the following system of four first-order differential equations:

$$2\,(A_1)_t = A_2\,\omega - \cot\left(\frac{p}{2}\right)A_1A_3\,\omega, \tag{9a}$$

$$2\,(A_2)_t = -A_1\,\omega - \cot\left(\frac{p}{2}\right)A_2A_3\,\omega, \tag{9b}$$

$$2\,(A_3)_t = -\cot\left(\frac{p}{2}\right)(A_3^2-1)\,\omega, \tag{9c}$$

$$p_t = A_3\,\omega. \tag{9d}$$

From the Eqs. (9c)–(9d) it follows

$$\frac{2\,A_3(A_3)_t}{A_3^2-1} = -\cot\left(\frac{p}{2}\right)p_t,$$

and hence,

$$(1 - A_3^2)\sin^2\left(\frac{p}{2}\right) = C, \quad C_t = 0. \tag{10}$$

Equation (10) immediately implies the following statement providing fulfillment of (7).

Proposition 2. *If $C \neq 0$, then $p(t) \in (0, 2\pi)$ for all $t \geq t_0$ if $p(t_0) \in (0, 2\pi)$.*

If one substitutes $A_3 = p_t/\omega$ from (9c) and replaces p with $q := \cos\left(\frac{p}{2}\right)$, then (10) takes the form

$$4\,q_t^2 = \omega^2\left(1 - q^2 - C\right).$$

This equation is easily solvable by the Maple routine *dsolve* which outputs four solutions. These solutions can be unified into the general solution

$$q = \sqrt{1-C}\sin\left(\frac{1}{2}\omega(C_1 - t)\right), \quad (C_1)_t = 0.$$

Then, the whole system (9a)–(9d) admits the following general analytic solution

$$A_1(s,t) = -\frac{\sqrt{C}\cdot sin(\frac{1}{2}\omega(C_2 - t))}{\sqrt{\omega^2\cos^2(\frac{1}{2}\omega(C_1 - t)) + C\sin^2(\frac{1}{2}\omega(C_1 - t))}}, \tag{11a}$$

$$A_2(s,t) = \frac{\sqrt{C}\cdot cos(\frac{1}{2}\omega(C_2 - t))}{\sqrt{\omega^2\cos^2(\frac{1}{2}\omega(C_1 - t)) + C\sin^2(\frac{1}{2}\omega(C_1 - t))}}, \tag{11b}$$

$$A_3(s,t) = \frac{\sqrt{\omega^2 - C}\cdot cos(\frac{1}{2}\omega(C_1 - t))}{\sqrt{\omega^2\cos^2(\frac{1}{2}\omega(C_1 - t)) + C\sin^2(\frac{1}{2}\omega(C_1 - t))}}, \tag{11c}$$

$$p(s,t) = 2\arccos\left(\frac{\sqrt{\omega^2 - C}\sin(\frac{1}{2}\omega(C_1 - t))}{\omega}\right), \tag{11d}$$

where C, C_1, C_2 are functions of s. These functions are determined by the following initial data:

$$C(s) := \omega^2\left(1 - A_3^2(s,t_0)\right)\sin^2\left(\frac{p(s,t_0)}{2}\right),$$

$$C_1(s) := t_0 + \frac{A_3(s,t_0)|\sin(p(s,t_0))|}{\sqrt{\omega^2 - C(s)}},$$

$$C_2(s) := t_0 + \frac{2}{\omega}\arctan\left(\frac{A_1(s,t_0)}{A_2(s,t_0)}\right).$$

Proposition 3. *$C(s) \equiv 0$ if and only if $A_3(s,t) = \pm 1$ which corresponds to a degenerated solution*

$$A_1(s,t) = 0, \quad A_2(s,t) = 0, \quad p(s,t) = p(s,t_0) \pm \omega t.$$

Computationally, this solution is not of interest, since it is unstable: a small (e.g. numerical) deviation of $A_3(s,t_0)$ from ± 1 converts the solution into a generic one.

Proof. The solution

$$p(s,t) = p(s,t_0) \pm \omega t, A_3(s,t) = \pm 1, A_2(s,t) = A_1(s,t) = 0$$

is singular. Unlike the generic solution, the value of $|p(t)|$ in this solution may increase indefinitely. However, it is unstable, since any small perturbation $\epsilon > 0$ to the initial value $A_3(s,t_0) = \pm(1-\epsilon(s))$ leads to the generic case when $p(t)$ remains bounded.

□

Our symbolic-numeric approach to the derivation of equations (9a)–(9d) and the construction of their explicit analytic solution (11a)–(11d) is in accord with the general principles of exponential integration (see e.g. [8]). The basic idea behind exponential integration is the identification of a prototypical differential system which has the stiffness properties similar to those in the original equation and which admits explicit solving.

In our case, the stiffness properties of the differential system (9a)–(9d) are similar to those in system (6a)–(6d). In doing so, the last system belongs to the second class of stiff problems (cf. [8], p. 210) whose stiffness is caused by the highly oscillatory behavior of their solutions. For such problems both explicit and implicit Euler schemes fail to provide the required stability unless the step size is strongly reduced to provide the resolution of all the oscillations in the solution. Thereby, the standard numerical treatment of the equations (1a)–(1d) and hence equations (6a)–(6d) is computationally inefficient. Just by this reason, special numerical solvers have been designed (cf. [10,20]) for the Eqs. (1a)–(1d).

Figure 2 illustrates the stiffness of the differential system (6a)–(6d). The behavior is shown, at $\omega = 1$, of the functions $p(s_0,t)$ and $A_3(s_0,t)$ as solutions of (9c)–(9d) for the initial conditions $p(s_0,0) = 1$ and $A_3(s_0,0) = 0.99$. As illustrated, the solution oscillates and changes drastically over time. A numerical reconstruction of such a behavior is possible only for very small step sizes of difference approximations.

4 Numerical Comparison with the Generalized α-Method

In 1993, Chung and Hulbert (cf. [4]) presented the generalized α-method as a new integration algorithm for problems from structural mechanics. It is characterized primarily by a controllable numerical dissipation of high-frequency components in the numerical solution. These occur, for example, in the context of finite element-based simulations, when the high-frequency states are too roughly resolved. Such methods usually improve the convergence behavior of iterative solving strategies for nonlinear problems.

The generalized α-method is well-established in the field of structural mechanics and has the major advantage of unconditional stability as well as

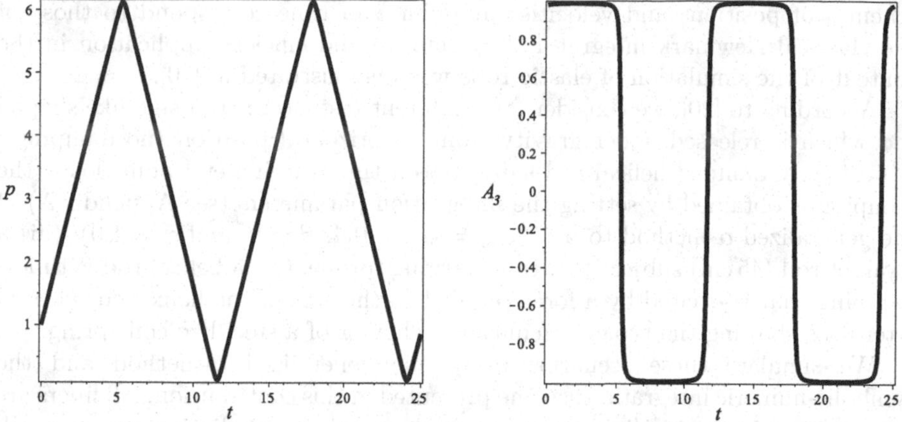

Fig. 2. Illustration of the temporal evolution of $p(s_0, t)$ (left) and $A_3(s_0, t)$ (right).

user controllable numerical damping. The idea of the introduction of a controllable numerical damping in the integration process is not new and found, among other things, realization in α-HHT (cf. [6]) or in the WBC-α-method (cf. [21]). The aim of the development of such methods is to maximize the attenuation of high-frequency components while preserving the important low-frequency components. Using the generalized α-method, this ratio is optimal, i.e., for a given attenuation of the high-frequency components, the attenuation of the low-frequency components is minimized. A brief specification of this method for general problems in structural mechanics is given in Appendix A.

Following [20], using a state vector

$$\boldsymbol{x}(s,t) = (\boldsymbol{v}(s,t), \boldsymbol{\omega}(s,t), \boldsymbol{\kappa}(s,t), \boldsymbol{n}(s,t))^{\mathsf{T}}$$

describing the rod, we can rewrite its equations of motion (1a)–(1d) in terms of a system

$$\hat{\mathbf{M}}\partial_t \boldsymbol{x}(s,t) + \hat{\mathbf{K}}\partial_s \boldsymbol{x}(s,t) + \boldsymbol{\Lambda}(s,t) = \mathbf{0}. \tag{12}$$

Here

$$\hat{\mathbf{M}} = \mathsf{diag}(\rho A, \rho A, \rho A, \rho I_1, \rho I_2, \rho I_3, 1, 1, 1, 0, 0, 0)$$

is the mass matrix and $\hat{\mathbf{K}} = -\mathsf{adiag}(\mathbf{1}, \mathbf{1}, \mathbf{K}, \mathbf{1})$ is the stiffness matrix with $\mathbf{K} = \mathsf{diag}(EI_1, EI_2, G\mu)$, in which the bending stiffness in the direction of the principal components of the cross section A is denoted by $EI_{1,2}$, and the torsional stiffness by $G\mu$. As above, the rod's density is given by ρ, the Young's modulus by E, and the shear modulus by G. The nonlinear terms are included in the nonlinearity $\boldsymbol{\Lambda}$. We do not explicitly write out the resulting equations here for brevity and refer to [20] for the explicit form of (12). In the generalized α-method, the update schemes of positions and velocities at point i in time correspond to those of the classical Newmark integrator. Its accurate and efficient application in the context of the simulation of elastic rods was demonstrated in [20].

According to [20], we consider four different test cases: (i) a sine-like shaped rod which is released under gravity from a horizontal position (no damping); (ii) a highly damped helical rod subject to a time-varying end point load (the damping is obtained by setting the integration parameters (see Appendix A) of the generalized α-method to $\alpha := \alpha_m = \alpha_f = 0.4$, $\beta = 0$, and $\gamma = 1.0$); (iii) a straight rod (45 cm) subject to a time-varying torque; (iv) a helical rod with low damping that is excited by a force parallel to the axis of the helix and released after 0.1 s showing the typical oscillating behavior of a steel-like coil spring.

We simulate these scenarios using the generalized α-method and the symbolic-numeric integration scheme presented in this contribution. All fibers are discretized using 100 individual segments. Using the symbolic-numeric method, damping is incorporated using the linear Rayleigh damping as described in [15]. We enforce a maximally tolerated relative L^2-error of 1% in the position and velocity space in order to ensure sufficient accuracy and measure the required computation time on a machine with an Intel(R) Xeon E5 with 3.5 GHz and 32 GB DDR-RAM without parallelization. For all test cases, we obtain significant speedups of the presented symbolic-numeric method ("snm") compared to the generalized α-method ("α"), in particular:

(i) speedup of a factor more than 20 (α: 4.1 s; snm: 0.2 s);
(ii) speedup of over 21× (α: 4.3 s; snm: 0.2 s);
(iii) speedup of approx. 19× (α: 3.8 s; snm: 0.2 s);
(iv) speedup of approx. 34× (α: 6.8 s; snm: 0.2 s).

Please note that the computation time of 0.2 s of the presented symbolic-numeric method is constant for all test cases (with identical duration of 8 s).

5 Conclusison

Based on the closed form solution to the kinematic part (1a)–(1b) of the governing Cosserat system (1a)–(1d) and with assistance of Maple, we have developed a new symbolic-numeric method for their integration. Our computational experiments demonstrate the superiority of the new method over the generalized α-method for the accurate and efficient integration of Cosserat rods. Its application prevents from numerical instabilities and allows for highly accurate and efficient simulations. This clearly shows the usefulness of the constructed analytic solution to the kinematic equations and should enable more complex and realistic Cosserat rod-based scenarios to be explored in scientific computing without compromising efficiency.

Acknowledgements. The authors appreciate the insightful comments of the anonymous referees. This work has been partially supported by the King Abdullah University of Science and Technology (KAUST baseline funding), the Max Planck Center for Visual Computing and Communication (MPC-VCC) funded by Stanford University and the Federal Ministry of Education and Research of the Federal Republic of Germany (BMBF grants FKZ-01IMC01 and FKZ-01IM10001), the Russian Foundation for Basic Research (grant 16-01-00080) and the Ministry of Education and Science of the Russian Federation (agreement 02.a03.21.0008).

A Generalized α-Method

In this appendix, we briefly explain the application of the generalized α-method for the common case of a system described by the standard equation from structural mechanics,

$$\boldsymbol{M}\ddot{\boldsymbol{x}} + \boldsymbol{D}\dot{\boldsymbol{x}} + \boldsymbol{K}\boldsymbol{x} + \boldsymbol{\Lambda}(t) = \mathbf{0}, \tag{13}$$

in which $\boldsymbol{M}$, $\boldsymbol{D}$, and $\boldsymbol{K}$ denote the mass, damping, and stiffness matrices. The time-dependent displacement vector is given by $\boldsymbol{x}(t)$, and its first- and second-order temporal derivatives describe velocity and acceleration. The vector $\boldsymbol{\Lambda}(t)$ describes external forces acting on the system at time t. We are searching for functions $\boldsymbol{x}(t)$, $\boldsymbol{v}(t) = \dot{\boldsymbol{x}}(t)$, and $\boldsymbol{a}(t) = \ddot{\boldsymbol{x}}(t)$ satisfying (13) for all t with initial conditions $\boldsymbol{x}(t_0) = \boldsymbol{x}_0$ and $\boldsymbol{v}(t_0) = \boldsymbol{v}_0$.

For the employment of the generalized α-method, we can write the integration scheme with respect to (13) as follows:

$$\boldsymbol{M}\boldsymbol{a}_{1-\alpha_m} + \boldsymbol{D}\boldsymbol{x}_{1-\alpha_f} + \boldsymbol{K}\boldsymbol{x}_{1-\alpha_f} + \boldsymbol{\Lambda}(t_{1-\alpha_f}) = \mathbf{0},$$

with the substitution rule $(\cdot)_{1-\alpha} := (1-\alpha)(\cdot)_i + \alpha(\cdot)_{i-1}$ and the approximations

$$\boldsymbol{x}_i = \boldsymbol{x}_{i-1} + \Delta t \boldsymbol{v}_{i-1} + \Delta t^2 \left(\left(\frac{1}{2} - \alpha \right) \boldsymbol{a}_{i-1} + \beta \boldsymbol{a}_i \right),$$

$$\boldsymbol{v}_i = \boldsymbol{v}_{i-1} + \Delta t \left((1-\gamma) \boldsymbol{a}_{i-1} + \gamma \boldsymbol{a}_i \right).$$

The parameters α_m, α_f, γ, and β are integration coefficients.

References

1. Antman, S.S.: Nonlinear Problems of Elasticity. Applied Mathematical Sciences, vol. 107. Springer, Heidelberg (1995). doi:10.1007/0-387-27649-1
2. Boyer, F., De Nayer, G., Leroyer, A., Visonneau, M.: Geometrically exact Kirchhoff beam theory: application to cable dynamics. J. Comput. Nonlinear Dyn. **6**(4), 041004 (2011)
3. Cao, D.Q., Tucker, R.W.: Nonlinear dynamics of elastic rods using the Cosserat theory: modelling and simulation. Int. J. Solids Struct. **45**, 460–477 (2008)
4. Chung, J., Hulbert, G.M.: A time integration algorithm for structural dynamics with improved numerical dissipation: the generalized-α method. J. Appl. Mech. **60**(2), 371–375 (1993)
5. Cosserat, E., Cosserat, F.: Théorie des corps déformables. Hermann, Paris (1909)
6. Hilber, H.M., Hughes, T.J.R., Taylor, R.L.: Improved numerical dissipation for time integration algorithms in structural dynamics. Earthq. Eng. Struct. Dyn. **5**, 283–292 (1977)

7. Hilfinger, A.: Dynamics of cilia and flagella. Ph.D. thesis, Technische Universität Dresden (2006)
8. Hochbruck, M., Ostermann, A.: Exponential integrators. Acta Numerica **19**, 209–286 (2010)
9. Ibragimov, N.H.: A Practical Course in Differential Equations and Mathematical Modelling. Classical and New Methods. Nonlinear Mathematical Models. Symmetry and Invariance Principles. Higher Education Press/World Scientific, Beijing (2009)
10. Lang, H., Linn, J., Arnold, M.: Multibody Dynamics Simulation of Geometrically Exact Cosserat Rods. Berichte des Fraunhofer ITWM, vol. 209. Fraunhofer, Munich (2011)
11. Luo, A.C.J.: Nonlinear Deformable-Body Dynamics. Springer, Heidelberg (2010). doi:10.1007/978-3-642-12136-4
12. Michels, D.L., Desbrun, M.: A semi-analytical approach to molecular dynamics. J. Comput. Phys. **303**, 336–354 (2015)
13. Michels, D.L., Lyakhov, D.A., Gerdt, V.P., Sobottka, G.A., Weber, A.G.: Lie symmetry analysis for cosserat rods. In: Gerdt, V.P., Koepf, W., Seiler, W.M., Vorozhtsov, E.V. (eds.) CASC 2014. LNCS, vol. 8660, pp. 324–334. Springer, Cham (2014). doi:10.1007/978-3-319-10515-4_23
14. Michels, D.L., Lyakhov, D.A., Gerdt, V.P., Hossain, Z., Riedel-Kruse, I.H., Weber, A.G.: On the general analytical solution of the Kinematic Cosserat equations. In: Gerdt, V.P., Koepf, W., Seiler, W.M., Vorozhtsov, E.V. (eds.) CASC 2016. LNCS, vol. 9890, pp. 367–380. Springer, Cham (2016). doi:10.1007/978-3-319-45641-6_24
15. Michels, D.L., Mueller, J.P.T., Sobottka, G.: A physically based approach to the accurate simulation of stiff fibers and stiff fiber meshes. Comput. Graph. **53B**, 136–146 (2015)
16. Michels, D.L., Sobottka, G.A., Weber, A.G.: Exponential integrators for stiff elastodynamic problems. ACM Trans. Graph. **33**, 7:1–7:20 (2014)
17. Olver, P.J.: Applications of Lie Groups to Differential Equations. Graduate Texts in Mathematics, vol. 107, 2nd edn. Springer, Heidelberg (1993). doi:10.1007/978-1-4684-0274-2
18. Rubin, M.B.: Cosserat Theories: Shells, Rods and Points. Kluwer Academic Publishers, Dordrecht (2000)
19. Seiler, W.M.: Involution: The Formal Theory of Differential Equations and its Applications in Computer Algebra. Algorithms and Computation in Mathematics, vol. 24. Springer, Heidelberg (2010). doi:10.1007/978-3-642-01287-7
20. Sobottka, G.A., Lay, T., Weber, A.G.: Stable integration of the dynamic Cosserat equations with application to hair modeling. J. WSCG **16**, 73–80 (2008)
21. Wood, W.L., Bossak, M., Zienkiewicz, O.C.: An alpha modification of Newmarks method. Int. J. Numer. Methods Eng. **15**, 1562–1566 (1981)

Algorithms for Zero-Dimensional Ideals Using Linear Recurrent Sequences

Vincent Neiger[1(✉)], Hamid Rahkooy[2], and Éric Schost[2]

[1] Department of Applied Mathematics and Computer Science, Technical University of Denmark, Lyngby, Denmark
vinn@dtu.dk

[2] Cheriton School of Computer Science, University of Waterloo, Waterloo, Canada

Abstract. Inspired by Faugère and Mou's sparse FGLM algorithm, we show how using linear recurrent multi-dimensional sequences can allow one to perform operations such as the primary decomposition of an ideal, by computing of the annihilator of one or several such sequences.

1 Introduction

In what follows, $\mathbb{K}$ is a perfect field. We consider the set $\mathscr{S} = \mathbb{K}^{\mathbb{N}^n}$ of n-dimensional sequences $\boldsymbol{u} = (u_m)_{m \in \mathbb{N}^n}$, and the polynomial ring $\mathbb{K}[X_1, \ldots, X_n]$, and we are interested in the following question. Let $I \subset \mathbb{K}[X_1, \ldots, X_n]$ be a zero-dimensional ideal. Given a monomial basis of $Q = \mathbb{K}[X_1, \ldots, X_n]/I$, together with the corresponding multiplication matrices $\mathsf{M}_1, \ldots, \mathsf{M}_n$, we want to compute the Gröbner bases, for a target order $>$, of pairwise coprime ideals $J_1, \ldots, J_K$ such that $I = \cap_{1 \le k \le K} J_k$.

Faugère *et al.*'s paper [11] shows how to solve this question with $K = 1$ (so J_1 is simply I) in time $O(nD^3)$, where $D = \deg(I)$; here, the degree $\deg(I)$ is the $\mathbb{K}$-vector space dimension of Q. More recently, algorithms have been given with the cost bound $O\tilde{~}(nD^\omega)$ [9,10,20], where the notation $O\tilde{~}$ hides polylogarithmic factors, still with $K = 1$. The algorithms in this paper allow splittings (so $K > 1$ in general) and assume that $>$ is a lexicographic order.

To motivate our approach, assume that the algebraic set $V(I)$ is in *shape position*, that is, the coordinate X_n separates the points of $V(I)$. Then, the Shape Lemma [14] implies that the Gröbner basis of the radical $\sqrt{I}$ for the lexicographic order $X_1 > \cdots > X_n$ has the form $\langle X_1 - G_1(X_n), \ldots, X_{n-1} - G_{n-1}(X_n), P(X_n) \rangle$, for some squarefree polynomial P, and some $G_1, \ldots, G_{n-1}$ of degrees less than $\deg(P)$. The polynomials P and $G_1, \ldots, G_{n-1}$ can be deduced from the values $(\ell(X_n^i))_{0 \le i \le 2D}$ and $(\ell(X_j X_n^i))_{1 \le j < n, 0 \le i < D}$, for a randomly chosen linear form $\ell : Q \to \mathbb{K}$, in time $O\tilde{~}(D)$ [4]. The algorithms in the latter reference use baby steps/giant steps techniques for the calculation of the values of ℓ.

Similar ideas were developed in [12]; the algorithms in this reference make no assumption on I but may fail in some cases, then falling back on the FGLM algorithm. For instance, if I itself (rather than $\sqrt{I}$) is known to have a lexicographic

V.P. Gerdt et al. (Eds.): CASC 2017, LNCS 10490, pp. 313–328, 2017.
DOI: 10.1007/978-3-319-66320-3_23

Gröbner basis of the form $\langle X_1 - H_1(X_n), \ldots, X_{n-1} - H_{n-1}(X_n), Q(X_n)\rangle$, the algorithms in [12] recover this basis, also by considering values of linear forms $\ell_i : Q \to \mathbb{K}$. A key remark made in that reference is that the values of the linear forms ℓ_i that we need can be computed efficiently by exploiting the sparsity of the multiplication matrices $\mathsf{M}_1, \ldots, \mathsf{M}_n$; this sparsity is then analyzed, assuming the validity of a conjecture due to Moreno-Socías [18]. These techniques are related as well to Rouillier's Rational Univariate Representation algorithm [21], which uses values of a specific linear form $Q \to \mathbb{K}$ called the *trace*. However, computing the trace (that is, its values on the monomial basis of Q) is non-trivial, and using random choices instead makes it possible to avoid this issue.

In this paper, we work in the continuation of [4]. Assuming $V(I)$ is in shape position, the results in that reference allow us to compute the Gröbner basis of $\sqrt{I}$, and our goal here is to recover Gröbner bases corresponding to a decomposition of I as stated above. Following [1,12], we discuss the relation of this question to instances of the following problem: given sequences $\boldsymbol{u}_1, \ldots, \boldsymbol{u}_s$ in $\mathscr{S}$, find the Gröbner basis of their *annihilator* $\mathrm{ann}(\boldsymbol{u}_1, \ldots, \boldsymbol{u}_s) \subset \mathbb{K}[X_1, \ldots, X_n]$, for a target order $>$. The annihilator, discussed in the next section, is a polynomial ideal corresponding to the linear relations which annihilate all sequences.

A direct approach to solve the FGLM problem using such techniques would be to pick initial conditions at random; knowing multiplication matrices modulo I allows us to compute the values of a sequence $\boldsymbol{u}$, for which I is contained in $\mathrm{ann}(\boldsymbol{u})$. If $I = \mathrm{ann}(\boldsymbol{u})$ holds, computing sufficiently many values of $\boldsymbol{u}$ and feeding them into an algorithm such as Sakata's [22] would solve our problem. This is often, but not always, possible: there exists a sequence $\boldsymbol{u}$ for which $I = \mathrm{ann}(\boldsymbol{u})$ if and only if $Q = \mathbb{K}[X_1, \ldots, X_n]/I$ is a *Gorenstein* ring, a notion going back to [15,16] (see e.g. [5, Proposition 5.3] for a proof of the above assertion). This is for instance the case if I is a complete intersection, or if I is radical over a perfect field [8]; however, an ideal such as $I = \langle X_1^2, X_1X_2, X_2^2\rangle \subset \mathbb{K}[X_1, X_2]$ is not Gorenstein.

To remedy this, we may have to use more than one sequence, so as to be able to recover I as $I = \mathrm{ann}(\boldsymbol{u}_1, \ldots, \boldsymbol{u}_s)$. However, proceeding directly in this manner, we do not expect the algorithm to be significantly better than applying directly the FGLM algorithm (the techniques we will use for computing annihilators follow essentially the same lines as the FGLM algorithm itself). We will see that starting from the Gröbner basis of $\sqrt{I}$, we will be able to decompose I into e.g. primary components (assuming we allow the use of factorization algorithms over $\mathbb{K}$), and that our approach is expected to be competitive in those cases where the multiple components of I have low degrees.

2 Generalities on Sequences and Their Annihilators

Define the shift operators $s_1, \ldots, s_n$ on $\mathscr{S}$ in the obvious manner, by setting $s_i(\boldsymbol{u}) = (u_{m+e_i})_{m \in \mathbb{N}^n}$, where $e_1, \ldots, e_n$ are the unit vectors. This makes $\mathscr{S}$ a $\mathbb{K}[X_1, \ldots, X_n]$-module, by setting $f \cdot \boldsymbol{u} = f(s_1, \ldots, s_n)(\boldsymbol{u})$. For $f = \sum_m f_m \boldsymbol{X}^m$, the entries of $f \cdot \boldsymbol{u}$ are thus $(\langle \boldsymbol{u} \mid \boldsymbol{X}^m f\rangle)_{m \in \mathbb{N}^n}$, where we write $\boldsymbol{X}^m = X_1^{m_1} \cdots X_n^{m_n}$

and $\langle \boldsymbol{u} | f \rangle = \sum_{m'} f_{m'} u_{m'}$. To a sequence $\boldsymbol{u} = (u_m)_{m \in \mathbb{N}^n}$ in $\mathscr{S}$, we can then associate its *annihilator* $\mathrm{ann}(\boldsymbol{u})$, defined as the ideal of all polynomials f in $\mathbb{K}[X_1, \ldots, X_n]$ such that $f \cdot \boldsymbol{u} = 0$. If we consider several sequences $\boldsymbol{u}_1, \ldots, \boldsymbol{u}_s$ in $\mathscr{S}$, we then define $\mathrm{ann}(\boldsymbol{u}_1, \ldots, \boldsymbol{u}_s) = \mathrm{ann}(\boldsymbol{u}_1) \cap \cdots \cap \mathrm{ann}(\boldsymbol{u}_s)$.

We will also occasionally discuss *kernels* of sequences. For $\boldsymbol{u} \in \mathscr{S}$, the kernel $\ker(\boldsymbol{u})$ is the $\mathbb{K}$-vector space formed by all polynomials f in $\mathbb{K}[X_1, \ldots, X_n]$ such that $\langle \boldsymbol{u} | f \rangle = 0$; this is not an ideal in general. If we consider several sequences $\boldsymbol{u}_1, \ldots, \boldsymbol{u}_s$, we will write $\ker(\boldsymbol{u}_1, \ldots, \boldsymbol{u}_s) = \ker(\boldsymbol{u}_1) \cap \cdots \cap \ker(\boldsymbol{u}_s)$.

Let I be a zero-dimensional ideal in $\mathbb{K}[X_1, \ldots, X_n]$. Define the residue class ring $Q = \mathbb{K}[X_1, \ldots, X_n]/I$ and let $D = \deg(I) = \dim_{\mathbb{K}}(Q)$. Consider also the dual $Q^* = \hom_{\mathbb{K}}(Q, \mathbb{K})$. To a linear form ℓ in Q^*, we associate the sequence $\boldsymbol{u}_\ell$ defined by $\boldsymbol{u}_\ell = (\ell(\boldsymbol{X}^m \bmod I))_{m \in \mathbb{N}^n}$.

For any linear form ℓ on Q, and any g in Q, define the linear form $g \cdot \ell \in Q^*$ by $(g \cdot \ell)(h) = \ell(gh)$. This induces a Q-module structure on Q^*, and we remark that we have the equality $g \cdot \boldsymbol{u}_\ell = \boldsymbol{u}_{(g \bmod I) \cdot \ell}$ for any g in $\mathbb{K}[X_1, \ldots, X_n]$. Following [23] (where it is described with $n = 1$), we call this operation *transposed product*.

For ℓ in Q^*, we can then define $\mathrm{ann}_Q(\ell)$ as the set of all g in Q such that $g \cdot \ell = 0$; this is an ideal of Q. The following lemma clarifies the relation between $\mathrm{ann}(\boldsymbol{u}_\ell) \subset \mathbb{K}[X_1, \ldots, X_n]$ and $\mathrm{ann}_Q(\ell) \subset Q$; it implies that $\mathrm{ann}(\boldsymbol{u}_\ell)$ is generated by I and any element of $\mathrm{ann}_Q(\ell)$ lifted to $\mathbb{K}[X_1, \ldots, X_n]$.

Lemma 1. *With notation as above, for f in $\mathbb{K}[X_1, \ldots, X_n]$, f is in $\mathrm{ann}(\boldsymbol{u}_\ell)$ if and only if $f \bmod I$ is in $\mathrm{ann}_Q(\ell)$.*

Proof. Take f in $\mathbb{K}[X_1, \ldots, X_n]$. Then f is in $\mathrm{ann}(\boldsymbol{u}_\ell)$ if and only if $f \cdot \boldsymbol{u}_\ell = 0$, that is, if and only if $\boldsymbol{u}_{(f \bmod I) \cdot \ell} = 0$, if and only if $(f \bmod I) \cdot \ell$ itself is zero. □

When Q^* is a free Q-module of rank one, we say that Q is a Gorenstein ring, and that I is Gorenstein. In this case, there exists a linear form λ such that $Q^* = Q \cdot \lambda$; by the previous lemma, $\mathrm{ann}(\boldsymbol{u}_\lambda) = I$. Conversely, if $\mathrm{ann}(\boldsymbol{u}_\lambda) = I$, $\mathrm{ann}_Q(\lambda) = \{0\}$, so that $Q^* = Q \cdot \lambda$ (and Q^* is free of rank one). For instance, it is known that if I is radical, or I a complete intersection, then I is Gorenstein. On the other hand, if $I = \langle X_1^2, X_1 X_2, X_2^2 \rangle$, the inclusion $I \subset \mathrm{ann}(\boldsymbol{u}_\ell)$ is strict for any linear form ℓ. Using several sequences, we can however always recover I.

Lemma 2. *Let $\ell_1, \ldots, \ell_D$ be linearly independent in Q^*, and let $\boldsymbol{u}_1, \ldots, \boldsymbol{u}_D$ be the corresponding sequences. Then $\mathrm{ann}(\boldsymbol{u}_1, \ldots, \boldsymbol{u}_D) = \ker(\boldsymbol{u}_1, \ldots, \boldsymbol{u}_D) = I$.*

Proof. Note first that the inclusion $I \subset \mathrm{ann}(\boldsymbol{u}_1, \ldots, \boldsymbol{u}_D) = \mathrm{ann}(\boldsymbol{u}_1) \cap \cdots \cap \mathrm{ann}(\boldsymbol{u}_D)$ is a direct consequence of Lemma 1, and that $\mathrm{ann}(\boldsymbol{u}_1, \ldots, \boldsymbol{u}_D)$ is contained in $\ker(\boldsymbol{u}_1, \ldots, \boldsymbol{u}_D)$. For the converse, let $\omega_1, \ldots, \omega_D$ be the basis of Q dual to $\ell_1, \ldots, \ell_D$. Suppose that f is in $\ker(\boldsymbol{u}_1, \ldots, \boldsymbol{u}_D)$, and assume without loss of generality that f has been reduced by I, so that f is a linear combination of the form $f_1 \omega_1 + \cdots + f_D \omega_D$. Fix i in $1, \ldots, D$ and apply ℓ_i to f; we obtain f_i. On the other hand, because f is in $\ker(\boldsymbol{u}_i)$, $\ell_i(f)$ must vanish. So we are done. □

We may however need less than D linear forms, as explained in the following discussion, which generalizes the comments we made in the Gorenstein case.

Let $B = (b_1, \ldots, b_D)$ be a monomial basis of Q. Given a linear form ℓ in Q^*, we define K_ℓ as the $D \times D$ matrix whose (i, j)th entry is $\ell(b_i b_j)$; this is the matrix of the mapping $f \in Q \mapsto f \cdot \ell \in Q^*$, so that its nullspace is $\mathrm{ann}_Q(\ell)$. More generally, given a positive integer s and linear forms $\ell_1, \ldots, \ell_s$, we define $K_{\ell_1,\ldots,\ell_s}$ as the $D \times sD$ matrix obtained as the concatenation of $K_{\ell_1}, \ldots, K_{\ell_s}$; this is the matrix of the mapping $(f_1, \ldots, f_s) \in Q^s \mapsto f_1 \cdot \ell_1 + \cdots + f_s \cdot \ell_s \in Q^*$.

Lemma 3. *For any linear forms $(\ell_1, \ldots, \ell_s)$, with all ℓ_i in Q^*,* $\mathrm{ann}(\boldsymbol{u}_{\ell_1}, \ldots, \boldsymbol{u}_{\ell_s}) = I$ *if and only if $(\ell_1, \ldots, \ell_s)$ are Q-module generators of Q^*.*

Proof. $(\ell_1, \ldots, \ell_s)$ are Q-module generators of Q^* if and only if $K_{\ell_1,\ldots,\ell_s}$ has rank D, if and only if $K^\perp_{\ell_1,\ldots,\ell_s}$ has a trivial nullspace. The nullspace of this matrix is the intersection of those of the matrices $K^\perp_{\ell_1}, \ldots, K^\perp_{\ell_s}$. All these matrices are symmetric, and we saw that for all i, the nullspace of $K^\perp_{\ell_i} = K_{\ell_i}$ is $\mathrm{ann}_Q(\ell_i)$; thus, the condition above is equivalent to $\mathrm{ann}_Q(\ell_1) \cap \cdots \cap \mathrm{ann}_Q(\ell_s) = \{0\}$. Lemma 1 shows that this is the case if and only if $\mathrm{ann}(\boldsymbol{u}_{\ell_1}) \cap \cdots \cap \mathrm{ann}(\boldsymbol{u}_{\ell_s}) = I$. □

Proposition 1. *There exists a unique integer $\tau \leq D$ such that for a generic choice of linear forms $(\ell_1, \ldots, \ell_\tau)$, with all ℓ_i in Q^*, the sequence of ideals $(\mathrm{ann}(\boldsymbol{u}_{\ell_1}, \ldots, \boldsymbol{u}_{\ell_t}))_{1 \leq t \leq \tau}$ is strictly decreasing, with* $\mathrm{ann}(\boldsymbol{u}_{\ell_1}, \ldots, \boldsymbol{u}_{\ell_\tau}) = I$.

Proof. Remark first that if τ exists with the properties above, it is necessarily unique. Let $(L_{1,1}, \ldots, L_{1,D}), \ldots, (L_{D,1}, \ldots, L_{D,D})$ be new indeterminates, let $\mathbb{L} = \mathbb{K}(L_{1,1}, \ldots, L_{D,D})$ and define the matrices $K_{L_1}, \ldots, K_{L_D}$ as follows. Let $Q_\mathbb{L} = Q \otimes_\mathbb{K} \mathbb{L}$; this allows us to define the linear forms $L_1, \ldots, L_D$ in $Q^*_\mathbb{L}$ by $L_t(b_j) = L_{t,j}$, for $1 \leq t \leq D$; then K_{L_t} is the matrix with entries $L_t(b_i b_j)$. The entries of K_{L_t} are linear forms in $L_{t,1}, \ldots, L_{t,D}$.

Define $K_{L_1,\ldots,L_t}$ as we did for $K_{\ell_1,\ldots,\ell_t}$. Then, for any linear forms $\ell_1, \ldots, \ell_t$ in Q^*, the matrix $K_{\ell_1,\ldots,\ell_t}$ is obtained by evaluating $K_{L_1,\ldots,L_t}$ at $L_{t,j} = \ell_t(b_j)$, for all t, j. The rank of $K_{\ell_1,\ldots,\ell_t}$ (over $\mathbb{K}$) is at most that of $K_{L_1,\ldots,L_t}$ (over $\mathbb{L}$).

We can then let τ be the smallest integer such that the matrix $K_{L_1,\ldots,L_\tau}$ has full rank D. Such an index exists, and is at most D, since by Lemma 2 (and by the remarks of the above paragraph) $K_{L_1,\ldots,L_D}$ has rank D.

Let $\ell_1, \ldots, \ell_\tau$ be such that $K_{\ell_1,\ldots,\ell_\tau}$ has rank D (this is our genericity condition); in this case, by the previous lemma, $\mathrm{ann}(\boldsymbol{u}_{\ell_1}, \ldots, \boldsymbol{u}_{\ell_\tau}) = I$. To conclude, it suffices to prove that the sequence of ideals $(\mathrm{ann}(\boldsymbol{u}_{\ell_1}, \cdots, \boldsymbol{u}_{\ell_t}))_{1 \leq t \leq \tau}$ is strictly decreasing. Suppose it is not the case, so that $\mathrm{ann}(\boldsymbol{u}_{\ell_1}, \ldots, \boldsymbol{u}_{\ell_t}) = \mathrm{ann}(\boldsymbol{u}_{\ell_1}, \ldots, \boldsymbol{u}_{\ell_{t+1}})$ for some $t < \tau$. Then, $\mathrm{ann}(\boldsymbol{u}_{\ell_1}, \ldots, \boldsymbol{u}_{\ell_t}, \boldsymbol{u}_{\ell_{t+2}}, \ldots, \boldsymbol{u}_{\ell_\tau}) = I$. Let us define $\ell'_1 = \ell_1, \ldots, \ell'_t = \ell_t, \ell'_{t+1} = \ell_{t+2}, \ldots, \ell'_{\tau-1} = \ell_\tau$. Then, we have $\mathrm{ann}(\boldsymbol{u}_{\ell'_1}, \ldots, \boldsymbol{u}_{\ell'_{\tau-1}}) = I$, so that $K_{\ell'_1,\ldots,\ell'_{\tau-1}}$ has rank D. This in turn implies (by the discussion above) that $K_{L_1,\ldots,L_{\tau-1}}$ has rank D, a contradiction. □

If Q is a local algebra with maximal ideal $\mathfrak{m}$, we can define the *socle* of Q as the $\mathbb{K}$-vector space of all elements f in Q such that $\mathfrak{m}f = 0$. For instance, if Q is local, the integer τ in the previous lemma is the dimension of the socle of Q. (we omit the proof, since we will not use this result in the rest of the paper).

3 Computing Annihilators of Sequences

Consider sequences $(\boldsymbol{u}_1,\ldots,\boldsymbol{u}_t)$ with $\boldsymbol{u}_i \in \mathscr{S}$ for all i, let J be the annihilator $\mathrm{ann}(\boldsymbol{u}_1,\ldots,\boldsymbol{u}_t) \subset \mathbb{K}[X_1,\ldots,X_n]$, and suppose that it has dimension zero; our goal is to compute a Gröbner basis of it. We first review an algorithm due to Marinari et al. [17], then introduce a modification of it that relaxes some of its assumptions. As a result, the algorithms in this section work under slightly different assumptions, and feature slightly different runtimes.

An algorithm with cost $(nt\deg(J))^{O(1)}$ would be highly desirable, but we are not aware of any such result. Most approaches (ours as well) involve reading a number of values of $\boldsymbol{u}_1,\ldots,\boldsymbol{u}_t$ and looking for dependencies between the columns of what is often called a generalized Hankel matrix, built using these values; the delicate question is how to control the size of the matrix.

Consider for instance the case $t=1$, $\langle \boldsymbol{u}_1 \mid X_1^{m_1}\cdots X_n^{m_n}\rangle = 1$ for $m_1+\cdots+m_n < \delta$ and $\langle \boldsymbol{u}_1 \mid X_1^{m_1}\cdots X_n^{m_n}\rangle = 0$ otherwise. The annihilator $J=\mathrm{ann}(\boldsymbol{u}_1)$ admits the lexicographic Gröbner basis $\langle X_1-X_n,\ldots,X_{n-1}-X_n,X_n^{\delta}\rangle$, so we have $\deg(J)=\delta$; on the other hand, this sequence takes $\binom{\deg(J)+n-1}{n}$ non-zero values, so taking them all into account leads us to an exponential time algorithm.

In the case $t=1$, Mourrain in [19] associates a Hankel operator to a sequence such that the kernel of the Hankel operator corresponds to the annihilator of the sequence. Algorithm 2 in that paper computes a border basis for the kernel of such a Hankel operator, taking as input its values over a finite set of monomials. As in the FGLM algorithm, this algorithm looks for linear dependencies between the monomials in the border of already computed linearly independent monomials. However, for examples as in the previous paragraph, we are not aware of how to avoid taking into account up to $\binom{\deg(J)+n-1}{n}$ values.

Several algorithms were also proposed in [1] for computing an annihilator $\mathrm{ann}(\boldsymbol{u}_1)$, and partly extended to arbitrary t in [2]. A first algorithm relies on the Berlekamp-Massey Algorithm, by means of a change of coordinates, which may require an exponential number of value of $\boldsymbol{u}_1$. The other algorithms extend the idea of FGLM, considering maximal rank sub-matrices of a truncated multi-Hankel matrix to compute a basis for the quotient algebra and a Gröbner basis. An algorithm with certified outcome (Scalar-FGLM) is presented; it considers the values of $\boldsymbol{u}_1$ at all monomials up to a given degree $\simeq \deg(J)$, so the issue pointed out above remains. An "adaptive" version uses fewer values of the sequence, but may fail in some cases (the conditions that ensure success of this algorithm seem to be close to the genericity assumptions we introduce in Subsect. 3.2). A comparison of Scalar-FGLM and Sakata's algorithm is presented in [3].

3.1 A First Algorithm

The first solution we discuss requires a strong assumption (written H_1 below): for any i and for any monomial b in $X_1,\ldots,X_n$, $b\cdot\boldsymbol{u}_i$ is in the $\mathbb{K}$-span of $(\boldsymbol{u}_1,\ldots,\boldsymbol{u}_t)$; as a result, the annihilator J of $(\boldsymbol{u}_1,\ldots,\boldsymbol{u}_t)$ equals the nullspace $\ker(\boldsymbol{u}_1,\ldots,\boldsymbol{u}_t)$. For this situation, Marinari et al. gave in [17] an algorithm that computes a Gröbner basis of J, for any order (for definiteness, we refer here to their second algorithm); it is an extension of both the Buchberger-Möller interpolation algorithm and the FGLM change of order algorithm.

Assumption H_1 above implies that $\deg(J) \leq t$, and the runtime of the algorithm, expressed in terms of n and t, is $O(nt^3)$ operations in $\mathbb{K}$, together with the computation of all values $\langle \boldsymbol{u}_i \mid b\rangle$, $1 \leq i \leq t$, for $O(nt)$ monomials b. These evaluations are done in incremental order, in the sense that for any monomial b for which we need all $\langle \boldsymbol{u}_i \mid b\rangle$, there exists $j \in \{1, \ldots, n\}$ such that $b = X_j b'$ and all $\langle \boldsymbol{u}_i \mid b'\rangle$ are known.

We will need the following property of this algorithm. Suppose that $(\boldsymbol{u}_1, \ldots, \boldsymbol{u}_t)$ is a subsequence of a larger family of sequences $(\boldsymbol{u}_1, \ldots, \boldsymbol{u}_{t'})$ that satisfies H_1, but that $(\boldsymbol{u}_1, \ldots, \boldsymbol{u}_t)$ itself may or may not, and that $(\boldsymbol{u}_1, \ldots, \boldsymbol{u}_t)$ and $(\boldsymbol{u}_1, \ldots, \boldsymbol{u}_{t'})$ have different $\mathbb{K}$-spans. Then, on input $(\boldsymbol{u}_1, \ldots, \boldsymbol{u}_t)$, the algorithm will still run its course, and at least one of the elements in the output will be a polynomial g that does not belong to $\mathrm{ann}(\boldsymbol{u}_1, \ldots, \boldsymbol{u}_{t'})$.

3.2 An Algorithm Under Genericity Assumptions

We now give a second algorithm for computing $J = \mathrm{ann}(\boldsymbol{u}_1, \ldots, \boldsymbol{u}_t)$, whose runtime is polynomial in $n, t, D = \deg(J)$ and an integer $B \leq \deg(J)$ defined below. We do not assume that H_1 holds, but we will require other assumptions; if they hold, the output is the lexicographic Gröbner basis G of J for the order $X_1 > \cdots > X_n$. Our first assumption is:

H_2. We are given an integer B such that the minimal polynomial of X_j in $\mathbb{K}[X_1, \ldots, X_n]/J$ has degree at most B for all j.

For j in $1, \ldots, n$, we will denote by J_j the ideal $\mathrm{ann}(\pi_j(\boldsymbol{u}_1), \ldots, \pi_j(\boldsymbol{u}_t)) \subset \mathbb{K}[X_j, \ldots, X_n]$, where for all i, $\pi_j(\boldsymbol{u}_i)$ is the sequence $\mathbb{N}^{n-j+1} \to \mathbb{K}$ defined by $\langle \pi_j(\boldsymbol{u}_i) \mid (m_j, \ldots, m_n)\rangle = \langle \boldsymbol{u}_i \mid (0, \ldots, 0, m_j, \ldots, m_n)\rangle$ for all $(m_j, \ldots, m_n)$ in $\mathbb{N}^{n-j+1}$; in particular, $J_1 = J$. We write $\deg(J_j) = D_j \leq D$, we let G_j be the lexicographic Gröbner basis of J_j, and we let $\mathscr{B}_j$ be the corresponding monomial basis of $\mathbb{K}[X_j, \ldots, X_n]/J_j$.

We can then introduce our genericity property; by contrast with H_2, we will not necessarily assume that it holds, and discuss the outcome of the algorithm when it does not. We denote this property by $\mathsf{H}_3(j)$, for $j = 1, \ldots, n-1$.

$\mathsf{H}_3(j)$. We have the equality $J_j \cap \mathbb{K}[X_{j+1}, \ldots, X_n] = J_{j+1}$.

Remark that the inclusion $J_j \cap \mathbb{K}[X_{j+1}, \ldots, X_n] \subset J_{j+1}$ always holds.

Suppose that for some j in $1, \ldots, n$, we have computed a sequence of monomials $\mathscr{B}'_{j+1}$ in $\mathbb{K}[X_{j+1}, \ldots, X_n]$ (if $j = n$, we let $\mathscr{B}'_{j+1} = (1)$). Since we will use them repeatedly, we define properties P and P' as follows, the latter being stronger than the former.

$\mathsf{P}(j+1)$. The cardinality D'_{j+1} of $\mathscr{B}'_{j+1}$ is at most D_{j+1}.
$\mathsf{P}'(j+1)$. The equality $\mathscr{B}'_{j+1} = \mathscr{B}_{j+1}$ holds.

We describe in the following paragraphs a procedure that computes a new family of monomials $\mathscr{B}'_j$, and we give conditions under which they satisfy $\mathsf{P}(j)$ and $\mathsf{P}'(j)$.

We call a family of monomials $\mathscr{B}$ in $\mathbb{K}[X_j, \ldots, X_n]$ *independent* if their images are $\mathbb{K}$-linearly independent modulo J_j (we call it *dependent* otherwise). We denote by $\mathsf{M}_{\mathscr{B}}$ the matrix with entries $\langle \boldsymbol{u}_i | bb' \rangle$, with rows indexed by $i = 1, \ldots, t$ and b' in $\mathscr{C}_{j+1} = \mathscr{B}'_{j+1} \times (1, X_j, \ldots, X_j^{B-1})$, and columns indexed by b in $\mathscr{B}$ (for any monomial b in $\mathbb{K}[X_j, \ldots, X_n]$, M_b is the column vector defined similarly).

Lemma 4. *If $\mathscr{B}$ is dependent, the right nullspace of $\mathsf{M}_{\mathscr{B}}$ is non-trivial. If both $\mathsf{P}'(j+1)$ and $\mathsf{H}_3(j)$ hold, the converse is true.*

Proof. Any $\mathbb{K}$-linear relation between the elements of $\mathscr{B}$ induces the same relation between the columns of $\mathsf{M}_{\mathscr{B}}$, and the first point follows.

By definition, a polynomial f in $\mathbb{K}[X_j, \ldots, X_n]$ belongs to J_j if and only if it annihilates $\pi_j(\boldsymbol{u}_1), \ldots, \pi_j(\boldsymbol{u}_t)$, that is, if $\langle \pi_j(\boldsymbol{u}_i) \mid X_j^{m_j} \ldots X_n^{m_n} f \rangle = 0$ for all $(m_j, \ldots, m_n)$ in $\mathbb{N}^{n-j+1}$ and all $i = 1, \ldots, t$. Now, assumptions $\mathsf{P}'(j+1)$, H_2 and $\mathsf{H}_3(j)$ imply that $\mathscr{C}_{j+1}$ generates $\mathbb{K}[X_j, \ldots, X_n]/J_j$, so that f is in J_j if and only if $\langle \boldsymbol{u}_i \mid bf \rangle = 0$, for all b in $\mathscr{C}_{j+1}$ and all $i = 1, \ldots, t$. □

The following lemma, that essentially follows the argument used in the proof of the FGLM algorithm [11], will be useful to justify our algorithm as well.

Lemma 5. *Suppose that $b_1 < \cdots < b_u < b_{u+1}$ are the first $u+1$ standard monomials of $\mathbb{K}[X_j, \ldots, X_n]/J_j$, for the lexicographic order induced by $X_j > \cdots > X_n$, with $b_1 = 1$. Then for any monomial b such that $b_u < b < b_{u+1}$, $\{b_1, \ldots, b_u, b\}$ is a dependent family.*

Proof. We prove the result by induction on $u \geq 0$, the case $u = 0$ being vacuously true. Assuming the claim is true for some index $u \geq 0$, we prove it for $u+1$. We proceed by contradiction, and we let b be the smallest monomial such that $b_u < b < b_{u+1}$ and $\{b_1, \ldots, b_u, b\}$ is an independent family (b exists by the well-ordering property of monomial orders).

We will use the fact that any monomial c less than b can be rewritten as a linear combination of $b_1, \ldots, b_i$, with $b_i < c$, for some $i \leq u$: if $c < b_u$, this is by the induction assumption; if $c = b_u$, this is obvious; if $b_u < c < b$, this is by the definition of b.

Now, either b is the leading term of an element in the Gröbner basis of J_j, or it must be of the form $b = X_e b'$, for some monomial b' not in $\{b_1, \ldots, b_u\}$. We prove that in both cases, b can be rewritten as a linear combination of $b_1, \ldots, b_u$, which is a contradiction. In the first case, b rewrites as a linear combination of smaller monomials, say $c_1, \ldots, c_v$, and by the previous remark, all of them can be rewritten as linear combinations of $b_1, \ldots, b_u$. Altogether, b itself can be rewritten as a linear combination of $b_1, \ldots, b_u$, a contradiction.

In the second case, $b = X_e b'$, for some monomial b' not in $\{b_1, \ldots, b_u\}$. As above, b' can be rewritten modulo J_j as a linear combination of monomials $b_1, \ldots, b_i$, for some $i \leq u$, with $b_i < b'$. Then, $b = X_e b'$ is a linear combination of $X_e b_1, \ldots, X_e b_i$. Since $b_i < b'$, we get $X_e b_1 < \cdots < X_e b_i < X_e b' = b$, so all of

$X_eb_1, \ldots, X_eb_i$ can be rewritten as linear combinations of $b_1, \ldots, b_u$. As a result, this is also the case for b itself, so we get a contradiction again. □

Suppose that $\mathsf{P}(j+1)$ holds. Then, the algorithm at step j proceeds as follows. We compute the reduced row echelon form of $\mathsf{M}_{\mathscr{C}_{j+1}}$. Using assumption $\mathsf{P}(j+1)$, this matrix has at most tBD_{j+1} rows and at most BD_{j+1} columns, and it has rank at most D_j (by the first item of Lemma 4). This computation can be done in time $O(tB^2D_{j+1}^2D_j) \in O(tB^2D^3)$. The column indices of the pivots allow us to define the monomials $\mathscr{B}'_j = (b'_1 < \cdots < b'_{D'_j})$, for some $D'_j \leq D_j$.

Lemma 6. *Property* $\mathsf{P}(j)$ *holds, and if* $\mathsf{P}'(j+1)$ *and* $\mathsf{H}_3(j)$ *hold, then* $\mathsf{P}'(j)$ *holds.*

Proof. The first item is a restatement of the inequality $D'_j \leq D_j$. To prove the second item, assuming that $\mathsf{P}'(j+1)$ and $\mathsf{H}_3(j)$ hold, we deduce from Lemma 4 that the columns indexed by the genuine $\mathscr{B}_j$ form a column basis of $\mathsf{M}_{\mathscr{C}_{j+1}}$, and we claim that it is actually the lexicographically smallest column basis (this will prove that $\mathscr{B}_j = \mathscr{B}'_j$). Indeed, write $\mathscr{B}_j = (b_1, \ldots, b_{D_j})$, and let $(f_1, \ldots, f_{D_j})$ be another subsequence of $\mathscr{C}_{j+1}$ whose corresponding columns form a column basis of $\mathsf{M}_{\mathscr{C}_{j+1}}$. Let m be the smallest index such that $b_m \neq f_m$. Then, applying Lemma 5 to $(b_1, \ldots, b_{m-1})$ and f_m, we deduce that $b_m < f_m$ (otherwise, since they are different, we must have $b_{m-1} < f_m < b_m$, which implies that f_m is a linear combination of $(b_1, \ldots, b_{m-1}) = (f_1, \ldots, f_{m-1})$, a contradiction). □

Thus, running this procedure for $j = n, \ldots, 1$, we maintain $\mathsf{P}(j)$; this implies that the running time is $O(ntB^2D^3)$, computing the values $\langle \boldsymbol{u}_i | b \rangle$, for $1 \leq i \leq t$, for $O(nB^2D^2)$ monomials b (with the same monotonic property as in the previous subsection). If $\mathsf{H}_3(j)$ holds for all j, the second item in the last lemma proves that $\mathscr{B}'_1 = \mathscr{B}_1$, the monomial basis of $\mathbb{K}[X_1, \ldots, X_n]/J$.

Once $\mathscr{B}'_1$ is known, we compute and return a family of polynomials G' defined as follows. We determine the sequence Δ of elements in $X_1\mathscr{B}'_1 \cup \cdots \cup X_n\mathscr{B}'_1 - \mathscr{B}'_1$, all of whose factors are in $\mathscr{B}'_1$ (finding them does not require any operation in $\mathbb{K}$; this can be done by using e.g. a balanced binary search tree with the elements of $\mathscr{B}'_1$, using a number of comparisons that is quasi-linear time in nD). Then, we rewrite each column M_b, for b in Δ, as a linear combination of the form $\sum_{1 \leq i \leq D'_1} c_i \mathsf{M}_{b'_i}$ and we put $b - \sum_{1 \leq i \leq D'_1} c_i b'_i$ in G'. If the reduction is not possible, the algorithm halts and returns fail. Using the reduced row echelon form of $\mathsf{M}_{\mathscr{C}_2}$, each reduction takes time $O(D_1^2) \in O(D^2)$ operations in $\mathbb{K}$, for a total of $O(nD^3)$.

If $\mathsf{H}_3(j)$ holds for all j, since $\mathscr{B}_1 = \mathscr{B}'_1$, the fact that $G' = G$ follows from Lemma 4. Assume now that G' differs from G; we prove that there exists an element in G not in J (we will use this in our main algorithm to detect failure cases). Indeed, in this case, $\mathscr{B}'_1$ must be different from $\mathscr{B}_1$, and since $\mathscr{B}'_1$ has cardinality at most equal to that of $\mathscr{B}_1$, there exists a monomial b in $\mathscr{B}_1$ not in $\mathscr{B}'_1$. This in turn implies that there exists an element g in G' that divides b, and thus with leading term in $\mathscr{B}_1$. Reducing g modulo G, we must then obtain a non-zero remainder, so that g does not belong to J.

4 Main Algorithm

4.1 Representing Primary Zero-Dimensional Ideals

Let I be a zero-dimensional ideal in $\mathbb{K}[X_1,\ldots,X_n]$; we assume that I is $\mathfrak{m}$-primary, for some maximal ideal $\mathfrak{m}$, and we write $D=\deg(I)$. In this paragraph, we briefly mention some possible representations for I (our main algorithm will compute either one of these representations).

The first, and main, option we will consider is simply the Gröbner basis G of I, for the lexicographic order induced by $X_1 > \cdots > X_n$. As an alternative, consider the following construction. Our assumption on I implies that the minimal polynomial R of X_n in $\mathbb{K}[X_1,\ldots,X_n]/I$ takes the form $R=P^e$, for some irreducible polynomial P in $\mathbb{K}[Z]$, of degree say f (remark that $R(X_n)$ is also the last polynomial in G). Let $\mathbb{L}=\mathbb{K}[Z]/\langle P\rangle$; this is a field extension of degree f of $\mathbb{K}$, and the residue class ζ of Z in $\mathbb{L}$ is a root of P. We then let I' be the ideal $I+\langle (X_n-\zeta)^e\rangle$ in $\mathbb{L}[X_1,\ldots,X_n]$, and let D' be its degree. Then, a second option is to compute the lexicographic Gröbner basis G' of I', for the order $X_1>\cdots>X_n$. The following lemma relates D and D'.

Lemma 7. *The ideal I' has degree $D'=D/f$.*

Proof. Let $\mathbb{M}$ be the splitting field of P and let $\zeta_1,\ldots,\zeta_f$ be the roots of P in $\mathbb{M}$. The ideals $J_i=I+\langle (X_n-\zeta_i)^e\rangle \subset [X_1,\ldots,X_n]$ are such that $\deg(J_1)+\cdots+\deg(J_f)=\deg(I)$. On the other hand, there exist f embeddings $\sigma_1,\ldots,\sigma_f$ of $\mathbb{L}$ into $\mathbb{M}$, with σ_i given by $\zeta\mapsto\zeta_i$; as a result, $\deg(I')=\deg(J_i)$ holds for all i, and the claim follows. □

The point behind this construction is to lower the degree of the ideal we consider, at the cost of working in a field extension of $\mathbb{K}$. This may be beneficial, as the cost of the main algorithm (which essentially relies on the one in the previous section) will be a polynomial of rather large degree with respect to the degree of the ideal, whereas computation in a field extension such as $\mathbb{K}\to\mathbb{L}$ is a well-understood task of cost ranging from quasi-linear to quadratic.

Our last option aims at producing a "simpler" Gröbner basis, by means of a change of coordinates. For this, we will assume that X_n separates the points of $V(\mathfrak{m})$ (over an algebraic closure of $\mathbb{K}$). As a result, the ideal $\mathfrak{m}$ being maximal, it admits a lexicographic Gröbner basis of the form $\langle X_1-G_1(X_n),\ldots,X_{n-1}-G_{n-1}(X_n),P(X_n)\rangle$. Define $\xi_1=G_1(\zeta),\ldots,\xi_{n-1}=G_{n-1}(\zeta),\xi_n=\zeta$, for $\zeta\in\mathbb{L}$ as above; then, $(\xi_1,\ldots,\xi_n)$ is the unique zero of I' (in fact, I' is $\mathfrak{m}'$-primary, with $\mathfrak{m}'=\langle X_1-\xi_1,\ldots,X_n-\xi_n\rangle$). We can then apply the change of coordinates that replaces X_i by $X_i+\xi_i$ in I', for all i, and call I'' the ideal thus obtained (so that I'' is generated by the polynomials $f(X_1+\xi_1,\ldots,X_n+\xi_n)$, for f in I, and X_n^e). Now, I'' is $\mathfrak{m}''$-primary, with $\mathfrak{m}''=\langle X_1,\ldots,X_n\rangle$; one of our options will be to compute the Gröbner basis G'' of I''.

Example 1. *Consider the polynomials in* $\mathbb{Q}[X_1, X_2]$

$$X_1^2 - 2X_1X_2 - 2X_1 + X_2^2 + 2X_2 + 1,$$
$$X_1X_2^2 + X_1X_2 + 2X_1 - X_2^3 - 2X_2^2 - 3X_2 - 2,$$
$$X_2^4 + 2X_2^3 + 5X_2^2 + 4X_2 + 4,$$

the last of them being $P(X_2)^2 = (X_2^2 + X_2 + 2)^2$, *and let* I *be the ideal they define. The polynomials above are the lexicographic Gröbner basis* G *of* I *for the order* $X_1 > X_2$. *Let* $\mathbb{L} = \mathbb{Q}[Z]/\langle Z^2 + Z + 2\rangle$, *and let* ζ *be the image of* Z *in* $\mathbb{L}$; *then, the ideal* $I' = I + \langle (X_2 - \zeta)^2\rangle$ *in* $\mathbb{L}[X_1, X_2]$ *admits the Gröbner basis* G'

$$X_1^2 - 2X_1\zeta - 2X_1 + \zeta - 1,$$
$$X_1X_2 - X_1\zeta - X_2\zeta - X_2 - 2,$$
$$X_2^2 - 2X_2\zeta - \zeta - 2.$$

Here, we have $e = 2$, $f = 2$, $D = 6$ *and* $D' = 3$. *The ideal* I *is* $\mathfrak{m}$*-primary, where* $\mathfrak{m}$ *admits the Gröbner basis* $\langle X_1 - X_2 - 1, X_2^2 + X_2 + 2\rangle$, *so that we have* $(\xi_1, \xi_2) = (\zeta + 1, \zeta)$, *and* I' *is* $\mathfrak{m}'$*-primary, with* $\mathfrak{m}' = \langle X_1 - \xi_1, X_2 - \xi_2\rangle$. *Applying the change of coordinates* $(X_1, X_2) \leftarrow (X_1 + \xi_1, X_2 + \xi_2)$, *the resulting ideal* I'' *admits the Gröbner basis* $G'' = \langle X_1^2, X_1X_2, X_2^2\rangle$, *from which we can readily confirm that it is* $\langle X_1, X_2\rangle$*-primary.*

4.2 The Algorithm

We consider a zero-dimensional ideal I in $\mathbb{K}[X_1, \dots, X_n]$. We assume that we know a monomial basis $B = (b_1, \dots, b_D)$ of $Q = \mathbb{K}[X_1, \dots, X_n]/I$, so that we let $D = \dim_{\mathbb{K}}(Q)$, together with the corresponding multiplication matrices $\mathsf{M}_1, \dots, \mathsf{M}_n$ of respectively $X_1, \dots, X_n$. We assume that the last variable X_n has been chosen generically; in particular, X_n separates the points of $V = V(I)$.

The algorithm in this section computes a decomposition of I into primary components $J_1, \dots, J_K$. Each such component J_k will be given by means of one of the representations described in the previous subsection; we will emphasize the first of them, the lexicographic Gröbner basis of J_k, and mention how to modify the algorithm in order to obtain the other representations. In order to find the primary components of I, we cannot avoid the use of factorization algorithms over $\mathbb{K}$; if desired, one may avoid this by relying on *dynamic evaluation techniques* [7], replacing for instance the factorization into irreducibles used below by a squarefree factorization (thus producing a decomposition of I into ideals that are not necessarily primary). In that case, if one wishes to compute descriptions such as the second or third ones introduced above, involving algebraic numbers as coefficients, one should take into account the possibility of splittings of the defining polynomials, as is usual with this kind of approach (a complete description of the resulting algorithm, along the lines of [6], is beyond the scope of this paper).

The Ideal I and Its Primary Decomposition. Let $P_{\min} \in \mathbb{K}[X_n]$ be the minimal polynomial of X_n in Q, let P be its squarefree part, and let polynomials

$G_1, \ldots, G_{n-1}$ in $\mathbb{K}[X_n]$, with $\deg(G_i) < \deg(P)$ for all i, be such that $\sqrt{I}$ admits the lexicographic Gröbner basis $\langle X_1 - G_1(X_n), \ldots, X_{n-1} - G_{n-1}(X_n), P(X_n)\rangle$. We write $P_{\min} = P_1^{e_1} \cdots P_K^{e_K}$, with the P_k's pairwise distinct irreducible polynomials in $\mathbb{K}[X_n]$ and $e_k \geq 1$ for all k. In particular, the factorization of P is $P_1 \cdots P_K$; we write $f_k = \deg(P_k)$ for all k.

Correspondingly, let $V_1, \ldots, V_K$ be the $\mathbb{K}$-irreducible components of V and for $k = 1, \ldots, K$, let $\mathfrak{m}_k$ be the maximal ideal defining V_k; hence, the reduced lexicographic Gröbner basis of $\mathfrak{m}_k$ is $\langle X_1 - (G_1 \bmod P_k), \ldots, X_{n-1} - (G_{n-1} \bmod P_k), P_k\rangle$. We can then write $I = J_1 \cap \cdots \cap J_K$, with J_k $\mathfrak{m}_k$-primary for all k; note that the ideal J_k is defined by $J_k = I + \langle P_k^{e_k}\rangle$. In what follows, we explain how to compute a Gröbner basis of this ideal by means of the results of the previous section. Without loss of generality, assume that L is such that $e_k = 1$ for $k > L$ and $e_k \geq 2$ for $k = 1, \ldots, L$. The fact that X_n is a generic coordinate implies that for $k > L$, $J_k = \mathfrak{m}_k$, so there is nothing left to do for such indices; hence, we are left with showing how to use the algorithms of the previous section to compute Gröbner bases of $J_1, \ldots, J_L$.

Data Representation. An element f of Q is represented by the column vector v_f of its coordinates on the basis B, whereas a linear form $\ell : Q \to \mathbb{K}$ is represented by the row vector $\mathsf{w}_\ell = [\ell(b_1), \ldots, \ell(b_D)]$. Computing $\ell(f)$ is then done by means of the dot product $\mathsf{w}_\ell \cdot \mathsf{v}_f$. Multiplying f by X_i amounts to computing $\mathsf{M}_i \mathsf{v}_f$, and the linear form $X_i \cdot \ell : g \mapsto \ell(X_i g)$ is obtained by computing the vector $\mathsf{w}_{X_i \cdot \ell} = \mathsf{w}_\ell \mathsf{M}_i$.

In terms of complexity, we assume that multiplying any matrix M_i by a vector (either on the left or on the right) can be done in m operations in $\mathbb{K}$. The naive bound on m is $O(D^2)$, but the sparsity properties of these matrices often result in much better estimates; see [12] for an in-depth discussion of this question. On the other hand, we assume $D \leq \mathsf{m}$.

Computing. $P_{\min}$ **and** $G_1, \ldots, G_{n-1}$. First, we compute generators of $\sqrt{I}$. We choose a random linear form $\ell_1 : Q \to \mathbb{K}$, and we compute the values $(\ell_1(X_n^i))_{0 \leq i < 2D}$ and $\ell_1(X_1 X_n^i), \ldots, \ell_1(X_{n-1} X_n^i)$, for $0 \leq i < D$. This is done by computing $1, X_n, \ldots, X_n^{2D-1}$ by repeated applications of M_n, which amounts to $O(D\mathsf{m})$ operations, and doing the corresponding dot products with $\ell, X_1 \cdot \ell, \ldots, X_{n-1} \cdot \ell$. For the latter, we have to compute the linear forms $X_i \cdot \ell$ in $O(n\mathsf{m})$ operations, then do a $D \times D$ by $D \times (n+1)$ matrix product, which costs $O(nD^2)$ operations (without using fast linear algebra).

Using the algorithm given in [4], given these values, we can compute the minimal polynomial $P_{\min}$, as well as the polynomials $G_1, \ldots, G_{n-1}$ describing $V(I)$ in $\tilde{O}(D)$ operations in $\mathbb{K}$. Then, as per the discussion in the preamble, we assume that we have an algorithm for factoring polynomials over $\mathbb{K}$, so that $(P_1, e_1), \ldots, (P_K, e_K)$ and P can be deduced from $P_{\min}$.

Constructing the Orthogonal of J_k**.** For $k = 1, \ldots, K$, we will write $Q_k = \mathbb{K}[X_1, \ldots, X_n]/J_k$. Any linear form $\ell : Q \to \mathbb{K}$ induces a linear form $\varphi_k(\ell) : Q_k \to \mathbb{K}$, defined as follows.

Let T_k be the polynomial $P_{\min}/P_k^{e_k}$. For f in Q_k, let $\hat{f}$ be any lift of f to $\mathbb{K}[X_1,\dots,X_n]$, and define $\varphi_k(\ell)(f) = \ell(T_k\hat{f} \bmod I)$. Notice that this expression is well-defined: indeed, any two lifts of f differ by an element δ of $J_k = I + \langle P_k^{e_k}\rangle$, so that $T_k\delta$ is in I, since $T_kP_k^{e_k} = P_{\min}$ is.

Lemma 1. *The mapping $\varphi_k : Q^* \to Q_k^*$ is $\mathbb{K}$-linear and onto.*

Proof. Linearity is clear by construction; we now prove that φ_k is onto. Let indeed A_k, B_k in $\mathbb{K}[X_n]$ be such that $A_kT_k + B_kP_k^{e_k} = 1$ (they exist by definition of T_k). Consider λ in Q_k^*, and define ℓ in Q^* by $\ell(f) = \lambda(A_kf \bmod J_k)$. Since $P_k^{e_k}$ vanishes modulo J_k, we have $A_kT_k = 1 \bmod J_k$, so $\ell(f) = \lambda(f \bmod J_k)$ holds for all f in Q; this in turn readily implies that $\varphi_k(\ell) = \lambda$. □

We saw in Subsect. 2 how to associate to an element $\ell \in Q^*$ a sequence $\boldsymbol{u}_\ell \in \mathscr{S}$, by letting $\langle \boldsymbol{u}_\ell \mid m\rangle = \ell(m \bmod I)$. The following tautological observation will then be useful below: for ℓ in Q^*, the sequences $\boldsymbol{u}_{T_k\cdot\ell}$ and $\boldsymbol{u}_{\varphi_k(\ell)}$ coincide, where $\boldsymbol{u}_{\varphi_k(\ell)}$ is defined starting from the linear form $\varphi_k(\ell) \in Q_k^*$. Indeed, take any monomial m in $X_1,\dots,X_n$; then, $\varphi_k(\ell)(m \bmod J_k)$ is defined as $\ell(T_km \bmod I)$, which is equal to $(T_k\cdot\ell)(m \bmod I)$. We will use this remark to compute values of $\varphi_k(\ell)$, through the computation of values of $T_k\cdot\ell$ instead.

In algorithmic terms, computing a single transposed product by a polynomial $T(X_n)$, that is, $T\cdot\ell$, can be done using Horner's rule, using d right-multiplications by M_n, with $d = \deg(T)$; this takes $O(d\mathsf{m})$ operations in $\mathbb{K}$. If several transposed products are needed, such as for instance computing $T_1\cdot\ell,\dots,T_L\cdot\ell$ as below, the cost becomes $O(LD\mathsf{m})$, using D as an upper bound on $\deg(T_1),\dots,\deg(T_L)$. One can actually do better, by computing inductively and storing the products $X_n^i\cdot\ell$, for $i = 0,\dots,D-1$. Then, the coefficients of $T_1\cdot\ell,\dots,T_L\cdot\ell$ can be computed as the product of the $D\times d'$ matrix of coefficients of $(X_n^i\cdot\ell)_{0\le i<D}$ by the matrix of coefficients of $T_1,\dots,T_L$; the cost is $O(D\mathsf{m} + LD^2)$.

One can improve this idea further using *subproduct tree* techniques, since the polynomials $T_1,\dots,T_L$ have a very specific structure. Recall that we defined $T_k = P_{\min}/P_k^{e_k}$. Hence, all of $T_1,\dots,T_L$ share a common factor $R = P_{L+1}^{e_{L+1}}\cdots P_K^{e_K}$. We can then treat the common factor R separately, by writing $T_k = RU_k$ for all these indices k, and computing $U_1\cdot\ell',\dots,U_L\cdot\ell'$ instead, with $\ell' = R\cdot\ell$. The cost to compute ℓ' is $O(D\mathsf{m})$.

The polynomials $U_1,\dots,U_L$ have no common factor anymore, but they are all of the form $P_1^{e_1}\cdots P_{k-1}^{e_{k-1}}P_{k+1}^{e_{k+1}}P_L^{e_L}$. We can then define a subproduct tree as in [13, Chap. 10], that is, a binary tree $\mathcal{T}$ having the polynomials $(P_k^{e_k})_{1\le k\le L}$ at its leaves, and where each node is labeled by the product of the polynomials at its two children. We proceed in a top-down manner: we associate ℓ' to the root of the tree, and recursively, if a linear form λ has been assigned to an inner node of $\mathcal{T}$, we associate to each of its children the transposed product of λ by the polynomial labelling the other child. At the leaves, this gives us $U_L\cdot\ell',\dots,U_K\cdot\ell'$, as claimed. The total cost at each level is $O(D\mathsf{m})$, for a total of $O(D\log(L)\mathsf{m})$.

The Main Procedure, Using the Algorithm of Subsect. 3.1. The first version of the main procedure determines the Gröbner bases of $J_L,\dots,J_K$ by applying the algorithm of Subsect. 3.1 to successive families of linear forms.

We maintain a list of "active" indices S, initially set to $S = (1, \dots, L)$; these are the indices for which we are not done yet. The algorithm proceeds iteratively; at step $i \geq 1$, we pick a random linear form $\ell_i \in Q^*$, and compute all $\ell_{k,i} = T_k \cdot \ell_i$, for k in S. We then apply the algorithm of Subsect. 3.1 to $(\boldsymbol{u}_{\ell_{k,1}}, \dots, \boldsymbol{u}_{\ell_{k,i}})$, for all k independently, and obtain families of polynomials $G_{k,i}$ as output. For verification purposes, we also choose a random $\ell_0 \in Q^*$, and compute the corresponding $\ell_{k,0}$.

Write $D_k = \deg(J_k)$, for $k \leq K$. Combining Lemma 2 and the equality $\boldsymbol{u}_{\ell_{k,i}} = \boldsymbol{u}(\varphi_k(\ell_i))$ seen above, we deduce that for a generic choice of $\ell_1, \dots, \ell_{D_k}$, $(\ell_{k,1}, \dots, \ell_{k,D_k})$ satisfies assumption H_1 needed for our algorithm, and that G_{k,D_k} is a Gröbner basis of J_k. In view of the discussion in Subsect. 3.1, for any $i < D_k$, $G_{k,i}$ contains a polynomial g not in J_k. Since ℓ_0 was chosen at random, $\ell_{k,0}$ will in general not vanish at g; hence, at every step i, we evaluate $\ell_{k,0}$ at all elements of $G_{k,i}$, and continue the algorithm for this index k if we obtain a non-zero value; else, we remove k from our list S, and append $G_{k,i}$ to the output.

In terms of complexity, we will have to apply the process in the previous paragraph to μ linear forms $\ell_{D_1}, \dots, \ell_\mu$, with $\mu = \max_{k \leq L}(D_k)$, for a cost $O(\mu D \mathsf{m} \log(L))$. Then, we will exploit a feature of Marinari-Möller-Mora's second algorithm: it is incremental in the number of linear forms given as input, so that the overall runtime of our D_k successive invocations is the same as if we called it once with $\ell_1, \dots, \ell_{D_k}$. For a given k, it adds up to $O(nD_k^2\mathsf{m} + nD_k^3) = O(nD_k^2\mathsf{m})$, where the first term describes the cost of the evaluations of the linear forms we need (since each new value requires the product by one of the M_i). Overall, the runtime is $O(\mu D \log(L)\mathsf{m} + n \sum_{k \leq L} D_k^2 \mathsf{m})$. This supports the comment made in the introduction: if the degrees of the multiple components are small, say $D_k = O(1)$ for all k, this is $O(nD\log(D)\mathsf{m})$.

Using the Algorithm of Subsect. 3.2. We can adapt our main procedure in order to use the algorithm of Subsect. 3.2 instead; the main difference is that we expect to use fewer linear forms.

For $k \leq K$, let indeed $t_k \leq D_k$ be the maximum of $\tau(Q_{k,\geq 1}), \dots, \tau(Q_{k,\geq n})$, with $Q_{k,\geq j} = \mathbb{K}[X_j, \dots, X_n]/J_k \cap \mathbb{K}[X_j, \dots, X_n]$, and with τ defined as in Proposition 1 (for instance, if I is a complete intersection ideal, $t_k = 1$ for all k). The main algorithm proceeds as in the previous variant: we choose random linear forms $\ell_1, \dots$ and deduce $\ell_{k,i} = T_k \cdot \ell_i$; we will compute the Gröbner basis G_k of J_k as $\mathrm{ann}(\boldsymbol{u}_{\ell_{k,1}}, \boldsymbol{u}_{\ell_{k,2}}, \dots)$. We claim that we only need t_k linear forms $\ell_1, \dots, \ell_{t_k}$ in order to recover G_k.

To confirm this, we consider again assumptions H_2 and H_3 made in Subsect. 3.2. The appendix of [4] implies that the minimal polynomial of any variable X_i in Q_k has degree at most e_k, except for X_n. We already know the minimal polynomial $P_k^{e_k}$ of X_n in Q_k, so we skip the first pass in the loop of the algorithm of Subsect. 3.2, and use the value $B = e_k$.

Regarding H_3, we prove that if $\ell_1, \dots, \ell_{t_k}$ are chosen generically, assumption $\mathsf{H}_3(j)$ holds for $j = 1, \dots, n$. For $i \geq 1$ and $j = 1, \dots, n$, define $\ell_{k,i,j}$ as the linear form in $Q_{k,\geq j}^*$ induced by restriction of $\varphi_k(\ell_i) \in Q_k^*$. Applying Proposition 1 to $Q_{k,\geq j}$ shows that there exists a Zariski open $\Omega_{k,j} \subset {Q_{k,\geq j}^*}^{t_k}$ such that if

$\ell_{k,1,j},\dots,\ell_{k,t_k,j}$ are in $\Omega_{k,j}$, they generate $Q^*_{k,\geq j}$ as a $Q_{k,\geq j}$-module, and thus (Lemma 3) $J_k \cap \mathbb{K}[X_j,\dots,X_n] = \mathrm{ann}(\boldsymbol{u}_{\ell_{k,1,j}},\dots,\boldsymbol{u}_{\ell_{k,t_k,j}})$. If this is true for some index k and all j, $\mathsf{H}_3(j)$ follows as well for these indices. Now, the mapping $\Delta_{k,j} : (\ell_1,\dots,\ell_{t_k}) \mapsto (\ell_{k,1,j},\dots,\ell_{k,t_k,j})$ is $\mathbb{K}$-linear and onto (we proved above that $(\ell_1,\dots,\ell_{t_k}) \mapsto (\varphi_k(\ell_1),\dots,\varphi_k(\ell_{t_k}))$ is onto, and the surjectivity of the projection is straightforward), so that the preimage $\Delta^{-1}_{k,j}(\Omega_{k,j})$ is Zariski open in Q^{*t_k} for all k, j. In other words, for generic $\ell_1,\dots,\ell_{t_k}$, $\mathsf{H}_3(j)$ holds for all j and all k, so the algorithm of Subsect. 3.2 computes G_k for all k.

We still need to discuss what happens when applying this algorithm to $\ell_{k,1},\dots,\ell_{k,i}$ for some $i < t_k$. In this case, as per the discussion in Subsect. 3.2, either we get generators of $\mathrm{ann}(\boldsymbol{u}_{\ell_{k,1}},\dots,\boldsymbol{u}_{\ell_{k,i}})$, which is a strict superset of J_k, or at least one of the polynomials in the output does not belong to $\mathrm{ann}(\boldsymbol{u}_{\ell_{k,1}},\dots,\boldsymbol{u}_{\ell_{k,i}})$. In any case, the output contains at least one polynomial g not in J_k, so we can use the same stopping criterion as in the previous paragraph, using a linear form ℓ_0 to test termination.

To control the complexity, at the ith step, we now use linear forms $\ell_1,\dots,\ell_{2^i}$; as a result, we need to go up to $i = t$, with $t = \max_k(t_k)$, and the overall runtime is proportional to that at $i = t$. The cost of preparing the linear forms $\ell_{k,i}$ is $O(tD\mathsf{m}\log(L))$, and the cost of computing annihilators is $O(nt\sum_{k\leq L} e_k^2 D_k^2 \mathsf{m})$. The first term is better than the equivalent term for our first algorithm, but the second one is obviously worse. On the other hand, the analysis in Subsect. 3.2 can be refined significantly, and possibly lead to improved estimates.

Using a Scalar Extension. To conclude, we discuss (without giving proofs) how to put to practice the idea introduced in Subsect. 4.1 of computing Gröbner bases of ideals of smaller degree over larger base fields, in the context (for definiteness) of the algorithm of the previous paragraph.

Let $\ell_{k,1},\dots,\ell_{k,t_k}$ be defined as before, let $\boldsymbol{u}_{\ell_{k,1}},\dots,\boldsymbol{u}_{\ell_{k,t_k}}$ be the corresponding sequences, and assume that these linear forms are such that the annihilator of $\boldsymbol{u}_{\ell_{k,1}},\dots,\boldsymbol{u}_{\ell_{k,t_k}}$ is J_k. Let further $\mathbb{L}_k$ be the field extension $\mathbb{K}[Z]/P_n(Z)$, and let ζ_k be the residue class of Z in $\mathbb{L}_k$ Then, the annihilator of $J'_k = J_k + \langle X_n - \zeta_k\rangle^{e_k}$ in $\mathbb{L}[X_1,\dots,X_n]$ has degree D_k/f_k by Lemma 7, so we might want to compute it instead of J_k. To accomplish this, we need sequences whose annihilator would be J'_k; we do this following the same strategy as above. Define $S_k = P_k/(X_n - \zeta_k) \in \mathbb{L}_k[X_n]$, as well as the linear form $\ell'_{k,i} = S_k^{e_k} \cdot \ell_{k,i} : \mathbb{L}[X_1,\dots,X_n]/I \to \mathbb{L}$, for $i \geq 1$. Then, one verifies that $\mathrm{ann}(\boldsymbol{u}_{\ell'_{k,1}},\dots,\boldsymbol{u}_{\ell'_{k,t_k}})$ is indeed J'_k.

Our last comment discusses the translation mentioned in Subsect. 4.1. The ideal J'_k is $\mathfrak{m}'$-primary, with $\mathfrak{m}' = \langle X_1 - \xi_1,\dots,X_n - \xi_n\rangle$, as in Subsect. 4.1. To replace J'_k by a $\langle X_1,\dots,X_n\rangle$-primary ideal, we need to modify the sequences $\boldsymbol{u}_{\ell'_{k,1}},\dots,\boldsymbol{u}_{\ell'_{k,t_k}}$. For $i \geq 1$, let $U_{k,i} \in \mathbb{L}[[X_1,\dots,X_n]]$ be the generating series of $\boldsymbol{u}_{\ell'_{k,i}}$, and let $\tilde{U}_{k,i} = \frac{1}{(1+\xi_1 X_1)\cdots(1+\xi_n X_n)} U_{k,i}(\frac{X_1}{1+\xi_1 X_1},\dots,\frac{X_n}{1+\xi_n X_n})$. Letting $\tilde{\boldsymbol{u}}_{k,i}$ be the sequence whose generating series is $\tilde{U}_{k,i}$, $\mathrm{ann}(\tilde{\boldsymbol{u}}_{k,1},\dots,\tilde{\boldsymbol{u}}_{k,t_k})$ is indeed the $\langle X_1,\dots,X_n\rangle$-primary ideal J''_k obtained by translation by $(\xi_1,\dots,\xi_n)$ in J'_k.

Acknowledgements. We thank the reviewers for their remarks and suggestions. The third author is supported by an NSERC Discovery Grant.

References

1. Berthomieu, J., Boyer, B., Faugère, J.-C.: Linear algebra for computing Gröbner bases of linear recursive multidimensional sequences. J. Symb. Comput. **83**, 36–67 (2016)
2. Berthomieu, J., Faugère, J.-C.: Guessing linear recurrence relations of sequence tuples and P-recursive sequences with linear algebra. In: ISSAC 2016, pp. 95–102. ACM (2016)
3. Berthomieu, J., Faugère, J.-C.: In-depth comparison of the Berlekamp-Massey-Sakata and the Scalar-FGLM algorithms: the non adaptive variants. hal-01516708, May 2017
4. Bostan, A., Salvy, B., Schost, É.: Fast algorithms for zero-dimensional polynomial systems using duality. AAECC **14**, 239–272 (2003)
5. Brachat, J., Comon, P., Mourrain, B., Tsigaridas, E.: Symmetric tensor decomposition. Linear Algebra Appl. **433**(11), 1851–1872 (2010)
6. Dahan, X., Moreno Maza, M., Schost, É., Xie, Y.: On the complexity of the D5 principle. In: Transgressive Computing, pp. 149-168 (2006)
7. Della Dora, J., Dicrescenzo, C., Duval, D.: About a new method for computing in algebraic number fields. In: Caviness, B.F. (ed.) EUROCAL 1985. LNCS, vol. 204, pp. 289–290. Springer, Heidelberg (1985). doi:10.1007/3-540-15984-3_279
8. Eisenbud, D.: Commutative Algebra: With a View Toward Algebraic Geometry, vol. 150. Springer Science & Business Media, New York (2013). doi:10.1007/978-1-4612-5350-1
9. Faugère, J.-C., Gaudry, P., Huot, L., Renault, G.: Polynomial Systems Solving by Fast Linear Algebra (2013). https://hal.archives-ouvertes.fr/hal-00816724
10. Faugère, J.-C., Gaudry, P., Huot, L., Renault, G.: Sub-cubic change of ordering for Gröbner basis: a probabilistic approach. In: ISSAC 2014, pp. 170-177. ACM (2014)
11. Faugère, J.-C., Gianni, P., Lazard, D., Mora, T.: Efficient computation of zero-dimensional Gröbner bases by change of ordering. J. Symb. Comput. **16**(4), 329–344 (1993)
12. Faugère, J.-C., Mou, C.: Sparse FGLM algorithms. J. Symb. Comput. **80**(3), 538–569 (2017)
13. von zur Gathen, J., Gerhard, J.: Modern Computer Algebra, 3rd edn. Cambridge University Press, Cambridge (2013)
14. Gianni, P., Mora, T.: Algebrric solution of systems of polynomirl equations using Groebher bases. In: Huguet, L., Poli, A. (eds.) AAECC 1987. LNCS, vol. 356, pp. 247–257. Springer, Heidelberg (1989). doi:10.1007/3-540-51082-6_83
15. Gröbner, W.: Über irreduzible Ideale in kommutativen Ringen. Math. Ann. **110**(1), 197–222 (1935)
16. Macaulay, F.S.: Modern algebra and polynomial ideals. Math. Proc. Camb. Philos. Soc. **30**(1), 27–46 (1934)
17. Marinari, M.G., Mora, T., Möller, H.M.: Gröbner bases of ideals defined by functionals with an application to ideals of projective points. AAECC **4**, 103–145 (1993)
18. Moreno-Socías, G.: Autour de la fonction de Hilbert-Samuel (escaliers d'ideaux polynomiaux). Ph.D. thesis, École polytechnique (1991)
19. Mourrain, B.: Fast algorithm for border bases of Artinian Gorenstein algebras. ArXiv e-prints, May 2017

20. Neiger, V.: Bases of relations in one or several variables: fast algorithms and applications. Ph.D. thesis, École Normale Supérieure de Lyon, November 2016
21. Rouillier, F.: Solving zero-dimensional systems through the rational univariate representation. AAECC **9**(5), 433–461 (1999)
22. Sakata, S.: Extension of the Berlekamp-Massey algorithm to N dimensions. Inform. Comput. **84**(2), 207–239 (1990)
23. Shoup, V.: A new polynomial factorization algorithm and its implementation. J. Symb. Comput. **20**(4), 363–397 (1995)

Symbolic-Numerical Analysis of the Relative Equilibria Stability in the Planar Circular Restricted Four-Body Problem

Alexander N. Prokopenya[1,2(✉)]

[1] Department of Applied Informatics, Warsaw University of Life Sciences – SGGW, Nowoursynowska Str. 159, 02-776 Warsaw, Poland
alexander_prokopenya@sggw.pl

[2] Collegium Mazovia Innovative Higher School, Sokolowska Str. 161, 08-110 Siedlce, Poland

Abstract. We study the stability of relative equilibrium positions in the planar circular restricted four-body problem formulated on the basis of the Euler collinear solution of the three-body problem. The stability problem is solved in a strict nonlinear formulation in the framework of the KAM theory. We obtained algebraic equations determining the equilibrium positions and showed that there are 18 different equilibrium configurations of the system for any values of the two system parameters μ_1, μ_2. Canonical transformation of Birkhoff's type reducing the Hamiltonian of the system to the normal form is constructed in a general symbolic form. Combining symbolic and numerical calculations, we showed that only 6 equilibrium positions are stable in Lyapunov's sense if parameters μ_1 and μ_2 are sufficiently small, and the corresponding points in the plane $O\mu_1\mu_2$ belong to the domain bounded by the second order resonant curve. It was shown also that the third order resonance results in instability of the equilibrium positions while in case of the fourth order resonance, either stability or instability can take place depending on the values of parameters μ_1 and μ_2. All relevant symbolic and numerical calculations are done with the aid of the computer algebra system Wolfram Mathematica.

1 Introduction

The circular restricted three-body problem is a well-known model of celestial mechanics (see, for example, [21]). Recall that we are interested in motion of the particle P_3 of negligible mass in the gravitational field of two massive particles P_0 and P_1 having masses m_0, m_1, respectively, and moving uniformly on circular Keplerian orbits around their common center of mass. A general solution of the problem cannot be written in symbolic form but there exist five exact particular solutions known as the homographic ones [22]. In the rotating frame of reference, where the particles P_0 and P_1 have rest, these solutions determine the equilibrium positions of the particle P_3 (or relative equilibrium positions)

V.P. Gerdt et al. (Eds.): CASC 2017, LNCS 10490, pp. 329–345, 2017.
DOI: 10.1007/978-3-319-66320-3_24

which are called the points of libration L_j $(j = 1, 2, \ldots, 5)$. The libration points are of great interest for applications and so their stability was a subject of many papers during the past two hundred years. As a result, it was proven that three points L_1, L_2, L_3 situated at the line P_0P_1 (collinear equilibrium positions) are unstable while the libration points L_4, L_5 (triangular equilibrium positions) may be stable if the mass ratio $\mu_1 = m_1/m_0$ is sufficiently small (see [14,21]).

The systems of particular interest in Celestial Mechanics and Cosmic Dynamics usually contain more than three bodies. So it makes sense to add the third particle P_2 of mass m_2 to the system $P_0P_1P_3$ and to analyze its influence on stability of equilibrium positions of the particle P_3. Note that the collinear and triangular equilibrium configurations exist also in a general case of the three-body problem. Moreover, both of them are realized in the Solar System (see [17]). So the particle P_2 may be situated in any of the five equilibrium positions of the corresponding three-body problem which coincide with the libration points $L_1, \ldots, L_5$ in case of $m_2 = 0$. In this way, we obtain the restricted four-body problem that has been a subject of many papers (see, for example, [1,5,10,18,19]). It should be emphasized that in the framework of this model, motion of the massive particles P_0, P_1, P_2 is given and is determined by the corresponding solutions of the three-body problem.

The case when the particle P_2 is situated in the vertex L_4 of the equilateral triangle $P_0P_1L_4$ was studied in detail in [2,4,7,18]. It was shown that four new equilibrium positions of particle P_3 arise from the point of libration L_4 if the second mass parameter $\mu_2 = m_2/m_0$ becomes greater than zero. The rest four libration points L_1, L_2, L_3, and L_5 only change their positions depending on the values of parameters μ_1 and μ_2. Besides, one or two new equilibrium positions may arise inside the triangle $P_0P_1P_2$. Stability of all these equilibrium positions was completely investigated in [5,6] on the basis of the KAM theory [3,15] that is widely used for solving similar problems, starting from the famous works [8, 13]. In particular, it was shown that all equilibrium positions situated near the collinear libration points L_1, L_2, L_3 remain unstable for any values of parameters μ_1, μ_2. A special case of the stability problem when $\mu_1 = \mu_2$ was investigated in [1,19]. Note that only two points (μ_1, μ_2) in the plane $O\mu_1\mu_2$ remain for which theorems of Arnold [3] and Markeev [14] cannot be applied, and the stability problem for the corresponding equilibrium positions has not been solved yet (see [5]).

To complete investigation of the influence of the particle P_2 on the motion of particle P_3 one needs to consider the case when particle P_2 is situated in one of the collinear libration points L_1, L_2, L_3, and this is the main research task of the present paper. Although such positions of particle P_2 are unstable, its quasi-periodic orbits near the collinear libration points may exist. So it is a matter of interest to investigate an influence of the particle P_2 mass on the equilibrium positions of particle P_3 and their stability. In addition, the problem is interesting from the theoretical point of view because it differs essentially from the case of triangular configuration of the particles P_0, P_1, P_2. Actually, in the latter case, the particles P_0, P_1, P_2 are fixed in the vertices of the equilateral triangle for any values of parameters μ_1, μ_2, while in the collinear case, mutual distances between particles depend on these parameters. Therefore, given the values of μ_1, μ_2 we have to look for both equilibrium configuration of the massive particles P_0, P_1, P_2 and equilibrium positions of the particle P_3 as solutions of the corresponding algebraic equations. Only afterwards we can analyze the Hamiltonian function in the neighborhood of the equilibrium configuration and conclude on stability or instability of equilibrium positions applying theorems of the KAM theory [3,15]. Realization of such an approach involves very advanced symbolic and numerical calculations (see [5,12,16]) which can be reasonably performed only with computers and modern software such as the computer algebra system *Wolfram Mathematica* [23], for example.

Note that if two particles P_1 and P_2 have the same mass and are situated symmetrically with respect to the particle P_0, the problem is simplified considerably because the system has only one parameter $\mu = \mu_1 = \mu_2$ and positions of the massive particles are fixed. Just such a case of the stability problem was considered earlier in linear approximation [11] and in a strict nonlinear formulation (see [9,20]). In case of $\mu_1 = \mu_2$, our results agree completely with the results of [20] and correct some inaccuracies in the computations performed in [9].

The paper is organized as follows. In Sect. 2, we obtain algebraic equations determining equilibrium configuration of the system and analyze their solutions for different values of parameters μ_1 and μ_2. Then in Sect. 3, we analyze the system stability in linear approximation and determine the domains of stability in the plane $O\mu_1\mu_2$ for different positions of the particle P_2. Section 4 is devoted to calculation of the third order term in the Hamiltonian expansion and analysis of the equilibrium positions' stability under the third order resonance. In Sect. 4, we consider the fourth order term of the Hamiltonian expansion and conclude on stability of the equilibrium positions applying theorems of Arnold and Markeev. At last, we conclude in Sect. 5.

2 Equilibrium Solutions

In the rotating frame of reference, where the particles P_0, P_1, P_2 have rest in the Oxy plane at the points $(0,0)$, $(1,0)$, $(a,0)$, respectively, the Hamiltonian function of the system can be written in the form

$$\mathcal{H} = \frac{1}{2}\left(p_x^2 + p_y^2\right) - xp_y + yp_x - \frac{1}{\kappa}\left(\frac{1}{\sqrt{x^2+y^2}} + \mu_1\left(\frac{1}{\sqrt{(x-1)^2+y^2}} - x\right) + \mu_2\left(\frac{1}{\sqrt{(x-a)^2+y^2}} - \frac{ax}{|a|^3}\right)\right), \quad (1)$$

where x, p_x and y, p_y are two pairs of canonically conjugate dimensionless coordinate and momentum, parameter κ is given by

$$\kappa = 1 + \mu_1 + \mu_2\left(\frac{a}{|a|^3} + \frac{1-a}{|1-a|^3}\right), \quad (2)$$

and dimensionless parameter a determining position of the particle P_2 at the Ox axis, is a real root of the equation

$$1+\mu_1+\mu_2\left(\frac{a}{|a|^3}+\frac{1-a}{|1-a|^3}\right)=\frac{1+\mu_2}{|a|^3}+\frac{\mu_1}{a}\left(1+\frac{a-1}{|a-1|^3}\right). \tag{3}$$

We assume here that both parameters μ_1 and μ_2 belong to the interval $0 < \mu_{1,2} \leq 1$, and this enables us to consider all physically different configurations of the system.

Equation (3) arises in the three-body problem (see [14,21]) and has three different real roots for any values of parameters μ_1, μ_2. Each of the three intervals $0 < a < 1$, $a > 1$ and $a < 0$ contains only one root, and in case of $\mu_2 = 0$, these roots determine the collinear points of libration L_1, L_2, L_3, respectively. For $\mu_2 > 0$, equilibrium positions of the particle P_2 are shifted from the points L_1, L_2, L_3 but the geometrical configuration of particles P_0, P_1, P_2 doesn't change. Note that symbolic solution of nonlinear equation (3) cannot be found but each of the three roots may easily be calculated numerically with necessary precision using the built-in function *FindRoot* (see [23]).

Using the Hamiltonian (1), one can easily write the equations of motion of particle P_3 and show that its equilibrium coordinates (x, y) are determined by the following system of two algebraic equations

$$\begin{aligned}\frac{x}{(x^2+y^2)^{3/2}}-\kappa x+\mu_1\left(1+\frac{x-1}{((x-1)^2+y^2)^{3/2}}\right)\\ +\mu_2\left(\frac{a}{|a|^3}+\frac{x-a}{((x-a)^2+y^2)^{3/2}}\right)=0,\end{aligned}$$

$$y\left(\frac{1}{(x^2+y^2)^{3/2}}-\kappa+\frac{\mu_1}{((x-1)^2+y^2)^{3/2}}+\frac{\mu_2}{((x-a)^2+y^2)^{3/2}}\right)=0. \tag{4}$$

In case of $\mu_2 = 0$, system (4) reduces to the equations determining the libration points in the restricted three-body problem (see [14,21]). If particle P_2 is situated at the collinear point of libration L_j, $(j = 1, 2, 3)$ then for $\mu_2 > 0$, two new collinear equilibrium positions arise from the point L_j, and the other two collinear libation points change their positions at the Ox axis. The x-coordinates of the four collinear equilibrium positions as functions of parameter μ_2 are shown in Fig. 1 for a fixed value of parameter μ_1 and the particle P_2 being located between P_0 and P_1 $(0 < a < 1)$. Note that position of P_2 also changes when parameter μ_2 grows (dashed curve in Fig. 1). Similar pictures are obtained for other positions of the particle P_2 and different values of μ_1.

If $\mu_2 = 0$ and particle P_2 is situated in one of the libration points L_1, L_2, L_3 the particle P_3 may also stay in equilibrium in each of the triangular points of libration L_4, L_5. Numerical analysis of the system (4) shows that increasing of parameter μ_2 results in shifting of the corresponding libration points in the Oxy plane (see Fig. 2). Three bold arrows starting at the points L_4, L_5 show equilibrium positions of the particle P_3 in the cases when particle P_2 is situated in the neighbourhood of one of the libration points L_1, L_2, L_3. The corresponding

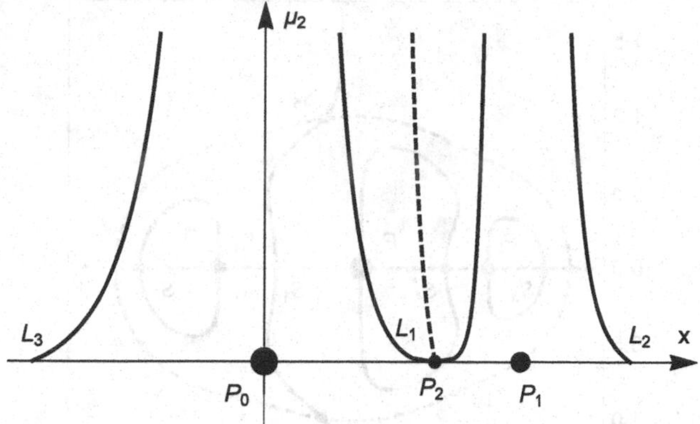

Fig. 1. Four collinear equilibrium positions of the particle P_3 arising from the libration points L_1, L_2, L_3 if particle P_2 is situated at the point L_1, $\mu_1 = 0.2$, $0 \leq \mu_2 \leq 1$.

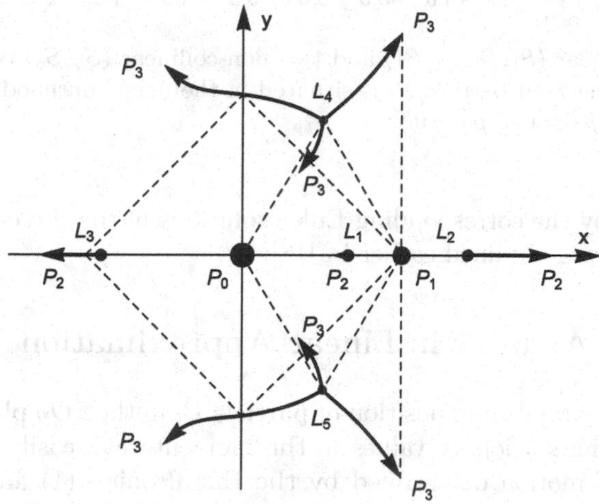

Fig. 2. Equilibrium positions of particle P_3 arising from the libration points L_4, L_5 if particle P_2 is situated at one of the libration points L_1, L_2, L_3, $\mu_1 = 0.2$, $0 \leq \mu_2 \leq 1$.

shifts of particle P_2 on the Ox axis are also shown by arrows starting at the points L_1, L_2, L_3.

Note that three cases of the particle P_2 localization in the neighbourhood of the libration points L_1, L_2, L_3 describe all physically different collinear geometrical configurations of the massive particles P_0, P_1, P_2. For each of these configurations, system (4) determines four collinear and two non-collinear equilibrium positions of particle P_3, geometrically they are represented as points of intersections of the solid and dashed curves (see Fig. 3) determined by the equations of system (4). Therefore, there are 18 different equilibrium solutions in the restricted four-body problem, where positions of the three massive particles

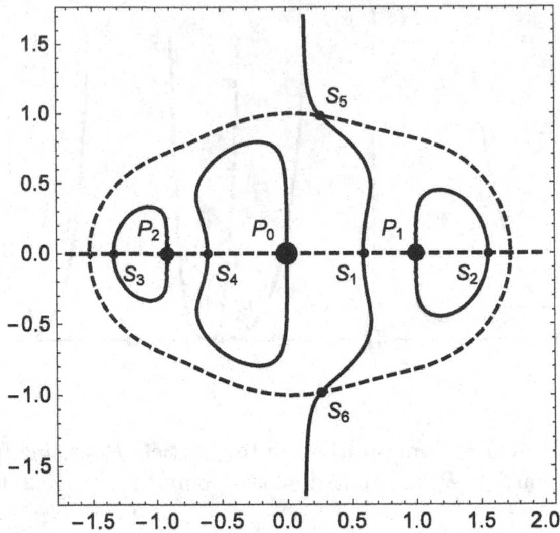

Fig. 3. Four collinear (S_1, S_2, S_3, S_4) and two non-collinear (S_5, S_6) equilibrium positions of the particle P_3 if particle P_2 is situated in the neighbourhood of the libration point L_3 $(a < 0)$, $\mu_1 = 0.4$, $\mu_2 = 0.2$.

are determined by the corresponding Euler solutions of the three-body problem. Similar results were obtained earlier in [18].

3 Stability Analysis in Linear Approximation

Let us denote an equilibrium position of particle P_3 in the xOy plane by (x_0, y_0). The corresponding stationary values of the momenta can easily be found from the equations of motion determined by the Hamiltonian (1) and are equal to $p_{x0} = -y_0$, $p_{y0} = x_0$. To analyze stability of this equilibrium solution we perform in the Hamiltonian (1) a substitution written in terms of the Wolfram language [23] as the following list of rules

$$rul1 = \{x \to x_0 + \delta x,\ y \to y_0 + \delta y,\ p_x \to -y_0 + \delta p_x,\ p_y \to x_0 + \delta p_y\}; \quad (5)$$

The variables x, y, p_x, p_y in the right-hand side of each rule in (5) are small deviations from to the equilibrium solution. To simplify calculation and to be able to extract in the Hamiltonian expansion in power series in terms of the perturbations the kth order term H_k $(k = 0, 1, \ldots)$, we add to each variable x, y, p_x, p_y a multiplier δ. Then it is sufficient to expand the Hamiltonian in the power series in terms of δ in the neighborhood of the point $\delta = 0$. The corresponding command written in the Wolfram Language is given by

$$H = Series[\mathcal{H}/.\ rul1,\ \{\delta, 0, 4\}]\ //\ Normal;$$

Here we obtain the Hamiltonian expansion up to the fourth order, and the kth order term is extracted by application of the built-in function $Coefficient$

$$H_k = Coefficient[H, \delta, k];$$

As a result, we calculate the Hamiltonian (1) in the form of power series in the neighborhood of equilibrium point, and it is represented in the form

$$\mathcal{H} = H_2 + H_3 + H_4 + \ldots, \tag{6}$$

where H_k is the kth order homogeneous polynomial with respect to canonical variables x, y, p_x, p_y. Note that zero-order term H_0 in (6) has been omitted as a constant, which does not influence the equations of motion, and the first-order term H_1 is equal to zero owing to Eq. (4) determining equilibrium positions. Therefore, the first non-zero term in expansion (6) is a quadratic one that has the form

$$H_2 = \frac{1}{2}\left(p_x^2 + p_y^2\right) - p_y x + p_x y + h_{20}x^2 + h_{11}xy + h_{02}y^2, \tag{7}$$

where coefficients h_{20}, h_{11}, and h_{02} are given by

$$h_{20} = -\frac{2x_0^2 - y_0^2}{2\kappa(x_0^2 + y_0^2)^{5/2}} - \frac{\mu_1}{2\kappa}\frac{2(x_0-1)^2 - y_0^2}{((x_0-1)^2 + y_0^2)^{5/2}} - \frac{\mu_2}{2\kappa}\frac{2(x_0-a)^2 - y_0^2}{((x_0-a)^2 + y_0^2)^{5/2}},$$

$$h_{11} = -\frac{3y_0}{\kappa}\left(\frac{x_0}{(x_0^2 + y_0^2)^{5/2}} + \frac{\mu_1(x_0-1)}{((x_0-1)^2 + y_0^2)^{5/2}} + \frac{\mu_2(x_0-a)}{((x_0-a)^2 + y_0^2)^{5/2}}\right), \tag{8}$$

$$h_{02} = \frac{x_0^2 - 2y_0^2}{2\kappa(x_0^2 + y_0^2)^{5/2}} + \frac{\mu_1}{2\kappa}\frac{(x_0-1)^2 - 2y_0^2}{((x_0-1)^2 + y_0^2)^{5/2}} + \frac{\mu_2}{2\kappa}\frac{(x_0-a)^2 - 2y_0^2}{((x_0-a)^2 + y_0^2)^{5/2}}.$$

One can readily check that the linearized equations of motion determined by the quadratic part H_2 of the Hamiltonian (6) form the fourth-order linear system of differential equations with constant coefficients. Characteristic exponents $\lambda_1, \ldots, \lambda_4$ for such a system can easily be found (see [5,6]) and may be represented in the form

$$\lambda_{1,2} = \pm i\sigma_1, \quad \lambda_{3,4} = \pm i\sigma_2, \tag{9}$$

where the frequencies σ_1 and σ_2 are given by

$$\sigma_{1,2} = \left(1 + h_{20} + h_{02} \pm \sqrt{h_{20}^2 + h_{02}^2 + h_{11}^2 - 2h_{20}h_{02} + 4h_{20} + 4h_{02}}\right)^{1/2}. \tag{10}$$

Recall that equilibrium position (x_0, y_0) may be stable for some values of parameters μ_1 and μ_2 only if the corresponding characteristic exponents (9) are different purely imaginary numbers or the frequencies σ_1 and σ_2 are different real numbers. As coefficients h_{20}, h_{11}, and h_{02} depend on both the mass parameters

μ_1 and μ_2 and geometrical parameters a, x_0, y_0 which are also functions of μ_1 and μ_2 (see (2), (3), (4)), it is very difficult to check the conditions of the equilibrium positions' stability in analytic form. To estimate the values of parameters μ_1 and μ_2 for which the equilibrium positions may be stable we can choose a grid with small step in the plane $O\mu_1\mu_2$ and calculate numerically the frequencies σ_1 and σ_2 in the grid nodes.

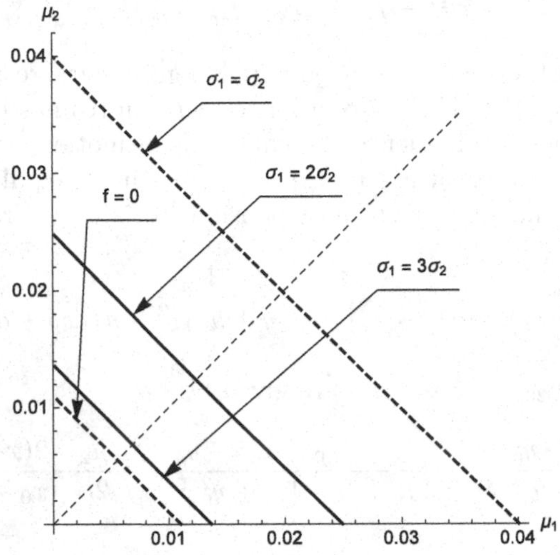

Fig. 4. Stability boundary and resonance curves for equilibrium positions S_5, S_6 if particle P_2 is situated in the neighbourhood of the libration points L_1 $(0 < a < 1)$ or L_2 $(a > 1)$. The curve $f = 0$ corresponds to the points for which condition (30) is not fulfilled.

Numerical analysis of the frequencies (10) in the domain $0 < \mu_{1,2} \leq 1$ has shown that for the collinear equilibrium points S_1, S_2, S_3, S_4, at least one of the frequencies σ_1 and σ_2 has an imaginary part for any values of parameters μ_1 and μ_2 and any of the three possible equilibrium positions of particle P_2. Therefore, all twelve collinear equilibrium positions are unstable in Lyapunov's sense. Due to the symmetry of the system with respect to the Ox axis, the equilibrium points S_5 and S_6 have the same properties, and they are stable in linear approximation if parameters μ_1 and μ_2 are smaller than their values on the stability boundaries which are determined from the condition $\sigma_1 = \sigma_2$. The corresponding curves for the equilibrium points S_5 and S_6 have been found numerically and are shown as dashed curves in the $O\mu_1\mu_2$ plane in Figs. 4 and 5. It should be noted that the stability domains are the same if particle P_2 is situated in the neighbourhood of the libration points L_1 and L_2. It is quite natural because one case is obtained from another one by means of mutual replacement $\mu_1 \to \mu_2$, $\mu_2 \to \mu_1$ and the scale transformation $a \to 1/a$. Besides, both domains of stability shown in Figs. 4 and 5 are symmetrical with respect to the line $\mu_2 = \mu_1$.

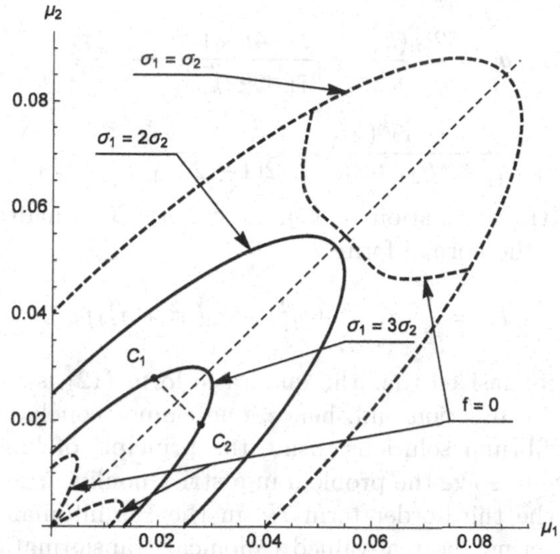

Fig. 5. Stability boundary and resonance curves for equilibrium positions S_5, S_6 if particle P_2 is situated in the neighbourhood of the libration point L_3 ($a < 0$).

4 Normalization of the Hamiltonian

The most-used method for studying the Hamiltonian systems of nonlinear differential equations is the Poincaré method of normal forms (see [14]). It includes constructing a real-valued canonical transformation, reducing the Hamiltonian of the system to the Birkhoff normal form, and applying theorems of the KAM theory [3,15]. As in the neighbourhood of the equilibrium position the Hamiltonian is represented in the form of expansion (6), we have to normalize successively the terms $H_2,\ H_3,\ H_4, \ldots$

Canonical transformation normalizing the quadratic part H_2 of the Hamiltonian may be constructed in symbolic form, the corresponding algorithm has been described in detail in [14,16]. Doing necessary symbolic calculations, we obtain the transformation in the form

$$\begin{aligned} x &= 2c_1p_1 + 2c_2p_2, \\ y &= -2\sigma_1u_1q_1 + 2\sigma_2u_2q_2 + h_{11}u_1p_1 + h_{11}u_2p_2, \\ p_x &= -v_1q_1 + v_2q_2 - h_{11}u_1p_1 - h_{11}u_2p_2, \\ p_y &= -h_{11}\sigma_1u_1q_1 + h_{11}\sigma_2u_2q_2 + g_1p_1 + g_2p_2, \end{aligned} \tag{11}$$

where p_1, q_1 and p_2, q_2 are two pairs of new canonically conjugated variables,

$$u_k = \frac{2c_k(1 - h_{20} + \sigma_k^2)}{h_{11}^2 + 4\sigma_k^2}, \; v_k = \frac{2c_k\sigma_k(-2 + 4h_{20} + h_{11}^2 + 2\sigma_k^2)}{h_{11}^2 + 4\sigma_k^2},$$

$$g_k = \frac{2c_k(h_{11}^2 + (2 + 4h_{20})\sigma_k^2 - 2\sigma_k^4)}{h_{11}^2 + 4\sigma_k^2},$$

$$c_k = \left(\frac{(-1)^k(h_{11}^2 + 4\sigma_k^2)}{4\sigma_k(-3 + h_{11}^2 + 4h_{20} + 4h_{20}^2 + 2(1 - 2h_{20})\sigma_k^2 + \sigma_k^4)}\right)^{1/2}, \; k = 1, 2.$$

Applying this transformation to (7), we reduce the quadratic part of the Hamiltonian H_2 to the normal form

$$H_2 = \frac{1}{2}\left(\sigma_1(p_1^2 + q_1^2) - \sigma_2(p_2^2 + q_2^2)\right). \tag{12}$$

It should be emphasized that the quadratic form (12) is neither a positive nor negative definite function and, hence, one cannot conclude on stability or instability of equilibrium solutions, using the principle of linearized stability. Therefore, we have to solve the problem in a strict nonlinear formulation.

To normalize the third-order term H_3 in the Hamiltonian (6) we use the method of constructing the real-valued canonical transformation of Birkhoff's type described in [5,6,9]. We start from the term H_3 in (6) given by

$$\begin{aligned} H_3 = \Bigg(&\frac{2x_0^3 - 3y_0^2}{(x_0^2 + y_0^2)^{7/2}} + \frac{\mu_1(x_0 - 1)(2(x_0 - 1)^2 - 3y_0^2)}{((x_0 - 1)^2 + y_0^2)^{7/2}} \\ &+ \frac{\mu_2(x_0 - a)(2(x_0 - a)^2 - 3y_0^2)}{((x_0 - a)^2 + y_0^2)^{7/2}}\Bigg)\frac{x^3}{2\kappa} \\ &+ \left(\frac{4x_0^3 - y_0^2}{(x_0^2 + y_0^2)^{7/2}} + \frac{\mu_1(4(x_0 - 1)^2 - y_0^2)}{((x_0 - 1)^2 + y_0^2)^{7/2}} + \frac{\mu_2(4(x_0 - a)^2 - y_0^2)}{((x_0 - a)^2 + y_0^2)^{7/2}}\right)\frac{3y_0x^2y}{2\kappa} \\ &- \Bigg(\frac{x_0^3 - 4x_0y_0^2}{(x_0^2 + y_0^2)^{7/2}} + \frac{\mu_1(x_0 - 1)((x_0 - 1)^2 - 4y_0^2)}{((x_0 - 1)^2 + y_0^2)^{7/2}} + \\ &+ \frac{\mu_2(x_0 - a)((x_0 - a)^2 - 4y_0^2)}{((x_0 - a)^2 + y_0^2)^{7/2}}\Bigg)\frac{3xy^2}{2\kappa} \\ &- \left(\frac{3x_0^3 - 2y_0^2}{(x_0^2 + y_0^2)^{7/2}} + \frac{\mu_1(3(x_0 - 1)^2 - 2y_0^2)}{((x_0 - 1)^2 + y_0^2)^{7/2}} + \frac{\mu_2(3(x_0 - a)^2 - 2y_0^2)}{((x_0 - a)^2 + y_0^2)^{7/2}}\right)\frac{y_0y^3}{2\kappa}. \end{aligned} \tag{13}$$

On substituting the coordinates x, y from (11) into (13) and simplifying the expression obtained, we reduce the third-order term H_3 to the form

$$H_3 = \sum_{i+j+k+l=3} h_{ijkl}^{(3)} q_1^i q_2^j p_1^k p_2^l. \tag{14}$$

As transformation (11) is linear, the coefficients $h_{ijkl}^{(3)}$ in (14) are determined only by the corresponding coefficients from (13) and coefficients of q_1, q_2, p_1, p_2

in (11). The expressions obtained for $h_{ijkl}^{(3)}$ are quite bulky, so we do not write them here. Similar, though more cumbersome calculations are performed with the fourth-order term H_4 and the transformation (11) reduces it to the form

$$H_4 = \sum_{i+j+k+l=4} h_{ijkl}^{(4)} q_1^i q_2^j p_1^k p_2^l. \tag{15}$$

Note that canonical transformation (11) does not mix the terms of different orders H_2, H_3, H_4 in the Hamiltonian (6).

The next step is to construct a canonical transformation that eliminates the third-order term H_3 in the Hamiltonian expansion. In addition to linear terms, such a transformation must contain the second-degree terms in new canonical variables and may be obtained with the aid of the following generating function

$$S(\tilde{p}_1, \tilde{p}_2, q_1, q_2) = q_1\tilde{p}_1 + q_2\tilde{p}_2 + \sum_{i+j+k+l=3} s_{ijkl}^{(3)} q_1^i q_2^j \tilde{p}_1^k \tilde{p}_2^l, \tag{16}$$

where coefficients $s_{ijkl}^{(3)}$ are to be found. The function (16) determines new momenta $\tilde{p}_1, \tilde{p}_2$ and coordinates $\tilde{q}_1, \tilde{q}_2$ according to the relationships

$$\tilde{q}_1 = \frac{\partial S}{\partial \tilde{p}_1},\ \tilde{q}_2 = \frac{\partial S}{\partial \tilde{p}_2},\ p_1 = \frac{\partial S}{\partial q_1},\ p_2 = \frac{\partial S}{\partial q_2}, \tag{17}$$

which form a system of algebraic equations with respect to the old canonical variables q_1, q_2, p_1, p_2. Solution of this system can be sought in the form of polynomials in terms of the new canonical variables $\tilde{q}_1$, $\tilde{q}_2$, $\tilde{p}_1$, $\tilde{p}_2$. To find an expression for the third-order term H_3 in new variables it is sufficient to consider the polynomials of second degree but the third-degree terms also should be taken into account because they influence the fourth-order term H_4 that will be normalized at the next step. In general, to calculate the term H_k in new variables one needs to consider solutions of system (17) in the form of polynomials of the $(k-1)th$ degree.

Applying the method of iterations and doing quite standard symbolic calculation, we solve system (17) and obtain the old canonical variables q_1, q_2, p_1, p_2 as functions of the new ones $\tilde{q}_1$, $\tilde{q}_2$, $\tilde{p}_1$, $\tilde{p}_2$ which in linear approximation determine an identical canonical transformation. Then we substitute these variables into the Hamiltonian expansion $\mathcal{H} = H_2 + H_3 + H_4$, where H_2, H_3, and H_4 are given by (12), (14), and (15), respectively, and expand the expression obtained into the power series in terms of new variables $\tilde{q}_1$, $\tilde{q}_2$, $\tilde{p}_1$, $\tilde{p}_2$ up to the fourth order. One can easily check that the second-order term $\tilde{H}_2$ in new variables retains the normal form (12) while coefficients $\tilde{h}_{ijkl}^{(3)}$ in the term $\tilde{H}_3$ represented in the form (14) become linear functions of $h_{ijkl}^{(3)}$ and unknown coefficients $s_{ijkl}^{(3)}$ determining the generating function (16). Equating the coefficients $\tilde{h}_{ijkl}^{(3)}$ to zero, we obtain a system of linear equations with respect to the coefficients $s_{ijkl}^{(3)}$. The calculations show that if the frequencies σ_1, σ_2 satisfy the conditions

$$\sigma_1 \pm 2\sigma_2 \neq 0,\ 2\sigma_1 \pm \sigma_2 \neq 0,\ \sigma_1 \pm \sigma_2 \neq 0, \tag{18}$$

and $\sigma_1 \neq 0$, $\sigma_2 \neq 0$ then this system has a unique solution. It means that realization of canonical transformation (17) with the found coefficients $s^{(3)}_{ijkl}$ eliminates the third-order term H_3 in the Hamiltonian expansion. The corresponding expressions for coefficients $s^{(3)}_{ijkl}$ are given in [5,6,16].

Analyzing frequencies (10), we obtain that there are such points (μ_1, μ_2) in the $O\mu_1\mu_2$ plane in the domain of linear stability of the equilibrium positions S_5, S_6 for which the condition of third-order resonance $\sigma_1 - 2\sigma_2 = 0$ is fulfilled (see Figs. 4 and 5). For such values of parameters μ_1 and μ_2, some coefficients $\tilde{h}^{(3)}_{ijkl}$ cannot be equal to zero and, therefore, it is not possible to eliminate the term $\tilde{H}_3$ completely (see [5,16]). There are six such coefficients and they should be chosen in such a way that the Hamiltonian $\mathcal{H}$ takes the form admitting application of Markeev's theorem on instability of equilibrium positions under the third-order resonance [14]. Requiring the following conditions to be fulfilled

$$\begin{aligned} \tilde{h}^{(3)}_{0012} &= \frac{B_1}{2\sqrt{2}}, \quad \tilde{h}^{(3)}_{0210} = -\frac{B_1}{2\sqrt{2}}, \quad \tilde{h}^{(3)}_{1101} = -\frac{B_1}{\sqrt{2}}, \\ \tilde{h}^{(3)}_{0111} &= \frac{B_2}{\sqrt{2}}, \quad \tilde{h}^{(3)}_{1002} = \frac{B_2}{2\sqrt{2}}, \quad \tilde{h}^{(3)}_{1200} = -\frac{B_2}{2\sqrt{2}}, \end{aligned} \tag{19}$$

where B_1, B_2 are some constants, and solving the system of equations (19), we obtain the corresponding coefficients $s^{(3)}_{ijkl}$ of the generating function (16) and find the constants B_1, B_2 as

$$B_1 = \frac{1}{\sqrt{2}}(h^{(3)}_{0012} - h^{(3)}_{0210} - h^{(3)}_{1101}), \quad B_2 = \frac{1}{\sqrt{2}}(h^{(3)}_{0111} + h^{(3)}_{1002} - h^{(3)}_{1200}). \tag{20}$$

Then the Hamiltonian (6) takes a form

$$\begin{aligned} \tilde{\mathcal{H}} = &\frac{1}{2}\sigma_1\left(\tilde{q}_1^2 + \tilde{p}_1^2\right) - \frac{1}{2}\sigma_2\left(\tilde{q}_2^2 + \tilde{p}_2^2\right) + \frac{B_1}{2\sqrt{2}}\left(\tilde{p}_1\tilde{p}_2^2 - \tilde{p}_1\tilde{q}_2^2 - 2\tilde{q}_1\tilde{q}_2\tilde{p}_2\right) \\ &+ \frac{B_2}{2\sqrt{2}}\left(\tilde{q}_1\tilde{p}_2^2 - \tilde{q}_1\tilde{q}_2^2 + 2\tilde{q}_2\tilde{p}_1\tilde{p}_2\right) + \tilde{H}_4 + \ldots. \end{aligned} \tag{21}$$

Numerical analysis of parameter $B = \sqrt{B_1^2 + B_2^2}$ for the equilibrium points S_5 and S_6 has shown that it is not equal to zero for all points (μ_1, μ_2) belonging to the resonance curves $\sigma_1 - 2\sigma_2 = 0$ for any of the three possible collinear configurations of the particles P_0, P_1, P_2. According to Markeev's theorem [14], we can conclude that equilibrium points S_5 and S_6 in the circular restricted four-body problem formulated on the basis of Euler's collinear configurations are unstable under third-order resonance of the form $\sigma_1 - 2\sigma_2 = 0$.

Let us consider the points (μ_1, μ_2) in the domain of linear stability of equilibrium positions S_5, S_6 for which the condition $\sigma_1 \neq 2\sigma_2$ is fulfilled, and there is no resonance in the system up to the third order inclusively. Then after normalization of the second and third order terms, we obtain the Hamiltonian (6) in the form

$$\tilde{\mathcal{H}} = \tilde{H}_2 + \tilde{H}_4 + \ldots, \tag{22}$$

where the second-order term

$$\tilde{H}_2 = \frac{1}{2}\left(\sigma_1(\tilde{p}_1^2 + \tilde{q}_1^2) - \sigma_2(\tilde{p}_2^2 + \tilde{q}_2^2)\right) \tag{23}$$

has the normal form, the third-order term $\tilde{H}_3$ is absent, and the fourth-order term $\tilde{H}_4$ may be written as

$$\tilde{H}_4 = \sum_{i+j+k+l=4} \tilde{h}^{(4)}_{ijkl}\tilde{q}_1^i\tilde{q}_2^j\tilde{p}_1^k\tilde{p}_2^l. \tag{24}$$

The sum (24) contains 35 terms but coefficients $\tilde{h}^{(4)}_{ijkl}$ are very cumbersome, and we do not write them here. Now we look for the fourth-degree polynomial

$$S(p_1^*, p_2^*, \tilde{q}_1, \tilde{q}_2) = \tilde{q}_1 p_1^* + \tilde{q}_2 p_2^* + \sum_{i+j+k+l=4} s^{(4)}_{ijkl}\tilde{q}_1^i\tilde{q}_2^j p_1^{*k} p_2^{*l}, \tag{25}$$

generating the canonical transformation reducing the fourth order term $\tilde{H}_4$ to the simplest form. New momenta p_1^*, p_2^* and coordinates q_1^*, q_2^* are determined by the relationships

$$q_1^* = \frac{\partial S}{\partial p_1^*},\ q_2^* = \frac{\partial S}{\partial p_2^*},\ \tilde{p}_1 = \frac{\partial S}{\partial \tilde{q}_1},\ \tilde{p}_2 = \frac{\partial S}{\partial \tilde{q}_2}. \tag{26}$$

Resolving (26) with respect to the old canonical variables $\tilde{q}_1$, $\tilde{q}_2$, $\tilde{p}_1$, $\tilde{p}_2$ in the neighborhood of the point $q_1^* = q_2^* = p_1^* = p_2^* = 0$ and substituting the solution into (22), we expand the Hamiltonian $\tilde{H}$ in the Taylor series in terms of q_1^*, q_2^*, p_1^*, p_2^*. Obviously, the second order term H_2^* in this expansion retains the form (23), the third order term H_3^* is absent, and the fourth order term H_4^* is a sum of 35 terms of the form

$$h^{*(4)}_{ijkl} q_1^{*i} q_2^{*j} p_1^{*k} p_2^{*l} \quad (i+j+k+l=4),$$

where new coefficients $h^{*(4)}_{ijkl}$ are linear functions of $\tilde{h}^{(4)}_{ijkl}$ and unknown coefficients $s^{(4)}_{ijkl}$, determining the generating function (25).

Analysis of the coefficients $h^{*(4)}_{ijkl}$ shows that they are divided into several independent groups, and each group forms a system of equations determining some coefficients $s^{(4)}_{ijkl}$. If the following conditions

$$\sigma_1 \neq 0, \quad \sigma_2 \neq 0, \quad \sigma_1 \pm \sigma_2 \neq 0, \quad \sigma_1 \pm 3\sigma_2 \neq 0, \quad 3\sigma_1 \pm \sigma_2 \neq 0, \tag{27}$$

are fulfilled we can solve the equations $h^{*(4)}_{ijkl} = 0$ and find coefficients $s^{(4)}_{ijkl}$ of the canonical transformation (26) eliminating the corresponding terms in (24). Nevertheless, there are ten terms in the expansion (24) which cannot be eliminated. They can be only simplified in such a way that the fourth order term $\tilde{H}_4$ takes the form (see [5,16])

$$H_4^* = \frac{1}{4}\left(c_{20}(p_1^{*2} + q_1^{*2})^2 + c_{11}(p_1^{*2} + q_1^{*2})(p_2^{*2} + q_2^{*2}) + c_{02}(p_2^{*2} + q_2^{*2})^2\right).$$

Then, using the standard canonical transformation

$$q_1^* = \sqrt{2\tau_1}\sin\varphi_1,\ p_1^* = \sqrt{2\tau_1}\cos\varphi_1,$$
$$q_2^* = \sqrt{2\tau_2}\sin\varphi_2,\ p_2^* = \sqrt{2\tau_2}\cos\varphi_2, \tag{28}$$

we rewrite the Hamiltonian (22) as

$$\mathcal{H}^* = \sigma_1\tau_1 - \sigma_2\tau_2 + c_{20}\tau_1^2 + c_{11}\tau_1\tau_2 + c_{02}\tau_2^2 + H_5^*(\varphi_1, \varphi_2, \tau_1, \tau_2) + \dots . \tag{29}$$

Recall that Arnold's theorem [3] states that in the case of absence of resonances up to the fourth order inclusively (conditions (18),(27) are fulfilled), equilibrium positions are stable if

$$f = c_{20}\sigma_2^2 + c_{11}\sigma_1\sigma_2 + c_{02}\sigma_1^2 \neq 0. \tag{30}$$

Numerical analysis of parameter f shows that for equilibrium points S_5 and S_6, there exist such values of parameters μ_1 and μ_2, for which $f = 0$ (see Figs. 4 and 5). For such μ_1, μ_2 the fifth and higher order terms in the Hamiltonian expansion (6) need to be analyzed to conclude on stability or instability of equilibrium solution. The corresponding calculations are very cumbersome, and this case will be analyzed in our next paper.

Besides, there are curves in the $O\mu_1\mu_2$ plane (see Figs. 4 and 5), where the condition of fourth-order resonance of the form $\sigma_1 = 3\sigma_2$ is fulfilled. In this case, eight additional terms appear in the expression for H_4^* because the following coefficients $h_{ijkl}^{*(4)}$ do not vanish and are expressed via two parameters A_1, A_2

$$h_{0013}^{*(4)} = -\frac{1}{3}h_{0211}^{*(4)} = -\frac{1}{3}h_{1102}^{*(4)} = h_{1300}^{*(4)} = \frac{A_1}{4},$$
$$h_{1003}^{*(4)} = \frac{1}{3}h_{0112}^{*(4)} = -\frac{1}{3}h_{1201}^{*(4)} = -h_{0310}^{*(4)} = \frac{A_2}{4}. \tag{31}$$

Solving system (31), we find the corresponding coefficients $s_{ijkl}^{(4)}$, and parameters A_1 and A_2 are obtained in the form

$$A_1 = \frac{1}{2}(\tilde{h}_{0013}^{(4)} - \tilde{h}_{0211}^{(4)} - \tilde{h}_{1102}^{(4)} + \tilde{h}_{1300}^{(4)}),$$

$$A_2 = \frac{1}{2}(\tilde{h}_{0112}^{(4)} - \tilde{h}_{0310}^{(4)} + \tilde{h}_{1003}^{(4)} - \tilde{h}_{1201}^{(4)}).$$

Finally, the Hamiltonian (22) is reduced to the form

$$\begin{aligned}\mathcal{H}^* = \frac{3\sigma_2}{2}\left(p_1^{*2} + q_1^{*2}\right) - \frac{\sigma_2}{2}\left(p_2^{*2} + q_2^{*2}\right)\\ + \frac{1}{4}\left(c_{20}(p_1^{*2} + q_1^{*2})^2 + c_{11}(p_1^{*2} + q_1^{*2})(p_2^{*2} + q_2^{*2}) + c_{02}(p_2^{*2} + q_2^{*2})^2\right)\\ + \frac{A_1}{4}\left(p_1^*p_2^{*3} - 3q_2^{*2}p_1^*p_2^* - 3q_1^*q_2^*p_2^{*2} + q_1^*q_2^{*3}\right)\\ + \frac{A_2}{4}\left(q_1^*p_2^{*3} - 3q_1^*q_2^{*2}p_2^* + 3q_2^*p_1^*p_2^{*2} - p_1^*q_2^{*3}\right).\end{aligned} \tag{32}$$

According to the theorem of Markeev [14], stability of the equilibrium solutions under the fourth-order resonance depends on the values of $c_{20}+3c_{11}+9c_{02}$ and $3\sqrt{3(A_1^2+A_2^2)}$. Our calculations show that if the particle P_2 is situated in the neighborhood of the points of libration L_1 and L_2, an inequality

$$c_{20}+3c_{11}+9c_{02}<3\sqrt{3(A_1^2+A_2^2)}$$

takes place for all points (μ_1,μ_2) belonging to the resonant curve $\sigma_1=3\sigma_2$ (see Fig. 4). It means that the fourth-order resonance results in instability of equilibrium positions S_5, S_6 for such collinear configurations of the massive particle. The same result is obtained in the case when the particle P_2 is situated in the neighborhood of the point L_3, and the point (μ_1,μ_2) is located at the resonant curve $\sigma_1=3\sigma_2$ to the left of the points C_1 and C_2 (see Fig. 5). However, for the points (μ_1,μ_2) belonging to the arc C_1C_2, we obtain

$$c_{20}+3c_{11}+9c_{02}>3\sqrt{3(A_1^2+A_2^2)},$$

and equilibrium positions S_5 and S_6 are stable in Liapunov's sense.

5 Conclusion

In the present paper, we have studied stability of the equilibrium positions in the planar circular restricted four-body problem formulated on the basis of the Euler collinear solutions of the three-body problem. We have proved that all collinear equilibrium positions of particle P_3 are unstable for any values of the system parameters μ_1 and μ_2. Similar results were obtained earlier in the restricted three-body problem (see [14,21]). The equilibrium positions S_5 and S_6 (see Fig. 3) are stable in Liapunov's sense if parameters μ_1 and μ_2 are sufficiently small and belong to the domains bounded by the curves $\sigma_1=\sigma_2$ (Figs. 4 and 5). However, in these domains, there are such values of parameters μ_1 and μ_2 for which the conditions of the third- or fourth-order resonances are fulfilled. The third-order resonance results in instability of equilibrium points S_5 and S_6 for any collinear position of the particle P_2 while in case of the fourth-order resonance, stability of these points may take place if the particle P_2 is situated in the neighborhood of the point of libration L_3, and the point (μ_1,μ_2) is located at the resonant curve between the points C_1 and C_2.

There are also such values of parameters μ_1 and μ_2 for which the conditions of Arnold's theorem are not fulfilled (curves $f=0$ in Figs. 4 and 5), and analysis of the fifth and higher order terms in the Hamiltonian expansion is required for the entire solution of the stability problem. Such analysis will be done in our next paper.

Note that all relevant calculations and visualization of the obtained results are performed with the computer algebra system *Wolfram Mathematica*.

References

1. Alvares-Ramirez, M., Skea, J.E.F., Stuchi, T.J.: Nonlinear stability analysis in a equilateral restricted four-body problem. Astrophys. Space Sci. (2015). doi:10.1007/s10509-015-2333-4
2. Arenstrof, R.E.: Central configurations of four bodies with one inferior mass. Celest. Mech. **29**, 9–15 (1982)
3. Arnold, V.I.: Small denominators and problems of stability of motion in classical and celestial mechanics. Russ. Math. Surv. **18**(6), 85–191 (1963)
4. Budzko, D.A., Prokopenya, A.N.: Symbolic-numerical analysis of equilibrium solutions in a restricted four-body problem. Program. Comput. Softw. **36**(2), 68–74 (2010)
5. Budzko, D.A., Prokopenya, A.N.: On the stability of equilibrium positions in the circular restricted four-body problem. In: Gerdt, V.P., Koepf, W., Mayr, E.W., Vorozhtsov, E.V. (eds.) CASC 2011. LNCS, vol. 6885, pp. 88–100. Springer, Heidelberg (2011). doi:10.1007/978-3-642-23568-9_8
6. Budzko, D.A., Prokopenya, A.N.: Stability of equilibrium positions in the spatial circular restricted four-body problem. In: Gerdt, V.P., Koepf, W., Mayr, E.W., Vorozhtsov, E.V. (eds.) CASC 2012. LNCS, vol. 7442, pp. 72–83. Springer, Heidelberg (2012). doi:10.1007/978-3-642-32973-9_7
7. Budzko, D.A., Prokopenya, A.N.: Symbolic-numerical methods for searching equilibrium states in a restricted four-body problem. Program. Comput. Softw. **39**(2), 74–80 (2013)
8. Deprit, A., Deprit-Bartholomé, A.: Stability of the triangular Lagrangian points. Astron. J. **72**(2), 173–179 (1967)
9. Gadomski, L., Grebenikov, E.A., Prokopenya, A.N.: Studying the stability of equilibrium solutions in the planar circular restricted four-body problem. Nonlinear Oscil. **10**(1), 66–82 (2007)
10. Grebenikov, E.A., Ikhsanov, E.V., Prokopenya, A.N.: Numeric-symbolic computations in the study of central configurations in the planar newtonian four-body problem. In: Ganzha, V.G., Mayr, E.W., Vorozhtsov, E.V. (eds.) CASC 2006. LNCS, vol. 4194, pp. 192–204. Springer, Heidelberg (2006). doi:10.1007/11870814_16
11. Kozak, D., Oniszk, E.: Equilibrium points in the restricted four-body problem. Sufficient conditions for linear stability. Rom. Astron. J. **8**(1), 27–31 (1998)
12. Kozera, R., Noakes, L., Klette, R.: External versus internal parameterizations for lengths of curves with nonuniform samplings. In: Asano, T., Klette, R., Ronse, C. (eds.) Geometry, Morphology, and Computational Imaging. LNCS, vol. 2616, pp. 403–418. Springer, Heidelberg (2003). doi:10.1007/3-540-36586-9_26
13. Leontovich, A.M.: On the stability of Lagrangian periodic solutions of the restricted three-body problem. Soviet Math. Dokl. **3**, 425–429 (1962)
14. Markeev, A.P.: Libration points in Celestial Mechanics and Cosmodynamics. Nauka, Moscow (1978). (in Russian)
15. Moser, J.: Lectures on the Hamiltonian systems. Mir, Moscow (1973). (in Russian)
16. Prokopenya, A.N.: Hamiltonian normalization in the restricted many-body problem by computer algebra methods. Program. Comput. Softw. **38**(3), 156–166 (2012)
17. Roy, A.E.: Orbital Motion, 4th edn. Institute of Physics Publishing, Bristol/ Philadephia (2005)
18. Simo, C.: Relative equilibrium solutions in the four body problem. Celest. Mech. **18**, 165–184 (1978)

19. Singh, J., Vincent, A.E.: Effect of perturbations in the Coriolis and centrifugal forces on the stability of equilubrium points in the restricted four-body problem. Few-Body Syst. **56**, 713–723 (2015)
20. Schmidt, D., Vidal, C.: Stability of the planar equilibrium solutions of a restricted $1 + N$ body problem. Regul. Chaotic Dyn. **19**(5), 533–547 (2014)
21. Szebehely, V.: Theory of Orbits. The Restricted Problem of Three Bodies. Academic Press, New York/London (1967)
22. Wintner, A.: The Analytical Foundations of Celestial Mechanics. Princeton Mathematical Series, vol. 5. Princeton University Press, Princeton (1941)
23. Wolfram, S.: An Elementary Introduction to the Wolfram Language, 2nd edn. Wolfram Media, Champaign (2017)

The Method of Collocations and Least Residuals Combining the Integral Form of Collocation Equations and the Matching Differential Relations at the Solution of PDEs

Vasily P. Shapeev[1,2] and Evgenii V. Vorozhtsov[1(✉)]

[1] Khristianovich Institute of Theoretical and Applied Mechanics, Russian Academy of Sciences, Novosibirsk 630090, Russia
{shapeev,vorozh}@itam.nsc.ru

[2] Novosibirsk National Research University, Novosibirsk 630090, Russia

Abstract. To increase the accuracy of computations by the method of collocations and least residuals (CLR) it is proposed to increase the number of degrees of freedom with the aid of the following two techniques: an increase in the number of basis vectors and the integration of the linearized partial differential equations (PDEs) over the subcells of each cell of a spatial computational grid. The implementation of these modifications, however, leads to the necessity of increasing the amount of symbolic computations needed for obtaining the work formulas of the new versions of the CLR method. The computer algebra system (CAS) *Mathematica* has proved to be successful at the execution of all these computations. It is shown that the proposed new symbolic-numeric versions of the CLR method possess a higher accuracy than the previous versions of this method. Furthermore, the version of the CLR method, which employs the integral form of collocation equations, needs a much lesser number of iterations for its convergence than the "differential" CLR method.

Keywords: Computer algebra system · Symbolic-numerical algorithm · Collocation of integral relations · Preconditioner · Krylov subspaces · Multigrid

1 Introduction

At present, the numerical simulation of various processes in technologies and industry with the aid of the numerical solution of the initial- and boundary-value problems for the systems of nonlinear partial differential equations (PDEs) has gained widespread acceptance. In particular, some applied tasks involving the solution of the Navier–Stokes equations are very computationally intensive and require a CPU time from several weeks to one year [2,14]. In this connection, the development of more efficient methods for the numerical solution of the PDE

V.P. Gerdt et al. (Eds.): CASC 2017, LNCS 10490, pp. 346–361, 2017.
DOI: 10.1007/978-3-319-66320-3_25

systems, which would enable a significant reduction of the needed CPU times, is urgent as before.

At the derivation of the work formulas of complex high-accuracy numerical methods, the errors are practically unavoidable in the cases when the above formulas are derived by a mathematician "by hand" with the aid of pen and paper. There are already by now fairly many works the authors of which have shown a substantial benefit from using computer algebra systems (CASs) in the process of deriving the formulas of new numerical algorithms, their realization and verification of the corresponding computer codes [1,7,8,25].

The CLR method, which was proposed in [15] and developed further in the subsequent works of other authors, is one of the methods which enable the efficient solution of PDEs [10,16,17,19–24]. The works [10,20–23] have shown the usefulness of the application of a CAS to the derivation of formulas of the different versions of the CLR method. The versions of the method were constructed, which have enabled the obtaining of the solutions of the 2D and 3D benchmark problems, which are among the most accurate ones at present [3,18].

During the last three decades, a class of the numerical techniques named LSFEM (Least-Squares Finite Element Method) [11,12] has gained a fairly wide acceptance. In this class of methods, the FEM (Finite Element Method) is combined with the method of least squares. In FEM, the PDEs to be solved are at first integrated over each finite element, which represents a subregion of the spatial computational region. This approach has stimulated the present authors to consider the following versions of the CLR method: (i) a version in which all the collocation equations derived from the PDE system are replaced with their integral counterparts, which are obtained at the integration of the PDEs over several subcells, into which each cell of the spatial grid is partitioned; (ii) a version in which both the collocation equations obtained from the PDEs and the equations obtained by integrating over the subcells are employed. All the analytic computations needed for obtaining the work formulas of the above modifications of the CLR method have been carried out with the aid of corresponding *Mathematica* codes to avoid any errors and to speed up all the needed jobs.

We describe below in the present work both the original "differential" CLR method and the CLR methods in which the "differential" version is combined with the integral form of collocation equations and the differential forms of matching conditions. The computational examples are presented, which show that the new modifications of the CLR method enable the obtaining of more accurate results than in the case of the "differential" versions of the CLR method.

2 The "Differential" CLR Method

2.1 Description of the Method

Consider a boundary-value problem for the system of Navier–Stokes equations

$$(\mathbf{V}\cdot\nabla)\mathbf{V}+\nabla p=\frac{1}{\mathrm{Re}}\Delta\mathbf{V}-\mathbf{f},\quad \operatorname{div}\mathbf{V}=0,\quad (x_1,x_2)\in\Omega, \tag{1}$$

$$\mathbf{V}\big|_{\partial\Omega}=\mathbf{g} \tag{2}$$

in the region Ω with the boundary $\partial\Omega$. In Eq. (1), x_1, x_2 are the Cartesian spatial coordinates, $\mathbf{V} = (v_1(x_1,x_2), v_2(x_1,x_2))$ is the velocity vector; $p = p(x_1,x_2)$ is the pressure, $\mathbf{f} = (f_1, f_2)$ is the given vector function, Re is the Reynolds number, $\Delta = \frac{\partial^2}{\partial x_1^2} + \frac{\partial^2}{\partial x_2^2}$, $(\mathbf{V}\cdot\nabla) = v_1\frac{\partial}{\partial x_1} + v_2\frac{\partial}{\partial x_2}$. System (1) is solved under the Dirichlet boundary conditions (2), where $\mathbf{g} = \mathbf{g}(x_1,x_2) = (g_1, g_2)$ is a given vector function. The pressure is determined from (1) and (2) with the accuracy up to a constant. We will choose this constant in the following in such a way that the following condition is satisfied:

$$\iint_\Omega p\, dx_1 dx_2 = 0. \tag{3}$$

The square

$$\Omega = \{(x_1,x_2),\ 0 \le x_i \le L,\ i = 1,2\}, \tag{4}$$

is taken as the problem solution region, where $L > 0$ is the given length of the square side. The quantity L was used in specific computations as the reference length at the non-dimensionalization of variables, and it enters the definition of the Reynolds number Re in (1) in a natural way. We will term in the following the boundary-value problem for the PDE the differential problem.

In the given problem (1)–(4), region (4) is discretized by a grid with square cells Ω_{ij}, $i,j = 1,\ldots,I$, $I \ge 1$. It is convenient to introduce the local coordinates y_1 and y_2 in each cell Ω_{ij}. The dependence of local coordinates on global spatial variables x_1 and x_2 is specified by relations $y_m = (x_m - x_{m,i,j})/h$, $m = 1,2$, where $x_{m,i,j}$ is the value of the coordinate x_m at the center of cell Ω_{ij}, and h is the halved length of the square cell side. Let $\mathbf{u}(y_1,y_2) = (u_1,u_2) = \mathbf{V}(hy_1 + x_{1,i,j}, hy_2 + x_{2,i,j})$, $q(y_1,y_2) = p(hy_1 + x_{1,i,j}, hy_2 + x_{2,i,j})$. The Navier–Stokes equations then take the following form:

$$\Delta u_m - \mathrm{Re}h\left(u_1\frac{\partial u_m}{\partial y_1} + u_2\frac{\partial u_m}{\partial y_2} + \frac{\partial q}{\partial y_m}\right) = \mathrm{Re}\cdot h^2 f_m, \quad m = 1,2; \tag{5}$$

$$\frac{1}{h}\left(\frac{\partial u_1}{\partial y_1} + \frac{\partial u_2}{\partial y_2}\right) = 0, \tag{6}$$

where $\Delta = \frac{\partial^2}{\partial y_1^2} + \frac{\partial^2}{\partial y_2^2}$. The Newton linearization of Eq. (5) gives the equation

$$\begin{aligned}\xi[\Delta u_m^{s+1} &- (\mathrm{Re}\cdot h)(u_1^s u_{m,y_1}^{s+1} + u_1^{s+1}u_{m,y_1}^s + u_2^s u_{m,y_2}^{s+1} \\ &+ u_2^{s+1}u_{m,y_2}^s + q_{y_m}^{s+1})] = \xi F_m,\end{aligned} \tag{7}$$

where $m = 1,2$, and s is the number of the iteration over the nonlinearity, $s = 0,1,2,\ldots$, u_1^s, u_2^s, q^s is the known approximation to the solution at the sth iteration starting from the chosen initial guess with index $s = 0$, $F_m = \mathrm{Re}\left[h^2 f_m - h\left(u_1^s u_{m,y_1}^s + u_2^s u_{m,y_2}^s\right)\right]$, $u_{m,y_l} = \partial u_m/\partial y_l$, $q_{y_m} = \partial q/\partial y_m$, $l,m = 1,2$. The user-specified parameter ξ has been introduced here as in [24] for the purpose of controlling the magnitude of the condition number of a system of linear algebraic equations (SLAE), which must be solved in each cell Ω_{ij}.

The approximate solution in each cell $\Omega_{i,j}$ is sought in the form of a linear combination of the basis vector functions φ_l:

$$(u_1^s, u_2^s, q^s)^T = \sum_{l=1}^{m_b} b_{i,j,l}^s \varphi_l, \tag{8}$$

where the superscript T denotes the transposition operation, and m_b is the user-specified number of the basis vector functions. In the given version of the method, the φ_l are the polynomials. Thus, the approximate solution is a piecewise polynomial. In the work [24], the second-degree polynomials in variables y_1, y_2 were employed for the approximation of velocity components, and the first-degree polynomial was used for the pressure approximation so that the total number of the basis vector functions in (8) amounted to $m_b = 12$.

It was shown previously in [10] that it is possible to increase the accuracy of the numerical solution obtained by the CLR method by using the polynomials of higher degrees. In this connection, we use in the present paper the second-degree polynomial also for the pressure approximation. In this case, there are eighteen basis functions in total. Since the coefficients are constant in the continuity equation, which has a simple form, it is easy to satisfy it at the expense of the choice of basis polynomials φ_l. It is not difficult to find that it is required to this end that they satisfy three linear relations. There will finally remain only fifteen independent basis polynomials from the original eighteen ones. They are presented in Table 1. One can term their set a solenoidal basis because $\operatorname{div} \varphi_l = 0$. The set of basis functions, which was employed in [24], is obtained from the set presented in Table 1 if one sets $m_b = 12$ in (8), that is if one retains in Table 1 only the first 12 basis vector functions.

Table 1. The form of basis functions φ_l

l	1	2	3	4	5	6	7	8	9	10	11	12	13	14	15
φ_l	1	y_1	y_2	y_1^2	$-2y_1y_2$	y_2^2	0	0	0	0	0	0	0	0	0
	0	$-y_2$	0	$-2y_1y_2$	y_2^2	0	1	y_1	y_1^2	0	0	0	0	0	0
	0	0	0	0	0	0	0	0	0	1	y_1	y_2	y_1^2	y_1y_2	y_2^2

The number of collocation points and their location inside the cell may vary in different versions of the method. In the given work, three versions of the specification of the collocation point coordinates have been implemented. Denote by N_c the number of collocation points inside each cell. In the case when $N_c = 2$, the coordinates of collocation points are as follows: (ω, ω), $(-\omega, \omega)$, where $0 < \omega < 1$. At $N_c = 4$, the local coordinates of collocation points have the form $(\pm\omega, \pm\omega)$. In the case of $N_c = 8$, the coordinates of collocation points were specified in the following way: the locations of the first four points were the same as at $N_c = 4$, and the coordinates of the next four points were specified by formulas $(\pm\omega, 0)$, $(0, \pm\omega)$.

Substituting (8) as well as the numerical values of the coordinates of each collocation point in (7) we obtain $2N_c$ linear algebraic equations:

$$\sum_{m=1}^{m_b} a_{\nu,m}^{(1)} \cdot b_m^{s+1} = f_\nu^s, \quad \nu = 1, \ldots, 2N_c. \tag{9}$$

By analogy with [24] let us augment the system of equations of the approximate problem in the Ω_{ij} cell by the conditions of matching with the solutions of the discrete problem, which are taken in all cells adhering to the given cell. We will write these conditions at separate points (called the matching points) on the sides of the Ω_{ij} cell, which are common with its neighboring cells. The matching conditions are taken here in the form

$$h\frac{\partial (u^+)^n}{\partial n} + \eta (u^+)^n = h\frac{\partial (u^-)^n}{\partial n} + \eta (u^-)^n, \tag{10}$$

$$h\frac{\partial (u^+)^\tau}{\partial n} + (u^+)^\tau = h\frac{\partial (u^-)^\tau}{\partial n} + (u^-)^\tau, \tag{11}$$

$$q^+ = q^-. \tag{12}$$

Here $h\frac{\partial}{\partial n} = h\left(n_1\frac{\partial}{\partial x_1} + n_2\frac{\partial}{\partial x_2}\right) = n_1\frac{\partial}{\partial y_1} + n_2\frac{\partial}{\partial y_2}$, $n = (n_1, n_2)$ is the external normal to the side of the Ω_{ij} cell, $(\cdot)^n$, $(\cdot)^\tau$ are the normal and tangential components of the velocity vector with respect to the cell side, u^+, u^- are the limits of the function u as its arguments tend to the matching point from inside and outside the Ω_{ij} cell. The user-specified parameter η has been introduced here as in [24] for the purpose of controlling the magnitude of the condition number of a SLAE, which must be solved in each cell Ω_{ij}.

For the uniqueness of the pressure determination in the solution, we either specify its value at a single point of the region or approximate condition (3) by the formula

$$\frac{1}{h}\left(\iint_{\Omega_{i,j}} q\, dy_1 dy_2\right) = \frac{1}{h}\left(-I^* + \iint_{\Omega_{i,j}} q^* dy_1 dy_2\right). \tag{13}$$

Here I^* is the integral over the entire region, which is computed as a sum of the integrals over each cell at the foregoing iteration, q^* is the pressure in a cell from the foregoing iteration.

Denote by N_m the number of matching points for the velocity vector components on the sides of each cell. At $N_m = 4$, the coordinates of these matching points are specified by the formulas $(\pm 1, 0)$, $(0, \pm 1)$. At $N_m = 8$, the coordinates of matching points are as follows: $(\pm 1, -\zeta)$, $(\pm 1, \zeta)$, $(-\zeta, \pm 1)$, $(\zeta, \pm 1)$, where $0 < \zeta < 1$. In the computational examples presented below, the value $\zeta = 1/2$ was used. The matching conditions for pressure (12) are set at four points with coordinates $(\pm 1, 0)$, $(0, \pm 1)$.

Using Eq. (8), we substitute the coordinates of these points in each of three matching conditions (10)–(12). We obtain from the first two conditions $2N_m$ linear algebraic equations for velocity components. The substitution of representation (8) in (12) also yields four linear algebraic (matching) equations.

In the present work, the pressure was specified at the vertex of the $\Omega_{1,1}$ cell or condition (13) was used. If the cell side coincides with the boundary of region

Ω, then the boundary conditions are written at the corresponding points instead of the matching conditions for the discrete problem solution: $u_m = g_m$, $m = 1, 2$.

Uniting the equations of collocations, matching, and the equations obtained form the boundary conditions, if the cell Ω_{ij} is the boundary cell, we obtain in each cell a SLAE of the form

$$A_{i,j} \cdot \boldsymbol{X}_{i,j}^{s+1} = \boldsymbol{f}_{i,j}^{s,s+1}, \tag{14}$$

where $\boldsymbol{X}_{i,j}^{s+1} = (b_{i,j,1}^{s+1}, \ldots, b_{i,j,m_b}^{s+1})^T$. In the versions studied in the present work, system (14) is overdetermined. The symbolic expressions for the coefficients of all equations of SLAE (14) were derived on computer in Fortran form by using symbolic computations with *Mathematica*. At the obtaining of the final form of the formulas for the coefficients of the equations, it is useful to perform the simplifications of the arithmetic expressions of polynomial form to reduce the number of the arithmetic operations needed for their numerical computation. To this end, we employed standard functions of the *Mathematica* system, such as SIMPLIFY[...] and FULLSIMPLIFY[...] for the simplification of complex symbolic expressions arising at the symbolic stages of the construction of the formulae of the method. Their application enabled a two-three-fold reduction of the length of polynomial expressions.

For the numerical solution of the SLAE of the discrete problem a process was applied which may be called conventionally the Gauss–Seidel iteration scheme. One global $(s+1)$th iteration meant that all the cells were considered sequentially in the computational region Ω. In each cell, SLAE (14) was solved by the orthogonal method (of Givens or Householder), and the values known at the solution construction at the $(s+1)$th iteration were taken in the right-hand sides of Eqs. (10)–(12) as the u^- and q^- in a given cell.

2.2 Preconditioners for the CLR Method

It is necessary to solve in each cell Ω_{ij} the SLAE of the form (14). Let us omit in (14) the superscripts and subscripts for the sake of brevity:

$$\mathrm{A}\boldsymbol{X} = \boldsymbol{f}. \tag{15}$$

The condition number of a rectangular matrix A is calculated by the formula

$$\kappa(\mathrm{A}) = \sqrt{\| \mathrm{A}_1 \| \cdot \| \mathrm{A}_1^{-1} \|}, \tag{16}$$

where it is assumed that matrix $\mathrm{A}_1 = \mathrm{A}^T\mathrm{A}$ is non-singular. In our case, we have a preconditioner involving the parameters ξ and η. A simple algorithm was described in [24] for finding the optimal values ξ_{opt} and η_{opt} in any cell from the requirement of minimizing the condition number $\kappa(\xi, \eta)$. The value $\kappa(\xi_{\mathrm{opt}}, \eta_{\mathrm{opt}})$ typically satisfied the inequalities $3 < \kappa(\xi_{\mathrm{opt}}, \eta_{\mathrm{opt}}) < 10$. It has turned out that the optimal values ξ_{opt} and η_{opt} depended weakly on the location of a specific cell in the spatial grid, at least in the cases of those test and benchmark problems which were considered in [24]. Some properties of the preconditioner were then investigated in [24]. In particular, it was shown that a reduction of N_c affects more significantly the value ξ_{opt} than the value η_{opt}.

2.3 Convergence Acceleration Algorithm Based on Krylov's Subspaces

To accelerate the convergence of the iterations used for the approximate solution construction we have used in all new versions of the CLR method, which are discussed in the present paper, a new variant of the well-known method [13] based on Krylov's subspaces, which was previously presented in detail in [22,26]. We present in the following a very brief description of the corresponding algorithm. Let the SLAE have the form $\boldsymbol{X} = \mathrm{T}\boldsymbol{X} + \boldsymbol{f}$, where the vector $\boldsymbol{X}$ is the sought solution, T is a square matrix, and $\boldsymbol{f}$ is a column vector. Let the matrix T have a full rank, and let the following iteration process converge: $\boldsymbol{X}^{n+1} = \mathrm{T}\boldsymbol{X}^n + \boldsymbol{f}$, $n = 0, 1, \ldots$, in which $\boldsymbol{X}^n$ is the approximation for the solution at the nth iteration. By the definition, $\boldsymbol{r}^n = \mathrm{T}\boldsymbol{X}^n + \boldsymbol{f} - \boldsymbol{X}^n = \boldsymbol{X}^{n+1} - \boldsymbol{X}^n$ is the residual of equations $\boldsymbol{X} = \mathrm{T}\boldsymbol{X} + \boldsymbol{f}$, and it is not difficult to obtain the following relation from the above formulas: $\boldsymbol{r}^{n+1} = \mathrm{T}\,\boldsymbol{r}^n$. Let us assume that $k+1$ iterations have been made starting from some initial guess $\boldsymbol{X}^0$, that is the quantities $\boldsymbol{X}^1$, $\boldsymbol{X}^2, \ldots, \boldsymbol{X}^{k+1}$ and $\boldsymbol{r}^0$, $\boldsymbol{r}^1, \ldots, \boldsymbol{r}^k$ have been computed. The value $\boldsymbol{X}^{k+1}$ is then refined by the formula $\boldsymbol{X}^{*k+1} = \boldsymbol{X}^{k+1} + \boldsymbol{Y}^{k+1}$. One employs the correction of the form

$$\boldsymbol{Y}^{k+1} = \sum_{i=1}^{k} \alpha_i\, \boldsymbol{r}^i \qquad (17)$$

with indefinite coefficients $\alpha_1, \ldots, \alpha_k$ that are found from the condition of the minimization of the residual functional $\Phi(\alpha_1, \ldots, \alpha_k) = \| \boldsymbol{X}^{*k+1} - \mathrm{T}\boldsymbol{X}^{*k+1} - \boldsymbol{f} \|_2^2$, which arises at the substitution of $\boldsymbol{X}^{*k+1}$ into the system $\boldsymbol{X} = \mathrm{T}\boldsymbol{X} + \boldsymbol{f}$. Here $||\boldsymbol{u}||_2$ is the Euclidean norm of the vector $\boldsymbol{u}$ of dimension N. The refined vector of the $k+1$th approximation $\boldsymbol{X}^{*k+1}$ is used as the initial approximation for further continuation of the sequence of iterations.

2.4 Convergence Acceleration by Using the Multigrid Algorithm

The main idea of multigrid is the selective damping of the error harmonics [5,27]. In the CLR method, as in other methods, the number of iterations necessary for reaching the given accuracy of the approximation to the solution depends on the initial guess. As a technique for obtaining a good initial guess for the iterations on the finest grid among the grids used in a multigrid complex we have applied the prolongation operations along the ascending branch of the V-cycle — the computations on a sequence of refining grids. The passage from a coarser grid to a finer grid is made with the aid of the prolongation operators. Let us illustrate the algorithm of the prolongation operation by the example of the velocity component $u_1(y_1, y_2, b_1, \ldots, b_{15})$. Let $h_1 = h$, where h is the half-step of the coarse grid, and let $h_2 = h_1/2$ be the half-step of the fine grid on which one must find the expansion of function u_1 over the basis.

Step 1. Let X_1 and X_2 be the global coordinates of the coarse grid cell center. We make the following substitutions into the polynomial expression for u_1:

$$y_l = (x_l - X_l)/h_1, \quad l = 1, 2. \qquad (18)$$

As a result, we obtain the polynomial

$$U_1(x_1, x_2, b_1, \dots, b_{15}) = u_1\left(\tfrac{x_1 - X_1}{h_1}, \tfrac{x_2 - X_2}{h_1}, b_1, \dots, b_{15}\right). \quad (19)$$

Step 2. Let $(\tilde{X}_1, \tilde{X}_2)$ be the global coordinates of the center of any of the four cells of the fine grid, which lie in the coarse grid cell. We make the substitution in (19) $x_l = \tilde{X}_l + \tilde{y}_l \cdot h_2$, $l = 1, 2$. As a result, we obtain the second-degree polynomial $\tilde{U}_1 = P(\tilde{y}_1, \tilde{y}_2, \tilde{b}_1, \dots, \tilde{b}_{15})$ in variables $\tilde{y}_1, \tilde{y}_2$ with coefficients $\tilde{b}_1, \dots, \tilde{b}_{15}$. After the collection of terms of similar structure it turns out that the coordinates X_1, X_2 and $\tilde{X}_1, \tilde{X}_2$ enter $\tilde{b}_l$ ($l = 1, \dots, 15$) only in the form of combinations $\delta x_l = (X_l - \tilde{X}_l)/h_1$. According to (18), the quantity $-\delta x_l = (\tilde{X}_l - X_l)/h_1$ is the local coordinate of the fine grid cell center in the coarse grid cell.

Let us present the expressions for coefficients $\tilde{b}_j$ $(j = 1, \dots, 15)$ of the solution representation in a fine grid cell with the half-step h_2 in terms of the coefficients $b_1, \dots, b_{15}$ of the solution representation in a cell with the half-step $h_1 = 2h_2$:

$$\begin{aligned}
&\tilde{b}_1 = b_1 - \delta x_1(b_2 - b_4\delta x_1) - \delta x_2(b_3 + 2b_5\delta x_1 - b_6\delta x_2),\ \tilde{b}_2 = \sigma_1(T_1 + b_5\delta x_2),\\
&\tilde{b}_3 = \sigma_1[b_3 + 2(b_5\delta x_1 - b_6\delta x_2)],\ \tilde{b}_4 = \sigma_2 b_4, \quad \tilde{b}_5 = \sigma_2 b_5,\ \tilde{b}_6 = \sigma_2 b_6,\\
&\tilde{b}_7 = b_7 - \delta x_1(b_8 - b_9\delta x_1) + \delta x_2 T_1,\ \tilde{b}_8 = \sigma_1(b_8 - 2b_9\delta x_1 + 2b_4\delta x_2),\\
&\tilde{b}_9 = \sigma_2 b_9,\ \tilde{b}_{10} = b_{10} - \delta x_1 T_2 - \delta x_2(b_{12} - b_{15}\delta x_2),\ \tilde{b}_{11} = \sigma_1(T_2 - b_{13}\delta x_1),\\
&\tilde{b}_{12} = \sigma_1(b_{12} - b_{14}\delta x_1 - 2b_{15}\delta x_2),\ \tilde{b}_{13} = \sigma_2 b_{13},\ \tilde{b}_{14} = \sigma_2 b_{14},\ \tilde{b}_{15} = \sigma_2 b_{15},
\end{aligned}$$

where $\sigma_1 = h_2/h_1$, $\sigma_2 = \sigma_1^2$, $T_1 = b_2 - 2b_4\delta x_1 + b_5\delta x_2$, $T_2 = b_{11} - b_{13}\delta x_1 - b_{14}\delta x_2$. The analytic expressions for coefficients $\tilde{b}_1, \dots, \tilde{b}_{15}$ were found efficiently with the aid of the *Mathematica* functions EXPAND[...], COEFFICIENT[...], SIMPLIFY[...]. To reduce the length of obtained coefficients we have applied a number of transformation rules as well as the *Mathematica* function FULLSIMPLIFY[...]. As a result, the length of the final expressions for $\tilde{b}_1, \dots, \tilde{b}_{15}$ proved to be three times shorter than the length of the original expressions. Note that the above expressions for $\tilde{b}_1, \dots, \tilde{b}_9$ coincide with the expressions which were presented in [24,26] for the case of $m_b = 12$ in (8).

3 The Use of the Integral Form of Collocation Equations

The "differential" version of the CLR method, in which the collocation equations (7) were obtained from the differential equations (5), was described in Sect. 2. By analogy with the LSFEM [11,12], one can use instead of collocation equations (7) their integral counterparts, which are obtained by integrating equations (5) over several subregions, and the consideration of collocation equations at several user-defined collocation points is replaced with the consideration of collocation relations, which account for the influence of the entire area of each cell Ω_{ij} of the spatial computational grid.

Furthermore, it was shown in [26] that the inclusion of the approximation (13) of the integral condition (3) in the overdetermined SLAE (14) instead of specifying the pressure at a single point speeds up considerably the iteration process convergence.

Let us at first introduce a uniform computational grid in each cell, which subdivides each cell face into $N_{\rm sub}$ intervals, where $N_{\rm sub} > 1$ is the user-specified number of cells along each local coordinate y_k, $k = 1, 2$. The lines of this grid subdivide the Ω_{ij} cell into $N_c = N_{\rm sub}^2$ subcells $\Omega_{ij}^{(l)}$, $l = 1, \ldots, N_c$. The $N_{\rm sub}$ value must be specified in such a way that the quantity $N_{\rm sub}^2$ be comparable with the number of unknown coefficients m_b in (8) or be higher than m_b. After that, the integration of Eq. (7) is carried out in each cell $\Omega_{ij}^{(l)}$:

$$\xi \iint_{\Omega_{ij}^{(l)}} [\Delta u_m^{s+1} - (\mathrm{Re} \cdot h)(u_1^s u_{m,y_1}^{s+1} + u_1^{s+1} u_{m,y_1}^s + u_2^s u_{m,y_2}^{s+1} + u_2^{s+1} u_{m,y_2}^s$$

$$+ q_{y_m}^{s+1})] \, dy_1 dy_2 = \xi \iint_{\Omega_{ij}^{(l)}} F_m dy_1 dy_2, \ m = 1, 2; \ l = 1, \ldots, N_c. \tag{20}$$

It is to be noted that the integration of the left-hand side of (20) can be performed in the analytic form because the integrand involves only the polynomial expressions according to Table 1, and the use of the *Mathematica* function INTEGRATE[...] proves to be efficient here.

It turns out that there are in the obtained $2 \cdot N_{\rm sub}^2$ collocation equations many common subexpressions. To reduce the CPU time needed for the numerical computation of the entries of the matrix A_{ij} in SLAE (14) it is reasonable to perform the common subexpression elimination (CSE) in (20). The basic idea is here to evaluate common subexpressions only once and put the results in temporary variables [6]. We have implemented an interactive CSE technique. As a result, 15 temporary variables have been introduced, and most of them are the functions of other temporary variables. This has resulted in a considerable (by factors from 2 to 5) reduction of the lengths of the expressions for the entries of the matrix A_{ij}.

The built-in *Mathematica* function REPLACEREPEATED(//.) repeatedly performs replacements until the expression no longer changes. But our practice shows that this function is not reliable, it performs not all replacements of common subexpressions with temporary variables. For illustration, let us consider the following very simple example: let us take the expression *expr* = `6*y1L - 6*y1R`. The transformation rule `y1L - y1R -> dy1` enables the replacement of the subexpression `y1L - y1R` with the temporary variable `dy1`. However, the application of the command *expr* = *expr*//. `{y1L - y1R -> dy1}` leaves the original expression unchanged. This situation can be rectified owing to the availability in CAS *Mathematica* of many other built-in functions performing elementary transformations, and their combination enables one, as a rule, to obtain the needed result. In the specific example considered above, the application of the built-in function FACTOR[...] has enabled us to obtain the needed result: *expr* = FACTOR[*expr*]`/.y1L - y1R -> dy1` yields the desired result *expr* = `6*dy1`.

At the symbolic implementation of the stages of deriving the work formulas of the versions of the CLR method, which are described in Sects. 2 and 3, we have used the following built-in *Mathematica* functions for symbolic computation and manipulation: AppendTo[...], Coefficient[...], D[...], Det[...], Expand[...], Factor[...], FortranForm[...], FullSimplify[...], InputForm[...], Integrate[...], Inverse[...], Length[...], Norm[...], ReplaceRepeated(//.), Sum[...], Table[...], ToExpression[...], ToString[...], Transpose[...].

It is to be noted that the analogs of the above functions are available also in such well-known CASs as Maple, REDUCE, and in a number of other general-purpose CASs. Thus, the researcher wishing to implement the above-described symbolic-numeric methods for solving the Navier–Stokes equations has a possibility to choose a specific CAS; this choice, in turn, depends on his personal experience in the matter of using one or other CAS. Although each of the CASs mentioned here has its individual advantages depending on a problem to be solved, the capabilities which *Mathematica* possesses were sufficient for doing with its aid the job presented in the given paper.

4 Results of Numerical Experiments

Consider the following exact solution of the Navier–Stokes equations (1) [4]:

$$u_1 = \frac{-2(1+x_1)}{(1+x_1)^2+(1+x_2)^2},\; u_2 = \frac{2(1+x_1)}{(1+x_1)^2+(1+x_2)^2},$$
$$p = -\frac{2}{(1+x_1)^2+(1+x_2)^2}, \quad 0 \le x_1, x_2 \le 1. \tag{21}$$

Note that the functions $u_1(x_1,x_2)$ and $u_2(x_1,x_2)$ describe the divergence-free velocity field. Furthermore,

$$\int_0^1\int_0^1 p\,dx_1dx_2 = 4G - \pi\ln 2 - 2i\left[\mathrm{Li}_2\left(-\frac{i}{2}\right) - \mathrm{Li}_2\left(\frac{i}{2}\right)\right]$$
$$\approx -0.462613146772815498 72,$$

where $i=\sqrt{-1}$, G is the Catalan's constant [28], $G \approx 0.9159655941772190150 5$, $\mathrm{Li}_2(z)$ is the polylogarithmic function. To ensure the satisfaction of Eq. (3) with an error not exceeding the error of machine computations, the pressure p in (3) was replaced with the quantity $\bar{p} = p + 0.4626131467728155$.

The root-mean-square solution errors were calculated as

$$Err(\mathbf{u}(h)) = \left[\frac{1}{2M^2}\sum_{i=1}^{M}\sum_{j=1}^{M}\sum_{\nu=1}^{2}(u_{\nu,i,j} - u^{ex}_{\nu,i,j})^2\right]^{\frac{1}{2}},$$
$$Err(p(h)) = \left[\frac{1}{M^2}\sum_{i=1}^{M}\sum_{j=1}^{M}(p_{i,j} - p^{ex}_{i,j})^2\right]^{\frac{1}{2}},$$

where M is the number of cells along each coordinate direction, $\mathbf{u}^{ex}_{i,j}$ and $p^{ex}_{i,j}$ are the velocity vector and the pressure according to the exact solution (21).

The quantities $\mathbf{u}_{i,j}$ and $p_{i,j}$ denote the numerical solution obtained by the CLR method described above. The convergence orders ν_u and ν_p are computed by the well-known formulas [21,23]. Let $b^s_{i,j,l}$, $s = 0, 1, \ldots$ be the values of the coefficients $b_{i,j,l}$ in (8) at the sth iteration. The following condition was used for termination of the iterations: $\delta b^{s+1} < \varepsilon$, where $\delta b^{s+1} = \max_{i,j}(\max_{1\le l\le m_b} |b^{s+1}_{i,j,l} - b^s_{i,j,l}|)$, and $\varepsilon < h^2$ is a small positive quantity. We will call the quantity δb^{s+1} the pseudo-error of the approximate solution.

Along with the criterion $\delta b^{s+1} < \varepsilon$, we have also applied the following criterion for the termination of iterations:

$$\delta \mathbf{u}^{n+1} = \| \mathbf{u}^{n+1} - \mathbf{u}^n \| < \varepsilon_2,$$

where $\| \cdot \|$ is the Euclidean norm of the vector, ε_2 is a user-specified small positive quantity.

In the work [26], we have introduced the definition of the overdetermination or underdetermination ratio of system (15) as the quantity $\chi(\mathrm{A}) = m_r/m_c$, where m_r and m_c are, respectively, the number of rows and the number of columns of the matrix A. We have investigated the influence of the quantity $\chi(\mathrm{A})$ on the condition number (16) when the "differential" CLR method is employed, and $m_b = 12$ in (8). This study was carried out by using the Dirichlet boundary conditions corresponding to the analytic solution (21).

For the sake of brevity, we omit the obtained tabular data and only enumerate the conclusions summarizing the above study.

1°. The convergence rate of the iteration process in the "differential" CLR method depends significantly on the condition numbers of the SLAEs to be solved in each cell.

2°. In cases where the matrix A includes only the rows corresponding to the collocation equations and one row corresponding to approximation (13) of the pressure integral, a very large condition number of the order of 10^5 is obtained depending on the number of collocation points, that is the matrix A is ill-conditioned. The inclusion in the matrix A of the rows corresponding to matching conditions results in a reduction of the condition number by three – five decimal orders depending on the number of grid cells and the number of collocation and matching points. It is this considerable reduction of the condition number which ensures the performance of the CLR method at the solution of boundary-value problems for linear and nonlinear partial differential equations.

3°. The condition number in boundary cells is always higher than in internal cells.

4°. At $N_c = 4$ and $N_{\mathrm{mat}} = 1$, the convergence of the CLR method slows down significantly in comparison with the case of $N_c = 8$ and $N_{\mathrm{mat}} = 2$.

5°. The data obtained present a practical proof of the fact that at the use of an overdetermined system in the approximate problem ($\chi(\mathrm{A}) > 1$), the corresponding SLAE of the approximate problem proves to be better conditioned than in the case when the SLAE is not overdetermined. This is an important advantage of the method of collocations and least squares (and the CLR method) over the collocation method, which does not use the overdetermined SLAE in the approximate problem.

To study the convergence and accuracy properties of the above-presented versions of the CLR method numerous numerical experiments were performed with the use of the analytic solution (21). The following names were used for different versions of the CLR method: CLRD_{12} is the "differential" CLR method of [24,26] employing twelve basis vectors, that is $m_b = 12$ in (8); CLRD_{15} is the "differential" CLR method with $m_b = 15$ in (8); and, finally, CLRI_{15} is the "integral" CLR method with $m_b = 15$ in (8).

Tables 2, 3, and 4 present the results of numerical experiments, in which only two of the above-described techniques for convergence acceleration were used: the two-parameter preconditioner and the Krylov subspace method. In these computations, it was assumed that the Reynolds number Re $= 1000$ and $L = 1$ in (4). These computations were done with $N_c = 8$ in the case of methods CLRD_{12} and CLRD_{15} and with $N_c = N_{\mathrm{sub}}^2 = 16$ in the case of method CLRI_{15}. The satisfaction of the inequality $\delta b^n < 10^{-9}$ was the criterion for termination of the computations by the versions of the CLR method. Comparing Table 2 with Tables 3 and 4 one can see that the 25% increase in the number of basis vector functions (from 12 to 15) reduces the error in the pressure and velocity in the numerical solution by the factors ranging from two to three orders of magnitude.

Figure 1 shows the solution obtained by the method CLRD_{15} by symbols $\triangle$ (v_1), $\circ$ (v_2), and ∇ (p); the curves of the exact solution are depicted by the

Table 2. The errors $Err(\mathbf{u}^n)$, $Err(p^n)$ and their convergence orders ν_u and ν_p on a sequence of grids, Re $= 1000$, $L = 1$, $N_c = 8$. Method CLRD_{12}

M	$Err(\mathbf{u}^n)$	$Err(p^n)$	ν_u	ν_p
10	1.309e−02	9.754e−03		
20	5.484e−03	3.664e−03	1.26	1.41
40	1.869e−03	1.135e−03	1.55	1.69
80	4.938e−04	2.870e−04	1.92	1.98

Table 3. The errors $Err(\mathbf{u}^n)$, $Err(p^n)$ and their convergence orders ν_u and ν_p on a sequence of grids, Re $= 1000$, $L = 1$, $N_c = 8$. Method CLRD_{15}

M	$Err(\mathbf{u}^n)$	$Err(p^n)$	ν_u	ν_p
10	2.947e−05	4.731e−05		
20	6.971e−06	2.168e−05	2.08	1.13
40	2.289e−06	1.046e−05	1.61	1.05
80	9.816e−07	4.950e−06	1.21	1.09

Table 4. The errors $Err(\mathbf{u}^n)$, $Err(p^n)$ and their convergence orders ν_u and ν_p on a sequence of grids, Re $= 1000$, $L = 1$, $N_c = 8$. Method CLRI_{15}

M	$Err(\mathbf{u}^n)$	$Err(p^n)$	ν_u	ν_p
10	2.910e−05	5.437e−05		
20	6.807e−06	2.249e−05	2.10	1.27
40	2.325e−06	1.076e−05	1.55	1.06
80	9.877e−07	4.893e−06	1.24	1.14

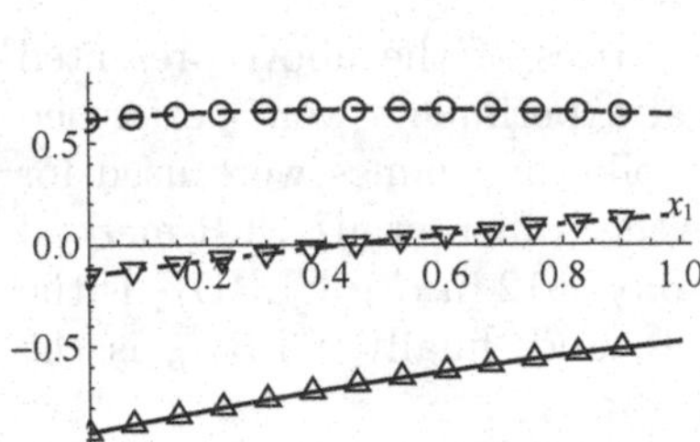

Fig. 1. Comparison of the approximate and exact solution profiles at $x_2 = L/2$. The 40 × 40 grid

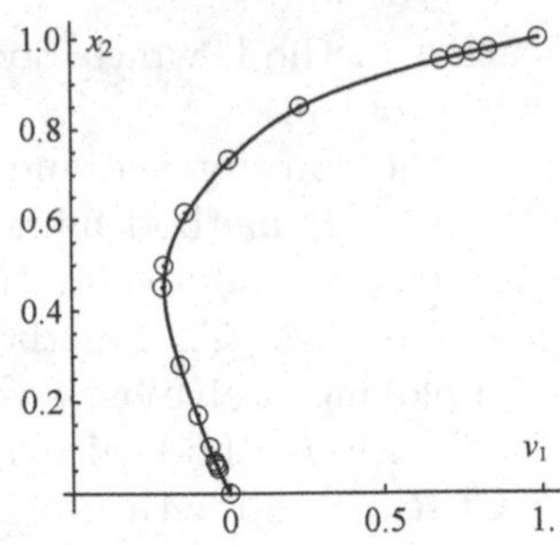

Fig. 2. Solution of the benchmark problem by the method CLRI_{15} at Re = 100 on the 40 × 40 grid: the profile of the velocity component v_1 along the centerline $x_1 = 0.5$ (solid line); (○ ○ ○) Ghia et al. [9]

solid, dashed, and dash-dot lines for v_1, v_2, and p, respectively. One can see here a good agreement between the numerical results and the analytic solution.

In the 2D driven cavity problem, the computational region is the cavity, which is a square (4) with side $L = 1$, the coordinate origin lies in its left lower corner. The upper lid of the cavity moves with unit velocity in dimensionless variables in the positive direction of the Ox_1 axis. The other sides of cavity (4) are at rest. The no-slip conditions are specified on all sides: $v_1 = 1$, $v_2 = 0$ at $x_2 = L$ and $v_m = 0$, $m = 1, 2$ on the remaining sides.

The lid-driven cavity flow has the singularities in the region upper corners. Their influence on the numerical solution accuracy enhances with increasing Reynolds number. Therefore, at high Reynolds numbers, it is necessary to apply adaptive grids for obtaining a more accurate solution: the grids with finer cells in the neighborhood of singularities. Only the uniform grids were applied here.

Table 5. The error δu_1 obtained at the use of different versions of the CLR method, Re = 100

Method	ξ	η	$\kappa(\xi,\eta)$	K_{mgr}	k	N_{it}	CPU time, s	δu_1
CLRD_{12}	2.0	3.5	8.536	4	8	2336	53.59	1.726e−02
CLRD_{15}	2.0	3.5	10.590	4	8	2540	99.59	1.150e−02
CLRI_{15}	0.25	1.75	6.774	4	9	1700	110.38	8.521e−03

We have compared the accuracy of different versions of the CLR method, which were presented in the foregoing sections, in the case when the Reynolds number Re = 100. For our comparisons, we have used the numerical results from [9]. Let us introduce the error $\delta u_1 = \max_j |u_{1,\mathrm{Ghia}}(0.5, x_{2j}) - u_{1,\mathrm{CLR}}(0.5, x_{2j})|$, where the coordinates x_{2j} were taken from [9]. In all computations whose results are presented in Table 5, we have used the combination of all three acceleration techniques, which were presented briefly above in Subsects. 2.2, 2.3, and 2.4. The grids, which were used in the multigrid complex, were as follows: 5×5, 10×10, 20×20, and 40×40 grids; K_{mgr} is the number of sequentially used grids in the multigrid complex. N_{it} is the number of iterations, which are necessary to satisfy the inequality $\delta b^n < 10^{-9}$. The value k is the number of residuals used in the Krylov's method (see Eq. (17)). The optimal values of the parameters ξ and η entering the two-parameter preconditioner were found using the algorithm of [24]. The condition number $\kappa(\xi, \eta)$ was computed for all three methods by formula (16) in the cell with indices (20, 20) on the 40×40 grid after the execution of 200 iterations on this grid. One can see from Table 5 that the method CLRI_{15} ensures the best accuracy among three considered versions of the CLR method (see also Fig. 2). In addition, it ensures the least condition number κ. This is the important property of the CLRI_{15} method, which extends the capabilities of the CLR method at the solution of ill-conditioned problems.

5 Conclusions

New versions of the CLR method have been presented. A large amount of symbolic computations, which arose at the derivation of the basic formulae of the new versions of the method, was done efficiently with *Mathematica*. It is very important that the application of CAS has facilitated greatly this work, reduced at all its stages the probability of errors usually introduced by the mathematician at the development of a new algorithm and also reduced the time needed for the development of new Fortran programs implementing the numerical stages of the CLR method. It is shown by examples of numerous computations that the new proposed versions of the CLR method produce more accurate numerical solutions than the previous "differential" version CLRD_{12} of the CLR method.

References

1. Amodio, P., Blinkov, Y., Gerdt, V., La Scala, R.: On consistency of finite difference approximations to the Navier-Stokes equations. In: Gerdt, V.P., Koepf, W., Mayr, E.W., Vorozhtsov, E.V. (eds.) CASC 2013. LNCS, vol. 8136, pp. 46–60. Springer, Cham (2013). doi:10.1007/978-3-319-02297-0_4
2. Bailly, O., Buchou, C., Floch, A., Sainsaulieu, L.: Simulation of the intake and compression strokes of a motored 4-valve SI engine with a finite element code. Oil Gas Sci. Technol. **54**, 161–168 (1999)
3. Botella, O., Peyret, R.: Benchmark spectral results on the lid-driven cavity flow. Comput. Fluids **27**, 421–433 (1998)
4. Chiu, P.H., Sheu, T.W.H., Lin, R.K.: An effective explicit pressure gradient scheme implemented in the two-level non-staggered grids for incompressible Navier-Stokes equations. J. Comput. Phys. **227**, 4018–4037 (2008)
5. Fedorenko, R.P.: The speed of convergence of one iterative process. USSR Comput. Math. Math. Phys. **4**(3), 227–235 (1964)

6. Fritzson, P., Engelson, V., Sheshadri, K.: MathCode: a system for C++ or Fortran code generation from Mathematica. Math. J. **10**, 740–777 (2008)
7. Ganzha, V.G., Mazurik, S.I., Shapeev, V.P.: Symbolic manipulations on a computer and their application to generation and investigation of difference schemes. In: Caviness, B.F. (ed.) EUROCAL 1985. LNCS, vol. 204, pp. 335–347. Springer, Heidelberg (1985). doi:10.1007/3-540-15984-3_290
8. Gerdt, V.P., Blinkov, Y.A.: Involution and difference schemes for the Navier–Stokes Equations. In: Gerdt, V.P., Mayr, E.W., Vorozhtsov, E.V. (eds.) CASC 2009. LNCS, vol. 5743, pp. 94–105. Springer, Heidelberg (2009). doi:10.1007/978-3-642-04103-7_10
9. Ghia, U., Ghia, K.N., Shin, C.T.: High-Re solutions for incompressible flow using the Navier-Stokes equations and a multigrid method. J. Comput. Phys. **48**, 387–411 (1982)
10. Isaev, V.I., Shapeev, V.P.: High-accuracy versions of the collocations and least squares method for the numerical solution of the Navier-Stokes equations. Comput. Math. Math. Phys. **50**, 1670–1681 (2010)
11. Jiang, B., Lin, T.L., Povinelli, L.A.: Large-scale computation of incompressible viscous flow by least-squares finite element method. Comput. Meth. Appl. Mech. Eng. **114**(3–4), 213–231 (1994)
12. Jiang, B.N.: The Least-Squares Finite Element Method: Theory and Applications in Computational Fluid Dynamics and Electromagnetics. Springer, Berlin (1998). doi:10.1007/978-3-662-03740-9
13. Krylov, A.N.: On the numerical solution of the equation, which determines in technological questions the frequencies of small oscillations of material systems. Izv. AN SSSR, Otd. matem. i estestv. nauk **4**, 491–539 (1931). (in Russian)
14. Li, K., Li, Q.: Three-dimensional gravity-jitter induced melt flow and solidification in magnetic fields. J. Thermophys. Heat Transf. **17**(4), 498–508 (2003)
15. Plyasunova, A.V., Sleptsov, A.G.: Collocation-grid method of solving the nonlinear parabolic equations on moving grids. Modelirovanie v mekhanike **18**(4), 116–137 (1987)
16. Semin, L., Shapeev, V.: Constructing the numerical method for Navier — Stokes equations using computer algebra system. In: Ganzha, V.G., Mayr, E.W., Vorozhtsov, E.V. (eds.) CASC 2005. LNCS, vol. 3718, pp. 367–378. Springer, Heidelberg (2005). doi:10.1007/11555964_31
17. Semin, L.G., Sleptsov, A.G., Shapeev, V.P.: Collocation and least -squares method for Stokes equations. Comput. Technol. **1**(2), 90–98 (1996). (in Russian)
18. Shapeev, A.V., Lin, P.: An asymptotic fitting finite element method with exponential mesh refinement for accurate computation of corner eddies in viscous flows. SIAM J. Sci. Comput. **31**, 1874–1900 (2009)
19. Shapeev, V.: Collocation and least residuals method and its applications. EPJ Web Conf. **108**, 01009 (2016). doi:10.1051/epjconf/201610801009
20. Shapeev, V.P., Isaev, V.I., Idimeshev, S.V.: The collocations and least squares method: application to numerical solution of the Navier-Stokes equations. In: Eberhardsteiner, J., Böhm, H.J., Rammerstorfer, F.G. (eds.) CD-ROM Proceedings of the 6th ECCOMAS, September 2012. Vienna University of Technology (2012). ISBN: 978-3-9502481-9-7
21. Shapeev, V.P., Vorozhtsov, E.V.: CAS application to the construction of the collocations and least residuals method for the solution of 3D Navier–Stokes equations. In: Gerdt, V.P., Koepf, W., Mayr, E.W., Vorozhtsov, E.V. (eds.) CASC 2013. LNCS, vol. 8136, pp. 381–392. Springer, Cham (2013). doi:10.1007/978-3-319-02297-0_31

22. Shapeev, V.P., Vorozhtsov, E.V., Isaev, V.I., Idimeshev, S.V.: The method of collocations and least residuals for three-dimensional Navier-Stokes equations. Vychislit. metody i programmirovanie **14**, 306–322 (2013). (in Russian)
23. Shapeev, V.P., Vorozhtsov, E.V.: Symbolic-numeric implementation of the method of collocations and least squares for 3D Navier–Stokes equations. In: Gerdt, V.P., Koepf, W., Mayr, E.W., Vorozhtsov, E.V. (eds.) CASC 2012. LNCS, vol. 7442, pp. 321–333. Springer, Heidelberg (2012). doi:10.1007/978-3-642-32973-9_27
24. Shapeev, V.P., Vorozhtsov, E.V.: Symbolic-numerical optimization and realization of the method of collocations and least residuals for solving the Navier–Stokes equations. In: Gerdt, V.P., Koepf, W., Seiler, W.M., Vorozhtsov, E.V. (eds.) CASC 2016. LNCS, vol. 9890, pp. 473–488. Springer, Cham (2016). doi:10.1007/978-3-319-45641-6_30
25. Valiullin, A.N., Ganzha, V.G., Meleshko, S.V., Murzin, F.A., Shapeev, V.P., Yanenko, N.N.: Application of Symbolic Manipulations on a Computer for Generation and Analysis of Difference Schemes. Prepr. Inst. Theor. Appl. Mech. Siberian Branch of the USSR Acad. Sci., Novosibirsk No. 7 (1981). (in Russian)
26. Vorozhtsov, E.V., Shapeev, V.P.: On combining the techniques for convergence acceleration of iteration processes during the numerical solution of Navier-Stokes equations. Vychislit. metody i programmirovanie **18**, 80–102 (2017). (in Russian)
27. Wesseling, P.: An Introduction to Multigrid Methods. Wiley, Chichester (1992)
28. Wolfram, S.: The Mathematica Book, 5th edn. Wolfram Media Inc., Champaign (2003)

A Special Homotopy Continuation Method for a Class of Polynomial Systems

Yu Wang[1], Wenyuan Wu[2(✉)], and Bican Xia[1]

[1] LMAM & School of Mathematical Sciences, Peking University, Beijing, China
yuxiaowang@pku.edu.cn, xbc@math.pku.edu.cn

[2] Chongqing Institute of Green and Intelligent Technology Chinese Academy of Sciences, Chongqing, China
wuwenyuan@cigit.ac.cn

Abstract. A special homotopy continuation method, as a combination of the polyhedral homotopy and the linear product homotopy, is proposed for computing all the isolated solutions to a special class of polynomial systems. The root number bound of this method is between the total degree bound and the mixed volume bound and can be easily computed. The new algorithm has been implemented as a program called LPH using C++. Our experiments show its efficiency compared to the polyhedral or other homotopies on such systems. As an application, the algorithm can be used to find witness points on each connected component of a real variety.

1 Introduction

In many applications in science, engineering, and economics, solving systems of polynomial equations has been a subject of great importance. The homotopy continuation method was developed in 1970s [1,2] and has been greatly expanded and developed by many researchers (see for example [3–7]). Nowadays, homotopy continuation method has become one of the most reliable and efficient classes of numerical methods for finding the isolated solutions to a polynomial system and the so-called *numerical algebraic geometry* based on homotopy continuation method has been a blossoming area. There are many famous software packages implementing different homotopy methods, including Bertini et al. [8], Hom4PS-2.0 [9], HOMPACK [10], PHCpack [11], etc.

Classical homotopy methods compute solutions in complex spaces, while in applications, it is quite common that only real solutions have physical meaning. Computing real roots of an algebraic system is a difficult and fundamental problem in real algebraic geometry. In the field of symbolic computation, there are some famous algorithms dealing with this problem. The cylindrical algebraic decomposition algorithm [12] is the first complete algorithm which has been implemented and used successfully to solve many real problems. However, in

The work is partly supported by the projects NSFC Grants 11471307, 11290141, 11271034, 61532019 and CAS Grant QYZDB-SSW-SYS026.

V.P. Gerdt et al. (Eds.): CASC 2017, LNCS 10490, pp. 362–376, 2017.
DOI: 10.1007/978-3-319-66320-3_26

the worst case, its complexity is doubly exponential in the number of variables. Based on the ideas of Seidenberg [13] and others, some algorithms for computing at least one point on each connected component of an real algebraic set were proposed through developing the formulation of critical points and the notion of polar varieties, see [14–17] and references therein. The idea behind is studying an objective function (or map) that reaches at least one local extremum on each connected component of a real algebraic set. For example, the function of square of the Euclidean distance to a randomly chosen point was used in [18,19]. On the other hand, some homotopy based algorithms for real solving have been proposed in [20–25]. For example, in [24], a numerical homotopy method to find the extremum of Euclidean distance to a point as the objective function was presented. More recently, the Euclidean distance to a plane was proposed as a linear objective function in [26].

Such critical point/plane approaches introduced above lead to a special class of polynomial systems. The main contribution of this paper is to give a special homotopy method for solving the system of that type efficiently by combining the polyhedral homotopy and the linear product homotopy. The root number bound of this method is not only easy to compute but also much smaller than the total degree bound and close to the BKK bound [27] when the polynomials defining the algebraic set are not very sparse. This key observation enables us to design an efficient homotopy procedure to obtain critical points numerically. The ideas and algorithms we proposed in this article avoid a great number of divergent paths to track compared with the total degree homotopy and save the great time cost for mixed volume computation compared with the polyhedral homotopy. The new algorithm has been implemented as a program called `LPH` using C++. Our experiments show its efficiency compared to the polyhedral or other homotopies on such systems.

The rest of this paper is organized as follows. Section 2 describes some preliminary concepts and results. Section 3 introduces a special type of polynomial systems we are considering. The new homotopy for these polynomial systems is also presented. It naturally leads to an algorithm which is described in Sect. 4. Based on this algorithm, in Sect. 5, we present a method to find real witness points of positive dimensional varieties, together with an illustrative example. The experimental performance of the software package `LPH`, which is an implementation of the method in C++, is given in Sect. 6.

2 Preliminary

2.1 Algebraic Sets and Genericity

For a polynomial system $f : \mathbb{C}^n \to \mathbb{C}^k$, let $V(f) = \{x \in \mathbb{C}^n | f(x) = 0\}$ and $V_{\mathbb{R}}(f) = V(f) \cap \mathbb{R}^n = \{x \in \mathbb{R}^n | f(x) = 0\}$ be the set of complex solutions and the set of real solutions of $f(x) = 0$, respectively. A set $X \subset \mathbb{C}^n$ is called an algebraic set if $X = V(g)$, for some polynomial system g.

An algebraic set X is irreducible if there does not exist a decomposition $X_1 \cup X_2 = X$ with $X_1, X_2 \neq X$ of X as a union of two strict algebraic subsets.

An algebraic set is reducible, if there exist such a decomposition. For example, the algebraic set $V(xy) \subset \mathbb{C}^2$ is consisting of the two coordinate axes, and is obviously the union of $V(x)$ and $V(y)$, hence reducible.

For an irreducible algebraic set X, the subset of smooth (or manifold) points X_{reg} is dense, open and path connected (up to the Zariski topology) in X. The dimension of an irreducible algebraic set X is the dimension of X_{reg} as a complex manifold.

Let $\mathcal{J}_f(x)$ denote the $n \times k$ Jacobian matrix of f evaluated at x. By the Implicit Function Theorem, for an irreducible algebraic set X defined by a reduced system f, $x \in X_{reg} \Leftrightarrow \text{rank}(\mathcal{J}_f(x)) = n - \dim X$. When $n = k$, the system f is said to be a square system. In this case, a point $x \in V(f)$ is nonsingular if $\det(\mathcal{J}_f(x)) \neq 0$, and singular otherwise.

On irreducible algebraic sets, we can define the notion of genericity, adapted from [6].

Definition 1. *Let X be an irreducible algebraic set. Property* P *holds generically on X, if the set of points in X that do not satisfy property* P *are contained in a proper algebraic subset Y of X. The points in Y are called* nongeneric points, *and their complements $X \backslash Y$ are called* generic points.

Remark 1. From the definition, one sees that the notion of generic is only meaningful in the context of property P in question.

Every algebraic set X has a (uniquely up to reordering) expression $X = X_1 \cup \ldots \cup X_r$ with X_i irreducible and $X_i \not\subset X_j$ for $i \neq j$. And X_i are the irreducible components of X. The dimension of an algebraic set is defined to be the maximum dimension of its irreducible components. An algebraic set is said to be pure-dimensional if each of its components has the same dimension.

2.2 Trackable Paths

In homotopy continuation methods, the notion of path tracking is fundamental, the following definition of trackable solution path is adapted from [28].

Definition 2. *Let $H(x,t) : \mathbb{C}^n \times \mathbb{C} \to \mathbb{C}^n$ be polynomial in x and complex analytic in t, and let x^* be nonsingular isolated solution of $H(x,0) = 0$, we say x^* is* trackable *for $t \in [0,1)$ from 0 to 1 using $H(x,t)$ if there is a smooth map $\xi_{x^*} : [0,1) \to \mathbb{C}^n$ such that $\xi_{x^*}(0) = x^*$, and for $t \in [0,1)$, $\xi_{x^*}(t)$ is a nonsingular isolated solution of $H(x,t) = 0$. The solution path started at x^* is said to be* convergent *if $\lim_{t \to 1} \xi_{x^*}(t) \in \mathbb{C}^n$, and the limit is called the* endpoint *of the path.*

2.3 Witness Set and Degree of an Algebraic Set

Let $X \subset \mathbb{C}^n$ be a pure i-dimensional algebraic set, given a generic co-dimension i affine linear subspace $L \subset \mathbb{C}^n$, then $W = L \cap X$ consists of a well-defined number d of points lying in X_{reg}. The number d is called the degree of X and denoted by $\deg(X)$. We refer to W as a set of witness points of X, and call L the associated $(n-i)$-slicing plane, or slicing plane for short [6].

It will be convenient to use the notations adapted from ([6], Chap. 8), when we prove the theorems in Sect. 3.

1. Let $\langle e_1, \ldots, e_n \rangle$ be the n dimensional vector space having basis elements $e_1, \ldots, e_n$ with complex coefficients. That is, a point in this space may be written as $\sum_{i=1}^{n} c_i e_i$, with $c_i \in \mathbb{C}$ for $i = 1, \ldots, n$. Note that we have not specified anything about the basis elements, it could be individual variables, monomials, or polynomials.
2. Let $\{p_1, \ldots, p_n\} \otimes \{q_1, \ldots, q_m\}$ be the product of two sets, that is, the set $\{p_i \cdot q_j | i = 1, \ldots, n; j = 1, \ldots, m\}$. In Sect. 3 we take this product as the image inside the ring of polynomials; that is, $x \otimes y = xy$ is just the product of two polynomials.
3. Define $P \times Q = \{pq | p \in P, q \in Q\}$. Accordingly, we have $\langle P \rangle \times \langle Q \rangle \subset \langle P \otimes Q \rangle$.
4. For repeated products, we use the shorthand notations $P^{(2)} = P \otimes P, \langle P \rangle^{(2)} = \langle P \rangle \times \langle P \rangle$, and similar for three or more products.
5. For a square polynomial system P, we denote by $MV(P)$ the mixed volume of the system P.

2.4 Critical Points

Let $X \subset \mathbb{C}^n$ be an algebraic set defined by a reduced polynomial system $f = \{f_1, \ldots, f_k\}$, and objective function Φ is polynomial function restricted to X.

Definition 3. *A point $x \in X$ is a* critical point *of Φ if and only if $x \in X_{reg}$ and* $\operatorname{rank}[\nabla\Phi(x)^T, \mathcal{J}_f(x)] = \operatorname{rank}[\nabla\Phi(x)^T, \nabla f_1^T, \ldots, \nabla f_k^T] \leqslant k$, *where $\nabla\Phi(x)$ is the gradient vector of Φ evaluated at x.*

Let Y denote the zero dimensional critical sets of Φ. One way to compute the critical points is to introduce auxiliary unknowns and consider a zero dimensional variety $\hat{Y}$ and then project $\hat{Y}$ onto Y. We use Lagrange Multipliers to define a squared system as follows

$$F(x, \lambda) := \begin{bmatrix} f \\ \lambda_0 \nabla\Phi(x)^T + \lambda_1 \nabla f_1^T + \ldots + \lambda_k \nabla f_k^T \end{bmatrix} \tag{1}$$

Note that if $x^* \in X$ is a critical point of Φ, then there exist $\lambda^* \in \mathbb{P}^k$, such that $F(x^*, \lambda^*) = 0$ by the Fritz John condition [29]. In the affine patch where $\lambda_0 = 1$, the system F becomes a square system, and its solution (x^*, λ^*) projects to the critical point x^*. We will use system (1) in Sect. 5 with an objective function Φ defined by a linear function, and consider the affine patch where $\lambda_0 = 1$, to find at least one point on each component of $V_{\mathbb{R}}(f)$.

3 Main Idea

In this section, we give a description of our idea. First we introduce a family of polynomial equations that we will be considering. In fact, such type of polynomial systems appears naturally when using the Method of Lagrange Multipliers in mathematical optimization, if the constraints are algebraic equations.

Thus, we consider the following class of polynomial systems:

$$F(x,\lambda) = \begin{cases} f \\ J \cdot \lambda - \beta \end{cases} \tag{2}$$

where

1. $f = \{f_1, \ldots, f_k\}$ are polynomials in $\mathbb{C}[x_1, \ldots, x_n]$, and $V(f_1, \ldots, f_k)$ is a pure $n-k$ dimension algebraic set in $\mathbb{C}^n$.
2. $J = \begin{pmatrix} g_{11} & \cdots & g_{1k} \\ \vdots & \ddots & \vdots \\ g_{n1} & \cdots & g_{nk} \end{pmatrix}$ and $g_{ij} (1 \leqslant i \leqslant n, 1 \leqslant j \leqslant k)$ are polynomials in $\mathbb{C}[x_1, \ldots, x_n]$ with $\max_{i,j} \deg(g_{ij}) = d$.
3. $\beta = (\beta_1, \ldots, \beta_n)^{\mathrm{T}}$ is a nonzero constant vector in $\mathbb{C}^n$, $\lambda = (\lambda_1, \ldots, \lambda_k)^{\mathrm{T}}$ are unknowns, and $n > k \geqslant 1$.

Remark 2. Note that, for any invertible $n \times n$ matrix A, $F(x,\lambda) = \{f, J \cdot \lambda - \beta\}$ and $F'(x,\lambda) = \{f, A \cdot (J \cdot \lambda - \beta)\}$ have the same solutions. It's easy to know that there exists an invertible matrix A such that $A \cdot \beta = (0, \ldots, 0, 1)^{\mathrm{T}}$. So without loss of generality, we may assume that $\beta = (0, \ldots, 0, 1)^{\mathrm{T}}$. Then, $J \cdot \lambda - \beta$ has $n-1$ equations in $\left\langle \{x_1, \ldots, x_n, 1\}^d \otimes \{\lambda_1, \ldots, \lambda_k\} \right\rangle$ and one equation in $\left\langle \{x_1, \ldots, x_n, 1\}^d \otimes \{\lambda_1, \ldots, \lambda_k, 1\} \right\rangle$.

Theorem 1. *Let $F(x,\lambda) = \{f, J \cdot \lambda - \beta\}$ be given as in (2), $\beta = (0, \ldots, 0, 1)^{\mathrm{T}}$, and $G = \{f, g\}$ where $g = \{g_1, \ldots, g_n\}$. For each $i = 1, \ldots, n-1$, $g_i = l_{i1} \cdots l_{id} h_i \in \langle x_1, \ldots, x_n, 1 \rangle^d \times \langle \lambda_1, \ldots, \lambda_k \rangle$, where $l_{i1}, \ldots, l_{id}$ are linear functions in $\mathbb{C}[x_1, \ldots, x_n]$ with randomly chosen coefficients, and h_i is a homogeneous linear function in $\mathbb{C}[\lambda_1, \ldots, \lambda_k]$ with randomly chosen coefficients. And $g_n = \sum_{i=1}^{k} \lambda_i g_{ni} - 1$, where g_{ni} is the $(n,i)^{th}$ entry of J.*

Let $H : \mathbb{C}^n \times \mathbb{C}^k \times \mathbb{C} \to \mathbb{C}^{n+k}$ be the homotopy defined by $H(x,\lambda,t) = G \cdot (1-t) + F \cdot \gamma \cdot t$ where γ is a randomly chosen complex number for Gamma Trick (see [6], Chap. 7 for details). Then, generically the following items hold,

1. *The set $S \subseteq \mathbb{C}^{n+k}$ of roots of $H(x,\lambda,0) = G(x,\lambda)$ is finite and each is a nonsingular solution of $H(x,\lambda,0)$.*
2. *The number of points in S is equal to the maximum number of isolated solutions of $H(x,\lambda,0)$ as coefficients of l_{ij}, h_i, $(i = 1, \ldots, n-1,\ j = 1, \ldots, d)$ and γ vary over $\mathbb{C}$.*

3. *The solution paths defined by H starting, with $t = 0$, at the points in S are trackable.*

Proof. As for item 1, since f has k equations only in x, and $V(f_1, \dots, f_k)$ is a pure $n-k$ dimension algebraic set in $\mathbb{C}^n$. To solve system G, it needs only $n-k$ linear functions L in g from different g_i with $i \in \{1, \dots, n-1\}$ to determine x. $\{f, L\}$ is a $n \times n$ square system, $V(f, L)$ is a finite witness set for the algebraic set $V(f_1, \dots, f_k)$, and each of the points is a nonsingular solution of $V(f, L)$ (see [6], Chap. 13 for details). And, we finally determine λ by solving a square system of linear equations. As for item 2, and item 3, it's a trivial deduction of Coefficient-Parameter Continuation [30]. □

Remark 3. From the proof of Theorem 1, the number of points of the finite set $V(f, L)$ is the degree of $V(f)$, and is independent of the choice of L. Thus, based on the number of different choices of L, and item 2, we can give a root count bound of system $F(x, \lambda)$ as in the following theorem, which is similar to the bound in [31].

Theorem 2. *For a system $F(x, \lambda) = \{f, J \cdot \lambda - \beta\}$ as in (2), the number of complex roots is bounded by*

$$\binom{n-1}{n-k} d^{n-k} D \tag{3}$$

where D is the degree of $V(f)$.

Due to Theorem 1, its proof and the remarks, we can design an efficient procedure to numerically find the isolated solutions of system $F(x, \lambda) = \{f, J \cdot \lambda - \beta\}$ in the form of (2). First, we solve a square system $\{f, L\}$, where L are $n-k$ randomly generated linear functions. Then for each group of $n-k$ linear functions L' chosen in g from different g_i with $i \in \{1, \dots, n-1\}$, we construct a linear homotopy from $\{f, L\}$ to $\{f, L'\}$, starting from points of $V(f, L)$, and solve the square linear equation of λ respectively. Let S be the set that consists of all the pairs of x and λ, *i.e.* (x, λ). Finally construct a linear homotopy $H(x, \lambda, t) = G \cdot (1-t) + F \cdot \gamma \cdot t$ starting from points in S, thus the endpoints of the convergent paths of homotopy $H(x, \lambda, t)$ are isolated solutions of system $F(x, \lambda) = \{f, J \cdot \lambda - \beta\}$. We put a specific description of this procedure in the next section.

4 Algorithm

From Theorem 1, its proof and Remarks 2 and 3, we propose an approach for computing isolated solutions of system $F(x, \lambda)$ as described in the end of last section. For consideration of the sparsity, we use the polyhedral homotopy method for solutions of the square system $\{f, L\}$. Actually, we use polyhedral homotopy method only once. Now we describe our algorithms.

Algorithm 1. LPH (Linear Product Homotopy)

input : $(n+k) \times (n+k)$ square polynomial system $F(x, \lambda) = \{f, J \cdot \lambda - \beta\}$ as in (2);

output: finite subset $V(F)$ of $\mathbb{C}^{n+k}$

1 Let $L = \{l_1, \ldots, l_{n-k}\}$ where l_i are linear equations with randomly chosen coefficients in $\mathbb{C}$;

2 Solve system $\{f, l\}$ by polyhedral homotopy method and denote the solution set as M;

3 Let $F'(x, \lambda) = \{f, A \cdot (J \cdot \lambda - \beta)\}$, $G = \{f, g\}$, $A \in GL_n(\mathbb{C})$ such that $A \cdot \beta = (0, \ldots, 0, 1)$, $g = \{g_1, \ldots, g_n\}$. $g_i = l_{i1} \cdots l_{id} \cdot h_i \in \langle x_1, \ldots, x_n, 1\rangle^d \times \langle \lambda_1, \ldots, \lambda_k\rangle$ for $i = 1, \ldots, n-1$ with coefficients randomly chosen in $\mathbb{C}$, and g_n is the last equation of $A \cdot (J \cdot \lambda - \beta)$;

4 Let $C = \left\{ I \,\middle|\, I = (\alpha_1, \ldots, \alpha_{n-1}) \in \{0,1\}^{n-1}, \sum_{i=1}^{n-1} \alpha_i = n-k \right\}$, and $\Omega = \emptyset$;

5 **repeat**

6   Pick one vector $I = (\alpha_1, \ldots, \alpha_{n-1})$ from C, and $C = C \backslash I$;

7   Let $L' = \emptyset$;

8   **for** *i from* 1 *to* $n-1$ **do**

9   **if** $\alpha_i = 1$ **then**

10   pick one linear equation l_i' from $\{l_{i1}, \ldots, l_{id}\}$ and $L' = L' \cup \{l_i'\}$.

11   **end**

12   **end**

13   Construct linear homotopy $H_1(x, t) = \{f, L\} \cdot (1-t) + \{f, L'\} \cdot \gamma_1 \cdot t$ starting at points in M. γ_1 is a randomly chosen complex number for gamma trick. Let the set of endpoints of the tracked paths be M';

14   Take every point $x^* = (x_1^*, \ldots, x_n^*)$ in M' into the system $G = \{f, g\}$ and resolve $\lambda^* = (\lambda_1^*, \ldots, \lambda_k^*)$. $\Omega = \Omega \cup \{(x^*, \lambda^*)\}$. ;

15 **until** $C = \emptyset$;

16 Construct linear homotopy $H_2(x, \lambda, t) = G \cdot (1-t) + F \cdot \gamma_2 \cdot t$ starting at points in Ω, γ_2 is a randomly chosen complex number for gamma trick. Let the set of convergent endpoints of the tracked paths be $V(F)$;

17 **return** $V(F)$;

Remark 4. $\#C = \binom{n-1}{n-k}$, and in Step 5, $I = (\alpha_1, \ldots, \alpha_{n-1})$ has exactly $n-k$ entries $\alpha_i = 1$. When $\alpha_i = 1$, we choose linear equation in g_i, and there are d candidates $\{l_{i1}, \ldots, l_{id}\}$ to choose. It adds up to be $\binom{n-1}{n-k} d^{n-k}$ different $\{f, L'\}$. Each $\{f, L'\}$ has the same number $D = \deg(V(f))$ of isolated roots as $\{f, L\}$, so the homotopy in Step 13 will have no path divergent. Thus we have $\binom{n-1}{n-k} d^{n-k} D$ points in Ω, which is the root bound we mention in Theorem

2. Should it so happen that some of the homotopy paths are divergent in Step 16, the method of end games for homotopy should be used [32–35].

5 Real Critical Set

In this section, we will combine the LPH algorithm in Sect. 4 and methods in [26] to compute a real witness set which has at least one point on each irreducible component of a real algebraic set, and give an illustrative example.

5.1 Critical Points on a Real Algebraic Set

We make the following assumptions (adapted from [26]). Let $f : \mathbb{C}^n \to \mathbb{C}^k$ be a polynomial system, and $f = (f_1, \dots, f_k)$ in $\mathbb{R}[x_1, \dots, x_n]$ satisfying the so-called Full Rank Assumption:

1. $V_{\mathbb{R}}(f_1, \dots, f_i)$ has dimension $n - i$ for $i = 1, \dots, k$;
2. the ideal $I(f_1, \dots, f_i)$ is radical for $i = 1, \dots, k$.

Under these assumptions, $(\nabla f_1^T, \dots, \nabla f_i^T)$ has rank i for a generic point $p \in V(f_1, \dots, f_i)$ for $i = 1, \dots, k$.

The main problem we consider is finding at least one real witness point on each connected component of $V_{\mathbb{R}}(f)$. For this purpose, we choose Φ in Definition 3 to be a linear function with $\Phi = x \cdot \beta + c$, where β is a random vector in $\mathbb{R}^n$, and c is a random real number. Then system (1) becomes

$$F = \left\{ f, \sum_{i=1}^{k} \lambda_i \nabla f_i - \beta \right\} = 0. \tag{4}$$

It may happen that there is no critical points of Φ in some connected component of $V_{\mathbb{R}}(f_1, \dots, f_k)$. In that case, we add Φ to f and construct a system with $k + 1$ equations

$$f^{(1)} = \{f, x \cdot \beta + c\}. \tag{5}$$

Then, recursively, we choose another linear function Φ_1, compute the critical points of Φ_1 with respect to $V(f^{(1)})$; and so on.

We give a concrete definition of the set of real witness points $W_{\mathbb{R}}(f)$ we are going to compute (see [26]).

Definition 4. *Let $f : \mathbb{C}^n \to \mathbb{C}^k$ be a polynomial system, $k \leqslant n$, and $f = (f_1, \dots, f_k)$ in $\mathbb{R}[x_1, \dots, x_n]$ satisfying Full Rank Assumption. F and $f^{(1)}$ defined as in (4) and (5). We define $W_{\mathbb{R}}(f)$ as follows:*

1. $W_{\mathbb{R}}(f) = V_{\mathbb{R}}(f)$ if $n = k$;
2. $W_{\mathbb{R}}(f) = V_{\mathbb{R}}(F) \cup W_{\mathbb{R}}(f^{(1)})$ if $k < n$.

It is obvious from the definition that we can recursively solve the square system (4), and apply plane distance critical points formulation of $f^{(1)}$ to finally get the set of witness points $W_{\mathbb{R}}(f)$ which contains finitely many real points on $V_{\mathbb{R}}(f)$, and there is at least one point on each connected component of $V_{\mathbb{R}}(f)$. Since the formulation introduces auxiliary unknowns, it increases the size of the system and leads to computational difficulties. For example, when $n = 15$ and $k = 10$, the size of system (4) becomes 25, which is challenging for general homotopy software. Combining the LPH algorithm, Theorems 1 and 2, we have the following algorithm and an upper bound of number of points in $W_{\mathbb{R}}(f)$, as in [31].

Algorithm 2. RWS (Real Witness Set)

input : a polynomial system $f = (f_1, \ldots, f_k)$, $k \leqslant n$, which satisfies the full rank assumption;

output: a finite subset $W_{\mathbb{R}}(f)$ of $\mathbb{R}^n$, which contains at least one point on each connected component of the real algebraic set $V_{\mathbb{R}}(f)$

```
Let W_R(f) = ∅;
while k ⩽ n do
    V_R ← LPH(f, J_f(x) · λ − β);
    W_R(f) ← W_R(f) ∪ V_R;
    f ← {f, x · β + c} where n is a random vector in R^n, and c is a
    random real number;
    k ← k + 1;
end
return W_R(f)
```

Remark 5. Algorithm 2 is essentially a recursive calling of Algorithm 1.

Theorem 3 *([31] Theorem 2.1). For a system $f = (f_1, \cdots, f_k)$ with n variables and degrees $d_i = \deg(f_i)$ for $i = 1, \ldots, k$, the number of complex roots of system (4) is bounded by*

$$\binom{n-1}{n-k}(d-1)^{n-k} D \tag{6}$$

where $d = \max\{d_1, \ldots, d_k\} > 1$ and $n > k > 0$, D is the degree of the pure $n-k$ dimensional component of $V = V(f)$.
Moreover, the total number of points in $W_{\mathbb{R}}(f)$ is bounded by

$$\sum_{j=0}^{n-k} \binom{n-1-j}{n-k-j}(d-1)^{n-k-j} D. \tag{7}$$

Obviously we have the following inequalities:

$$MV(F) \leqslant \binom{n-1}{k-1}(d-1)^{n-k} D \leqslant \binom{n-1}{k-1}(d-1)^{n-k} \prod_{i=1}^{k} d_i \leqslant d^n \prod_{i=1}^{k} d_i.$$

If f is dense, the equalities hold. And if f is sparse, they vary considerably most of the time. For example, let $f=\{-62xy+97y-4xyz-4, 80x-44xy+71y^2-17y^3+2\}$ with $d=3, n=3, k=2$. We have $MV(F)=11$, $\binom{n-1}{k-1}(d-1)^{n-k}D=28$, $\binom{n-1}{k-1}(d-1)^{n-k}\prod_{i=1}^{k}d_i=36$, and $d^n\prod_{i=1}^{k}d_i=243$.

5.2 Illustrative Example

In this subsection, we present an illustrative example for Algorithm 2.

Example 1. Consider the hypersurface defined by $f=(y^2-x^3-ax-b)\cdot((x-y+e)^3+x+y)$, $e=6, a=-4, b=-1$. Clearly, $V_{\mathbb{R}}(f)$ is the combination of a cubic ellipse (y^2-x^3-ax-b), and a cubic curve $(x-y+e)^3+x+y$, as plotted in Fig. 1. We show how to compute $W_{\mathbb{R}}(f)$ by Algorithm 2.

- For computing $V_{\mathbb{R}}=\mathtt{LPH}(f)$, we randomly choose a line l in $\mathbb{C}^2$ and solve $L=\{f,l\}$ by polyhedral homotopy, which follows $D=6$ paths. Then to compute Ω by linear homotopy, we follow $\binom{2-1}{2-1}(6-1)^{2-1}6=30$ convergent paths, and for $V_{\mathbb{R}}$ by linear homotopy, we follow 30 paths, of which 6 are convergent and 19 divergent. Then
$$V_{\mathbb{R}}=\{(-1.44299,-1.32941),(-0.781143,1.28371)\}.$$
- For computing $W_{\mathbb{R}}(f)$, we solve $f^{(1)}=\{f,x\cdot\beta+c\}$ by polyhedral homotopy, with $x\cdot\beta+c=0.874645x+1.0351y-3.9825$ and
$$W_{\mathbb{R}}(f^{(1)})=\{(2.4052801,1.815026),(-1.992641,5.531208)\}.$$
So $W_{\mathbb{R}}(f)=W_{\mathbb{R}}(f^{(1)})\cup V_{\mathbb{R}}$, which has at least one point in each connected component of $V_{\mathbb{R}}(f)$ as in Fig. 1.

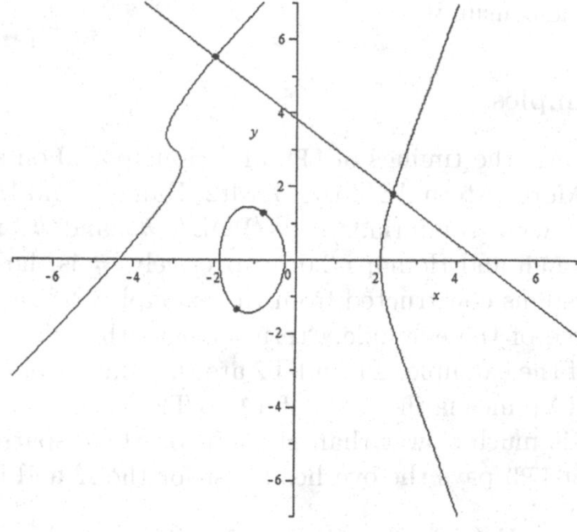

Fig. 1. $n=10, k=4, \deg=2$

6 Experiment Performance

As shown in Sect. 5, to compute the set $W_{\mathbb{R}}(f)$, the key and most time consuming steps are solving the system $F = \left\{f, \sum_{i=1}^{k} \lambda_i \nabla f_i - \beta\right\}$ in Algorithm 1. In this section, given $f = \{f_1, \ldots, f_k\}$, we solve the square system $F = \left\{f, \sum_{i=1}^{k} \lambda_i \nabla f_i - \beta\right\}$. We compare our program LPH which implements Algorithm 1 to Hom4PS-2.0 (available at http://users.math.msu.edu/users/li/). All the examples were computed on a PC with Intel Core i5 processor (2.5GHz CPU, 4 Cores and 6 GB RAM) in the Windows environment. We mention that LPH is a program written in C++, available at http://arcnl.org/PDF/LHP.zip, and an interface of Maple is provided on this site.

6.1 Dense Examples

In Table 1, we provide the timings of LPH and Hom4ps-2.0 for solving systems $F = \left\{f, \sum_{i=1}^{k} \lambda_i \nabla f_i - \beta\right\}$, where $f = (f_1, \cdots, f_k)$ consists of dense polynomials of degree 2, $n = 2, \ldots, 14$ and $1 \leqslant k \leqslant n-1$. T1 ,T2 are the the timings for LPH and Hom4ps-2.0, respectively, and RAT is the ratio of T1 to T2. 'overflow' means running out of memory. When T2=overflow, we set RAT=ε.

It may be observed that LPH is much faster than Hom4ps-2.0 when $k > 1$. Note also that LPH is a little bit slower than Hom4ps-2.0 when $k = 1$. The main reason is obvious. That is, the root number bound of LPH, *i.e.*

$$\binom{n-1}{n-k}(d-1)^{n-k}D,$$

is close to the mixed volume $MV(F)$ when F is dense but the computation of $MV(F)$ is very time-consuming.

6.2 Sparse examples

In Table 2, we provide the timings of LPH and Hom4ps-2.0 on sparse examples: Czapor Geddes2, Morgenstern AS(3or), Gerdt2, Hairer1, and Hawes2 which are available at: http://www-sop.inria.fr/saga/POL/. #1 and #2 is the number of curves followed by LPH and Hom4ps-2.0, respectively. # is the number of roots of the Jacobian systems constructed from the examples. '*d*' means the minimal and maximal degree of the example. "term" means the minimal and maximal number of terms of the example. T1 and T2 are the timings of LPH and Hom4ps-2.0, respectively. RAT means the ratio of T1 to T2.

Note that LPH is much slower than Hom4ps on these sparse examples. The main reason is that LPH pays the overhead cost for the Ω and homotopy

$$H_2(x, \lambda, t) = G \cdot (1-t) + F \cdot \gamma_2 \cdot t.$$

Table 1. Dense Examples (n: number of variables; k: number of polynomials; T1: time for LHP; T2: time for Hom4PS-2.0; RAT: ratio of T1 to T2.)

n	k	T1	T2	RAT	n	k	T1	T2	RAT
2	1	0.125s	0.094s	1.32	11	1	0.28s	0.23s	1.37
3	1	0.125s	0.109s	1.14	11	2	1.20s	2.85s	0.42
3	2	0.125s	0.109s	1.14	11	3	8.7s	41.5s	0.209
4	1	0.125s	0.109s	1.14	11	4	51.5s	5m25s	0.158
4	2	0.156s	0.156s	1.00	11	5	3m15s	25m30s	0.128
4	3	0.202s	0.265s	0.76	11	6	9m1s	76m32s	0.118
5	1	0.125s	0.109s	1.14	11	7	16m14s	2h.45m30s	0.098
5	2	0.187s	0.202s	0.93	11	8	18m34s	3h52m52s	0.079
5	3	0.390s	0.655s	0.59	11	9	16m22s	3h34m38s	0.076
5	4	0.687s	1.280s	0.54	11	10	6m37s	overflow	ε
6	1	0.140s	0.109s	1.28	12	1	0.29s	0.218s	1.360
6	2	0.281s	0.328s	0.86	12	2	1.27s	4.3s	0.294
6	3	0.76s	1.68s	0.45	12	3	13.5s	1m10s	0.191
6	4	1.61s	4.36s	0.37	12	4	1m25s	10m0.2s	0.142
6	5	1.90s	6.59s	0.29	12	5	6m17s	53m58s	0.116
7	1	0.14s	0.10s	1.29	12	6	20m1s	3h15m10s	0.102
7	2	0.344s	0.56s	0.61	12	7	44m23s	8h7m1s	0.091
7	3	1.3s	3.82s	0.34	12	8	1h8m30s	14h38m24s	0.0779
7	4	3.7s	14.1s	0.27	12	9	1h8m42s	overflow	ε
7	5	6.318s	27.9s	0.23	12	10	46m21s	overflow	ε
7	6	6.006s	27.6s	0.21	12	11	21m54s	overflow	ε
8	1	0.15s	0.15s	1.00	13	1	0.343s	0.218s	1.573
8	2	0.54s	0.73s	0.74	13	2	1.716s	6.193s	0.277
8	3	2.29s	7.2s	0.31	13	3	18s	1m51s	0.61
8	4	8.018s	34.1s	0.23	13	4	2m15s	18m35s	0.121
8	5	17.6s	91.2s	0.19	13	5	11m10s	1h52m27s	0.099
8	6	23.7s	153s	0.15	13	6	40m6s	7h16m34s	0.092
8	7	19.4s	128s	0.15	13	7	1h39m40s	21h25m14s	0.078
9	1	0.2s	0.18s	1.08	13	8	2h58m48s	overflow	ε
9	2	0.7s	1.2s	0.58	13	9	3h59m32s	overflow	ε
9	3	4.1s	13.1s	0.31	13	10	3h40m3s	overflow	ε
9	4	16.1s	1m19s	0.20	13	11	2h13m9s	overflow	ε
9	5	46.5s	4m29s	0.17	13	12	56m48.309s	overflow	ε
9	6	1m20s	9m38s	0.138	14	2	2.5s	9.6s	0.264
9	7	1m30s	11m10s	0.135	14	3	24.3s	3m0.4s	0.134
9	8	59.9s	7m52s	0.126	14	4	3m19s	37m19s	0.089
10	1	0.23s	0.18s	1.25	14	5	19m28s	8h24m29s	0.038
10	2	0.98s	1.9s	0.51	14	6	1h16m20s	15h58m59s	0.079
10	3	5.8s	24.8s	0.24	14	7	3h34m52s	overflow	ε
10	4	31.1s	2m53s	0.18	14	8	7h50m34s	overflow	ε
10	5	1m46s	11m30s	0.15	14	9	12h43m8s	overflow	ε
10	6	3m41s	29m15s	0.13	14	10	16h48m4s	overflow	ε
10	7	5m10s	48m32s	0.107	14	11	13h9m8s	overflow	ε
10	8	4m57s	48m27s	0.102	14	12	6h29m37s	overflow	ε
10	9	3m2s	30m7.553s	0.1	14	13	2h18m27s	overflow	ε

Table 2. Sparse examples

Ex	n	k	d	term	#1	#2	#	T1	T2	RAT
C2	5	4	4–5	7–32	2*1767	692	383	16.6s	19.4 s	0.85
M3	9	5	2	2–7	2*2240	368	32	69 s	6 s	11
G2	5	2	4	8–9	2*1080	17	15	8.4 s	0.2 s	31.8
H1	8	6	1–3	2–4	2*400	15	15	9.8 s	0.18 s	52
H2	8	5	2–4	3–5	2*12320	148	80	5m49 s	1.2s	267

Moreover, LPH executes $2 * \binom{n-1}{k-1} (d-1)^{n-k} D$ times of curve following, while Hom4ps does only $MV(F)$ times of curve following. $\binom{n-1}{k-1} (d-1)^{n-k} D$ is not tight for these sparse examples and much greater than $MV(F)$.

6.3 RAT/Density

In Fig. 2, we present the changes of ratio of T1 to T2 as terms increase. We randomly generate $f = (f_1, \ldots, f_k)$ with different n, k and degrees, and increase the number of terms from 2 to dense.

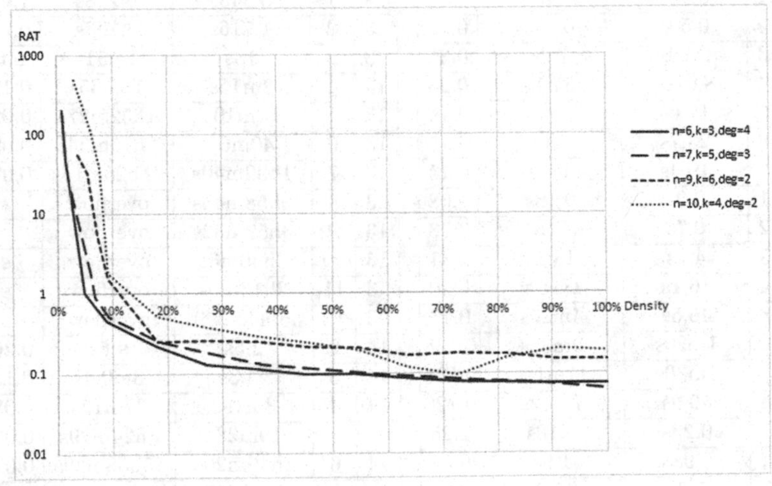

Fig. 2. LPH vs Hom4ps-2.0 with the growth of density

It can be observed that, when the polynomials are not very sparse, *e.g.* the number of terms are more than 10% of $\binom{n+d}{d}$, LPH is faster than Hom4ps-2.0.

Actually, when the polynomials are not very sparse, the root number bound $\binom{n-1}{k-1}(d-1)^{n-k}D$ is close to $MV(F)$.

Acknowledgement. We gratefully acknowledge the very helpful suggestions of Hoon Hong on this paper with emphasis on Sect. 6. We also thank Changbo Chen for his helpful comments. And the authors would like to thank the anonymous reviewers for their constructive comments that greatly helped improving the paper.

References

1. Garcia, C.B., Zangwill, W.I.: Finding all solutions to polynomial systems and other systems of equations. Math. Program. **16**(1), 159–176 (1979)
2. Drexler, F.J.: Eine Methode zur Berechnung sämtlicher Lösungen von Polynomgleichungssystemen. Numer. Math. **29**(1), 45–58 (1977)
3. Sommese, A.J., Verschelde, J., Wampler, C.W.: Numerical algebraic geometry. In: The Mathematical of Numerical Analysis. Lectures in Applied Mathematics, vol. 32, pp. 749–763. AMS (1996)
4. Allgower, E.L., Georg, K.: Introduction to numerical continuation methods. Reprint of the 1979 original. Society for Industrial and Applied Mathematics (2003)
5. Li, T.: Numerical solution of polynomial systems by homotopy continuation methods. In: Handbook of Numerical Analysis, vol. 11, pp. 209–304 Elsevier (2003)
6. Sommese, A.J., Wampler, C.W.: The Numerical Solution of Systems of Polynomials Arising in Engineering and Science. World Scientific, Singapore (2005)
7. Morgan, A.: Solving Polynominal Systems Using Continuation for Engineering and Scientific Problems. Society for Industrial and Applied Mathematics, Philadelphia (2009)
8. Bates, D.J., Haunstein, J.D., Sommese, A.J., Wampler, C.W.: Numerically Solving Polynomial Systems with Bertini. Society for Industrial and Applied Mathematics, Philadelphia (2013)
9. Lee, T.L., Li, T.Y., Tsai, C.H.: Hom4ps-2.0: a software package for solving polynomial systems by the polyhedral homotopy continuation method. Computing **83**(2), 109 (2008)
10. Morgan, A.P., Sommese, A.J., Watson, L.T.: Finding all isolated solutions to polynomial systems using hompack. ACM Trans. Math. Softw. **15**(2), 93–122 (1989)
11. Verschelde, J.: Algorithm 795: Phcpack: a general-purpose solver for polynomial systems by homotopy continuation. ACM Trans. Math. Softw. **25**(2), 251–276 (1999)
12. Collins, G.E.: Quantifier elimination for real closed fields by cylindrical algebraic decompostion. In: Brakhage, H. (ed.) GI-Fachtagung 1975. LNCS, vol. 33, pp. 134–183. Springer, Heidelberg (1975). doi:10.1007/3-540-07407-4_17
13. Seidenberg, A.: A new decision method for elementary algebra. Ann. Math. **60**(2), 365–374 (1954)
14. El Din, M.S., Schost, É.: Polar varieties and computation of one point in each connected component of a smooth real algebraic set. In: Proceedings of ISSAC 2003, pp. 224–231. ACM, New York (2003)
15. El Din, M.S., Spaenlehauer, P.J.: Critical point computations on smooth varieties: degree and complexity bounds. In: Proceedings of ISSAC 2016, pp. 183–190. ACM, New York (2016)

16. Bank, B., Giusti, M., Heintz, J., Pardo, L.M.: Generalized polar varieties and an efficient real elimination. Kybernetika **40**(5), 519–550 (2004)
17. Bank, B., Giusti, M., Heintz, J., Pardo, L.: Generalized polar varieties: geometry and algorithms. J. Complex. **21**(4), 377–412 (2005)
18. Rouillier, F., Roy, M.F., El Din, M.S.: Finding at least one point in each connected component of a real algebraic set defined by a single equation. J. Complex. **16**(4), 716–750 (2000)
19. El Din, M.S., Schost, É.: Properness defects of projections and computation of at least one point in each connected component of a real algebraic set. Discrete Comput. Geom. **32**(3), 417–430 (2004)
20. Li, T.Y., Wang, X.: Solving real polynomial systems with real homotopies. Math. Comp. **60**(202), 669–680 (1993)
21. Lu, Y., Bates, D.J., Sommese, A.J., Wampler, C.W.: Finding all real points of a complex curve. Technical report. In: Algebra, Geometry and Their Interactions (2006)
22. Bates, D.J., Sottile, F.: Khovanskii-rolle continuation for real solutions. Found. Comput. Math. **11**(5), 563–587 (2011)
23. Besana, G.M., Rocco, S., Hauenstein, J.D., Sommese, A.J., Wampler, C.W.: Cell decomposition of almost smooth real algebraic surfaces. Numer. Algorithms **63**(4), 645–678 (2013)
24. Hauenstein, J.D.: Numerically computing real points on algebraic sets. Acta Appl. Math. **125**(1), 105–119 (2013)
25. Shen, F., Wu, W., Xia, B.: Real root isolation of polynomial equations based on hybrid computation. In: Feng, R., Lee, W., Sato, Y. (eds.) Computer Mathematics, pp. 375–396. Springer, Heidelberg (2014). doi:10.1007/978-3-662-43799-5_26
26. Wu, W., Reid, G.: Finding points on real solution components and applications to differential polynomial systems. In: Proceedings of ISSAC 2013, pp. 339–346. ACM, New York (2013)
27. Bernshtein, D.N.: The number of roots of a system of equations. Funct. Anal. Appl. **9**(3), 183–185 (1975)
28. Hauenstein, J.D., Sommese, A.J., Wampler, C.W.: Regeneration homotopies for solving systems of polynomials. Math. Comp. **80**(273), 345–377 (2011)
29. John, F.: Extremum problems with inequalities as subsidiary conditions. In: Giorgi, G., Kjeldsen, T.H. (eds.) Traces and Emergence of Nonlinear Programming, pp. 197–215. Springer, Basel (2014). doi:10.1007/978-3-0348-0439-4_9
30. Morgan, A.P., Sommese, A.J.: Coefficient-parameter polynomial continuation. Appl. Math. Comput. **29**(2), 123–160 (1989)
31. Wu, W., Reid, G., Feng, Y.: Computing real witness points of positive dimensional polynomial systems. Theoretical Computer Science (2017). http://doi.org/10.1016/j.tcs.2017.03.035. Accessed 31 Mar 2017
32. Morgan, A.P., Sommese, A.J., Wampler, C.W.: A power series method for computing singular solutions to nonlinear analytic systems. Numer. Math. **63**(1), 391–409 (1992)
33. Morgan, A.P.: A transformation to avoid solutions at infinity for polynomial systems. Appl. Math. Comput. **18**(1), 77–86 (1986)
34. Huber, B., Verschelde, J.: Polyhedral end games for polynomial continuation. Numer. Algorithms **18**(1), 91–108 (1998)
35. Bates, D.J., Hauenstein, J.D., Sommese, A.J.: A parallel endgame. Contemp. Math. **556**, 25–35 (2011). AMS, Providence, RI

Penalty Function Based Critical Point Approach to Compute Real Witness Solution Points of Polynomial Systems

Wenyuan Wu[1,2], Changbo Chen[1,2(✉)], and Greg Reid[3]

[1] Chongqing Key Laboratory of Automated Reasoning and Cognition, Chongqing Institute of Green and Intelligent Technology, Chinese Academy of Sciences, Chongqing, China
{wuwenyuan,chenchangbo}@cigit.ac.cn

[2] University of Chinese Academy of Sciences, Beijing, China

[3] Applied Mathematics Department, Western University, London, Canada
reid@uwo.ca

Abstract. We present a critical point method based on a penalty function for finding certain solution (witness) points on real solutions components of general real polynomial systems. Unlike other existing numerical methods, the new method does not require the input polynomial system to have pure dimension or satisfy certain regularity conditions.

This method has two stages. In the first stage it finds approximate solution points of the input system such that there is at least one real point on each connected solution component. In the second stage it refines the points by a homotopy continuation or traditional Newton iteration. The singularities of the original system are removed by embedding it in a higher dimensional space.

In this paper we also analyze the convergence rate and give an error analysis of the method. Experimental results are also given and shown to be in close agreement with the theory.

1 Introduction

Computational real algebraic geometry is the study of the global structure of real solution sets of polynomial systems, including positive dimensional solution components (see [2] for a background text on algorithms for exact real algebraic geometry). This paper is a contribution to the development of numerical algorithms for computational real algebraic geometry directed at numerically describing such global structure. In contrast, conventional numerical methods seek local solutions which are points, and generally do not give information on positive dimensional solution components.

Numerical algebraic geometry [16,29] was pioneered by Sommese, Wampler, Verschelde and others (see the texts [9,27] for references and background). They first considered the easier characterization of complex solution components of each possible dimension, by slicing the solution set with appropriate random

V.P. Gerdt et al. (Eds.): CASC 2017, LNCS 10490, pp. 377–391, 2017.
DOI: 10.1007/978-3-319-66320-3_27

planes, that intersected the solution components in complex points called *witness points*. The complex points are computed by homotopy continuation solvers. For example, a one dimensional circle, $x^2+y^2-1=0$ in $\mathbb{C}^2$ is intersected by a random line in two such witness points, but this method obviously fails for $(x,y) \in \mathbb{R}^2$ since a real line can miss the circle.

Instead the method in [31,32] yields real witness points as critical points of the distance from a random hyperplane to the real variety. The reader can easily see this yields two real witness points for the circle example. An alternative numerical approach where the witness points are critical points of the distance from a random point to the real variety has been developed in [15]. The works [15,31,32] use Lagrange multipliers to set up the critical point problem.

A contribution of our current paper is to remove the assumptions in [15, 31,32] by developing a penalty function based critical point method where the singularities are removed by embedding systems in a higher dimensional space. The method has two stages. In the first stage it finds approximate solution points of the input system such that there is at least one real point on each connected solution component. In the second stage it refines the points by a homotopy continuation or traditional Newton iteration. We also analyze the convergence rate and give an error analysis of our method. Experimental results are given and shown to closely agree with the theory.

Critical point methods in Lagrange form appeared previously in important symbolic works [24–26]. In those works, the systems are analyzed using Gröbner Bases. Ultimately numerical methods have to be used to approximate points on components, but only after application of symbolic algorithms to the systems, instead of the fully numerical methods used here and in [15,31,32]. Also see the early related symbolic works [1,6] and the recent work [5,11].

More distantly related symbolic approaches for computational real algebraic geometry include cylindrical algebraic decomposition (CAD) introduced by Collins [13] and improved by many others. Recent improvement of CAD by using triangular decompositions are given in [12] for solving semi-algebraic systems. But the double exponential cost of the CAD algorithm [14] is the main barrier to its application.

Numerical methods based on moment matrices and semi-definite programming techniques have been developed to approximate real radical ideals of zero dimensional systems, e.g. [20,21]. For a positive dimensional system, an approach is given in [7] which combines numerical algebraic geometry and sums of squares programming to test whether the input is real radical or not. Also see [23,33], based on moment matrices, and [22].

As a development of critical point approaches [15,24–26,31], this article will propose an approximation method to compute real witness points of polynomial systems without any regularity assumption [31,32] or pure dimension assumption [15]. In Sect. 2, we will describe how the polynomial systems are embedded in a higher dimensional space. In Sect. 3, we will describe error control with a rank assumption. In Sect. 4, this rank assumption is removed and error control is provided for general systems. In Sect. 5, our method is illustrated with examples and concluding remarks are given in Sect. 6.

2 Augmented System

In this section we introduce our augmented system, that involves adding a variable to each equation, so the original system is obtained when these slack variables are set to zero. The resulting augmented system has solution set that is a smooth real manifold. We alert the reader that this is different to the embedding systems of (complex) numerical algebraic geometry. To avoid confusion with the well known embedding systems of the complex case we have used a different name for our systems, Augmented System.

Let $x = (x_1, \ldots, x_n)$. Let $f = \{f_1, \ldots, f_k\}$ be a set of polynomials in the ring $\mathbb{R}[x]$. We construct the following augmented system g for f with slack variables $z = (z_1, \ldots, z_k)$:

$$g = \{f_1 + z_1, f_2 + z_2, \ldots, f_k + z_k\}. \tag{1}$$

Note that $g \subset \mathbb{R}[x, z]$ holds.

Lemma 1. *The Jacobian matrix of g w.r.t. the variables $(x_1, \ldots, x_n, z_1, \ldots, z_k)$ has rank k at any point of $V_{\mathbb{R}}(g)$ and $V_{\mathbb{R}}(g)$ is a smooth submanifold of $\mathbb{R}^{n+k}$ with dimension n.*

Proof. Firstly, $V_{\mathbb{R}}(g) \neq \emptyset$ since $\{x_1 = 0, \ldots, x_n = 0, z_1 = -f_1(0), \ldots, z_k = -f_k(0)\}$ is a real solution. Secondly, it is easy to see that the Jacobian matrix $\frac{\partial g}{\partial(x,z)}$ has full rank k at any solution $(x^*, z^*) \in V_{\mathbb{R}}(g)$, which implies that $V_{\mathbb{R}}(g)$ is a smooth submanifold of $\mathbb{R}^{n+k}$ with dimension n by the regular level set theorem (see pp. 113–114 of [19]). □

By Lemma 1, the augmented system g satisfies the regularity assumptions A_1 and A_2 of [31]. Moreover, any point on $V_{\mathbb{R}}(g)$ being smooth is a crucial property for numerical stability of numerical methods applied to $V_{\mathbb{R}}(g)$.

Using the critical point technique [24], we choose a random point $\mathfrak{a} = (\mathfrak{a}_1, \ldots, \mathfrak{a}_n)$, where $\mathfrak{a} \notin V_{\mathbb{R}}(f)$, in x-space and consider the minimal distance from $V_{\mathbb{R}}(f)$ to this point. As the norm of the slack variables z approaches zero, the corresponding point of $V_{\mathbb{R}}(g)$ approaches $V_{\mathbb{R}}(f)$. To force the slack variables z to be very small, we introduce a penalty function $\beta \cdot (z_1^2 + \cdots + z_k^2)/2$ with penalty factor $\beta \gg 0$ and formulate the following optimization problem

$$\min \mu = (\beta \cdot (z_1^2 + \cdots + z_k^2) + \sum\nolimits_{i=1}^{n} (x_i - \mathfrak{a}_i)^2)/2 \qquad s.t. \quad g = 0. \tag{2}$$

To solve the optimization problem above, we can use Lagrange multiplier techniques:

$$\begin{pmatrix} x_1 - \mathfrak{a}_1 \\ \vdots \\ x_n - \mathfrak{a}_n \\ \beta\, z_1 \\ \vdots \\ \beta\, z_k \end{pmatrix} = \begin{pmatrix} \partial f_1/\partial x_1 & \cdots & \partial f_k/\partial x_1 \\ \vdots & \ddots & \vdots \\ \partial f_1/\partial x_n & \cdots & \partial f_k/\partial x_n \\ 1 & & \\ & \ddots & \\ & & 1 \end{pmatrix}_{(n+k)\times k} \cdot \begin{pmatrix} \lambda_1 \\ \vdots \\ \lambda_k \end{pmatrix}. \tag{3}$$

Then (3) implies that $\lambda_i = \beta z_i = -\beta f_i$. Substituting this solution back into the Eq. (3) above yields a square system with n variables

$$\begin{pmatrix} x_1 \\ \vdots \\ x_n \end{pmatrix} + \beta \cdot \mathcal{J}^t \cdot \begin{pmatrix} f_1 \\ \vdots \\ f_k \end{pmatrix} = \begin{pmatrix} \mathfrak{a}_1 \\ \vdots \\ \mathfrak{a}_n \end{pmatrix}. \tag{4}$$

where the $n \times k$ matrix $\mathcal{J}^t$ is the transpose of the Jacobian of f.

The optimization problem (2) is equivalent to the following unconstrained optimization problem (will be used in the next two sections):

$$\min\ \mu = (\beta \cdot (f_1^2 + \cdots + f_k^2) + \textstyle\sum_{i=1}^n (x_i - \mathfrak{a}_i)^2)/2. \tag{5}$$

Setting the gradient of μ to be zero, we also obtain Eq. (4).

Note that the left hand side of Eq. (4) defines a smooth mapping $M : \mathbb{R}^n \to \mathbb{R}^n$.

Lemma 2. *For a random point $\mathfrak{a} = (\mathfrak{a}_1, \ldots, \mathfrak{a}_n) \notin V_{\mathbb{R}}(f)$, Problem (2) has solutions and $M^{-1}(\mathfrak{a}) \neq \emptyset$. Moreover, every point of the real variety $M^{-1}(\mathfrak{a})$ is a regular point of M with probability* 1.

Proof. Let $z_i = w_i/\sqrt{\beta}$, $i = 1, \ldots, k$, and substitute them into (2). Let $h = \{\sqrt{\beta} f_1 + w_1, \ldots, \sqrt{\beta} f_k + w_k\}$. We obtain another equivalent form of Problem (2), where the objective function is now formulated as a distance function:

$$\min\ (w_1^2 + \cdots + w_k^2 + \sum_{i=1}^n (x_i - \mathfrak{a}_i)^2)/2 \qquad s.t. \quad h = 0. \tag{6}$$

By Lemma 1, $V_{\mathbb{R}}(g)$ is a smooth submanifold of $\mathbb{R}^{n+k}$ with dimension n. So is $V_{\mathbb{R}}(h)$. For any $\mathfrak{a} \notin V_{\mathbb{R}}(f)$, the point $(x = \mathfrak{a}, w = 0)$ does not belong to $V_{\mathbb{R}}(h)$. Thus, (6) always has minimum distance from $(\mathfrak{a}, 0)$ to $V_{\mathbb{R}}(h)$ by completeness of the real numbers, which implies that the minimal value of (2) can always be attained. Since the Jacobian matrix of g has full rank at any point of $V_{\mathbb{R}}(g)$, a solution $\{x^*, z^*\}$ of Problem (2) must be a solution of Eq. (3), which implies that x^* is a solution of Eq. (4).

Thus $M^{-1}(\mathfrak{a}) \neq \emptyset$. By Sard's Theorem [30], for almost all $\mathfrak{a}$, every point of the real variety $M^{-1}(\mathfrak{a})$ is regular point of M. □

Lemma 2 implies that all the solutions of Eq. (4) can be obtained by applying homotopy continuation methods.

Among these solutions, we look for solutions with small residuals i.e. $\|z\| \ll 1$. It is possible that such points do not exist, which then provides strong evidence that $V_{\mathbb{R}}(f)$ is empty. Intuitively, this is because if $V_{\mathbb{R}}(f)$ is not empty, increasing the penalty factor β will force $\|z\|$ to be close to zero.

A theoretical study on the relationship between the magnitude of the residual $\|z\|$ and the emptiness of $V_{\mathbb{R}}(f)$ is out of the scope of this paper and will be treated in a future work. **In the rest of this paper, we always assume that $V_{\mathbb{R}}(f) \neq \emptyset$.** A natural question is how to estimate the distances of the local minima of Problem (5) to $V_{\mathbb{R}}(f)$. We divide this problem into two cases w.r.t. the rank of the Jacobian and will address them in the next two sections.

Note that, throughout this paper, the norm $\|\cdot\|$ always means the 2-norm.

3 Error Control with Rank Assumption

Since Problem (5) with penalty function is different from the goal of finding real witness points of the original system $f = 0$, it is of great importance to study the difference between their solutions. In this section, we will give an error estimate of the approximate answer given by solving Problem (5) under a rank assumption. In the next section, we will remove this assumption and give an error estimate for general systems.

For a smooth point x on $V_{\mathbb{R}}(f)$, let the local dimension of $V_{\mathbb{R}}(f)$ at point x be ℓ. The Jacobian matrix at x is a $k \times n$ matrix denoted by $\mathcal{J}_x$. Suppose its rank is m, where $m \leq \min\{k, n\}$. Then we say that x satisfies the **rank condition**, if $m = n - \ell$, i.e.

$$rank \mathcal{J}_x = n - \dim V_{\mathbb{R}}(f)_x \tag{7}$$

If any smooth point on $V_{\mathbb{R}}(f)$ satisfies the rank condition, then we say the system f satisfies the rank condition. For example this occurs if $f = \{x - y, x^2 - y^2\}$. This means that f can be an over-determined system and even generate a non-radical ideal (e.g. consider $f = \{(x^2+1)^2(x-y)\}$). Note that $f = (x-y)^2$ does not satisfy the rank condition, although its graph is a smooth line. Such systems will be discussed in the next section.

For a random point $\mathfrak{a} \in \mathbb{R}^n$, there is at least one point on each connected component of $V_{\mathbb{R}}(f)$ with minimal distance to $\mathfrak{a}$ satisfying the following problem:

$$\begin{aligned} &\min \sum_{i=1}^{n} (x_i - \mathfrak{a}_i)^2 \\ &s.t. \quad x \in V_{\mathbb{R}}(f). \end{aligned} \tag{8}$$

Let us consider such a point p of (8) with local minimal distance to $\mathfrak{a}$. For this point, there exists a constant c and we have $\|f_i(p + \Delta x)\| < c\|\Delta x\|$ for each polynomial f_i when Δx is sufficiently small. The value of the target function μ at p of (5) is $D^2/2$ where $D = \|p - \mathfrak{a}\|$. If we move p towards $\mathfrak{a}$ with a sufficiently small distance Δx to p' then

$$\mu(p') = \beta\|f(p')\|^2/2 + (p' - \mathfrak{a})^2/2 < \beta c^2\, \Delta x^2/2 + (D - \Delta x)^2/2 < D^2/2. \quad (9)$$

It means that p of Problem (8) is not a local minimum of Problem (5). Let p' be a local minimum of (5) for a given β. Consequently, $p' \notin V_{\mathbb{R}}(f)$. But we have the following result.

Corollary 1. *Let p be a local minimum of (8). There exists a local minimum p' of (5) for sufficiently large β, such that $\|p - p'\|$ can be arbitrarily small.*

Proof. For any small $\delta > 0$, consider the sphere S of a ball centered at p with radius δ. Let $D = \|p - \mathfrak{a}\|$, where $\mathfrak{a} \notin V_{\mathbb{R}}(f)$ is the given point for both problems (5) and (8). The sphere S can be divided into two sets: $S_1 = \{x \in S : \|x - \mathfrak{a}\| \leq D\}$ and $S_2 = \{x \in S : \|x - \mathfrak{a}\| > D\}$. Since p obtains the local minimum distance from $V_{\mathbb{R}}(f)$ to $\mathfrak{a}$, we have $S_1 \cap V_{\mathbb{R}}(f) = \emptyset$ for a small enough δ. Let $s = \min_{x \in S_1}(\sum_j f_j(x)^2)$. So $s > 0$. When $\beta s + (D - \delta)^2 > D^2$, i.e. $\beta > \frac{2\delta D - \delta^2}{s}$, we have $\mu(x) > D^2/2 = \mu(p)$ for any point x on the sphere S. Since the ball is a compact set, the local minimal value of μ must be attained at p' inside this ball. □

We now consider how to estimate the error $\|p - p'\|$ for a given β. First let us consider a simple case when a local minimum p of (8) satisfies the rank condition (7). Then, we have the following result.

Theorem 2. *Suppose p is a local minimum of (8) satisfying the rank condition (7). Then there is at least one real solution p' of Eq. (4) such that $\|p - p'\| < \frac{D}{\beta\sigma_m^2 + 1}$, where $D = \|p - \mathfrak{a}\|$ and σ_m is the smallest nonzero singular value of $\mathcal{J}_p$.*

Proof. Without loss of generality, we assume that p is the origin o. Because of the rank condition, the local dimension at p is equal to $n - \mathit{rank}\mathcal{J}_p = n - m$. Moreover, the null-space of $\mathcal{J}_p$ is the tangent space T at p. Let N be the orthogonal complement of T in $\mathbb{R}^n$. Since $p \in T$ has the minimum distance to $\mathfrak{a}$, the vector $\mathfrak{a}$ belongs to N.

Let $U^T \mathcal{J}_p V = \Sigma_{k \times n} = \mathit{diag}(\sigma_1, \ldots, \sigma_m, 0, \ldots, 0)$ be the singular value decomposition of the Jacobian matrix at p, where $U = ([u_1|\cdots|u_k]) \in \mathbb{R}^{k \times k}$ and $V = ([v_1|\cdots|v_n]) \in \mathbb{R}^{n \times n}$. Then, the space N is spanned by $\{v_1, \ldots, v_m\}$.

By Corollary 1 for sufficiently large β, there exists a local minimum p' of (5) such that $\|p' - p\| = \delta \ll 1$ and $f(p') = f(p) + \mathcal{J}_p \cdot p' + O(\delta^2)$ since p is the origin.

Let $p' = t + b$, where $t \in T, b \in N$. Recall that $\mathfrak{a} \in N$ and $N \perp T$. Then we have $\mathcal{J}_p \cdot p' = \mathcal{J}_p \cdot b$ and $\|p' - \mathfrak{a}\|^2 = \|t\|^2 + \|\mathfrak{a} - b\|^2$. Since $\mathfrak{a}, b \in N$, we choose $\{v_1, \ldots, v_m\}$ as the coordinates of N and suppose $\mathfrak{a} = (a_1, \ldots, a_m)$, $b = (b_1, \ldots, b_m)$. Thus, ignoring high order errors we have

$$\|f(p')\|^2 = \|\mathcal{J}_p \cdot b\|^2 = b^T V \Sigma^2 V^T b = \sum_{i=1}^{m} \sigma_i^2 b_i^2.$$

Hence, p' is a point near p satisfying the following problem

$$\min_{t,b} \mu = (\beta(\sum_{i=1}^{m} \sigma_i^2 b_i^2) + \sum_{i=1}^{m}(a_i - b_i)^2 + \|t\|^2)/2.$$

It is straightforward to show that when $b_i = \frac{a_i}{\beta\sigma_i^2+1}$ and $t = 0$, the function μ attains the minimum $\sum_i \frac{\beta\sigma_i^2}{\beta\sigma_i^2+1} a_i^2/2$, which is less than $\mu(p) = \sum_i a_i^2/2 = D^2/2$.

Therefore,

$$\|p' - p\|^2 = \sum_i b_i^2 = \sum_i (\frac{a_i}{\beta\sigma_i^2 + 1})^2 \leq \sum_i (\frac{a_i}{\beta\sigma_m^2 + 1})^2 = (\frac{D}{\beta\sigma_m^2 + 1})^2.$$

Moreover, p' can be found by solving Eq. (4). □

Example 1. Consider the system $f = \{x^2 + y^2 - 2x, 2x^2 + 2y^2 - 4x\}$ and $\mathfrak{a} = (-0.8, 0.6)$. In this case $m < k$ holds. The real variety is a circle centered at $(1, 0)$ with radius 1. The point $p = (1 - \frac{3\sqrt{10}}{10}, \frac{\sqrt{10}}{10})$ has the minimal distance to $\mathfrak{a}$. Consequently, we have $D = 1, \sigma_m = 4.472$ and $r = \|p - p'\| \leq \mathfrak{e} = \frac{1}{20\beta+1}$. The behaviors of both the actual error r and the estimated error $\mathfrak{e}$ with increasing of β are given in Fig. 1, where the differences between the log of estimated errors and the log of actual errors are greater than 0.047 and less than 0.048. That is $0.895\mathfrak{e} < r < 0.897\mathfrak{e}$. Thus, the theoretical estimation is quite sharp.

Here increasing the value of β and producing more and more accurate roots aim to verify Theorem 2. Since the local minimum satisfies the rank condition, we can simply apply Gauss-Newton iteration [4] to improve the accuracy.

Remark 1. Since we only have a local minimum p' which is an approximation of p, a good estimate of σ_m can be obtained by $\mathcal{J}_{p'}$ for sufficiently small $\|p - p'\|$ because of Weyl's theorem [28].

This theorem only works for $\sigma_m > 0$. However, if σ_m is close to zero because of singularity of p or non-radicalness of the system f, the convergence will be very slow as $\beta \to \infty$. But Corollary 1 still applies.

Example 2. In an example of [31], $f = \{x_2^2 + x_3^2 - (2x_1 - x_1^2)^3\}$ and $\mathfrak{a} = (-0.5, -1, 0.1)$. The point $p = (0, 0, 0)$ with the minimal distance to $\mathfrak{a}$ is singular in $V_{\mathbb{R}}(f)$. To see the asymptotic behavior of the error given in Corollary 1, we plot the magnitude of the actual error against the magnitude of the penalty factor β in Fig. 2. Applying the `CurveFitting[LeastSquares]` command in Maple yields $\log(r) \doteq -0.538 - 0.202\log(\beta)$.

4 Error Control for General Systems

Previously, we know that if a local minimum of (8) satisfies the rank condition $m = n - \ell$, the estimated error is of order $O(1/\beta)$ as can be observed in Fig. 1.

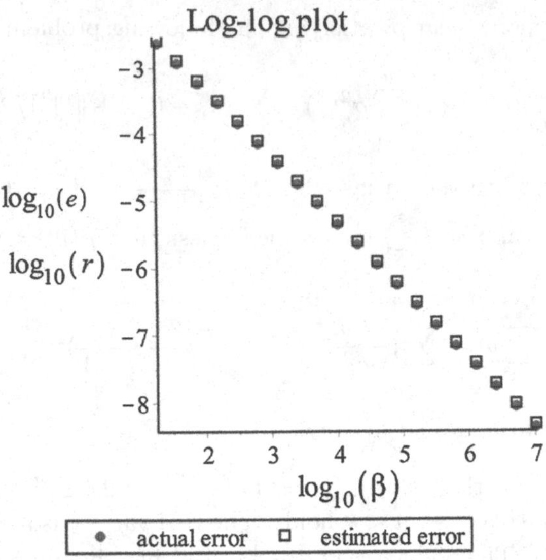

Fig. 1. For Example 1, the log of the estimated error, $log(\mathfrak{e})$ (resp. actual error, $log(r)$) is decreasing linearly with the increase of the magnitude of penalty factor β.

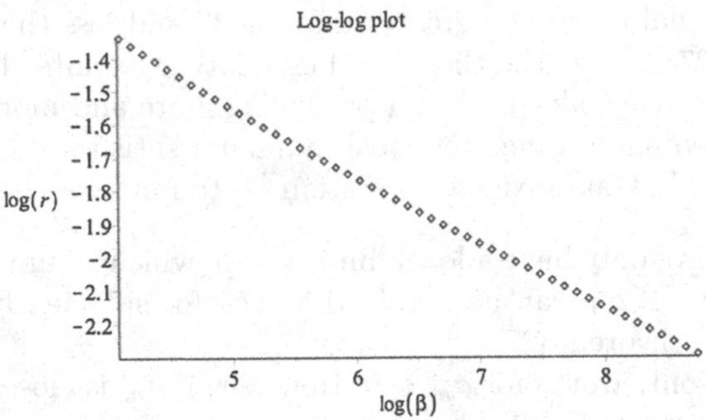

Fig. 2. The actual error $log(r)$ for Example 2 is decreasing slowly with slope -0.202.

For this case, although increasing the value of β can improve the accuracy, the Gauss-Newton iteration will improve convergence.

However, whether the input system satisfies the regularity assumptions or not, a local minimum of Problem (5) could be close to singularities (as in Example 2) which means $m < n - \ell$. Since such singularities are unavoidable, we will address the convergence for general systems with no assumptions in this section.

4.1 Degree Index

In Problem (5), the residual of f is amplified by a penalty factor β when x does not belong to $V_{\mathbb{R}}(f)$. In this section we will give a lower bound of the residual first. Then it leads to an error control of our method for general systems.

Let $f \in \mathbb{R}[x_1, \ldots, x_n]$. Suppose $f(0) = 0$. If we write $f = f_m + f_{m+1} + \ldots$ with f_α homogeneous of degree α and $f_m \neq 0$, then m is the multiplicity of f at the origin. For example $f = x^2 + y^3$ has a cusp at the origin with multiplicity 2.

Here we will consider the value of f near the origin. Let a direction be $v = (a, b)$ with $a^2 + b^2 = 1$. Then the bivariate polynomial $f = x^2 + y^3$ becomes a univariate polynomial $a^2t^2 + b^3t^3 = t^2(a^2 + b^3t)$ by substituting $(x = at, y = bt)$. For a generic direction v, the magnitude of f will be $O(t^2)$ as $t \to 0$. Here the degree 2 coincides with the multiplicity. But a lower bound is obtained for the direction $v = (0, 1)$ where the value of $f = t^3$ is even smaller of order 3.

In general we define **degree index** to study multiplicity discussed above.

Definition 3. *Let $f_v = f(vt)$ which is a polynomial in $\mathbb{R}[t]$ by substituting $x = vt$ into f with $v \neq 0 \in \mathbb{R}^n$. The lowest degree of nonzero terms of f_v is denoted by $\deg_{\min}(f_v)$. We define the* ***degree index*** *of f to be*

$$\deg_{ind}(f) = \max_v \deg_{\min}(f_v) \tag{10}$$

Furthermore, for any polynomial f and a point $p \in \mathbb{R}^n$, if $f(p) = 0$ then we define the ***degree index*** *of f at p to be $\deg_{ind}(f(x + p))$.*

For instance, $\deg_{ind}(x^2 - y^2) = 2$, $\deg_{ind}(x^2 + y^2) = 2$, and $\deg_{ind}(x^2 + y^3) = 3$, etc. But it is difficult to compute the degree index of an arbitrary multivariate polynomial f. It can be reduced to finding a nonzero common real root of the sequence $\{f_m = 0, f_{m+1} = 0, \ldots\}$. However, by definition, if $\deg(f) = d > 0$, then we have $1 \leq \deg_{ind}(f) \leq d$, which gives a simple bound.

Suppose $f_v(t) = a_0 t^{\alpha_0} + a_1 t^{\alpha_1} + \cdots + a_k t^{\alpha_k}$ is not a zero polynomial and $\deg_{\min}(f_v) = \alpha_0 < \alpha_1 < \cdots < \alpha_k$. The lowest degree term is $a_0 t^{\alpha_0}$ which is the dominant term when $t \ll 1$. Thus, we have the following result.

Proposition 4. *Let $f \in \mathbb{R}[x_1, \ldots, x_n]$ and $f(0) = 0$. For any direction $v \neq 0 \in \mathbb{R}^n$, if $f_v(t) = f(vt)$ is not a zero polynomial, then there is a constant $c > 0$ such that $|f_v(t)| > c\, t^{\deg_{ind}(f)}$ for sufficiently small $t > 0$.*

As in Sect. 3, suppose the point $p \in V_{\mathbb{R}}(f)$ minimizing distance to a random point $\mathfrak{a}$ is singular. Let p' be the corresponding local minimum close to p of Problem (5). We have the following estimation for $\|p - p'\|$.

Theorem 5. *For a random point $\mathfrak{a} \in \mathbb{R}^n$ and a sufficiently large β, suppose $p \in V_{\mathbb{R}}(f)$ attains the local minimal distance to $\mathfrak{a}$. Then there is a solution p' of Eq. (4) such that $\|p' - p\| \leq O(\sqrt[2I-1]{1/\beta})$, where $I = \max\{\deg_{ind}(f_i(x + p)), i = 1, \ldots, k\}$.*

Proof. By Eq. (5), $\mu(p') = (\beta \sum_i f_i(p')^2 + \|p' - \mathfrak{a}\|^2)/2$. Let $D = \|p - \mathfrak{a}\|$. The relationship between points p and p' is shown in Fig. 3, where $r = \|p' - p\|$. Since D is the local minimal distance from $V_{\mathbb{R}}(f)$ to $\mathfrak{a}$ and $\mu(p') < \mu(p) = D^2/2$, we have $\|p' - \mathfrak{a}\| < D$ and $p' \notin V_{\mathbb{R}}(f)$ and the angle θ between $\overline{pp'}$ and $\overline{p\mathfrak{a}}$ is less

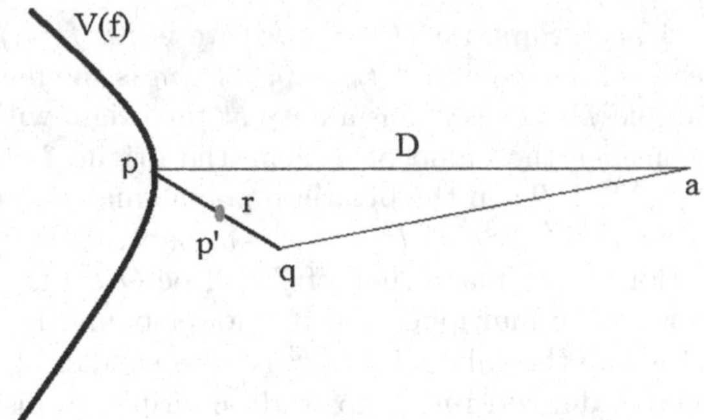

Fig. 3. The minimum value μ is attained at p' which is not a zero of f.

than $\pi/2$. Then $2\mu(p') = (D - r\cos\theta)^2 + r^2\sin^2\theta + \beta\sum_i f_i(p')^2 < D^2 = 2\mu(p)$, which is equivalent to $\beta\sum_i f_i(p')^2 + r^2 < 2r\cos\theta D$.

Thus $\beta\sum_i f_i(p')^2 < 2rD$ holds. Since $f(p') \neq 0$, there is at least one nonzero $f_i(p')$. Recall that $r = \|p' - p\|$, thus for the direction $v = (p' - p)/\|p - p'\|$, $f_i(vr + p) = f_i(p') \neq 0$, which implies that $f_i(vt + p)$ is a nonzero polynomial in t. By Proposition 4, when r is small enough, we have

$$f_i(p')^2 = f_i(vr + p)^2 > c^2\, r^{2\deg_{ind}(f_i(x+p))} \geq c^2\, r^{2I}.$$

Thus, we get $\beta c^2 r^{2I} < 2rD \Rightarrow r^{2I-1} < O(1/\beta)$. □

Remark 2. If the input system f satisfies the rank condition, then we have $I = 1$ and $\|p' - p\| \leq O(1/\beta)$ in Theorem 5, which is in consistent with Theorem 2. But Theorem 2 provides a more precise estimation in this case.

Recall Example 2 in Sect. 3, $I = \deg_{ind}(f) = 3$. By Theorem 5, $r = C\sqrt[2I-1]{1/\beta}$ for some constant C. Then $\log(r) = \log(C) - \frac{1}{2I-1}\log(\beta) = \log(C) - 0.2\log(\beta)$ which is in close agreement with the experimental results.

4.2 Improve Accuracy

In contrast to Sect. 3, the input polynomial system may not satisfy the rank assumption. Consequently it is difficult to apply local methods such as Newton iteration to improve accuracy. For example if $f = x^2 + y^2$ and an approximate root of $f = 0$ close to 0 is given, it is still difficult to determine how to update the root because there is only one equation in f.

By Corollary 1, theoretically we can use Eq. (4) to update the approximate root x' by increasing β. But introducing a very large β will lead to numerical instability. To ease this difficulty, we substitute $\beta = 1/t$ into Eq. (4). Multiplying by t gives

$$t\begin{pmatrix} x_1 - a_1 \\ \vdots \\ x_n - a_n \end{pmatrix} + \mathcal{J}^t \cdot \begin{pmatrix} f_1 \\ \vdots \\ f_k \end{pmatrix} = 0. \tag{11}$$

which can be considered as a homotopy with initial points in the form of (t_0, x_0), where $t_0 = 1/\beta$ with $\beta \gg 0$ and x_0 is a real solution of Eq. (4). When $t \to 0$, the homotopy path $x(t)$ will approach $V_{\mathbb{R}}(f)$. The invertibility of the Jacobian along the homotopy path is guaranteed by Lemma 2.

Let us reduce the value of t by a half at each step i.e. $t = \frac{1}{2^s \beta}$ after s steps. Combining with the result of Theorem 5, we have the following result.

Corollary 6. *Let $\tau = 2^{-1/(2d-1)} < 1$. After s steps of path tracking, the error of root is reduced to $O(\tau^s\, r)$, where r is the initial error $\|p' - p\|$.*

Example 3. Next we consider a sum of squares $f = x^2 + y^2$ with $\mathfrak{a} = (-1, 0.5)$. When $\beta = 1000$, the solution of (4) is the real point $(x = -0.0719, y = 0.0359)$ with a small residual $f(x, y) = 0.00646$. By tracking the path of the homotopy (11), it yields a sequence of points shown in Fig. 4. After 30 steps, we obtain the point $(x = 0.0000517, y = 0.000103)$ with residual $f(x, y) = 1.34 \times 10^{-8}$.

The `CurveFitting[LeastSquares]` command in Maple gives the formula $\log(r) \doteq -0.101 - 0.331 \log(\beta)$, where the coefficient -0.331 is very consistent with the formula $-\frac{1}{2I-1} = -0.333$ in Theorem 5, where $I = \deg_{ind}(x^2 + y^2) = 2$.

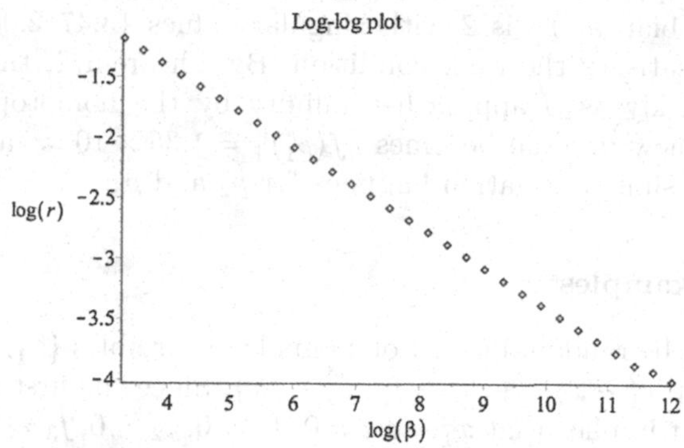

Fig. 4. The error for $f = x^2 + y^2$ in Example 3.

5 Examples

In this section, we demonstrate the generalized critical point method for finding real witness points of a general system on several examples. The numerical tool for solving the zero dimensional system (4) can be found in [18].

5.1 ISSAC 2016 System

Let us consider an interesting example $f = \{xyz, z(x^2 + y^2 + z^2 + y), y(y + z)\}$ in [7]. The real variety consists of a line $\{y = 0, z = 0\}$ and an isolated point $(0, -1/2, 1/2)$. We verify this result by `RealTriangularize` [12] in Maple 18.

To obtain the initial approximation, we set $\beta = 10000, \mathfrak{a} = (1, 0.5, 2)$ and solve the corresponding system (4) by Hom4ps2 [18] with the output $\{[x = 0.00159, y = -0.499, z = 0.500], [x = 0.999, y = 0.0286, z = 0.000169]\}$. The first is an isolated solution with full rank Jacobian, so it can be refined simply. But the Jacobian at the second point is close to singular with singular values $\{1.03, 0.057, 0.0000045\}$. Using the homotopy (11), we can refine this point to $(1.0, 0.000044, 0.0)$ after 30 steps of path tracking. At the exact solution $(1, 0, 0)$ the degree index is 2. Hence, by Corollary 6 we have $\tau = 0.794$ and $\tau^{30} \times 0.0286 = 0.000028$ which has the same magnitude with 0.000044.

5.2 Seiler System

The system is $f = \{x_3^2 + x_2x_3 - x_1^2, x_1x_3 + x_1x_2 - x_3, x_2x_3 + x_2^2 + x_1^2 - x_1\}$ whose real variety is a curve. We use RealTriangularize to obtain a triangular set

$$\{(x_2 + x_3)\, x_1 - x_3, x_2^3 + 3\, x_3x_2^2 + 3\, x_3^2x_2 + x_3^3 - x_3\}.$$

Our method gives three initial points when $\beta = 10000$ and $\mathfrak{a} = (1, 1, 1)$, namely $p_1 = (0.233, 0.37, 0.113)$, $p_2 = (0.0546, -0.22, -0.013)$ and $p_3 = (1.12, 0.13, -1.19)$ with residuals $\|f(p_1)\| = 0.000096$, $\|f(p_2)\| = 0.00013$, $\|f(p_3)\| = 0.000014$. The rank of the Jacobian at p_1 is 2 with singular values $\{3.47, 2.06, 4.16 \times 10^{-6}\}$ which means f satisfies the rank condition. By Theorem 2, the accuracy can be improved quickly as β approaches infinity by the homotopy (11). In our experiment, the new residual becomes $\|f(p_1')\| = 1.35 \times 10^{-11}$ after 20 steps of path tracking. A similar situation happens for p_2 and p_3.

5.3 Larger Examples

Let f_1, f_2 and f_3 be random linear polynomials in variables $\{x_1, \ldots, x_n\}$ (where $n \geq 4$) and $f = \{(x_1^2 - x_2)^2 + f_1^2 + f_2^2, f_2^2 - f_3^2\}$. Since the first polynomial is a sum of squares, it implies that $x_1^2 - x_2 = 0, f_1 = 0, f_2 = 0, f_3 = 0$ which defines an $n - 4$ dimensional real variety of degree two. Let $f_1 = 97x_1 - 67x_2 + 58x_3 + 29x_4 + 37, f_2 = 5x_1 - 36x_2 - 57x_3 + 85x_4 + 80, f_3 = 90x_1 + 74x_2 + 27x_3 + 9x_4 - 91$.

Applying RealTriangularize to f yields a triangular set T in 3.6 s: $\{876997x_1 + 665882x_4 + 70645 = 0, 876997x_2 - 321399x_4 - 932414 = 0, 876997x_3 - 1046403x_4 - 635783 = 0, 443398837924x_4^2 - 187783491023x_4 - 812733564733 = 0\}$. Thus, there are two isolated solutions:

$$(0.799123750840176, 0.638598769156873, -0.657417515791706, -1.15857484076095),$$
$$(-1.28179034942671, 1.64298649988345, 2.61264348189493, 1.58208404954058).$$

Let $\beta = 10000$ and $\mathfrak{a} = (0, 1, 0.21, -0.053)$ and solve the corresponding square system (4) numerically by Hom4ps2 in 3.09 s to obtain 4 real solutions, which are $(0.775, 0.642, -0.613, -1.12)$, $(0.775, 0.642, -0.613, -1.12)$, $(0.781, 0.650, -0.614, -1.12)$ and $(-1.24, 1.60, 2.53, 1.49)$.

We cannot refine these roots directly since f consists of only two equations. To improve the accuracy we apply the techniques introduced in Sect. 4.2 and after 30 steps of path tracking the refined roots are

$$(0.799096390527587, 0.638616146725934, -0.657352004837882, -1.15851537630920),$$
$$(0.799085595521195, 0.638625438788475, -0.657323985904906, -1.15848904058360),$$
$$(0.799096848516397, 0.638616471399574, -0.657351954804033, -1.15851596503370),$$
$$(-1.28174450424170, 1.64292754794384, 2.61255064161436, 1.58197011638993).$$

Apparently, the first three are multiple roots.

When $n = 5$, numerical solving for approximate points costs 17.6 s and refinement costs 0.26 s. It gives two real witness points. On the other hand, if we still use `RealTriangularize` to compute the triangular set of f, after 2902 s Maple 18 displayed an Error message and indicated that "Maple was unable to allocate enough memory to complete this computation".

Moreover, numerical solving stage costs 129 and 559 s for $n = 6$ and $n = 7$ respectively. Since it is difficult to compute the exact solutions, we verify the output by substituting the refined witness points back to f and the residuals are less than 10^{-7}.

6 Conclusions

This paper is part of a series in which we develop algorithms for numerical algebraic geometry. In current paper, we present a new formulation with penalty function, which is a development of critical point techniques. Comparing with existing numerical critical point methods, this method does not require the input system to have pure dimension or satisfy regularity assumptions. It leads to a kind of penalty function approximation method. The convergence rate of the method is given and it is in close agreement with our experimental results.

We plan to apply our method to larger systems, and those with approximate coefficients, which are beyond limitations of current (e.g. symbolic computation) based approaches. Since a non-radical polynomial system does not satisfy the rank condition, it is still difficult to move the obtained approximate real witness points on positive dimensional components to detect more geometric information. Extracting such information will be a focus of future work. In the second stage of our method, we assume local convergence of Newton iteration. An interesting research problem is to compare our approach with the certified path tracking approach [10].

Acknowledgements. The authors would like to thank the anonymous reviewers for their constructive comments that greatly helped improving the paper. This work is partially supported by the projects NSFC (11471307, 11671377, 61572024), cstc2015jcyjys40001, and the Key Research Program of Frontier Sciences of CAS (QYZDB-SSW-SYS026).

References

1. Aubry, P., Rouillier, F., El Din, M.S.: Real solving for positive dimensional systems. J. Symb. Comput. **34**(6), 543–560 (2002)
2. Basu, S., Pollack, R., Roy, M.-F.: Algorithms in Real Algebraic Geometry. Algorithms and Computation in Mathematics, vol. 10, 2nd edn. Springer, Heidelberg (2006). doi:10.1007/3-540-33099-2
3. Besana, G.M., DiRocco, S., Hauenstein, J.D., Sommese, A.J., Wampler, C.W.: Cell decomposition of almost smooth real algebraic surfaces. Numer. Algorithms **63**(4), 645–678 (2013)
4. Bjorck, A.: Numerical Methods for Least Squares Problems. SIAM, Philadelphia (1996)
5. Bank, B., Giusti, M., Heintz, J.: Point searching in real singular complete intersection varieties - algorithms of intrinsic complexity. Math. Comput. **83**(286), 873–897 (2014)
6. Bank, B., Giusti, M., Heintz, J., Mbakop, G.-M.: Polar varieties, real equation solving, and data structures: the hypersurface case. J. Complex. **13**, 5–27 (1997)
7. Brake, D.A., Hauenstein, J.D., Liddell, A.C.: Numerically validating the completeness of the real solution set of a system of polynomial equations. ISSAC **2016**, 143–150 (2016)
8. Bates, D.J., Hauenstein, J.D., Sommese, A.J., Wampler, C.W.: Adaptive multiprecision path tracking. SIAM J. Numer. Anal. **46**(2), 722–746 (2008)
9. Bates, D.J., Hauenstein, J.D., Sommese, A.J., Wampler, C.W.: Numerically Solving Polynomial Systems with the Software Package Bertini. SIAM, Philadelphia (2013)
10. Beltrán, C., Leykin, A.: Robust Certified Numerical Homotopy Tracking. Found. Comput. Math. **13**(2), 253–295 (2013)
11. Basu, S., Roy, M.-F., El Din, M.S., Schost, É.: A baby step-giant step roadmap algorithm for general algebraic sets. Found. Comput. Math. **14**(6), 1117–1172 (2014)
12. Chen, C., Davenport, J.H., May, J.P., Moreno Maza, M., Xia, B., Xiao, R.: Triangular decomposition of semi-algebraic systems. J. Symb. Comput. **49**, 3–26 (2013)
13. Collins, G.E.: Quantifier elimination for real closed fields by cylindrical algebraic decompostion. In: Brakhage, H. (ed.) GI-Fachtagung 1975. LNCS, vol. 33, pp. 134–183. Springer, Heidelberg (1975). doi:10.1007/3-540-07407-4_17
14. Davenport, J.H., Heintz, J.: Real quantifier elimination is doubly exponential. J. Symb. Comp. **5**, 29–35 (1988)
15. Hauenstein, J.: Numerically computing real points on algebraic sets. Acta Appl. Math. **125**(1), 105–119 (2013)
16. Hauenstein, J., Sommese, A.: What is numerical algebraic geometry? J. Symb. Comp. **79**, 499–507 (2017). Part 3
17. Hong, H.: Improvement in CAD-Based Quantifier Elimination. Ph.D. thesis. Ohio State University, Columbus, Ohio (1990)
18. Li, T.Y., Lee, T.L.: Homotopy method for solving Polynomial Systems software. http://www.math.msu.edu/~li/Software.htm
19. Lee, J.M.: Introduction to Smooth Manifolds, vol. 218. Springer, Heidelberg (2003). doi:10.1007/978-0-387-21752-9
20. Lasserre, J.B., Laurent, M., Rostalski, P.: Semidefinite characterization and computation of zero-dimensional real radical ideals. Found. Comput. Math. **8**(5), 607–647 (2008)

21. Lasserre, J.B., Laurent, M., Rostalski, P.: A prolongation-projection algorithm for computing the finite real variety of an ideal. Theoret. Comput. Sci. **410**(27–29), 2685–2700 (2009)
22. Lu, Y.: Finding all real solutions of polynomial systems. Ph.D thesis. University of Notre Dame (2006). Results of this thesis appear. In: (with Bates, D.J., Sommese, A.J., Wampler, C.W.), Finding all real points of a complex curve, Contemp. Math. vol. 448, pp. 183–205 (2006)
23. Ma, Y., Zhi, L.: Computing Real Solutions of Polynomial Systems via Low-rank Moment Matrix Completion. In: ISSAC, pp. 249–256 (2012)
24. Rouillier, F., Roy, M.-F., El Din, M.S.: Finding at least one point in each connected component of a real algebraic set defined by a single equation. J. Complex. **16**(4), 716–750 (2000)
25. El Din, M.S., Schost, É.: Polar varieties and computation of one point in each connected component of a smooth real algebraic set. In: ISSAC 2013, pp. 224–231 (2003)
26. El Din, M.S., Schost, É.: Properness defects of projection functions and computation of at least one point in each connected component of a real algebraic set. J. Discrete Comput. Geom. **32**(3), 417–430 (2004)
27. Sommese, A.J., Wampler, C.W.: The Numerical Solution of Systems of Polynomials Arising in Engineering and Science. World Scientific Press (2005)
28. Stewart, G.W.: Perturbation theory for the singular value decomposition. In: SVD and Signal processing, II: Algorithms, Analysis and Applications, pp. 99–109. Elsevier (1990)
29. Sommese, A.J., Verschelde, J., Wampler, C.W.: Introduction to numerical algebraic geometry. In: Bronstein, M., et al. (eds.) Solving Polynomial Equations. AACIM, vol. 14, pp. 339–392. Springer, Heidelberg (2005). doi:10.1007/3-540-27357-3_8
30. Sternberg, S.: Lectures on Differential Geometry. Prentice-Hall, Englewood Cliffs (1964)
31. Wu, W., Reid, G.: Finding points on real solution components and applications to differential polynomial systems. In: ISSAC, pp. 339–346 (2013)
32. Wu, W., Reid, G., Feng, Y.: Computing real witness points of positive dimensional polynomial systems. Accepted by Theoretical Computer Sciences (2017). http://doi.org/10.1016/j.tcs.2017.03.035
33. Yang, Z., Zhi, L., Zhu, Y.: Verified error bounds for real solutions of positive-dimensional polynomial systems. In: ISSAC, pp. 371–378 (2013)

Computing Multiple Zeros of Polynomial Systems: Case of Breadth One (*Invited Talk*)

Lihong Zhi[1,2(✉)]

[1] Key Laboratory of Mathematics Mechanization, Academy of Mathematics and System Sciences, Beijing, China
[2] School of Mathematical Sciences, University of Chinese Academy of Sciences, Beijing 100190, China
lzhi@mmrc.iss.ac.cn

Abstract. Given a polynomial system f with a multiple zero x whose Jacobian matrix at x has corank one, we show how to compute the multiplicity structure of x and the lower bound on the minimal distance between the multiple zero x and other zeros of f. If x is only given with limited accuracy, we give a numerical criterion to guarantee that f has μ zeros (counting multiplicities) in a small ball around x. Moreover, we also show how to compute verified and narrow error bounds such that a slightly perturbed system is guaranteed to possess an isolated breadth-one singular solution within computed error bounds. Finally, we present modified Newton iterations and show that they converge quadratically if x is close to an isolated exact singular solution of f. This is joint work with Zhiwei Hao, Wenrong Jiang, Nan Li.

1 Introduction

Let I_f be an ideal generated by polynomials $f = \{f_1, \ldots, f_n\}$, where $f_i \in \mathbb{C}[X_1, \ldots, X_n]$. An isolated zero of multiplicity μ for f is a point $x \in \mathbb{C}^n$ such that

1. $f(x) = 0$,
2. there exists a ball $B(x,r)$ of radius $r > 0$ such that $B(x,r) \cap f^{-1}(0) = \{x\}$,
3. $\mu = \dim(\mathbb{C}[X]/Q_{f,x})$,

where

$$B(x,r) := \{y \in \mathbb{C}^n : \|y - x\| < r\},$$

and $Q_{f,x}$ is a primary component of the ideal I_f whose associate prime is

$$m_x = (X_1 - x_1, \ldots, X_n - x_n).$$

This research was supported in part by the National Key Research Project of China 2016YFB0200504 (Zhi) and the National Natural Science Foundation of China under Grants 11571350 (Zhi).

V.P. Gerdt et al. (Eds.): CASC 2017, LNCS 10490, pp. 392–405, 2017.
DOI: 10.1007/978-3-319-66320-3_28

Let $\mathbf{d}_x^{\boldsymbol{\alpha}} : \mathbb{C}[X] \to \mathbb{C}$ denote the differential functional defined by

$$\mathbf{d}_x^{\boldsymbol{\alpha}}(g) = \frac{1}{\alpha_1! \cdots \alpha_n!} \cdot \frac{\partial^{|\boldsymbol{\alpha}|} g}{\partial x_1^{\alpha_1} \cdots \partial x_n^{\alpha_n}}(x), \quad \forall g \in \mathbb{C}[X], \tag{1}$$

where $x \in \mathbb{C}^n$ and $\boldsymbol{\alpha} = [\alpha_1, \ldots, \alpha_n] \in \mathbb{N}^n$. We have

$$\mathbf{d}_x^{\boldsymbol{\alpha}}\left((X - x)^{\boldsymbol{\beta}}\right) = \begin{cases} 1, \text{ if } \boldsymbol{\alpha} = \boldsymbol{\beta}, \\ 0, \text{ otherwise.} \end{cases} \tag{2}$$

The local dual space of I_f at a given isolated singular solution x is a subspace $\mathcal{D}_{f,x}$ of $\mathfrak{D}_x = \operatorname{span}_{\mathbb{C}}\{\mathbf{d}_x^{\boldsymbol{\alpha}}\}$ such that

$$\mathcal{D}_{f,x} = \{\Lambda \in \mathfrak{D}_x \mid \Lambda(g) = 0, \ \forall g \in I_f\}. \tag{3}$$

When the evaluation point x is clear from the context, we write $d_1^{\alpha_1} \cdots d_n^{\alpha_n}$ instead of $\mathbf{d}_x^{\boldsymbol{\alpha}}$ for simplicity.

Let $\mathcal{D}_{f,x}^{(k)}$ be the subspace of $\mathcal{D}_{f,x}$ with differential functionals of orders bounded by k, we define

1. breadth $\kappa = \dim\left(\mathcal{D}_{f,x}^{(1)} \setminus \mathcal{D}_{f,x}^{(0)}\right)$,
2. depth $\rho = \min\left(\left\{k \mid \dim\left(\mathcal{D}_{f,x}^{(k+1)} \setminus \mathcal{D}_{f,x}^{(k)}\right) = 0\right\}\right)$,
3. multiplicity $\mu = \dim\left(\mathcal{D}_{f,x}^{(\rho)}\right)$.

If x is an isolated singular solution of f, then $1 \leq \kappa \leq n$ and $\rho < \mu < \infty$.

We recall α-theory below according to [1] and refer to [16,37–41,43] for more details.

Let $Df(x)$ denote the Jacobian matrix of f at x. Suppose $Df(x)$ is invertible, x is called a simple (regular) zero of f. The Newton's iteration is defined by

$$N_f(x) = x - Df(x)^{-1} f(x). \tag{4}$$

Shub and Smale [37] defined

$$\gamma(f, x) = \sup_{k \geq 2} \left\| Df(x)^{-1} \cdot \frac{D^k f(x)}{k!} \right\|^{\frac{1}{k-1}}, \tag{5}$$

where $D^k f$ denotes the k-th derivative of f which is a symmetric tensor whose components are the partial derivatives of f of order k, $\|\cdot\|$ denotes the classical operator norm.

According to [1, Theorem 1], if

$$\|z - x\| \leq \frac{3 - \sqrt{7}}{2\gamma(f, x)}, \tag{6}$$

then Newton's iterations starting at z will converge quadratically to the simple zero x.

If y is another zero of f, according to [1, Corollary 1], we have

$$\|y - x\| \geq \frac{5 - \sqrt{17}}{4\gamma(f, x)}, \tag{7}$$

which separates the simple zero x from other zeros of f.

Furthermore, according to [1, Theorem 2], if only a system f and a point x are given such that

$$\alpha(f,x) \leq \frac{13-3\sqrt{17}}{4} \approx 0.157671, \tag{8}$$

where $\alpha(f,x) = \beta(f,x)\gamma(f,x)$ and

$$\beta(f,x) = \|x - N_f(x)\| = \|Df(x)^{-1}f(x)\|,$$

then Newton's iterations starting at x will converge quadratically to a simple zero ξ of f and

$$\|x-\xi\| \leq 2\beta(f,x).$$

It is a challenge to extend α-theory for polynomial systems with singular solutions. When $Df(x)$ is not invertible, many modifications of Newton's iteration to restore the quadratic convergence for singular solutions have been proposed in [2,6–8,12–14,29–33,36,46]. Recently, some symbolic-numeric methods based on deflated systems have also been proposed for refining approximate isolated singular solutions to high accuracy [3–5,10,11,18–20,25]. For example, as shown in [19], let $r = \operatorname{rank}(Df(x))$, with probability one, there exists a unique vector $\lambda = (\lambda_1, \lambda_2 \ldots, \lambda_{r+1})^T$ such that (x, λ) is an isolated solution of a deflated polynomial system, i.e.,

$$\begin{cases} f(x) = 0, \\ Df(x)B\lambda = 0, \\ h^T\lambda = 1, \end{cases} \tag{9}$$

where $B \in \mathbb{C}^{n\times(r+1)}$ is a random matrix, $h \in \mathbb{C}^{r+1}$ is a random vector. If (x, λ) is still a singular solution of (9), the deflation is repeated. Furthermore, they proved that the number of deflations needed to derive a regular solution of an augmented system is strictly less than the multiplicity of x. Dayton and Zeng showed that the depth of $\mathcal{D}_{f,x}$ is a tighter bound for the number of deflations [5].

In [44,45], we present a method based on the reduction to geometric involutive form to compute the primary component and a basis of the local dual space of a polynomial system at an isolated singular solution. We also present an algorithm based on correctly computed multiplicity structure such as index and multiplicity at an approximate singular solution to restore the quadratic convergence of Newton's iterations.

In this paper, we introduce some recent contributions related to extending α-theory for polynomial systems with singular zeros satisfying $f(x) = 0$, $\dim\ker Df(x) = 1$. It is also called breadth-one singular zero in [5] as

$$\dim(\mathcal{D}_{f,x}^{(k)} \setminus \mathcal{D}_{f,x}^{(k-1)}) = 1,\ k = 1\ldots,\rho,\ \rho = \mu - 1. \tag{10}$$

Therefore, the local dual space of I_f at x is

$$\mathcal{D}_{f,x} = \operatorname{span}_{\mathbb{C}}\{\Lambda_0, \Lambda_1, \ldots, \Lambda_{\mu-1}\},$$

where $\deg(\Lambda_k) = k$ and $\Lambda_0 = 1$.

As pointed out in [11], the breath one case is the least degenerate one and therefore most likely to be of practical significance. Moreover, it is also the worst case for the deflation method [5,19,29,30] since the deflation always terminates at step $\mu - 1$. Hence the size of the matrices grows extremely fast with the multiplicity.

2 Local Dual Space

Let us introduce a morphism $\Phi_\sigma : \mathfrak{D}_x \to \mathfrak{D}_x$ which is an anti-differentiation operator defined by

$$\Phi_\sigma(d_1^{\alpha_1} \cdots d_n^{\alpha_n}) = \begin{cases} d_1^{\alpha_1} \cdots d_\sigma^{\alpha_\sigma - 1} \cdots d_n^{\alpha_n}, & \text{if } \alpha_\sigma > 0, \\ 0, & \text{otherwise.} \end{cases}$$

Computing a closed basis of the local dual space is done essentially by matrix-kernel computations based on the stability property of $\mathcal{D}_{f,x}$ [26,28,42]:

$$\forall \Lambda \in \mathcal{D}_{f,x}^{(k)},\ \Phi_\sigma(\Lambda) \in \mathcal{D}_{f,x}^{(k-1)},\quad \sigma = 1, \ldots, n. \tag{11}$$

Let $\mathcal{D}_{f,x}^{(k)}$ be the subspace of $\mathcal{D}_{f,x}$ with differential functionals of orders bounded by k. Let $\Psi_\sigma : \mathfrak{D}_x \to \mathfrak{D}_x$ be a differential operator defined by

$$\Psi_\sigma(d_1^{\alpha_1} \cdots d_n^{\alpha_n}) = \begin{cases} d_\sigma^{\alpha_\sigma + 1} \cdots d_n^{\alpha_n}, & \text{if } \alpha_1 = \cdots = \alpha_{\sigma-1} = 0, \\ 0, & \text{otherwise.} \end{cases}$$

We deal with multiple zeros satisfying $f(x) = 0$, $\dim\ker Df(x) = 1$. The local dual space of I_f at a given isolated singular solution x is

$$\mathcal{D}_{f,x} = \mathrm{span}_{\mathbb{C}}\{\Lambda_0, \Lambda_1, \ldots, \Lambda_{\mu-1}\},$$

where $\deg(\Lambda_k) = k$ and $\Lambda_0 = 1$.

As shown in [23, Theorem 3.4], suppose $\Lambda_1 = a_{1,1}d_1 + \cdots + a_{1,n}d_n$, without loss of generality, we assume $a_{1,1} = 1$, $a_{k,1} = 0$, $k = 2, \ldots, n$. Then for $k = 2, \ldots, \mu - 1$, we have

$$\Lambda_k = \Delta_k + a_{k,2}d_2 + \cdots + a_{k,n}d_n, \tag{12}$$

where

$$\Delta_k = \sum_{\sigma=1}^{n} \Psi_\sigma(a_{1,\sigma}\Lambda_{k-1} + \cdots + a_{k-1,\sigma}\Lambda_1), \tag{13}$$

and $a_{k,2}, \ldots, a_{k,n}$ are determined by solving the linear system obtained from setting $\Lambda_k(f_i) = 0, i = 1, \ldots, n$:

$$\begin{pmatrix} d_2(f_1) & \cdots & d_n(f_1) \\ \vdots & \ddots & \vdots \\ d_2(f_n) & \cdots & d_n(f_n) \end{pmatrix} \begin{pmatrix} a_{k,2} \\ \vdots \\ a_{k,n} \end{pmatrix} = - \begin{pmatrix} \Delta_k(f_1) \\ \vdots \\ \Delta_k(f_n) \end{pmatrix}. \tag{14}$$

Definition 1 [15]. *For a polynomial function* $f : \mathbb{C}^n \rightarrow \mathbb{C}^n$, *suppose* $f(x) = 0, \dim\ker Df(x) = 1$. *Then* $Df(x)$ *has a normalized form if*

$$Df(x) = \begin{pmatrix} 0 & D\hat{f}(x) \\ 0 & 0 \end{pmatrix}, \tag{15}$$

$D\hat{f}(x)$ *is the nonsingular Jacobian matrix of polynomials* $\hat{f} = \{f_1, \ldots, f_{n-1}\}$ *with respect to variables* $X_2, \ldots, X_n$.

If x is a multiple zero of multiplicity μ for f and $Df(x)$ has the normalized form (15), which is always possible to obtain by performing unitary transformations when $\dim\ker Df(x) = 1$, see [15, Sect. 2.3], then we have $\Delta_k(f_n) = 0$, for $k = 2, \ldots, \mu - 1$, $\Delta_\mu(f_n) \neq 0$, and the linear system (14) for getting the values of $a_{k,2}, \ldots, a_{k,n}$ can be simplified to:

$$\begin{pmatrix} d_2(f_1) & \cdots & d_n(f_1) \\ \vdots & \ddots & \vdots \\ d_2(f_{n-1}) & \cdots & d_n(f_{n-1}) \end{pmatrix} \begin{pmatrix} a_{k,2} \\ \vdots \\ a_{k,n} \end{pmatrix} = - \begin{pmatrix} \Delta_k(f_1) \\ \vdots \\ \Delta_k(f_{n-1}) \end{pmatrix}. \tag{16}$$

3 Local Separation Bound and Cluster Location

In [9], Dedieu and Shub gave quantitative results for simple double zeros satisfying $f(x) = 0$ and

(A) $\dim\ker Df(x) = 1$,
(B) $D^2 f(x)(v, v) \notin \mathrm{im} Df(x)$,

where $\ker Df(x)$ is spanned by a unit vector $v \in \mathbb{C}^n$. They generalized the definition of γ in (5) to

$$\gamma_2(f, x) = \max\left(1, \sup_{k\geq 2} \left\| A(f, x, v)^{-1} \cdot \frac{D^k f(x)}{k!} \right\|^{\frac{1}{k-1}}\right), \tag{17}$$

where

$$A(f, x, v) = Df(x). + \frac{1}{2} D^2 f(x)(v, \Pi_v), \tag{18}$$

is a linear operator which is invertible at the simple double zero x, and Π_v denotes the Hermitian projection onto the subspace $[v] \subset \mathbb{C}^n$.

In [9, Theorem 1], Dedieu and Shub also presented a lower bound for separating simple double zeros x from the other zeros y of f,

$$\|y - x\| \geq \frac{d}{2\gamma_2(f, x)^2}, \tag{19}$$

where $d \approx 0.2976$ is a positive real root of

$$\sqrt{1 - d^2} - 2d\sqrt{1 - d^2} - d^2 - d = 0. \tag{20}$$

In [9, Theorem 4], Dedieu and Shub showed that if the following criterion is satisfied at a given point x and a given vector v

$$\|f(x)\| + \|Df(x)v\| \frac{d}{4\gamma_2(f,x,v)^2} < \frac{d^3}{32\gamma_2^4\|B(f,x,v)^{-1}\|}, \tag{21}$$

then f has two zeros in the ball of radius

$$\frac{d}{4\gamma_2(f,x)^2}, \tag{22}$$

around x. Let us set

$$B(f,x,v) = A(f,x,v) - L,$$

where $L(v) = Df(x)v$, $L(w) = 0$ for $w \in v^{\perp}$, and

$$\gamma_2(f,x) = \max\left(1, \sup_{k\geq 2}\left\|B(f,x,v)^{-1}\cdot\frac{D^k f(x)}{k!}\right\|^{\frac{1}{k-1}}\right). \tag{23}$$

Based on the multiplicity structure of the singular zero x of f computed in the last section, we generalize Dedieu and Shub's results to multiple zeros with arbitrary large multiplicity.

Let $f : \mathbb{C}^n \to \mathbb{C}^n$, and x be a singular zero of f of multiplicity μ, where $Df(x)$ has the normalized form $Df(x) = \begin{pmatrix} 0 & D\hat{f}(x) \\ 0 & 0 \end{pmatrix}$, $D\hat{f}(x)$ is invertible and

$$\Delta_k(f_n) = 0, \text{ for } k = 2, \ldots, \mu - 1, \quad \Delta_\mu(f_n) \neq 0. \tag{24}$$

Let y be another vector in $\mathbb{C}^n$ and $y \neq x$. Recall that $\varphi = d_P(v, y - x)$, $v = (1, 0, \ldots, 0)^T$ and $w = x - y = (\zeta, \eta_2, \ldots, \eta_n)^T$, $\eta = (\eta_2, \ldots, \eta_n)^T$, then we have $|\zeta| = \|w\| \sin\varphi$, $\|\eta\| = \|w\| \cos\varphi$. Let

$$\mathcal{A} = \begin{pmatrix} \sqrt{2}D\hat{f}(x) & 0 \\ 0 & \frac{1}{\sqrt{2}}\Delta_\mu(f_n) \end{pmatrix},$$

and $\gamma_\mu = \max(\hat{\gamma}_\mu,\ \gamma_{\mu,n})$, where

$$\hat{\gamma}_\mu = \hat{\gamma}_\mu(f,x) = \max\left(1,\ \sup_{k\geq 2}\left\|D\hat{f}(x)^{-1}\frac{D^k\hat{f}(x)}{k!}\right\|^{\frac{1}{k-1}}\right), \tag{25}$$

where $D^k\hat{f}(x)$ for $k \geq 2$ denote the partial derivatives of $\hat{f}$ of order k with respect to $X_1, X_2, \ldots, X_n$ evaluated at x, and

$$\gamma_{\mu,n} = \gamma_{\mu,n}(f,x) = \left(1,\ \sup_{k\geq 2}\left\|\frac{1}{\Delta_\mu(f_n)}\cdot\frac{D^k f_n(x)}{k!}\right\|^{\frac{1}{k-1}}\right), \tag{26}$$

Definition 2 [15, Defintion 3]. *We define* $d = \min(d_1, d_2, d_3)$, *where*

$$d_1 = \sqrt{\frac{1}{c_{\mu-1,1}^2 + 1}}, \; d_2 = \sqrt{\frac{1}{\mu - 1}},$$

and d_3 *is the smallest positive real root of the polynomial*

$$p(d) = (1 - d^2)^{\frac{\mu}{2}} - \sum_{i+j=\mu, j>0} c_{i,j} d (1 - d^2)^{\frac{i}{2}} d^{j-1} \tag{27}$$
$$- d \left(\sum_{1 \le i \le \mu-2} t_{i,0} + \sum_{1 \le i+j \le \mu-2, j>0} t_{i,j} (1 - d^2)^{\frac{i}{2}} d^j + 1 \right),$$

where $c_{i,j}$ *and* $t_{i,j}$ *can be obtained by the method given in [15, Case 2].*

Theorem 1 [15, Theorem 5]. *Let* x *be a multiple zero of* f *of multiplicity* μ, $\dim \ker Df(x) = 1$, *and* y *be another zero of* f, *then*

$$\|y - x\| \ge \frac{d}{2\gamma_\mu^\mu}.$$

Remark 1. For $\mu = 2$, we have [15, Sect. 3.3]

$$p(d) = 1 - 2d^2 - 2d\sqrt{1 - d^2} - d. \tag{28}$$

The smallest positive real root of $p(d)$ is

$$d \approx 0.2865.$$

For $\mu = 3$, we have [15, Lemma 3]

$$p(d) = (1 - 2d - 8d^2)\sqrt{1 - d^2} - 9d - d^2 + 6d^3. \tag{29}$$

The smallest positive root of $p(d)$ is

$$d \approx 0.08507.$$

Theorem 2 [15, Theorem 8]. *Given* $f : \mathbb{C}^n \to \mathbb{C}^n$, $x \in \mathbb{C}^n$, *such that* $D\hat{f}(x)$ *is invertible, and* $\Delta_\mu(f_n) \neq 0$. *Let*

$$H_1 = \begin{pmatrix} \frac{\partial \hat{f}(x)}{\partial X_1} & 0 \\ \frac{\partial f_n(x)}{\partial X_1} & \frac{\partial f_n(x)}{\partial \hat{X}} \end{pmatrix},$$

$$H_k = \left(\begin{pmatrix} 0 & 0 \\ \Delta_k(f_n) & 0 \end{pmatrix} \mathbf{0}_{\underbrace{n \times \cdots \times n}_{k} \times (n-1)} \right), \; 2 \le k \le \mu - 1,$$

and polynomials

$$g(X) = f(X) - f(x) - \sum_{1 \le k \le \mu-1} H_k(X-x)^k.$$

Let $\gamma_\mu = \gamma_\mu(g,x)$, *if*

$$\|f(x)\| + \sum_{1 \le k \le \mu-1} \|H_k\| \left(\frac{d}{4\gamma_\mu^\mu}\right)^k < \frac{d^{\mu+1}}{2\,(4\gamma_\mu^\mu)^\mu \,\|\mathcal{A}^{-1}\|}, \tag{30}$$

then f *has* μ *zeros (counting multiplicities) in the ball of radius* $\frac{d}{4\gamma_\mu^\mu}$ *around* x.

4 Verified Error Bound

Let $\mathbb{IR}$ be the set of real intervals, and let $\mathbb{IR}^n$ and $\mathbb{IR}^{n\times n}$ be the set of real interval vectors and real interval matrices, respectively. Standard verification methods for nonlinear systems are based on the following theorem.

Theorem 1 [17,27,34]. *Let* $f : \mathbb{R}^n \to \mathbb{R}^n$ *be a system of nonlinear equations. Suppose* $x \in \mathbb{R}^n$, $\mathbf{X} \in \mathbb{IR}^n$ *with* $0 \in \mathbf{X}$ *and* $R \in \mathbb{R}^{n\times n}$ *are given. Let* $\mathbf{M} \in \mathbb{IR}^{n\times n}$ *be given such that*

$$\{Df_i(y) : y \in x + \mathbf{X}\} \subseteq \mathbf{M}_{i,:}, i = 1, \ldots, n. \tag{31}$$

Denote by I_n *the* $n \times n$ *identity matrix and assume*

$$-Rf(x) + (I_n - R\mathbf{M})\mathbf{X} \subseteq \mathrm{int}(\mathbf{X}). \tag{32}$$

Then there is a unique $\tilde{x} \in x + \mathbf{X}$ *satisfying* $f(\tilde{x}) = 0$. *Moreover, every matrix* $\tilde{M} \in \mathbf{M}$ *is nonsingular. In particular, the Jacobian matrix* $Df(\tilde{x})$ *is nonsingular.*

Theorem 1 is restricted to verifying the existence of a simple solution of a square and regular system. Notice that Theorem 1 is valid over complex numbers with the necessary modifications. In [35], by introducing a smoothing parameter, Rump and Graillat developed a verification method for computing verified and narrow error bounds, such that a slightly perturbed system is proved to possess a double root within computed error bounds.

In [23], by adding a univariate polynomial in one selected variable with some smoothing parameters to one selected equation of the original system, we generalized the algorithm in [35] to compute guaranteed error bounds such that a slightly perturbed system is proved to have a breadth-one isolated singular solution within computed error bounds.

For a polynomial function $f : \mathbb{C}^n \to \mathbb{C}^n$, where $f_i \in \mathbb{C}[X_1, \ldots, X_n]$, and suppose x is a zero of f of multiplicity μ and satisfying $\dim \ker Df(x) = 1$. Suppose the i-th column of $Df(x)$ can be written as a linear combination of the other $n-1$ columns, then we choose x_i as the variable. Similarly, suppose the j-th row of $Df(x)$ can be written as a linear combination of the other $n-1$

linearly independent rows, then we add the perturbed univariate polynomial in x_i to f_j. Finally, we permute

$$x_1 \leftrightarrow x_i \text{ and } f_1 \leftrightarrow f_j$$

to construct a deflated system below.

We introduce $\mu - 1$ smoothing parameters $b_0, b_1, \ldots, b_{\mu-2}$ and construct a deflated system $G(X, b, a)$ with μn variables and μn equations:

$$G(X,b,a) = \begin{pmatrix} F_1(X,b) = f(X) - \left(\sum_{\nu=0}^{\mu-2} \frac{b_\nu x_1^\nu}{\nu!}\right) e_1 \\ F_2(X,b,a_1) \\ F_3(X,b,a_1,a_2) \\ \vdots \\ F_\mu(X,b,a_1,\ldots,a_{\mu-1}) \end{pmatrix}, \tag{33}$$

where $e_1 = (1, 0, \ldots, 0)^T$, $b = (b_0, b_1, \ldots, b_{\mu-2})$, $a = (a_1, a_2, \ldots, a_{\mu-1})$, $a_1 = (1, a_{1,2}, \ldots, a_{1,n})^T$, $a_k = (0, a_{k,2}, \ldots, a_{k,n})^T$ for $1 < k \leq \mu$, and

$$F_k(X, b, a_1, \ldots, a_{k-1}) = L_{k-1}(F_1), \tag{34}$$

where L_k are differentiation operators corresponding to Λ_k defined by (12).

Theorem 2 [23, Theorem 4.3]. *Suppose $G(x, \tilde{b}, \tilde{a}) = 0$. If the Jacobian matrix of the deflated polynomial system $G(X, b, a)$ at $(x, \tilde{b}, \tilde{a})$ is nonsingular, then x is an isolated root of the perturbed polynomial system $F(X) = F_1(X, \tilde{b})$ with multiplicity μ and the corank of $DF(x)$ is one.*

Theorem 3 [23, Theorem 4.5]. *Suppose Theorem 1 is applicable to $G(X, b, a)$ in (33) and yields inclusions for x, $\tilde{b}$ and $\tilde{a}$ such that $G(x, \tilde{b}, \tilde{a}) = 0$. Then x is an isolated breadth-one root of $F(X) = F_1(X, \tilde{b})$ with multiplicity μ.*

5 Modified Newton Iterations

In [22], we presented a symbolic-numeric method to refine an approximate isolated singular solution $\tilde{x} = (x_1, \ldots, x_n)$ of a polynomial system $f = \{f_1, \cdots, f_n\}$ when the Jacobian matrix of f evaluated at $\tilde{x}$ has corank one approximately. Our approach is based on the regularized Newton iteration and the computation of differential conditions satisfied at the approximate singular solution. The size of matrices involved in our algorithm is bounded by $n \times n$. The algorithm converges quadratically if $\tilde{x}$ is close to the isolated exact singular solution of f.

Theorem 4 [22, Theorem 3.16]. *If the Jacobian matrix of f evaluated at x has corank one and the approximate singular solution $\tilde{x}$ of f satisfying*

$$\|\tilde{x} - x\| = \varepsilon \ll 1,$$

where the positive number ϵ is small enough such that there are no other solutions of f nearby, then the refined singular solution $\tilde{x}$ returned by Algorithm 1 satisfies

$$\|N_f(\tilde{x}) - x\| = O(\varepsilon^2).$$

Algorithm 1. Modified Newton's Iterations for Breadth-one Multiple Zero

Input:

f: a polynomial system;

$\tilde{x}$: an approximate singular zero of f;

μ: the multiplicity

Output:

$N_f(\tilde{x})$: a refined solution after one iteration;

1: solve the regularized least squares problem

$$(Df(\tilde{x})^* Df(\tilde{x}) + \sigma_n I_n)\tilde{y} = Df(\tilde{x})^* b,$$

where $b = -f(\tilde{x})$, I_n is the $n \times n$ identity matrix and σ_n is the smallest singular value of $Df(\tilde{x})$;

2: compute the singular value decomposition of $Df(\tilde{x} + \tilde{y}) = U \cdot \Sigma \cdot V^*$, let

$$g(X) = f(W \cdot X), \quad W = (v_n, v_1, \ldots, v_{n-1}),$$

and set $\tilde{z} \leftarrow W^*(\tilde{x} + \tilde{y})$;

3: construct Δ_μ and a closed approximate basis of the local dual space

$$\mathcal{D}_{g,\tilde{z}} = \mathrm{Span}(\Lambda_0, \Lambda_1, \ldots, \Lambda_{\mu-1}),$$

by Algorithm MultiplicityStructureBreadthOneNumeric in [21];

4: solve the linear system

$$\left[\Delta_\mu(g), \frac{\partial g(\tilde{z})}{\partial z_2}, \ldots, \frac{\partial g(\tilde{z})}{\partial z_n}\right] \delta = -\Lambda_{\mu-1}(g)$$

5: update the zero of g

$$\tilde{z}_1 \leftarrow \tilde{z}_1 + \frac{\delta_1}{\mu}, \quad \tilde{z}_i \leftarrow \tilde{z}_i, \; 2 \leq i \leq n$$

and

$$N_f(\tilde{x}) \leftarrow W \cdot \tilde{z}.$$

The proof of Theorem 4 in [22] is based on studying zeros of deflated systems. It is difficult to quantify the quadratical convergence of Algorithm 1. In [15], we present a new algorithm for refining an approximate singular zero whose Jacobian matrix has corank one. The main idea is to perform the unitary transformations to both variables and equations defined at the approximate singular solutions, then define the modified Newton's iteration which are very similar to Step 4 in Algorithm 1.

Theorem 3. *Given an approximate zero z of a polynomial system f associated to a multiple zero ξ of multiplicity μ and satisfying* $f(\xi) = 0$, $\dim \ker Df(\xi) = 1$. *Suppose*

$$\hat{\gamma}_\mu(f, z)\|z - \xi\| < \frac{1}{2},$$

Algorithm 2. Modified Newton's Iteration for Breadth-one Multiple Zeros

Input:

f: a polynomial system;

z: an approximate singular zero of f;

μ: the multiplicity;

Output:

$N_f(z)$: a refined solution after one iteration;

1: compute the singular value decomposition

$$Df(z) = U \cdot \begin{pmatrix} \Sigma_{n-1} & 0 \\ 0 & \sigma_n \end{pmatrix} \cdot V^*, \; W_\dagger = (v_n, v_1, \ldots, v_{n-1});$$

2: perform the unitary transformations to equations and variables

$$f(X) \leftarrow U^* \cdot f(W_\dagger \cdot X), \quad z \leftarrow W_\dagger^* z;$$

3: update the last $n-1$ elements of the approximate zero

$$N_1(\hat{f}, \hat{z}) \leftarrow \hat{z} - D\hat{f}(z)^{-1}\hat{f}(z), \quad y = (y_1, \hat{y}) \leftarrow (z_1, N_1(\hat{f}, \hat{z}));$$

4: compute the singular value decomposition

$$Df(y) = U \cdot \begin{pmatrix} \Sigma_{n-1} & 0 \\ 0 & \sigma_n \end{pmatrix} \cdot V^*, W_\ddagger = (v_n, v_1, \ldots, v_{n-1});$$

5: perform the unitary transformations to equations and variables:

$$g(X) \leftarrow U^* \cdot f(W_\ddagger \cdot X), \quad w = (w_1, \hat{w}) \leftarrow W_\ddagger^* y;$$

6: update the first element of the approximate zero

$$N_2(g_n, w) \leftarrow w_1 - \frac{1}{\mu}\Delta_\mu(g_n)^{-1}\Delta_{\mu-1}(g_n), \quad x = (x_1, \hat{x}) \leftarrow (N_2(g_n, w), \hat{w});$$

7: update the zero of f

$$N_f(z) \leftarrow W_\dagger \cdot W_\ddagger \cdot x.$$

where $\hat{\gamma}_\mu(f, z)$ *is defined by (25), then the refined singular solution* $N_f(z)$ *returned by Algorithm 2 satisfies*

$$\|N_f(z) - \xi\| = O(\|z - \xi\|^2). \tag{35}$$

In [15, Theorem 12], we give a quantified quadratic convergence proof of the Algorithm 2 for simple triple zeros. There is no significant obstacle to extend the proof to multiple zeros of higher multiplicities. However, the computation will become more complicated.

Theorem 4 [15, Theorem 12]. *Given an approximate zero* z *of a system* f *associated to a simple triple zero* ξ *of multiplicity* 3 *and satisfying* $f(\xi) = 0$, $\dim\ker Df(\xi) = 1$. *Let* $u = \max\{\gamma_3(f, \xi)^3\|\xi - z\|, L\gamma_3(f, \xi)^2\|\xi - z\|\}$, *where* L *is the Lipschitz constant of the function* $Df(X)$.

(1) If $u < u_3 \approx 0.0137$,
then the output of Algorithm 2 satisfies:

$$\|N_f(z) - \xi\| < \|z - \xi\| .$$

(2) If $u < u_3' \approx 0.0098$ *then after k times of iteration we have*

$$\left\|N_f^k(z) - \xi\right\| < \left(\frac{1}{2}\right)^{2^k - 1} \|z - \xi\| .$$

6 Conclusion

The Maple code of algorithms mentioned in the paper and test results are available http://www.mmrc.iss.ac.cn/~lzhi/Research/hybrid.

Although the algorithms and proofs of quadratic convergence given in the paper are for polynomial systems with exact multiple zeros, examples are given to demonstrate that our algorithms are also applicable to analytic systems and polynomial systems with a cluster of simple roots.

References

1. Blum, L., Cucker, F., Shub, M., Smale, S.: Complexity and Real Computation. Springer, New York (1998)
2. Chen, X., Nashed, Z., Qi, L.: Convergence of Newton's method for singular smooth and nonsmooth equations using adaptive outer inverses. SIAM J. Optim. **7**(2), 445–462 (1997)
3. Corless, R.M., Gianni, P.M., Trager, B.M.: A reordered Schur factorization method for zero-dimensional polynomial systems with multiple roots. In: Küchlin, W.W. (ed) Proceedings of ISSAC 1997, pp. 133–140. ACM, New York (1997)
4. Dayton, B., Li, T., Zeng, Z.: Multiple zeros of nonlinear systems. Math. Comput. **80**, 2143–2168 (2011)
5. Dayton, B., Zeng, Z.: Computing the multiplicity structure in solving polynomial systems. In: Kauers, M. (ed) Proceedings of ISSAC 2005, pp. 116–123. ACM, New York (2005)
6. Decker, D.W., Kelley, C.T.: Newton's method at singular points I. SIAM J. Numer. Anal. **17**, 66–70 (1980)
7. Decker, D.W., Kelley, C.T.: Newton's method at singular points II. SIAM J. Numer. Anal. **17**, 465–471 (1980)
8. Decker, D.W., Kelley, C.T.: Convergence acceleration for Newton's method at singular points. SIAM J. Numer. Anal. **19**, 219–229 (1982)
9. Dedieu, J.P., Shub, M.: On simple double zeros and badly conditioned zeros of analytic functions of n variables. Math. Comput. **70**(233), 319–327 (2001)
10. Giusti, M., Lecerf, G., Salvy, B., Yakoubsohn, J.C.: On location and approximation of clusters of zeros of analytic functions. Found. Comput. Math. **5**(3), 257–311 (2005)
11. Giusti, M., Lecerf, G., Salvy, B., Yakoubsohn, J.C.: On location and approximation of clusters of zeros: case of embedding dimension one. Found. Comput. Math. **7**(1), 1–58 (2007)

12. Griewank, A.: On solving nonlinear equations with simple singularities or nearly singular solutions. SIAM Rev. **27**(4), 537–563 (1985)
13. Griewank, A.: Analysis and modification of Newton's method at singularities. Australian National University, thesis (1980)
14. Griewank, A., Osborne, M.R.: Newton's method for singular problems when the dimension of the null space is >1. SIAM J. Numer. Anal. **18**, 145–149 (1981)
15. Hao, Z., Jiang, W., Li, N., Zhi, L.: Computing simple multiple zeros of polynomial systems (2017). https://www.arxiv.org/pdf/1703.03981.pdf
16. Hauenstein, J.D., Sottile, F.: Algorithm 921: AlphaCertified: certifying solutions to polynomial systems. ACM Trans. Math. Softw. **38**(4), 28:1–28:20 (2012)
17. Krawczyk, R.: Newton-Algorithmen zur Bestimmung von Nullstellen mit Fehlerschranken. Computing **4**(3), 187–201 (1969)
18. Lecerf, G.: Quadratic Newton iteration for systems with multiplicity. Found. Comput. Math. **2**(3), 247–293 (2002)
19. Leykin, A., Verschelde, J., Zhao, A.: Newton's method with deflation for isolated singularities of polynomial systems. Theoret. Comput. Sci. **359**(1), 111–122 (2006)
20. Leykin, A., Verschelde, J., Zhao, A.: Higher-order deflation for polynomial systems with isolated singular solutions. In: Dickenstein, A., Schreyer, F.O., Sommese, A.J. (eds.) Algorithms in Algebraic Geometry. IMA, vol. 146, pp. 79–97. Springer, New York (2008)
21. Li, N., Zhi, L.: Compute the multiplicity structure of an isolated singular solution: case of breadth one. J. Symb. Comput. **47**, 700–710 (2012)
22. Li, N., Zhi, L.: Computing isolated singular solutions of polynomial systems: case of breadth one. SIAM J. Numer. Anal. **50**(1), 354–372 (2012)
23. Li, N., Zhi, L.: Verified error bounds for isolated singular solutions of polynomial systems: case of breadth one. Theoret. Comput. Sci. **479**, 163–173 (2013)
24. Li, N., Zhi, L.: Verified error bounds for isolated singular solutions of polynomial systems. SIAM J. Numer. Anal. **52**(4), 1623–1640 (2014)
25. Mantzaflaris, A., Mourrain, B.: Deflation and certified isolation of singular zeros of polynomial systems. In: Leykin, A. (ed.) Proceedings of ISSAC 2011, pp. 249–256. ACM, New York (2011)
26. Marinari, M.G., Mora, T., Möller, H.M.: Gröbner duality and multiplicities in polynomial system solving. In: Proceedings of ISSAC 1995, pp. 167–179. ACM, New York (1995)
27. Moore, R.E.: A test for existence of solutions to nonlinear systems. SIAM J. Numer. Anal. **14**(4), 611–615 (1977)
28. Mourrain, B.: Isolated points, duality and residues. J. Pure Appl. Algebra **117**, 469–493 (1996). 117
29. Ojika, T.: Modified deflation algorithm for the solution of singular problems. i. a system of nonlinear algebraic equations. J. Math. Anal. Appl. **123**(1), 199–221 (1987)
30. Ojika, T., Watanabe, S., Mitsui, T.: Deflation algorithm for the multiple roots of a system of nonlinear equations. J. Math. Anal. Appl. **96**(2), 463–479 (1983)
31. Rall, L.B.: Convergence of the Newton process to multiple solutions. Numer. Math. **9**(1), 23–37 (1966)
32. Reddien, G.W.: On Newton's method for singular problems. SIAM J. Numer. Anal. **15**(5), 993–996 (1978)
33. Reddien, G.W.: Newton's method and high order singularities. Comput. Math. Appl. **5**(2), 79–86 (1979)

34. Rump, S.M.: Solving algebraic problems with high accuracy. In: Proceedings of the Symposium on A New Approach to Scientific Computation, pp. 51–120. Academic Press Professional Inc., San Diego (1983)
35. Rump, S.M., Graillat, S.: Verified error bounds for multiple roots of systems of nonlinear equations. Numer. Algorithms **54**(3), 359–377 (2010)
36. Shen, Y.Q., Ypma, T.J.: Newton's method for singular nonlinear equations using approximate left and right nullspaces of the Jacobian. Appl. Numer. Math. **54**(2), 256–265 (2005)
37. Shub, M., Smale, S.: Complexity of bezout's theorem IV: probability of success; extensions. SIAM J. Numer. Anal. **33**(1), 128–148 (1996)
38. Shub, M., Smale, S.: Computational complexity: on the geometry of polynomials and a theory of cost: I. Ann. Sci. Éc. Norm. Supér. **18**(1), 107–142 (1985)
39. Shub, M., Smale, S.: Computational complexity: on the geometry of polynomials and a theory of cost: II. SIAM J. Comput. **15**(1), 145–161 (1986)
40. Smale, S.: The fundamental theorem of algebra and complexity theory. Bull. Amer. Math. Soc. **4**(1), 1–36 (1981)
41. Smale, S.: Newton's method estimates from data at one point. In: Ewing, R.E., Gross, K.I., Martin, C.F. (eds.) The Merging of Disciplines: New Directions in Pure, Applied, and Computational Mathematics. Springer, New York (1986)
42. Stetter, H.: Numerical Polynomial Algebra. SIAM, Philadelphia (2004)
43. Wang, X., Han, D.: On dominating sequence method in the point estimate and smale theorem. Sci. China Ser. A **33**(2), 135–144 (1990)
44. Wu, X., Zhi, L.: Computing the multiplicity structure from geometric involutive form. In: Jeffrey, D. (ed) Proceedings of ISSAC 2008, pp. 325–332. ACM, New York (2008)
45. Wu, X., Zhi, L.: Determining singular solutions of polynomial systems via symbolic-numeric reduction to geometric involutive forms. J. Symb. Comput. **47**(3), 227–238 (2012)
46. Yamamoto, N.: Regularization of solutions of nonlinear equations with singular Jacobian matrices. J. Inf. Process. **7**(1), 16–21 (1984)

Author Index

Printed in the United States
By Bookmasters